T0310165

Principles of Forensic Engineering Applied to Industrial Accidents

Principles of Forensic Engineering Applied to Industrial Accidents

Luca Fiorentini
TECSA S.r.l.
Italy

Luca Marmo
Politecnico di Torino
Italy

This edition first published 2019

Registered Offices
John Wiley & Sons, Inc., 111 River Street, Hoboken, NJ 07030, USA
John Wiley & Sons Ltd, The Atrium, Southern Gate, Chichester, West Sussex, PO19 8SQ, UK

Editorial Office
The Atrium, Southern Gate, Chichester, West Sussex, PO19 8SQ, UK

For details of our global editorial offices, customer services, and more information about Wiley products visit us at www.wiley.com.

Wiley also publishes its books in a variety of electronic formats and by print-on-demand. Some content that appears in standard print versions of this book may not be available in other formats.

Library of Congress Cataloging-in-Publication Data

Names: Fiorentini, Luca, 1976- author. | Marmo, Luca, 1967- author.
Title: Principles of forensic engineering applied to industrial accidents / Luca Fiorentini, Prof. Luca Fiorentini, TECSA S.r.l., IT, Luca Marmo, Prof. Luca Marmo, Politecnico di Torino, IT.
Description: First edition. | Hoboken, NJ, USA : Wiley, 2019. | Includes bibliographical references and index. |
Identifiers: LCCN 2018034915 (print) | LCCN 2018037469 (ebook) | ISBN 9781118962787 (Adobe PDF) | ISBN 9781118962794 (ePub) | ISBN 9781118962817 (hardcover)
Subjects: LCSH: Forensic engineering. | Industrial accidents. | Accident investigation–Case studies.
Classification: LCC TA219 (ebook) | LCC TA219 .F57 2018 (print) | DDC 363.11/65–dc23
LC record available at https://lccn.loc.gov/2018034915

Cover Design: Wiley
Cover Image: © Phonix_a/GettyImages

Set in 10/12pt WarnockPro by SPi Global, Chennai, India
Printed in Singapore by C.O.S. Printers Pte Ltd

10 9 8 7 6 5 4 3 2 1

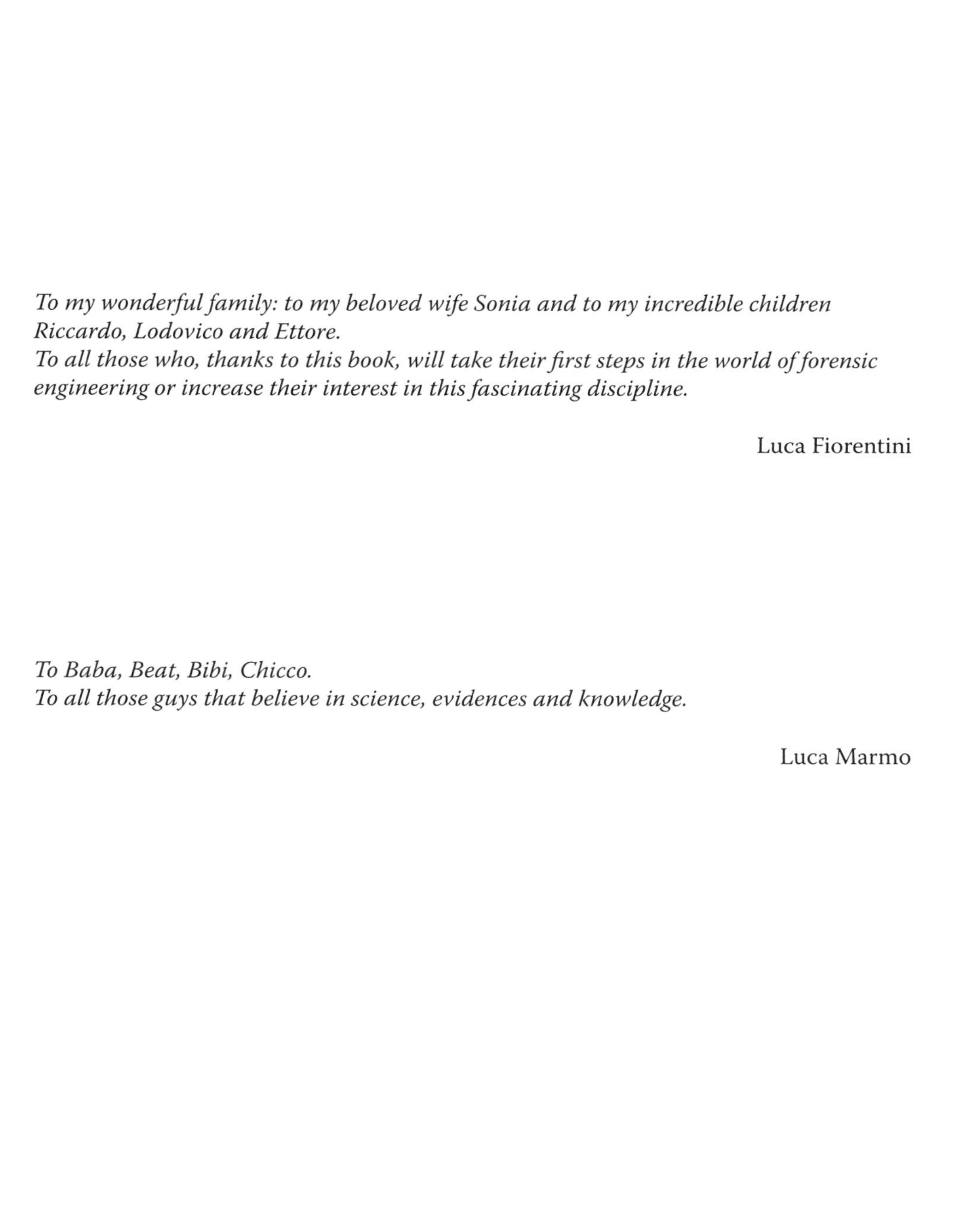

To my wonderful family: to my beloved wife Sonia and to my incredible children Riccardo, Lodovico and Ettore.
To all those who, thanks to this book, will take their first steps in the world of forensic engineering or increase their interest in this fascinating discipline.

Luca Fiorentini

To Baba, Beat, Bibi, Chicco.
To all those guys that believe in science, evidences and knowledge.

Luca Marmo

Contents

Foreword by Giomi

Fires and explosions, by their very nature, tend to delete any evidence of their causes, destroying it or making it unrecognizable. Establishing the origins and causes of fire, as well as the related responsibilities, therefore requires significantly complex investigations.

Simple considerations illustrate these difficulties. In the case of arson retarding devices may be used to delay the phenomenon, or accelerating substances, such as petroleum derivatives, alcohols and solvents, by pouring them on combustible materials present on site. The use of flammable and/or combustible liquids determines a higher propagation velocity, the possible presence of several outbreaks of diffuse type – which do not occur in accidental fires that usually start from single points, in addition temperatures are higher than those that would result from just solid fuels, such as paper, wood or textiles.

Generally, in accidental fires, burning develops slowly with a rate that varies according to the type and quantity of combustible materials present, as well as to the ventilation conditions of involved buildings. In addition, temperatures are, on the average, lower than those reached in malicious acts.

Obviously, these considerations must be applied to the context: the discovery of a container of flammable liquid is not in itself a proof of arson, on the other hand, the absence of traces of ignition at the place of the fire is not evidence that the fire is of an accidental nature!

Forensic Engineering, science and technology at the same time, interprets critically the results of an experiment in order to explain the phenomena involved, borrowing from science the method of investigation, replacing the experimental results with the evidence collected in the investigation, to understand how a given phenomenon took place and what were its causes, and also any related responsibility.

The reconstruction takes place through reverse engineering to establish the possible causes of the event.

The same scientific and engineering methodologies are used for the analysis of failures of particular elements (failure analysis) as well as the procedures for the review of what happened, researching the primary causes (root causes analysis).

The accident is seen as the unwanted final event of a path that starts from organizational and contextual conditions with shortcomings, due to inefficiencies and errors of design and actual conditions in which individuals find themselves working, and continues by examining the unsafe actions, human errors and violations that lead to the occurrence of the accident itself.

The assessment of the scientific skills and abilities of the forensic engineer should not be limited, as often happens, to just ascertaining the existence of the specialization, but should also include the verification of an actual qualified competence, deducting it from previous experiences of a professional, didactic, judicial, etc. nature.

In this context, the book "Principi di ingegneria forense applicati ad incidenti industriali" (Principles of forensic engineering applied to industrial accidents) by Prof. Luca Fiorentini and Prof. Luca Marmo constitutes an essential text for researchers and professionals in forensic engineering, as well as for all those, including technical consultants, who are preparing to systematically approach the discipline of the so-called "industrial forensic engineering".

The authors, industrial process safety experts and recognised "investigators" on fires and explosions, starting from the analysis of accidents or quasi-accidents that actually occurred in the industrial field, offer, among other things, an overview of the methodologies to be adopted for collecting evidence and storing it by means of an appropriate measurement chain, illustrate some analysis methodologies for the identification of causes and dynamics of accidents and provide guidance for the identification of the responsibilities in an industrial accident.

The illustration of some highly complex cases requiring the use of specialist knowledge ensures that this text can also be a useful reference for the Investigative Police, that, as is well known, in order to validate the sources of evidence must be able to understand the progress of the events.

Gioacchino Giomi
Head, National Fire Brigade, Italy

Foreword by Chiaia

The number and the magnitude of industrial accidents worldwide has risen since the 70s and continues to grow in both frequency and impact on human wellbeing and economic costs. Several major accidents (see, e.g. the Seveso disaster in 1976, the Bhopal gas tragedy in 1984, the Chernobyl accident in 1986, and Deepwater Horizon oil spill in 2010) and the increased number of hazardous substances and materials have been under the lens of the United Nations Office for Disaster Risk Reduction (UNISDR), which puts great effort in developing safety guidelines within the Sendai Framework for Disaster Risk Reduction 2015–2030.

On the other hand, man-made and technological accidents still represent a major concern in both the advanced countries and in under-developed ones. In the first case, risk is related not only to possible human losses but also to the domino effects, in terms of fires, explosions and possible biological effects in highly populated areas. Indeed, as pointed out by a great number of forensic engineering cases, the safety regulations for industries in developed countries are usually very strict and demanding. On the contrary, in underdeveloped countries, there is clear evidence that industrial regulations are less strict and that a general lack of the "culture of safety" which generally results in a looser application of the rules, thus providing higher frequency of industrial accidents.

Quite often, the default of a plant component or a human error are individuated as the principal causes of an accident. However, in most cases the picture is not so simple. For instance, the *intrinsic probability* of experiencing a human error within a certain industrial process is a crucial factor that should be kept in mind when designing the process *ex-ante* and, inversely, during a forensic investigation *ex-post,* to highlight correctly responsibilities and mistakes. Another source of complexity is represented by the so-called *black swans,* i.e. the negative events which were not considered before their occurrence (i.e. neither during the plant design, nor during functioning of the plant) simply because no one had never encountered such events (black swans are also called the *unknown unknowns*).

In this complex framework, Forensic Engineering, as applied in the realm of industrial accidents, plays the critical and fundamental role of knowledge booster. As pointed out by Fiorentini and Marmo in this excellent and comprehensive book, application of the structured methods of *reverse engineering* coupled with the specific intuition of the smart, experienced consultant, permits the reader to reconstruct the fault event tree, to individuate the causes of defaults and even to identify, *a posteriori*, possible black swan events. In this way, a well-conducted Forensic Engineering activity not only aims at solving the specific investigation problem but, in many cases, provides significant advancements for science, technology, and industrial engineering.

Bernardino Chiaia
Vice Rector, Politecnico di Torino, Italy

Foreword by Tee

It is my pleasure and privilege to write the foreword for this book, titled *Principles of Forensic Engineering Applied to Industrial Accidents*. I was invited to do so by one author of this book, Luca Fiorentini, who is the editorial board member of the International Journal of Forensic Engineering published by Inderscience Publishers.

Forensic engineering is defined as the application of engineering methods in determination and interpretation of causes of damage to, or failure of, equipment, machines or structures. Despite prevention and mitigation efforts, disasters still occur everywhere around the world. Nothing is so certain as the unexpected. Engineering failures and disasters are quite common and occur because of flaws in design, human error and certain uncontrollable situations, for instance, collapse of the I-35 West bridge in Minneapolis, crash of Air France Flight 447, catastrophic pipe failure in Weston, Fukushima nuclear disaster, just to name a few. Forensic engineering has played increasingly important roles in discovering the root cause of failure, determining whether the failure was accidental or intentional, lending engineering rationale to dispute resolution and legal processes, reducing future risk and improving next generation technology.

Nevertheless, forensic engineering investigations are not widely published, partly because most of the investigations are confidential. It then denies others the opportunity to learn from failure so as to reduce the risk of repeated failure. As forensic engineering is continuing to develop as a mature professional field, the launch of this book is timely. The topics of this book are well balanced and provide a good example of the focus and coverage in forensic engineering. The scope of this book includes all aspects of industrial accidents and related fields. Its content includes, but is not limited to, investigation methods, real case studies and lessons learned. This book was motivated by the author's experience as an expert witness and forensic engineer. It is appropriate for use to raise awareness of current forensic engineering practices both to the forensic community itself and to a wider audience. I believe this book has great value to students, academician and practitioners from world-wide as well as all others who are interested in forensic engineering.

Kong Fah Tee
Editor-in-Chief: International
Journal of Forensic Engineering;
Reader in Infrastructure Engineering,
Department of Engineering Science,
University of Greenwich,
Kent, United Kingdom

Preface

If you read this book, you are forensic engineers, or you would like to become one. Or you are simply curious. We hope this reading will stimulate your curiosity. A forensic engineer must be curious. He/she must look for answers to facts, give them scientific proof and above all he/she must not stop at the first explanation of the facts, even when it may seem the most obvious and solid.

A forensic engineer collects fragments, and, with these, he/she builds a mosaic where each tessera has one and only one natural location. Why do we do it? The reasons may be different. You could work on behalf of justice, or for the defence of an accused, or for an insurance company called to compensate an accident, just to name a few. Whatever your principle, you have a responsibility that goes beyond the professional one. A scientific responsibility. By reconstructing the mosaic of the facts that led to the disaster you are investigating or will investigate, you will give your explanation of the facts and the causes that determined them. If our explanation is based irrefutably on scientific arguments and the evidence, free from considerations related to the standards and desires of our principle, we will have made a contribution, sometimes small, sometimes significant, to progress. How much did the fire of the Deepwater Horizon, the release of Methyl Isocyanate of Bhopal or the fire of the ThyssenKrupp of Turin or the explosion of Chernobyl cost to the human community? Sometimes we find it difficult to estimate exactly the tribute of human lives; it is even more challenging to estimate material, image and environmental damage. If in the profession of the forensic engineer there is a mission, it is to contribute so that these facts are not repeated, so that the community learns from its mistakes, so that our well-being is increasingly based on sustainable activities, respectful of the rights of those who are more vulnerable or more exposed.

Galileo Galilei said: "Philosophy is written in this great book that is constantly open in front of our eyes (I say the universe), but we cannot understand it if we do not learn to understand the language first and know the characters in which it is written. It is written in mathematical language, and the characters are triangles, circles, and other geometric figures, without which it is impossible to understand them on a human scale; without these, it is a vain wandering through an obscure labyrinth." In our opinion, it also applies to the Forensic Engineer. The facts and their causes are written in the universe of the scene of the disaster, but we must understand the language and the characters of the writing. In reconstructing the dynamics and causes of an accident we must apply science to the facts, we must reconcile the reconstruction based on objective evidence with its explanation based on scientific evidence. In this way, in our opinion, one can ultimately achieve a precious result, that is expanding knowledge, drawing lessons from adverse facts so that they do not repeat themselves. We believe this is the highest mission that a forensic engineer can pursue in his/her professional life. Professor Trevor Kletz showed us how important it is to learn from accidents. This belief is the basis of the large space given in this book to the case studies. Obviously, we need a systematic and orderly method of work, which is what we have tried to describe in the text. And then we need a team. The forensic engineer cannot, in our opinion, have such a large baggage to deal with a complex case like the Thyssen Krupp case described in Chapter 7. We need specialists with very different characteristics to retrieve the data of a control system and interpret them, to simulate a jet fire and to determine the chemical-physical properties of the substances involved. We believe that a forensic engineer should never be afraid to seek the help of a specialist, but rather should fear to possess not the technical and scientific skills to dialogue with the many specialists who will contribute in his/her investigations. We hope that reading this text can help you build some of these bases.

Luca Fiorentini
Luca Marmo

Acknowledgement

Writing a book on the principles of forensic engineering represented a double challenge. First of all, the writing activity, whatever is written, requires moments of reflection to be devoted solely to the composition and in today's life this may mean taking a few hours from sleep. But such a large work, although limited to the principles of this discipline, could not be achieved without the precious contribution of those people who helped us to gather the necessary information for some topics of this text, as well as for the various case studies mentioned in Chapter 7.

In particular, we would like to thank MFCforensic for the valuable help provided in the preparation of this book. Clarifying that the objective of this book is not to publicise an investigative tool, but to provide a wide knowledge about the main methodologies used, a special thank you, however, goes to those who have allowed us to enrich the volume with a broad examination of the main instruments at the service of the forensic investigator. We therefore thank CGE Risk Management Solution for providing important support with its images on the main investigative tools, such as BSCAT™, Tripod Beta and BFA, which have undoubtedly embellished this text. Special thanks also to Fadi E. Rahal for providing the necessary material for the knowledge of Apollo RCA™; Mark Paradies and Barbara Carr for TapRoot®; and Jason Elliot Jones for Reason© RCA.

One of the most important contributions comes from those who have shared with us the information necessary for drafting the case studies reported in Chapter 7, often offering themselves for writing them. Proceeding in the order in which the case studies are presented in the book, we wish to thank Norberto Piccinini, former professor of Industrial Safety at the Turin Polytechnic, for his invaluable collaboration on the ThyssenKruup case; ARCOS Engineering s.r.l., in the person of Rosario Sicari, Alessandro Cantelli Forti, CNIT researcher at the Radar and Surveillance Systems National Laboratory of Pisa, and Simone Bigi by Tecsa s.r.l. for their help in drafting the case on the Norman Atlantic; Giovanni Pinetti and Pasquale Fanelli by Tecsa s.r.l. for having shared the material concerning a LOPC of flammable substance; Salvatore Tafaro, commander of the provincial command of Vibo Valentia of Italian National Fire Brigade, for valuable information on the case study of a refinery pipeway fire; Vincenzo Puccia, director of the provincial command of the Padua National Fire Brigade, and Serena Padovani for their contribution about the flash fire at silo and the explosion of a rotisserie van case studies; a special thanks to Vincenzo also for his example about the value of the digital evidence, shown in Paragraph 4.4.3.1; Numerics GmbH, in the person of Ernst Rottenkolber and Stefan Greulich, for the valuable collaboration on the case study of the fragment projection; Iplom S.p.A., in the person of Gianfranco

Peiretti, for the material relating to the fire of a process unit; ARCOS Engineering s.r.l., in the person Bernardino Chiaia and Stefania Marello, and TECSA S.r.l., in the person of Federico Bigi, for the support in the case study of an oil pipeline cracking; Giovanni Manzini for information regarding the case study on storage building on fire.

The authors give a special thanks to Rosario Sicari who oversaw the drafting of the work with care, precision and dedication, qualities that distinguish his activity as a forensic engineer and that we have been able to appreciate on several occasions of shared professional activity, from which have made Rosario not only an esteemed colleague to entrust the management of this complex and important work, but also an excellent friend with whom to share in the future, with great confidence, a growing number of assignments in the forensic field.

List of Acronyms

AHJ	Authorities Having Jurisdiction
AI	Accident Investigation
AIT	Auto Ignition Temperature
AIChE	American Institute of Chemical Engineers
ALARP	As Low As Reasonably Practicable
ANSI	American National Standards Institute
API	American Petroleum Institute
ASME	American Society of Mechanical Engineers
ATG	Automatic Tank Gauging
BBS	Behavior Based Safety
BFA	Barrier Failure Analysis
BFD	Block Flow Diagram
BLEVE	Boiling Liquid Expanding Vaprs Explosion
BOP	Blow Out Preventer
BPCS	Basic Process Control System
BRF	Basic Risk Factor
BSCAT	Barrier-based Systematic Cause Analysis Technique
CAC	Critical Administrative Control
CAS	Chemical Abstracts Service (number)
CCDM	Cause-Consequence Diagram Method
CCPS	Centre for Chemical Process Safety
CEO	Chief Executive Officer
CFD	Computational Fluid Dynamics
COMAH	Control Of Major Accident Hazards
CSB	US Chemical Safety Board
CPU	Central Processing Unit
DCS	Distributed Control System
E/E/PE	Electrical/Electronic/Programmable Electronic
EFV	Excessive Flow Valve
EIV	Emergency Isolation Valve
EPA	U.S. Environmental Protection Agency
EPG	Equipment Performance Gaps
ERT	Emergency Response Team
ESReDA	European Safety Reliability and Data Association
ETA	Event Tree Analysis

FLPPG	Front-Line Personnel Performance Gaps
FMEA	Failure Mode and Effect Analysis
FMECA	Failure Modes, Effects and Criticality Analysis
FDS	Fire Dynamics Simulator
FPT	Flash Point Temperature
FRC	Flow Recorder Controller
FTA	Fault Tree Analysis
GIGO	Garbage In Garbage Out
HAZID	HAZard IDentification
HAZOP	HAZard and OPerability Analysis
HD	Hard Disk
HDA	HydroDeAlkylation
HEMP	Hazard and Effects Management Process
HIRA	Hazard Identification and Risk Analysis
HPEP	Human Performance Evaluation Process
HR	Human Resources
HRR	Heat Release Rate
HSE	Health, Safety and Environmental
HSSE	Health, Safety, Security and Environmental
ICT	Information Computer Technology
IE	Initiating Event
IEC	International Electrotechnical Commission
IHLS	Independent High-Level Switch
IPL	Individual Protection Layer
ISO	International Organization for Standardization
IT	Information Technology
JA	Job Ability
JD	Job Demand
LEL	Lower Explosive Limit
LFE	Learning From Experience
LFL	Lower Flammability Limit
LI	Level Indicator
LLA	Low-Level Alarm
LOC	Limiting Oxygen Concentration
LOPA	Layer Of Protection Analysis
LOPC	Loss Of Primary Containment
LPG	Liquefied Petroleum Gases
LTA	Less Than Adequate
MARS	Major Accident Reporting System
MIC	Methyl-IsoCyanate
MIE	Minimum Ignition Energy
MOC	Management Of Change
MOC	Minimum Oxygen Concentration
MOOC	Management Of Organizational Change
MORT	Management Oversight Risk Tree
MSDS	Material Safety Data Sheet
MTO	Man, Technology and Organization

NFPA	National Fire Protection Association
NIST	U.S. National Institute for Standards and Technology
OCM	Organizational Change Management
OE	Operational Excellence
OSHA	Occupational Safety and Health Administration
PAH	Polycyclic Aromatic Hydrocarbons
P&A	Pickling and Annealing
P&ID	Piping and Instrumentation Diagram
PFD	Probability of Failure on Demand
PFS	Process Flow Sheet
PFH	Probability of Failure per Hour
PHA	Process Hazard Analysis / Preliminary Hazard Analysis
PLC	Programmable Logic Controller
PM	Project Manager
PPE	Personal Protective Equipment
PRP	Primary Responsible Party
PSI	Process Safety Information
PSM	Process Safety Management
PSV	Pressure Safety Valve
PV	PhotoVoltaic
QIQO	Quality In Quality Out
QRA	Quantitative Risk Assessment
RA	Risk Assessment
RCA	Root Cause Analysis
RCV	Remote Controller isolation Valve
R&D	Research & Development
RMP	Risk Management Program
ROI	Return On Investment
RPN	Risk Priority Number
RV	Relief Valve
SCE	Safety Critical Equipment
SIF	Safety Instrumented Functions
SIL	Safety Integrity Level
SIS	Safety Instrumented System
SLC	Safety Life Cycle
SMS	Safety Management System
SPAC	Standard, Policies and Administrative Control
SRK	Skill-Rule-Knowledge
STEP	Sequentially Timed Events Plotting
SWOT	Strengths, Weaknesses, Opportunities and Threats analysis
TCDD	TetraChloroDibenzoDioxin
TCP	TriChloroPhenol
TIC	Temperature Indicator Controller
TRV	Thermal Relief Valve
UEL	Upper Explosive Limit

UFL	Upper Flammability Limit
UVCE	Unconfined Vapor Cloud Explosion
VCE	Vapor Cloud Explosion
VDR	Voyage Data Recorder
VGS	Vent Gas Scrubber

1

Introduction

1.1 Who Should Read This Book?

"Principles of forensic engineering applied to industrial accidents" is intended to be an introductory volume on the investigation of industrial accidents. Forensic engineering should be seen as a rigorous approach to the discovery of root causes that lead to an accident or a near-miss. The approach should be suitable to identify both the immediate causes as well as the underlying factors that affected, amplified or modified the events (regarding consequences, evolution, dynamics), and the contribute by an eventual "human error".

A number of books have already been published on similar topics. The idea behind this book is not to replace those important volumes but to obtain a single concise and introductory volume (also for students and authorities) to the forensic engineering discipline that helps understand the link among those critical but very functional aspects of the same problem in the global strategy of learning from accidents (or near-misses). The reader, in this sense, will benefit from a single point of access to this vast technical literature that can be only accessed with proficiency having the right terms, definitions, and links in mind. On the contrary, the reader could get lost in all the quoted literature that day by day increases due to the speed of the research in this complex field.

The intent of the book is:

- Presenting simple real cases as well as give an overview of more complex ones, each of them investigated with the same framework;
- giving the readers the bibliography to access more in-depth specific aspects;
- giving them an overview of the most and commonly used methodologies and techniques to investigate accidents;
- giving them a summary of the evidence, which should be collected to define the cause, dynamics, and responsibilities of an industrial accident;
- giving them an overview of the most appropriate methods to collect and to preserve evidence through an appropriate chain of custody; and
- giving an overview of the main mistakes that can lead to misjudgment or loss of proof.

The book is an introductory volume for readers in academia as well as professionals who want to know more about the forensic engineering methodologies to be applied to discover more about the causes of industrial accidents in order to derive lessons. Among those professionals, we can identify process and safety managers, risk managers,

Principles of Forensic Engineering Applied to Industrial Accidents, First Edition.
Luca Fiorentini and Luca Marmo.

industrial risks consultants, attorneys, authorities having jurisdiction, judges and prosecutors, and so on.

It is particularly addressed to those who would like to approach the fundamentals of forensic engineering discipline without directly going to specialised already available volumes and handbooks that need a sound background to be read. Nonetheless, reading this book may help professionals (e.g. loss adjusters, risk engineers, safety professionals, safety management systems consultants.) and students who want to have a concise book as prompt reference towards the main important recognised resources available (e.g. CCPS©-AIChE© books also edited by Wiley, NFPA© 921 Standard, etc.) or as a bridge between risk assessment and accidents investigation (as a tool to learn from real accidents or near-misses in order to improve safety).

1.2 Going Beyond the Widget!

When investigating an industrial accident or a near miss, it should be well kept in mind that the primary goal to be reached is not to find a concise fault of a well-defined widget, confined to a distinct domain. A rigorous approach to the forensic discipline requires going much deeper in the investigation, not stopping at the main relevant evidence, even if properly gathered and analysed. It often happens that accident reports are one-dimensional [1]: in simple words, they identify only a single cause, usually corresponding to the outer layer of the complexity that surrounds the reconstruction of the incidental dynamics. Even when multiple causes are discovered, the investigator seldom looks beyond them.

In the industrial context, a complex system of relations, information, and people is present, with its peculiarity and hierarchy, creating a structured entity that needs to be considered when investigating an accident or a near miss. Thus, it becomes necessary to consider as an element of investigation the management systems as well, as some causes of the accident may be related to management failure, so to take the corrective actions and to prevent a further similar failure. A good investigator does not find culprits, does not blame. A good investigator collects evidence, analyses it and finds the root causes and the relations among them that lead to the accident, whilst also considering the managerial duties and, as usually happens, then provides suggestions about corrective actions to avoid the reoccurrence of the undesired event.

Focusing on the system, rather than the individual, represents the right way to face an investigation, at least for two reasons [2]. Firstly, if equipment and systems provided to persons reveal to be not effective, thus it is not the individual responsibility that has to be pointed out as the fault cause. Secondly, it is much easier to change a managerial choice rather than a person or his/her behavior, which is susceptible to vary daily. Third, human errors may often be the consequence of insufficient training, motivation or attention to safety, all being aspects that the management should promote and monitor. It is a matter of controllability and reliability, as they are the two most essential ingredients to ensure that the lesson learnt will guarantee an increasing, or a restoration at least, of the safety level accepted in the industry at the corporate, field and line levels. Metaphorically speaking, an accident investigation is like peeling an onion: this concept, cited in [3], gives us a live image of what we are called to solve (see Figure 1.1). Technical problems and mechanical failures are the outer layers of the onion: they are the immediate causes.

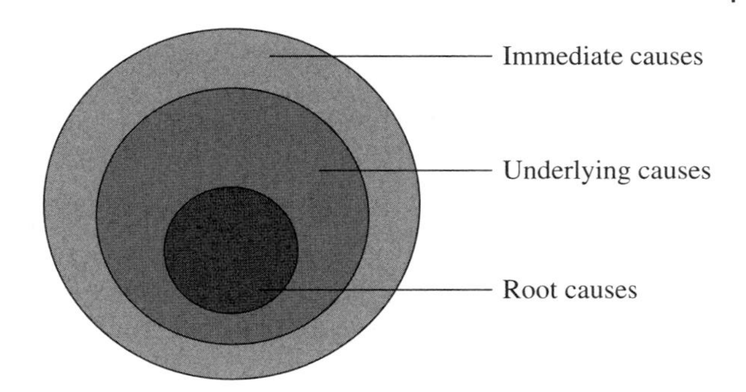

Figure 1.1 The onion-like structure between immediate causes and root causes.

Only once you peel them you can find the inner layers, thus the underlying causes like those involving the management weaknesses.

Going beyond the widget is what a professional investigator does. Let us consider a relief valve that fails, causing harm and loss (thus an accident) also involving some injuries to the line operators. A neophyte may conclude: "It was a fault in the relief valve. Case is closed, people". On the contrary, a good investigator may wonder: "Is it a consequence of an unexpected running condition, exceeding the operational limits? Was there an erroneous maintenance procedure? Was it installed correctly? Is it a result of an entire procurement of damaged relief valves?". The differences in the two extreme examples are clear: it is highly recommended to investigate spanning at least over the following three levels: line, field, and corporate levels. This good practice should suggest what a proper investigation requires: a project management and a variously skilled team of investigators.

Conducting an investigation means to plan the activities, to organise meetings, to schedule recognitions of the accident area, to inform and to be informed, to commission tests to external laboratories, to manage resources, mainly time and budget. But most of all conducting an investigation means to link the collected elements in a multidisciplinary network. To do this you need many different skills to work together. Many people get confused about how to conduct an investigation. The best way to face such a complex challenge is to consider it as an ordinary project: organisational and managerial skills, listening capacity in addition to a problem-solving attitude, are the desirable features of the investigator.

The recent approach in accident investigation reflects the simple concept discussed in this Paragraph. Indeed, over the past decade, a transition has occurred not only in the way accidents are investigated, but also in the way they are perceived [4]. One more time, the transition has shown an increasing focus on the organisational context rather than on the technical failures and human errors. This transition is also felt by the public opinion that forms after an industrial accident and is broadcasted by media. It is interesting to observe that such a transition can also be noted from the legal point of view, with an evolution of national laws and international technical standards and codes supporting a progressive shift of liability from the worker to the contractor and, more recently, to the top management of the company or, in some countries like Italy, to the Company itself. It is possible to claim that there is a sort of alignment among the technical aspects implicated in the forensic science, including the procedural way to conduct an investigation, and the legal issues. This transition has given rise to new methods to analyse an industrial accident, whose attention is primarily focused on the so-called "organisational network" and whose objective is to reconstruct empirically the real accidental phenomenon exploring the theoretical organisational structures. The goal is very ambitious and hard. It requires a multiplicity of transversal scientific skills, attitude, intuition and managerial capabilities. It requires ground competencies to find, gather and analyse that evidence that may be the trace of some precursor events, thus helping directly in the search for the root causes, or that may be weak signal, thus requiring a much more in-depth analysis to be referred into the organisational network and put in position, just like a puzzle piece, both in time and space.

The approach here described is also encouraged by some recent studies, like the one reported in [5], which analyses the phenomenon and the request for a different methodological approach taking inspiration from complexity theory. After having observed that single-factor explanations usually prevail and that also the language used in the accident reports reveals a historical trend in finding in individual human actions and failures the single leading cause of an accident, it is possible to identify the limits of the Cartesian-Newtonian worldview. According to these studies, the classical accident investigation is based on the Newtonian vision of the world, where a chain of causes–effects is the trace to identify everything since everything is deterministic and materialistic. Following this investigation methodology, the time becomes reversible. In other words, it is always possible to cross the time domain in both its directions, because of the bi-unique relationship between cause and effect. The knowledge is complete, and the perceived complexity is only apparent because of the human incapability in thoroughly reading this world. However, if you insert the idea of a failure in the theory of complexity, then conclusions change. The attention is now focused not only on the individual components of the system but also on their relationships. The rising complexity, which is an intrinsic feature of the whole system – not of its parts –, implicates the time irreversibility (thus the link between a cause and effect is not always bi-directional because of the sufficient or necessary nature of the condition that links the two). The Newtonian certainties collapse leaving the field to the uncertainty of knowledge and the foreseeability of probabilities, nothing more. These implications of complexity theory for safety investigations represent an interesting topic that needs to be further studied deeply, especially regarding the consequences it may have on the daily activities of the forensic engineers, the judges, the attorneys, and all the people called into the forensic path, whose need–primarily the legal need–might not accept such a loss of knowledge. What it can be doubtless taken into account is a broad look at the relations, thus to interactions at all levels including management. Facing complexity is a challenge requiring a strong capability to deal with sociotechnical systems, system safety, resilience engineering: these are the main ingredients of a more in-depth accident analysis [6]. According to what just said, the reader is asked to not confuse the attribute "complex" with "complicated" for the rest of the book.

However, in some cases "going beyond the widget" could not be necessary: these situations represent some (fortunately) rare uncontrollable events, because they are the consequences of deliberately malicious acts, dereliction of duty, working under the influence of drugs or alcohol and so on. If one of these events occurs, then blaming is legitimated. This is why these examples of industrial accidents or near misses are not considered in this book. Moreover, the analysis of Natural Hazard Triggering Technological Disasters (NaTech) is not treated here.

1.3 Forensic Engineering as a Discipline

The arising of forensic disciplines in the modern era can be considered as a consequence of several factors. The most important one is the constant needing of skilled professionals called upon to deal with judicial cases, thus providing a tangible help to the complex machinery of Justice, whichever it is the role they assume in the context of the judicial parties. What emerges concisely is the need for an expert and competent help to judges and attorneys: this is another reason that led to the necessity to regulate the field, not only from a legal standpoint but firstly from the methodological one. Indeed, the rights to prosecute and to defend when called to participate in the discussions of the Court can be exerted only if these rights are soundly based on facts. No ideas, no principles and no intuition: only facts. As a natural consequence, the gathering of evidence and its analysis – being the focal point of the entire judgment – are steps that need to be regulated. It is now that forensic engineering arises as a discipline, just like forensic psychology, criminology, and other related fields.

Forensic engineering becomes a discipline when it meets a method. In forensic engineering, the scientific method by Bacon and Galilei is the one followed to ensure comparability, shared methodologies and proven results. These are the basic conditions to trigger a favorable discussion when facts are cited in the Court, with the primary goal of presenting the Truth. A forensic engineer should well keep in mind its role: you find the Truth, not the Blame. Prosecuting is not in the tasks; you do not investigate to search the culprits, but to discover the facts and to reconstruct the dynamics of the event. A Forensic Engineer should also be capable to speak to and with the legal professionals, to ensure that all the technical facets of the accident will be properly considered in the judgment process. This may be one of the most challenging tasks for the Forensic Engineer.

Forensic science is a challenging mix of science, law, and management. What makes it in this way are the continuous changing legislation and legal decisions which push for constant research for new methods, protocols, and sciences [7]. In the previous Paragraph, it was briefly mentioned that an accident investigation requires the typical structure of multidisciplinary project management: this is because of the multidisciplinary approach usually adopted. After the first step is concluded, consisting in analysing the problem, the synthesis is then required. This path is typical of a problem-solving approach and a project management attitude is the only way to ensure a standard quality, in terms of a guaranteed chain of custody of the collected evidence, reproducibility of tests–when repeatable–soundly obtained results based on scientific method, logic, and cause-and-effect analysis. The final objective is to ensure an incontestable outcome capable of reconstructing the Truth. In simple words, a project management attitude is required because of the scientific complexity combined with the bureaucratic administrative path imposed by the legal context in which the accident investigation is conducted. The consequence is that very often the investigator assumes the role of leader of a multidisciplinary team that works following a holistic approach.

The basis of the rigorous method required is logic. Distinguishing between inductive methods and deductive methods is possible. The inductive method goes back to Aristotle, and it is based on the reconstruction of general principles starting from peculiar evidence. A mistake in generating the conclusion can be made when the collection of proof is not wide enough to ensure a robust logic sequence. There are some methods (described in Chapter 5) based on this logic path. However, ancient Greeks are also famous for the deductive method, whose frame of logical argumentation – the syllogism – represents one of its primary achievement. The interested reader can go deep into the historical background of the scientific method by consulting [7].

Nowadays the scientific method is worldly recognised as the core layer on which humanity has created its scientific – and then social – achievements. As well known, the scientific method is not the unique method on which humans relied. At the time of Bacon, the doctrine of *apriorism* was the only accepted: according to this doctrine, a selection of *a priori* assumptions was the only starting point – thus the only cause – of the entire Universe. It was not possible to overcome these assumptions since they were perceived as a religious dogma [8]. This brief passage is necessary to understand the power, as well the courage, of the revolution of Roger Bacon and Galileo Galilei (Figure 1.2). According to the scientific method, which refuses the *apriorism*, only a close observation and experimentation can ensure a complete knowledge of Nature. Centuries were necessary to guarantee a solid establishment of the scientific method.

Forensic engineering spans many fields. The necessity to share standard models and approaches has brought about the formation of international associations. Their purpose is to ensure an advantageous exchange of expertise, experience and capability about how to generally face an accident investigation and how to properly treat a peculiar case (like a bombing scene investigation – see [9] for details –, an industrial accident, a ship disaster, a fire investigation). When the accident implies severe injuries to humans, then the application of forensic pathology may be required [10]. Being a discipline, just like forensic engineering, the application of the scientific method is mandatory. This

Figure 1.2 Galileo Galilei (left) and Roger Bacon (right): two of the brightest scientists of the world who supported the scientific method. Source: Attribution 4.0 International (CC BY 4.0) https://en.wikipedia.org/wiki/Wikipedia:Text_of_Creative_Commons_Attribution-ShareAlike_3.0_Unported_License.

feature allows the reconstruction of the accidental dynamics, starting from the study of the penetrating and perforating shrapnel, the dust tattooing, the burns from heat and so on: these are all elements, here taken as a mere example, necessary to the medico-legal opinion at autopsy.

Being a forensic engineer implies a multidisciplinary approach and therefore a sound proficiency in physics, chemistry, mechanics, metallurgy, computer science regardless of whether you decide to work in a team or not. The rigorously adopted approach, relying on the scientific method, is the unique assurance of doing this job in the right way.

References

1 Kletz, T. (2002) Accident investigation - Missed opportunities. Hazards XVI: Analysing the Past, Planning the Future. Manchester: Institution of Chemical Engineers. pp. 3–8.
2 Sutton, I. (2010) *Process Risk and Reliability Management*. Burlington: William Andrew, Inc.
3 Kletz, T. (2012) *Learning from accidents*. 3rd ed. New York: Taylor & Francis
4 Dien, Y., Llory, M., and Montmayeul, R. (2004) Organisational accidents investigation methodology and lessons learned. *Journal of Hazardous Materials*, 111(1-3):147–153.
5 Dekker, S., Cilliers, P., and Hofmeyr, J. (2011) The complexity of failure: Implications of complexity theory for safety investigations. *Safety Science*, 49(6):939–945.
6 Pasman, H. (2015) *Risk analysis and control for industrial processes*. 1st ed. Oxford: Elsevier Butterworth-Heinemann.
7 Noon, R. (2009) *Scientific method*. Boca Raton, FL: CRC Press.
8 Noon, R. (2001) *Forensic engineering investigation*. 1st ed. Boca Raton, FL: CRC Press,
9 La, A. (2001) Guide for explosion and bombing scene investigation. https://www.ncjrs.gov National Criminal Justice Reference Service.
10 Beveridge, A. (2012) *Forensic investigation of explosions*. 1st ed. Boca Raton, Fla.: CRC Press.

Further Reading

CCPS (Center for Chemical Process Safety). (2003) *Guidelines for investigating chemical process incidents*. 2nd ed. New York: American Institute of Chemical Engineers.
ESReDA Working Group on Accident Investigation. (2009) *Guidelines for Safety Investigations of Accidents*. 1st ed. European Safety and Reliability and Data Association.

2

Industrial Accidents

2.1 Accidents

Industrial accidents include some of the saddest events in the history of the humans on Earth. Regardless of the effort to limit their consequences, this particular type of event has always had a significant impact on the society, the public opinion and the industry as well. Two aspects are peculiar to an incident: being low-probability and having high-consequences [1]. This characteristic relies on the process industry risk sources, which expose the environment, the people and the business to acute effects. Even a person who is not an expert may agree about the hazard of dealing with gasoline, natural gas (LNG), ammonia, liquefied petroleum gas (LPG), hydrogen, and so on. Indeed, public opinion often overestimates the risk, having the consequence, very frequently seen in the country of the Authors of this book, to refuse *a priori* the idea of a new plant in the vicinity. The interested reader can find additional historical information about propellants and explosives – both military and commercial – in [2]. Every day, many industries in the world deal with these major hazards and are exposed to their risks, which can remotely cause an accident. The last 50 years experienced important business, industry, and energy trends [1]. Operations have been enlarged and diversified, the globalization and the increased competition affected the priority goals of industrial managers, driving them towards a cost-cutting strategy which pushed on more efficient technology and automation, saving energy but delaying the investments in safeguards. At the same time, plants became more complex, and people continuously changed their duties, in a reorganised complex structure. Also, the way accidents are perceived significantly changed, because of the reduced sensibility of the younger generations towards fires and explosions, due to its exposure to digital reality (like video games and films) where accidents are seen, but not physically experienced. This reflects in more significant efforts for companies to carry out an effective training about safety-related issues. Moreover, performance pressures (i.e. time and cost pressure) have increased. All these tasks may hamper safe working. Management is the available tool to face this scenario. It has to re-think itself continuously, in order to ensure a reliable and resilient work environment.

This Chapter is intended to provide the basic knowledge about the industrial accidents, the chemistry and physics at their base, together with an introduction to the process safety and the instruments that may allow the monitoring of safety-related performance. Some of the most important industrial accidents are presented in

Principles of Forensic Engineering Applied to Industrial Accidents, First Edition.
Luca Fiorentini and Luca Marmo.

Paragraph 2.4, just to introduce the reader to the investigator's mindset. However, at this stage of the book, a structured approach to "solve" them is not provided voluntarily. In the end, the role of "Uncertainty" and "Risk" is discussed, giving some useful definitions.

Before discussing the principles of forensic engineering applied to industrial accidents, some definitions need to be provided. A unique definition of accident does not exist since different explanations have been given in time. The most straightforward definition of accident is "an event that results in injury or ill health" [3]. This definition limits the impact of an accident to the health sphere, so other definitions have to be explored. In [4], an accident is defined as an undesired event that causes injury or property damage, recalling [5]. Similarly, according to [6], an accident is "the final event in an unplanned process, resulting in injury or illness to an employee and possibly property damage. It is the final effect of multiple causes". This definition introduces a larger view of what causes an incident, and it immediately establishes how there can be more than a single cause. A further definition is provided by [7], describing an incident as "an unplanned event or sequence of events that either resulted in or had the potential to result in adverse impacts". This definition covers not only safety and environmental harm but also economic loss. Finally, the incident sequence can be defined as a series of events composed of an initiating cause and intermediate events leading to an undesirable outcome.

The terms "accident" and "incident" are used differently by many companies, and also many books on the argument make the same distinction [8]. Both the two words describe an event that causes harm or loss. The main difference between them is that an incident, by definition, can be preventable thanks to the use of the facility's normal management systems (including the process safety culture, when talking about industrial accidents); an accident, instead, implies uncontrollability, misfortune, and surprise. As we have already pointed out in the previous Chapter, these types of events are not discussed in this book. This is why we use the term "accident" as a synonymous with "incident", being the difference irrelevant in the context of this book. Thus they are used interchangeably. Many authors agree with the approach here adopted about the definition of terms [4].

Accidents occur because failures occur. And even if there are many ways to be safe, failures seem to have a single path [9]. Having a single path does not mean that only a single cause exists. The problem is usually in the relations, causalities, or spaces around the single detected immediate cause: it relies on the complexity of the system. The equation between accident and failure requires defining what a failure is. In our context, the failure concerns the incapability of a set of barriers to stop the incident sequence before the occurrence of the incident itself. It is an important term since its meaning is shared among different professionals (quality engineers, production engineers, maintenance engineers, front-line operators, and managers share the same idea on what a failure is). According to [9], failures can be mainly of two types: individual failure and organizational/system failure. The former happens when the worker is not protected from the dangers, and it includes cuts, slips, falls, and chemical exposure. The consequences of an individual failure affect the worker or workers in the event. The latter has the potentiality to have a consequence extended to many people. They typically occur when several layers of protection have been broken. Every failure can be divided into three parts: the context, the consequence, and the retrospective understanding. The context is everything that led up to the actual failure event; the consequence is the failure itself; the retrospective understanding is everything that happens after the failure happens (i.e. the organizational reaction). The understanding of a failure requires:

- An explanation of the failure (it does not mean to have a root cause analysis, a fault tree analysis, or a timeline. Just an explanation);
- to know what went wrong and what went right;
- to understand why barriers failed or were not present at all;
- to be aware that the consequence size does not determine the importance of a failure; and
- to be aware that unwanted events do not discriminate between good and bad people.

Analyses of accidents revealed that they are generated by immediate causes (technical failures and/or human error), which are induced, facilitated or accelerated by underlying organisational conditions (root causes) [10]. According to this accident causation model, an accident happens after an incubation period, during which the latent preconditions give signals that are not adequately perceived as potentially dangerous for the safety. From this standpoint, an accident is a materialised risk: it is now more evident why it is fundamental to deal with the concepts of hazards, risks and their identification and assessment.

Several techniques, developed to face safety-related incidents, can also be used to investigate an environmental harm or an economic loss: this is implicit in the definition of "incident", which spans over these three different typologies of risk: safety, environmental and business. Safety-related incidents involve harms to human life, like injuries or death directly correlated with the crucial event (e.g. a fire, an explosion of an item) or with some of its immediate consequences, like a Loss Of Primary Containment (LOPC) being toxic, flammable, or generally harmful for humans (LOPC are often, but not always, a consequence of the reached structural resistance of an item, attributable to overpressure, over temperature, over level and other typical deviations from the standard process). Environmental incidents concern an environmental harm due to a leakage, a spill, an LOPC arising after the main event, an increasing of the wasted gas released into the atmosphere, and so on. A business incident happens when a loss of production, a reduced efficiency, or a loss of equipment (requiring high costs to perform its maintenance) occurs.

Having clarified that an accident is a predictable event causing harm or loss, the next step in our approach to forensic engineering is to go deeper in their analysis. The objective of the book, at this stage, is to provide a sound knowledge to the reader, in order to face the concepts presented further in the book and to obtain the best lesson from reading the case studies discussed in Chapter 6. When an incident meets some particular features, it may be classified as "potential incident" or "high potential incident". A "potential incident" is an incident where nothing happened at all [8]. This definition may generate confusion, so it is better to explain it by an example: let us suppose that a worker is on the upper deck of a four stories scaffold. He loses equilibrium and the wrench he was using falls to the ground. If it hits another person walking or working on the same construction site, then it is an accident. However, if the wrench stops its downfall at a lower deck without touching the ground, thus having the possibility to create a potential loss, then it is only a potential incident. A similar example will be used to explain the concept of "near miss", described in the next Paragraph. A "high potential incident" is a potential incident with the possibility to generate a severe major loss.

An example of high potential incident is a toxic gas release from a flange, which does not cause any consequences solely because nobody was present nearby because the area was restricted for a maintenance issue. Typically, a high potential incident occurs when the last Individual Protection Layer (IPL) is used by the system. In other words, all the safeguards, put in place to mitigate the risks related to the occurrence of an undesired event promoted by an Initiating Event (IE), fail except one. Obviously, if all the safeguards fail, then the incident is no more potential but actual.

To sum up, an incident may be defined as an unusual or unexpected event which either resulted in or had the reasonable potential to cause an injury, release, fire, explosion, environmental impact, damage to property, interruption of operations, adverse quality affecting or security breach or irregularity.

Commonly, the incident scenarios that affect the process industry are classified into three types:

- Fires (any combustion regardless of the presence of flame; this includes smouldering, charring, smoking, singeing, scorching, carbonising or the evidence that any of these have occurred);
- explosions (including thermal deflagrations, physical bursts and detonations); and
- toxic releases (mainly gas/vapors but also liquids).

The consequences of an accident span from fatal to minor injury and damage only (economic loss), in a scale of magnitude which is not uniquely predefined. Similarly, the likelihood that an adverse event will happen again spans from certain to rare [3]. Talking about fires, explosions, and toxic releases, their likelihood and magnitude are summed up in Table 2.1.

Fires and explosions are generally among the most common typologies and the most potentially dangerous in industry: indeed, the case studies discussed in Chapter 6 belong to these categories. Consequentially, close attention is focused primarily on the description of the peculiarities and modalities of evolution related to fires and explosions, leaving out a bit the toxic releases, which in most cases are the consequence of the first two.

The analysis of the general unit operations and their failure modes [12] shows some typical mechanical failures that contribute to triggering the adverse sequence of events resulting in an accident. Pumps, compressors, fans, heat exchange equipment, reactors and reactive hazards, tanks and storage issues, operations concerning mass transfer, distillation, leaching and extraction, adsorption, mechanical separation: they are only the outer surface of the complex (i.e. full of relations) system which can be identified in the process industry. Hierarchies, procedures, sets of accountabilities and responsibilities, operative instructions, alarm management, functional safety, are only a few elements that are not "tangible" but significantly take part in the definition of the eventual

Table 2.1 Incident typologies and correlated potentiality and magnitude.

Type of incident	Likelihood of occurrence	Death potentiality	Economic loss potentiality
Fire	High	Low	Medium
Explosion	Medium	Medium	High
Release of toxic	Low	High	Low

Source: Adapted from [11].

occurrence of an incident. We have anticipated how important is going beyond the widget, and here this concept starts to be enforced. Plants are not merely a group of mechanical items. Rather, they are complex sociotechnical systems. Concluding this reasoning, it is no longer true that accidents are always unplanned [6]. However, this concept is underlined throughout the book, with the aim of providing a growing awareness about this essential mental step to carry out a more in-depth accident investigation.

It may be of interest to cite the results of some statistics about the most common causes at the origin of industrial accidents since they highlight the critical factors on which the attention of the consultant should have to be primarily addressed. It is essential to keep in mind that these results are purely indicative (See Figures 2.1 and 2.2).

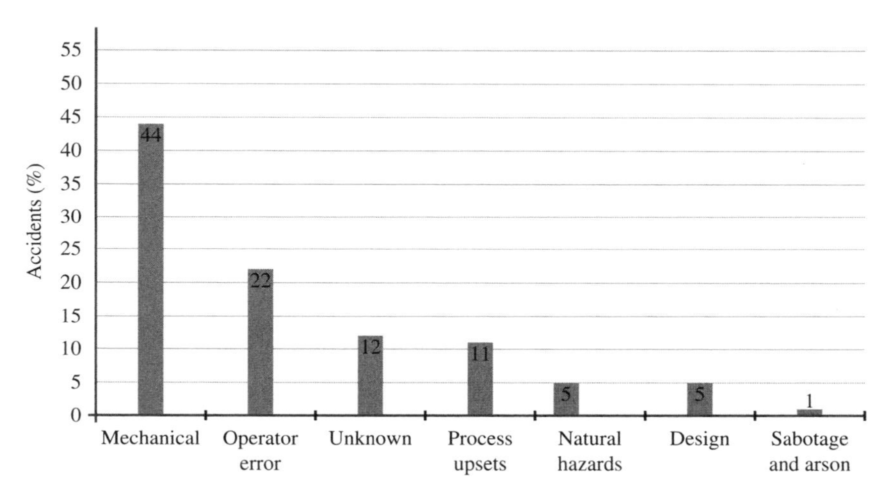

Figure 2.1 Causes of industrial accidents in chemical and petrochemical plants in the United States in 1998. Source: Data elaborated from [11].

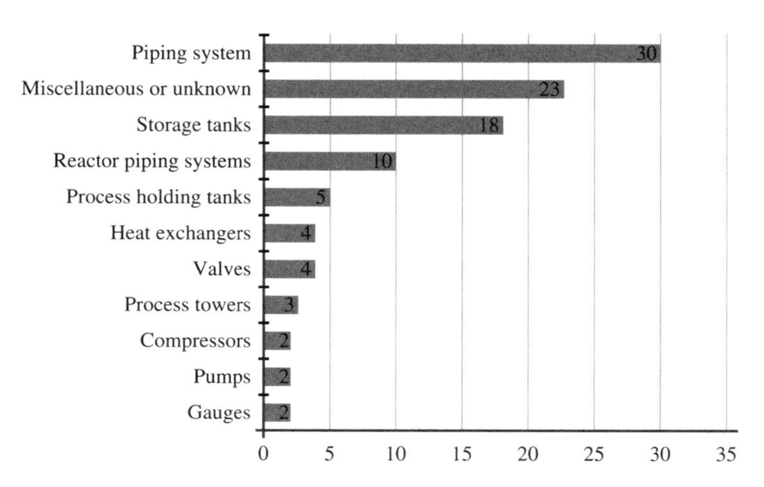

Figure 2.2 Components related to the industrial accidents in chemical and petrochemical plants in the United States in 1998. Source: Data elaborated from [11].

The phenomenon at the base of fires and explosions scenarios is the combustion: the principles of this phenomenon are now treated.

2.1.1 Principles of Combustion

Many fires are investigated by law enforcement since they are often the consequence of negligence, criminal activity (which are not within the scope of this book) and man-related incidents. The objectives of a fire investigation are two: determining the origin and the cause of the fire. A proper scene observation is the starting point to reach such goals. It requires tracking the fire back to its origin, by studying the fire patterns and identifying the ignition source. To do so, the fundamental theories in fire chemistry and physics are used. They are the preliminary background knowledge that a forensic engineer must have, since they govern important phenomena like the transfer of heat, the fire propagation, and its interaction with fuels. This Paragraph intends to provide some basic notions about the chemistry of fire, taking inspiration from [13].

Combustion is a chemical reaction between a substance, named fuel, with an oxidant, usually oxygen, characterised by heat development and, usually, by a visible flame. The fire triangle (Figure 2.3) is the graphical representation of the three conditions that must be present in order to have a fire:

- Presence of fuel (combustible material);
- presence of an oxidising agent; and
- presence of a source of ignition (like heat).

Only the contemporary presence of the three sides of the triangle will ensure the birth and propagation of a fire. Indeed, the suppression of only one of the three conditions will cause the extinguishment of the fire. Breaking the connection between two sides of the fire triangle is actually the strategy adopted to extinguish every fire. For instance, adding water to the fire, one removes the heat; using carbon dioxide fire extinguishers, one removes the oxygen, and thus the fire dies.

Usually, the oxidising agent is the oxygen contained in the air. Besides, other substances already have, as part of their molecular structure, a sufficient quantity of oxygen to start combustion (if triggered): this is the case with explosives or with cellulose. Alogens (like Chlorine) are also excellent oxidisers. Therefore, a fire investigator must be aware that, especially in chemical plants, the presence of oxygen is often not a crucial factor to start a fire or sustain a combustion, because of the particular substances that are used.

Figure 2.3 The Fire Triangle.

Fuels are almost everywhere. They can be solid, liquid or gas. The possibility of the presence of fuels in all the three states may have a significant impact on a fire investigation, becoming more complicated. Typically, some elements like carbon, sulphur or hydrogen are part of the molecular composition of fuel, and they tend to combine with the oxygen. A substance is defined flammable if it triggers easily and burns rapidly.

In the fire triangle, heat is the activation energy or the source of ignition. However, to have combustion, it is necessary to meet some further conditions. Depending on the fuels, examples of these conditions include reaching the flash point (i.e. the lowest temperature for a volatile material at which vapors ignite, in the presence of an ignition source) as they may be inside the flammability limits. Indeed, many fuels are in daily contact with oxygen, but they do not cause a fire. Moreover, it is important to understand that the capability of the source of ignition to trigger a fire depends on the fuel. For example, a lit cigarette may ignite a newspaper, but not gasoline. Five sources of heat exist in nature: electrical, mechanical, chemical, biological, and nuclear. Heat is also the primary cause for the propagation of a fire. Depending on the thermal effects induced on the structures, it may produce a specific harm. The different mechanisms of heat transfer (convection, conduction and radiation) are illustrated in Figure 2.4 Convection is typical of fluids (liquids and gases): energy is transferred by the motion of fluids, activated by the different density of warmer and colder fluids. This phenomenon is at the base of the layering of the hot gas at the ceiling of a compartment. Conduction is typical of solids, and it is due to the interaction of vibrating atoms from particle to particle. A higher temperature causes a wider vibration of the atoms, which transfer their energy to the atoms nearby, thus transferring heat. The capability of a solid to transfer heat by conduction is called "conductivity".

Finally, with radiation the heat transfers through electromagnetic waves. The peculiarity of this last mechanism is that heat can be transferred in all directions even through void. Radiation is the dominant heat transfer mechanism for development of fire. Figure 2.5 shows how radiation was responsible for the propagation of fire from deck 4 to deck 3 inside the Norman Atlantic ferryboat [14].

Combustion is a redox reaction: electrons are exchanged such that the oxidiser is reduced, and the fuel is oxidised. The most important feature of this reaction is the high exothermicity. The combustion reaction is commonly written according to stoichiometry, where the exact amount of oxydiser and fuel react to give completely oxidised products, as below:

$$CH_4 + 2O_2 \rightarrow CO_2 + 2H_2O \tag{2.2}$$

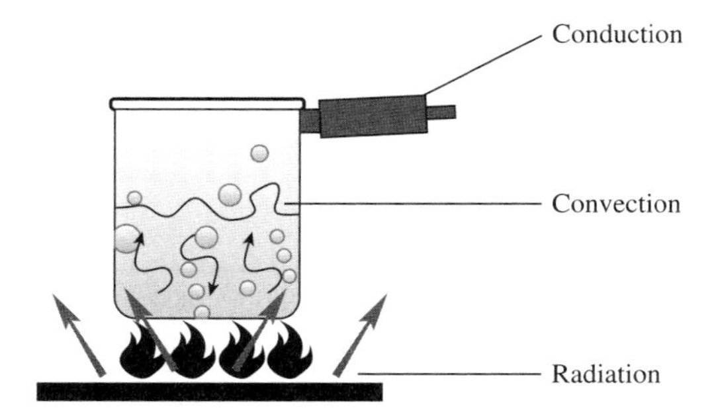

Figure 2.4 The different mechanisms of heat transfer.

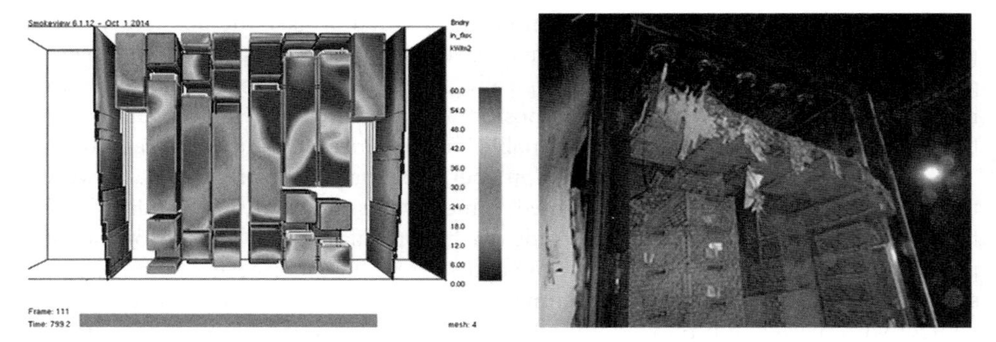

Figure 2.5 The involvement of deck no. 3 of the Norman Atlantic into the fire, due to radiation: simulation and evidence (plastic boxes, melted at the top). Source: [14].

This is what is called a complete combustion. Complete combustion is often the result of a controlled burn, where the full oxidization of the fuel is reached. If hydrocarbons are the fuel, in a complete combustion they are fully transformed into carbon dioxide and water. Complete combustion rarely occurs. In general it eqiores a huge excess of oxidiser. In practice, only a few fires meet this requirement. In general the the combustion is incomplete, such producing significant amounts of partly oxidised products: Carbon monoxide (CO), soot, tar, charcoal. Carbon monoxide is a colorless and odourless toxic gas. Carbon monoxide is the top cause of asphyxiation in fires. Moreover, it is also a combustible gas, usually resulting in explosions. When the level of oxygen is not enough to oxidise carbon, the incomplete combustion generates soot. Typically, we are used to seeing orange flames during a fire: they are a typical indicator of incomplete combustion. Actually, it is possible, for gaseous fuel, to correlate the temperature of combustion with a chromatic scale of the flames (Figure 2.6).

In case of shortage of oxygen, the fire may have limited power.: it is called an oxygen-controlled fire, in opposition to a fuel-controlled fire, which is governed by the

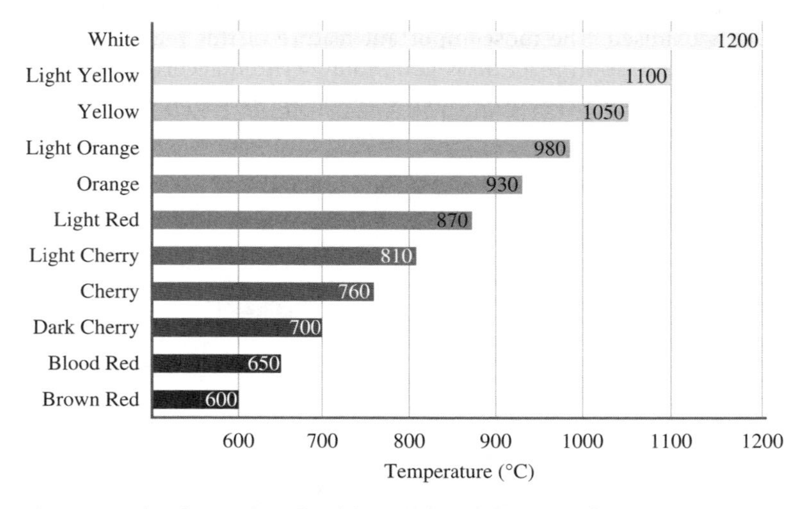

Figure 2.6 The chromatic scale of the temperatures in a gas fuel.

availability of fuel. In case of strong shortage of oxygen, the fire will extinguish, often leaving a hot, reactive environment. Oxygen-controlled fires are extremely hazardous to the firefighters since a sudden supply of oxygen can transform a gentle situation into an explosion. For example, this happens when openings are created to access a closed room, exposing firefighters to severe risks.

The smoke production during fires, in the form of microscopic solid and liquid particles, may lead to the production of highly toxic products. It very often results in a significant obstacle to the emergency operations, especially in confined areas.

During a fire, if the oxygen concentration drops below a certain percentage, flames disappear, and the fire smoulders. In this condition, the combustion reaction is between the solid fuel and the gaseous oxidiser in its surroundings. Differently from smouldering fires, which can only occur with solids, a flaming fire involves a gas-to-gas reaction. This means that only the combustion of gases can result into flames; thus, solids and liquids must transform into gas before creating a flaming fire. This is a key concept of the fire investigation: solids and liquids do not burn, and they need to be transformed into gases to burn in flaming fire. For a fire investigator, this result in a preliminary step: determining how the fuel transformed in its proper form before being ignited. The phase change analysis therefore becomes crucial.

In addition to the well-known phase changes, fire investigation requires the knowledge of a further one: the pyrolysis. Pyrolysis is the chemical decomposition of a larger molecule, belonging to a solid, into smaller molecules at the gaseous phase. The chemical decomposition is due to heat, but no oxidation occurs. This phenomenon is the main cause of transformation of a solid fuel into ignitable gases.

2.1.1.1 Flammable Gases and Vapors

The flammability of gases and vapors is connected with the value of their composition in the mixture with the oxidising agent. Their quantity must be within two flammability limits that identify a range of values.

The Lower Flammability Limit (LFL) is the lowest concentration of a gas or vapor in the air that can sustain combustion at a given temperature and pressure. In the context of this book, the LFL is synonymous with Lower Explosive Limit (LEL). If the concentration is lower than the LFL, then combustion does not occur even if a source of ignition is present. In simple words, this happens because the quantity of fuel is not sufficient.

The Upper Flammability Limit (UFL), also referred to as Upper Explosive Limit (UEL), is the maximum acceptable concentration to have a flammable gas/vapor cloud that can sustain combustion. Over this limit, there is a lack of the oxidiser, hampering the combustion. These basic concepts are shown in Figure 2.7.

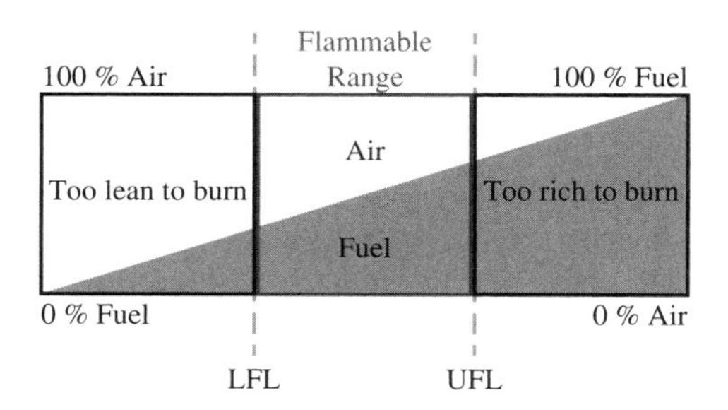

Figure 2.7 Graphical representation of the concepts of LFL and UFL.

The LFL and the UFL vary with temperature and pressure: generally speaking, a higher temperature produces an increase in the UFL and a decrease in the LFL: this means that the gas/vapor mixture has a greater range of possible concentration values falling inside the flammable region. Instead, the effect of pressure is usually hard to predict, since it may vary depending on the specific mixture. In Table 2.2, the flammability limits of some substances are listed.

When the fuel is a mix of components, the flammability limits are evaluated through additivity criteria, starting from the values of the single elements, using the empiric relation of Le Châtelier, known as "the mixing rule".

$$LFL_{mix} = \frac{1}{\sum \frac{x_i}{LFL_i}} \tag{2.1}$$

In equation (2.1), LFL_{mix} is the LFL of the mixture, LFL_i is the LFL of the generic single component, and x_i is the molar fraction of the single generic component of the mixture of flammable species. The results based upon the additivity criterion have to be considered prudently, especially if there are components of different chemical structures that tend to react differentially and to influence each other. In order to calculate the effect of temperature and pressure on the limits, other empirical relations may be used to estimate them starting from the stoichiometric concentration in air. The temperature influences the characteristics of flammability of both gases and vapors significantly, acting on the vapor pressure, the reaction rate, the flammability limits, and the speed of flames propagations. Generally an increasing of the temperature produces an enlargement of the flammability range, with a particular focus on the upper limit. This growth may require the introduction of a larger quantity of inert substance to transform the mixture into a non-flammable one. Pressure is also capable to vary the flammability range, but these modifications are less evident and their effect is not easy to identify because a single trend does not exist, but it depends on the specific mixture. Generally, a significant decrease of pressure causes a reduction of the flammability range.

Another important parameter governing the flammability of gas and vapors is the **Minimum Oxygen Concentration** (MOC), also known as Limiting Oxygen Concentration (LOC). It identifies a limiting value of the oxygen concentration below which

Table 2.2 Flammability limits of some gas and vapors.

	Flammability limits	
Substance	**LFL[% in volume]**	**UFL[% in volume]**
Hydrogen	4	75
Methane	5	15
Butane	1.8	8.4
n-Hexane	1.1	7.5
Gasoline	1.7	7.6
Toluene	1.2	7.1
Methanol	6.7	36
Ethanol	3.3	19

Source: Data taken from [11].

combustion cannot occur. This value is expressed in units of volume percent of oxygen. It depends on pressure, temperature, and the type of inert (non-flammable) gas. Some values of this parameter are in Table 2.3. The knowledge of this value is particularly important in fire safety engineering, since severe risks of explosion can be eliminated by adding the inert gas (like nitrogen or carbon dioxide) in the compartment, forcing the O_2 concentration to drop below the MOC.

Among the flammability properties of gases and vapors, there is the **Auto Ignition Temperature** (AIT). It is defined as the lowest temperature at which a substance starts to burn spontaneously when an oxidiser is present, without a direct source of ignition. In this case, the temperature is itself an efficient source of ignition to start the combustion. It is important to note that autoignition is not synonymous with instantaneous ignition. Indeed, a period, named "induction period" or "ignition delay", exists and it varies according to the specific mixture and temperature. Typically, this period decreases as the

Table 2.3 MOC values (volume percent oxygen concentration above which combustion can occur).

Gas or vapor	N_2/Air	CO_2/Air	Gas or vapor	N_2/Air	CO_2/Air
Methane	12	14.5	Kerosene	10 (150°C)	13 (150°C)
Ethane	11	13.5	JP-1 fuel	10.5 (150°C)	14 (150°C)
Propane	11.5	14.5	JP-3 fuel	12	14.5
n-Butane	12	14.5	JP-4 fuel	11.5	14.5
Isobutane	12	15	Natural gas	12	14.5
n-Pentane	12	14.5	*n*-Butyl chloride	14	–
Isopentane	12	14.5		12 (100°C)	–
n-Hexane	12	14.5	Methylene chloride	19 (30°C)	–
n-Heptane	11.5	14.5		17 (100°C)	–
Ethylene	10	11.5	Ethylene dichloride	13	–
Propylene	11.5	14		11.5 (100°C)	–
1-Butene	11.5	14	Methyl chloroform	14	–
Isobutylene	12	15	Trichloroethylene	9 (100s°C)	–
Butadiene	10.5	13	Acetone	11.5	14
3-Methyl-butene	11.5	14	*t*-butanol	NA	16.5 (150°C)
Benzene	11.4	14	Carbon disulphide	5	7.5
Toluene	9.5	–	Carbon monoxide	5.5	5.5
Styrene	9	–	Ethanol	10.5	13
Ethylbenzene	9	–	2-Ethyl butanol	9.5 (150°C)	–
Vinyltoluene	9	–	Ethyl ether	10.5	13
Diethylbenzene	8.5	–	Hydrogen	5	5.2
Cyclopropane	11.5	14	Hydrogen sulphide	7.5	11.5
Gasoline			Isobutyl formate	12.5	15
(73/100)	12	15	Methanol	10	12
(100/130)	12	15	Methyl acetate	11	13.5
(115/145)	12	14.5			

Source: Data from [11].

temperature is much higher than the AIT and it increases as the temperature is close to the AIT. For instance, it is possible to expose a combination methane-air, whose AIT is 580°C, to a jet of gas at a higher temperature, but only for a very short time. Autoignition provoked by contact of a flammable atmosphere with a hot surface often triggers an explosion, as described in [15, 16]. The relations among the flammability limits and the properties of gases and vapors are shown in Figure 2.8. Table 2.4 lists the values of AIT for some substances.

The AIT is not an intrinsic parameter of the material since it depends on the same factors that influence the reaction rate in the gaseous phase and on the peculiar system used for its measurement. They include:

- Volume and geometry of the container, in particular the surface/volume ratio;
- presence of inert (N_2, CO_2, water steam, and so on);
- pressure;
- presence of additives (inhibitors or promoters);

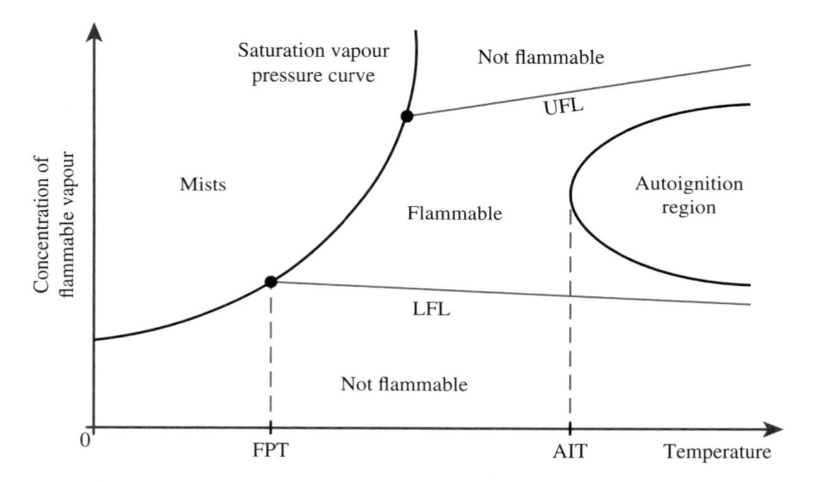

Figure 2.8 Relations among the flammability properties of gas and vapors. Source: Adapted from [11].

Table 2.4 Approximate values of the Auto Ignition Temperature for some substances.

Substance	Auto Ignition Temperature (AIT) [°C]
Methane	537
Gasoline	246
Hydrogen	570
Hexane	220
Paper	230
Wood	220–250
Synthetic rubber	300
Wool	205

- physical state of the fuel (fog, vapor);
- cold flames;
- ignition delay; and
- superficial effects (correlated to the material of the container).

It is possible to classify the flammable gases depending on the way they are stored. According to this classification, it is possible to distinguish among:

- Compressed gasses. They are stored at the gaseous state at a higher pressure than the atmospheric one. They are usually stored in cylinders at the pressure listed in Table 2.5;
- liquefied gases. They are liquefied by compression, at the room temperature. Hydrocarbons and their mixtures are typically stored in this way. The main advantage is the space saving, generally in a ratio of 1/800, meaning that from 1 litre of liquefied gas, 800 litres at the gaseous state are obtained (propane, ammonia, chlorine, LPG are some examples);
- dissolved gases. They are stored at the gaseous state, dissolved in a liquid at a certain pressure (acetylene is a case); and
- refrigerated gases. They are liquefied gases, by compression and low temperatures. Using low temperature allows storing them at a lower pressure than the compressed gas (liquid nitrogen is an example whose pressure is comparable to the atmospheric one).

2.1.1.2 Flammable Liquids

The flammability of liquids is correlated to their capacity of generating vapors in a sufficient quantity to create a combustible mixture with the air. Every liquid is characterised by a value of the vapor pressure. Its increase with the temperature determines the capacity of the liquid to create a higher concentration of flammable vapors in air. This explains why the parameter governing the flammability of liquids is the temperature.

The **Flash Point Temperature** (FPT) is the lowest temperature at which a liquid develops, at equilibrium conditions, a sufficient quantity of vapor to create a flammable mixture with air (i.e. a mixture that inflames under the action of a source of ignition). In other words, at the flashpoint temperature the vapor pressure (i.e. the vapor concentration found at the surface of a liquid at equilibrium) equals the lower flammability limit (LFL). The value of the flash point temperature can be derived from the boiling point temperature using some numerical constants, available in the literature. A different parameter is the **Fire Point Temperature**: it is the lowest temperature at

Table 2.5 Storage pressure of some compressed gasses.

Gas	Storage pressure [bar]
Hydrogen	250
Oxygen	250
Air	250
Methane	300
CO_2	20

which a combustible substance, contained in an open recipient, burns with a sustained combustion after ignition. This distinction is required because at the flash point temperature, the air/vapor mixture above the liquid burns quickly (flash flame) and fire does not sustain. The reader should keep in mind that Flash Point and Fire Point holds at ambient pressure. A deviation from ambient pressure may vary significantly both the parameters.

In practice, it is hard to deal with pure substances, whose value of Flash Point Temperature is known. Industrial processes very often work with multicomponent mixtures. Predicting the Flash Point of a mixture is a complex issue as in principle it may depend upon several variables:

- The composition
- Whether or not some components are not flammable
- The interaction beween the components
- The vapor pressure of the components
- The flammability limits of the components.

Attempts can be made to calculate the vapor pressure and vapor composition in equilibrium with the liquid at a given temperature, and then compare it with the flammability limit of the same mixture as the vapor phase. Most mixtures do not behave as ideal so to calculate the vapor composition may be quite challenging.

As a rule of thumb the reader may consider that a small amount of a highly volatile liquid (hence low Flash Point liquid) may significantly lower the flash point of the mixture. Furthermore, it may happen that some liquid mixtures, classified as not flammable, become flammable in time, maybe after stagnation in open receptacles as is the case with the mix benzene-titanium tetrachloride. In the context of firefighting protection, industries usually take advantage of the inhibitory power of the components in the mixture that are not flammable.

Liquids are divided into different categories of flammability, depending on the interval of temperature to which their Flash Point Temperature and boiling point belongs. For example, the three different categories in Table 2.6 can be identified.

The reader who wants to quickly acquire the basis of combustion and liquid flammability may refer to [17].

The FPT values of some substances and their classification can be consulted in Table 2.7.

2.1.1.3 The Ignition

Ignition is a complex phenomenon that may occur in many ways. Perhaps the most complete treaty on ignition phenomena is the handbook by Dr. Babrauskas [18]. The ignition

Table 2.6 Classification of flammable liquids according to CLP Rule (EU Directive 1272/08).

Category	Flash Point Temperature (FPT)
1	FPT < 23°C and Tb<35°C
2	FPT < 23°C and Tb>35°C
3	23<FPT < 60°C
4	FPT>60°C

Table 2.7 Classification and FPT of some common flammable liquids.

Substance	Flash Point Temperature [°C]	Category
Gasoline	−20	1
Methyl alcohol	11	2
Ethyl alcohol	13	2
Diesel	55	3
Lube oil	149	4

is the side of the fire triangle that provides the needed energy to start the combustion of a mixture fuel-air, within the flammability limits. The **Minimum Ignition Energy** (MIE) is the smallest quantity of energy required to an electric arc (or an electrostatic spark) to ignite a propagating combustion. Its value varies with the mixture composition: it is minimum at about the stoichiometric concentration and maximum in correspondence of the flammability limits. There are many sources of ignition, each with its different typology, provided energy, and duration. They may be classified as follow:

- Autoignition. It occurs when the material is heated to the point at which it ignites. For example, a material inside an oven will ignite at a certain temperature. No external sources of energy, like flames, have been used to pilot the ignition. This phenomenon is quite frequent in fires. For example, the radiant heat transferred from the ceiling of a burning room may increase the temperature of other materials on the floor, until they reach their autoignition temperature and start burning, propagating the fire;
- Spontaneous ignition. It is a particular type of auto ignition, where the heat is produced by the fuel itself. The pyrophoric substances that spontaneously burn at the room temperature are an example;
- Piloted ignition. In this case, a direct external ignition source exists, such as flames or sparks. This means that a portion of the material is heated up to the AIT thanks to an external source, even if the rest of the material is below this temperature. Examples of external sources include free flames, incandescent materials, electric sparks, cigarette stubs, matches, cutting and welding, friction or impact. Also, electrostatic charges are an external source of ignition: this is why they are not wanted. They are usually correlated with transferring, decanting, mixing and shaking operations. Their danger becomes real when they recombine themselves rapidly, after having been separated, or discharged; indeed, the energy developed by electrostatic charges is comparable with the MIE of most gases and vapors [11], as shown in Figure 2.9.

2.1.2 Fires

Fire is an uncontrolled combustion that develops without limits in time and space. The most visible part of fire is the flame, whose color depends on the temperature and the chemistry of the fire, and the smoke, which is an unwanted by-product of fire. The distinct characteristics of fires are:

- Decomposition. The combustion reaction "consumes" the reactants, burning them into other products;

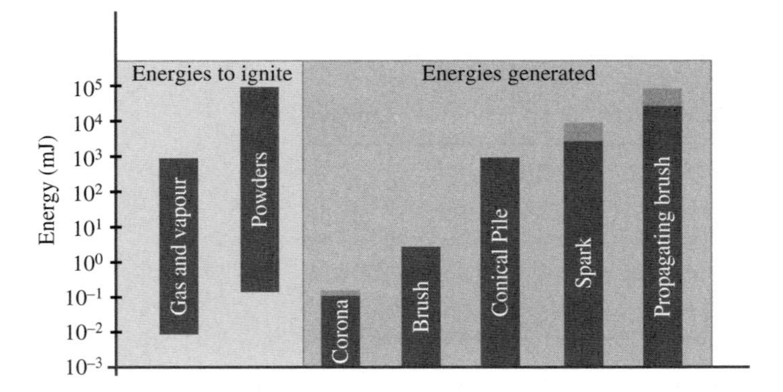

Figure 2.9 Comparison among the MIE of gases and vapors and the energy of electrostatic sparks. Adapted from [11].

Table 2.8 Extinguishers and their actions.

Extinguisher	Action	Features
Water	Temperature reduction Suffocation Dilution of flammable substances Inhibition of solid fuels	Good availability Cheap To not use on electric parts It does not work on light hydrocarbons
Foam	Separation between fuel and oxidiser Suffocation	Classified in high, medium or low expansion To not use on electric parts
Powder	Separation Chemical inhibition Cooling	
Inert gases	Temperature reduction Reduction of oxidising concentration	Suitable on electric fire

- heat. Fire is an exothermic reaction; and
- transfer of heat. The produced heat is transferred through conduction, convection and radiation, as already discussed.

When the extinguishment of the fire does not occur spontaneously, because all the fuel was burnt, then the usage of particular substances, named "extinguishers", is required. They adopt different actions, summarised in Table 2.8.

Fires may be classified in different ways, according to the purpose of the classification. A typical distinction, also promoted by several national technical regulations, concerns the material of the fuel. Taking inspiration from [19], the identified classes are:

- Class A: Fires of solid materials, typically of organic nature (they leave ashes when they combust). Wood, paper, rubber, trash, and cloth are examples. Class A fires are faced removing the "heat" side in the fire triangle, e.g. using water;
- Class B: Fires of liquid substances or liquefiable solids, like petroleum, oils, greases, paints. They do not leave ashes when burning. The most used firefighting strategy to

face Class B fires is to eliminate the oxidiser or, at least, its possibility to react with the fuel, using foam or carbon dioxide extinguishers;
- Class C: Fires in live electrical materials. Circuit breakers, wirings, electrical outlets are examples of Class C combustible materials. If Class A extinguishers are used for a Class C fire, then severe hazards of electrocution can occur. Indeed, Class C fires are extinguished through dry chemical or powder. If the fire involves electronics, such as TVs and PCs, the residual damages likely caused by the powder may be avoided using carbon dioxide or halon extinguishers; and
- Class D: Fires of combustible metallic substances. They include titanium, magnesium, aluminium, sodium, and so on. The best way to face these fires is to use a special powder, handled by trained personnel, to cover the combusting material, excluding the contact between the oxygen and the combusting metal. Typically, Class D fire extinguishers contain sodium chloride, or sodium carbonate.

Depending on the national context, other classes can also be identified. The Class F fires, involving cooking oil or fat, are common in the UK, even if in the US these materials are listed as fuels in Class B fires.

Fires are also classified according to the spatial attribute in which they originate. Indeed, it is possible to distinguish between confined and not confined fires (a third category is the semi-confined fire, being in the middle between the two extreme cases). This classification points out if fires occur outdoors or inside a closed space, like a room or industrial building. On the one hand, confinement may limit the availability of oxygen, hampering the fire propagation; on the other hand, a confined fire takes into account the radiation from hot elements (smokes, walls, ceiling) towards other combustible materials, propagating the fire. The initial evolution is similar for both confined and not confined fires, but when the power increases in time and the fire propagates in space, the confinement effect starts to become relevant and may generate two different conditions. The two subsequent regimes of combustion have been already discussed previously in this book. They are:

- Oxygen-controlled fire. It is typical of confined fire, and the combustion rate depends on the availability of oxygen. An oxygen-controlled fire is the one occurred at deck 3 of the Norman Atlantic. The significant presence of black soot is a key indicator of the lack of oxygen (Figure 2.10); and
- fuel-controlled fire. It is typical of typical of not confined fire. Here, the combustion rate depends mainly on the fuel availability. A fuel-controlled fire is the one occurred at deck 4 of the Norman Atlantic, where the lateral openings supplied a continuous flow of fresh air, allowing the fuel to fully burn. In Figure 2.10 it is possible to see the different colors on the walls of the garage ramp of the ferryboat: on the right, the access to deck 3 is covered by black soot, while these signs are not present on the left, where there is the access to the windowed deck 4.

The main parameters governing a fire are:

- The maximum temperature and the temperature rate of the combustion products;
- the quantity of heat being generated and the rate of its development;
- the duration of the fire;
- the required time to reach the maximum temperature;

Figure 2.10 Different colors at the access of deck 3 and 4 of the Norman Atlantic, suggesting two different typologies of fire. The oxygen-controlled fire at deck 3 (on the right) and fuel-controlled fire at deck 4 (on the left).

- the fire load, which is defined as the ratio between the heat developed by a complete combustion of the fuels and the surface in the plan view of the considered space. It is expressed in kJ/m^2. Conventionally, kilograms of equivalent wood (with a predefined value of 4400 kcal/kg) is used as an alternative measure of heat: and
- the availability of oxygen, which affects semiconfined fires and mainly depends upon the size of the openings.

The evolution in time of fire is usually represented through the Heat Release Rate (HRR) curve, shown in Figure 2.11. It shows the typical trend of a fire, focusing on the variation of the generated power (energy per seconds) in time.

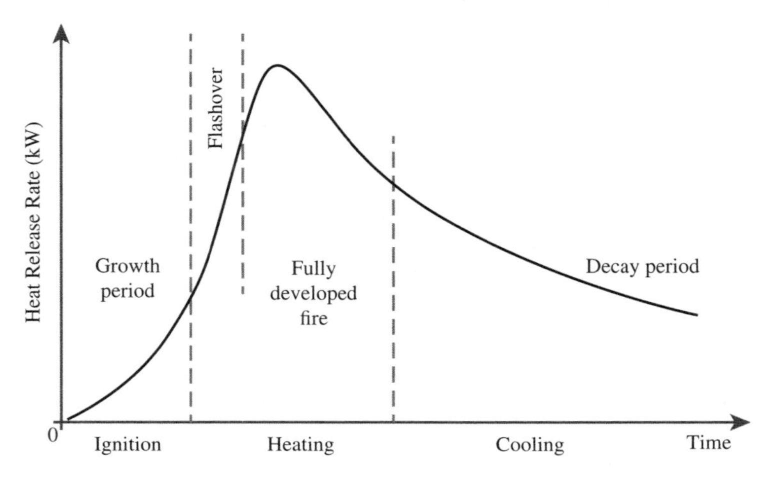

Figure 2.11 Evolution of a fire.

Three main stages may be found:

- Development stage (or pre-flashover). The average temperature of gases is low, and the fire is localised in its origin point. The temperature slowly varies in time because the heat is mainly used to increase the temperature of the fuel materials above their AIT, and to warm the surrounding air and combustible materials next to the origin of the fire;
- complete development stage (or **flashover**). The temperature rapidly increases because the number of materials involved in the fire grows and the increasing temperature causes a higher combustion rate. The fire propagates, and the heat release rate reaches the maximum values; and
- decay stage: the fuel is almost all burned and combustion rate lowers. The temperature decreases because of the heat dispersion through the smokes and the irradiation toward the coolest zones.

In Table 2.9, the growth rate of fire has been classified into categories, depending on the time t_1 required to reach the power threshold of 1 MW. The following Table 2.10 collects the values of t_1 for some commonly used materials. These values have to be intended as approximated, thus purely as a guidance, because of the several variables that affect them.

2.1.3 Explosions

Many types of explosions may be encountered by a forensic engineer. By far the most of them are the consequence of a rapid chemical reaction that delivers a huge amount of energy in a very short time. In many cases an explosion is the result of an uncontrolled combustion occurring at a very high rate. To better appreciate the role of the rate of reaction you may consider the following example. Consider a typical Italian use to have a coffee for breakfast. Imagine being at home and to using gas for cooking. You may imagine two procedures for getting your coffee ready as quickly as possible in the morning (we are always in a hurry…). Procedure one: late in the evening you prepare the coffee machine (moka type, Italian style), open the gas and light the flame. Then you go to sleep. When you wake up in the morning you just put your coffee machine on the cooking pot and (provided the gas cooking pot lighted thorough the night did not make the air in your kitchen toxic with Carbon Monoxide) you get a warm coffee in about one to two minutes. Second procedure: similar to the first but you open the gas valve of your cooking pot but do not light the flame. When you get back to the kitchen in

Table 2.9 Categories of growth velocity of fire.

Category	t_1 [s]
Slow	600
Medium	300
Fast	150
Very fast	75

Source: Data taken from [20].

Table 2.10 Values of t_1 for some materials commonly used.

Material	t_1 [s]
Wooden pallets, stacked, height 45 cm	155 ÷ 310
Wooden pallets, stacked, height 1.5 m	92 ÷ 187
Wooden pallets, stacked, height 3 m	77 ÷ 115
Wooden pallets, stacked, height 4.8 m	72 ÷ 115
Rolls of paper, vertical, stacked, height 6 m	16 ÷ 26
Clothing, cotton and polyester, shelves, height 4 m	21 ÷ 42
Paper, densely packed in cardboard boxes, stocked in shelf, height 6 m	461
Canisters of wasted polyethylene, stacked, height 4.5 m	53
Polyethylene bottles packed in cardboard boxes, height 4.5 m	82
Pallets made of polyethylene, stacked, height 1 m	145
Pallets made of polyethylene, stacked, height 2 m	31 ÷ 55
Single mattress in polyurethane, horizontal	115
Polystyrene tubs stacked in cardboard boxes, stacked, height 4.5 m	115
Polyethylene and polypropylene films in rolls, stacked, height 4 m	38
Insulating panels in rigid foam in polystyrene, stacked, height 4 m	6

Source: Data taken from [21].

the morning, the gas has accumulated in the room so when you try to light the flame an explosion occurs and your home is destroyed. The difference between fire and explosion here is the *combustion rate*. In the first case gas burns slowly throughout the night. In the second case the gas burns in about 2 to 3 seconds. Same amount of gas, hence same energy delivered, but different duration hence different power.

The reader should also consider that not all combustion phenomena may result in an explosion. Combustion may occur in three ways:

- Diffusive flame;
- premixed flame; or
- smoldering (no flame).

When fuel and oxidant move to come into contact to make a reaction, you get a diffusive flame. Examples are a candle, the burning of a pool of liquid or a pile of wood. In general, all fires, even of a very huge size such the fire of a huge building, occur by diffusive flame. This is a rather slow phenomenon as the limiting phenomenon is mass transfer which drives the motion of fuel and oxidant that must come in contact. Diffusive flames cannot produce explosions.

When fuel and oxidant are mixed together before the combustion starts, after ignition you will get a premixed flame. Examples are an oxyacetylene torch, a gasoline engine, or the flame you get after an unwanted release of flammable gas or vapor in air. Flame speed is not more limited by mass transfer and the flame front may reach very high speed. Premixed flames may cause an explosion.

Smoldering is the burning of a solid without flame. A typical example is the burning of a cigarette. Smoldering itself cannot produce explosions but may produce a huge quantity of smoke containing unburnt flammable substances. It may, sometimes, cause the accumulation of a flammable atmosphere in a closed environment. An example is the smoldering combustion of wood chip in a silo which often ends with the silo explosion.

An explosion is the ignition of a mixture of one or more flammable substances in air, with a consequent rapid volume expansion or a pressure increase, depending on the space, confined or not, in which the event takes place. The ignition starts the chemical reaction, producing heat that is then transferred to the adjacent mixture, thus generating a reaction (or flame) front that moves from the combusted gases to the fresh mixture. The propagation rate of the reaction front depends on the velocity of the heat conduction. Explosions can be distinguished, according to their conformation and the reached propagation rate value, in:

- Deflagration. It is an explosion characterised by a flame front that proceeds with subsonic speed. Usually, explosions of gases and airborne powders are part of this category, even if a blast inside a long duct, like a tunnel, may transform it into a detonation. A deflagration is a reaction propagated by heat transfer [22]; and
- detonation. It is an explosion whose flame front has a marked turbulent structure, proceeding at supersonic speed. Compression waves are generated, and they precede the reaction front, propagating in the fuel mixture like a shock wave. Therefore, a detonation can be defined as a reaction propagated by a shock wave [20]. This phenomenon is typical of those real explosive substances that detonate.

The difference between detonation and deflagration are shown in Figure 2.12.

An explosion is defined:

- Mechanical, if it is due to the rupture of a recipient in pressure not containing a reactive gas;
- chemical, if it is generated by the rapidly expanding gases produced by a chemical reaction;

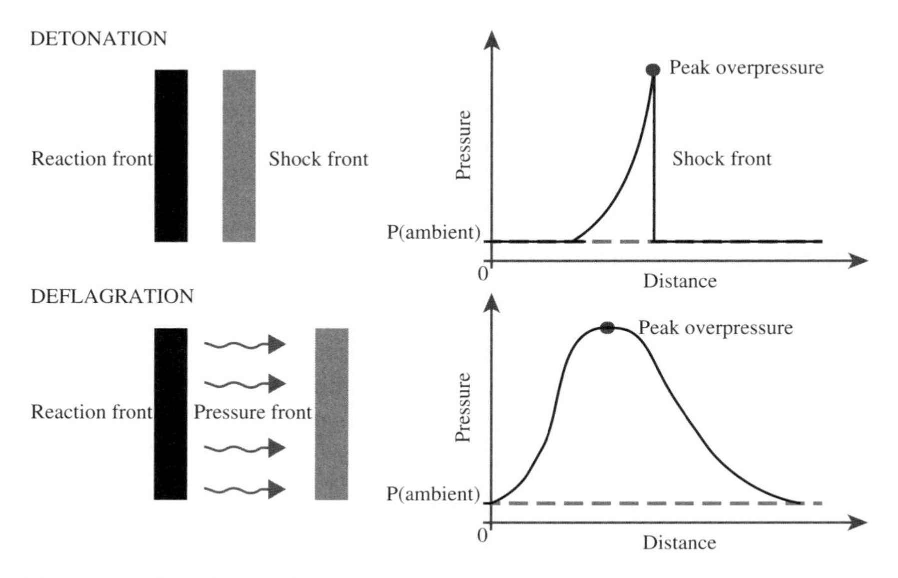

Figure 2.12 Shock front and pressure front in detonations and deflagrations. Source: Adapted from [23].

- physical, if it is due to the expansion of a liquefied gas, even if stored at room pressure;
- confined, if it occurs inside a container, a vessel, a building. The confinement is responsible for a significant increase in pressure; or
- not confined, if it occurs outdoors. It may occur because of a leak of flammable gases or vapors.

Depending on the type of the explosion, mechanical or chemical energy is produced. The energy released by the explosion is dissipated by various phenomena, like a shock wave, radiation, acoustic energy, throw of fragments. Its estimation can be performed by calculating the Gibbs's free energy. The free energy of a compressed gas may be estimated by different methods available in the literature [23].

Process plants may be affected by a number of different types of explosions [23]:

- Condensed phase explosion;
- physical explosion (hydraulic or pneumatic);
- confined gas explosion;
- vapor Cloud Explosion (VCE);
- boiling Liquid Expanding Vapor Explosion (BLEVE); and
- dust explosion.

In the context of this book, only a few of them are discussed, since the other ones, like the condensed phase explosions (e.g. TNT) are far from being of interest for the process plant context, even if they stimulated the creation of several well-established investigation methods. The interested reader can find additional information in [22]. The physical explosions are typically the result of overpressure during a fire. If the vessel is full of liquid, then a hydraulic explosion occurs, otherwise it is a pneumatic explosion, which is more violent. The failure modality depends on the weakest feature of the containment. A confined gas explosion may result in a deflagration or detonation, depending on the nature of the confined space. With a single vessel the explosion is likely to be a deflagration, with a uniform stress over the volume. But with a multi-compartment volume the flame speed can accelerate, such as with piping, resulting in a detonation. In a slow deflagration, bursting occurs at a pressure equal to the structural strength of the item. Instead, with a detonation, the bursting occurs at a higher pressure. In simple words, this happens because the slow deflagration is likely to cause tears, thus a brittle failure will affect the structural response of the item. Conversely, a detonation does not generally cause brittle fracture and the item can use its ductile reserve. Distinguishing the structural response (brittle or ductile) according to the type of detonation is one way that investigators have to determine the type of explosion and the initial point. With this regard, the pattern of the bursting of a single vessel may not reveal too much; alternatively, in a multi-component configuration, the maximum damage is usually observed far from the source of ignition. However, it is harder to find the ignition source for a vapor cloud explosion rather than for solid materials [23]. Also, injury to humans is a potential source of information in the investigation of an explosion. This information is discussed in [24]. A useful guide for explosion and bombing scene investigation is in [25].

Different measurable physical dimensions can be adopted to represent the effects of an explosion. They depend on the specific material and are experimentally determinable by using specific equipment. The maximum pressure of explosion and the deflagration index (a measure of the explosibility) are two examples of these physical dimensions.

Some peculiar values, available by using specific test apparatus for acquiring vapor explosion data [23], are in Table 2.11 and Table 2.12. Table 2.11 is referred to in [23], where data have been selected from the three referenced sources.

Dust explosions deserve a separate discussion because of their features. With the term "dust", we mean solid combustible materials with a diameter lower than 500 μm. They are dispersed in air, forming a cloud that can rapidly burn if ignited and generate an explosion like a gas cloud.

The dynamic of a dust explosion is very peculiar, since the following two distinct phenomena develop:

- A primary explosion. It involves those portions of dust that cause direct structural damages and, by expanding and generating convective motions, lift the dust eventually dispersed in ducts or generally present nearby; and
- a secondary explosion. The dust lifted by the first explosion participates, enlarging significantly the destroying effects of the event.

The primary and secondary explosions are shown in Figure 2.13.

To determine the dangerousness of the explosive mixture dust-air, some tests can be carried out, following the international technical standards. In order to have ignition and explosion of a dust cloud, it is necessary that the concentration of the fuel is between

Table 2.11 Characteristic explosion indexes for gasses and vapors.

	Maximum pressure P_{max} [barg]			Deflagration index K_G [bar · m/s]		
Chemical	**NFPA 68 (1997)**	**Bartknecht (1993)**	**Senecal and Beaulieu (1998)**	**NFPA 68 (1997)**	**Bartknecht (1993)**	**Senecal and Beaulieu (1998)**
Acetylene	10.6			109		
Ammonia	5.4			10		
Butane	8.0	8.0		92	92	
Carbon disulphide	6.4			105		
Diethyl ether	8.1			115		
Ethane	7.8	7.8	7.4	106	106	78
Ethyl alcohol	7.0			78		
Ethylbenzene	.6.6	7.4		94	96	
Ethylene			8.0			171
Hydrogen	6.9	6.8	6.5	659	550	638
Hydrogen sulphide	7.4			45		
Isobutane			7.4			67
Methane	7.05	7.1	6.7	64	55	46
Methyl alcohol		7.5	7.2		75	94
Methylene chloride	5.0			5		
Pentane	7.65	7.8		104	104	
Propane	7.9	7.9	7.2	96	100	76
Toluene		7.8			94	

Source: Data taken from [23].

Table 2.12 Characteristic explosion indexes for powders.

	Deflagration index K_{St} [bar·m/s]	St class			
	0	St-0			
	1–200	St-1			
	200–300	St-2			
	>300	St.3			

Dust	Median particle size [μm]	Minimum explosive dust concentration [g/m^3]	P_{max} [barg]	K_{St} [bar·m/s]	Minimum Ignition Energy [mJ]
Cotton, wood, peat					
Cotton	44	100	7.2	24	–
Cellulose	51	60	9.3	66	250
Wood dust	33	–	–	–	100
Wood dust	80	–	–	–	7
Paper Dust	<10	–	5.7	18	–
Feed, food					
Dextrose	80	60	4.3	18	–
Fructose	200	125	6.4	27	180
Fructose	400	–	–	–	>4000
Wheat grain dust	80	60	9.3	112	–
Milk powder	165	60	8.1	90	75
Rice flour	–	60	7.4	57	>100
Wheat flour	50	–	–	–	540
Milk sugar	10	60	8.3	75	14
Coal, coal products					
Activated carbon	18	60	8.8	44	–
Bituminous coal	<10	–	9.0	55	–
Plastics, resins, rubber					
Polyacrylamide	10	250	5.9	12	–
Polyester	<10	–	9.0	55	–
Polyethylene	72	–	7.5	67	–
Polyethylene	280	–	6.2	20	–
Polypropylene	25	30	8.4	101	–
Polypropylene	162	200	7.7	38	–
Polystyrene (copolymer)	155	30	8.4	110	–
Polystyrene (hard foam)	760	–	8.4	23	–
Polyurethane	3	<30	7.8	156	–
Intermediate products, auxiliary materials					
Adipinic acid	<10	60	8.0	97	–
Naphthalene	95	15	8.5	178	<1
Salicylic acid	–	30	–	–	–

Source: Data taken from [23].

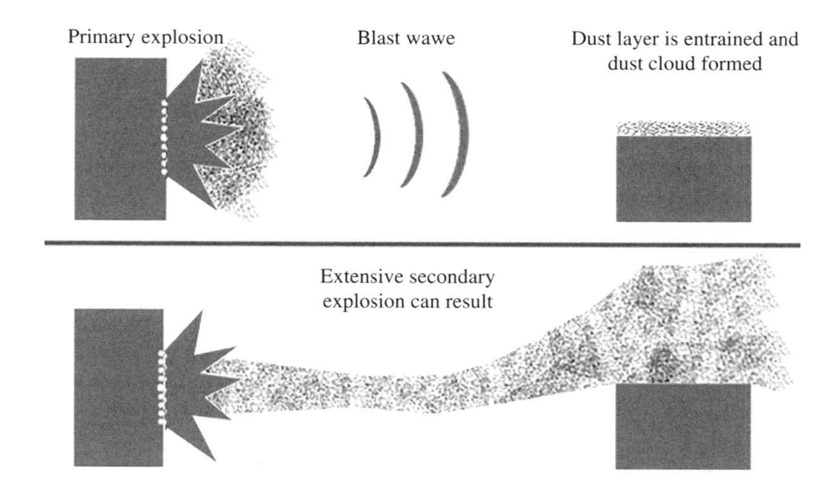

Figure 2.13 Primary and secondary dust explosion. Source: Adapted from [26].

the flammability limits already discussed: the LEL and the UEL. These boundaries of minimum and maximum concentration depend on several variables, including:

- Condition of the superficial layer of the dust particle;
- dimensions of the particle;
- temperature and pressure;
- presence of inert gases; and
- presence of inert dust.

Differentially from gases, UEL does not represent a solid reference for dust: this is not only because of the marked influence numerous variables listed above but also because the upper bound is quite difficult to be found experimentally. Thus it remains undetermined. Moreover, it is almost impossible to obtain a homogeneous system dust-air with a uniform composition, because segregation phenomena usually happen. This results in hard measures. The reader should keep in mind that the data listed in Table 2.12 should be regarded to as order of magnitude rather than definitive values. In any investigation the forensic engineer should collect samples and make new measurements whenever it is possible.

VCE and BLEVE, being significant in the industrial context, are discussed in the next Paragraph, dedicated to the most frequent incidental scenarios in the process industry.

2.1.4 Incidental Scenarios

Industrial accidents involving fires or explosions can be different in nature, as outlined in [27]; consequentially their evolution follows different dynamics. Generally, process plants handle a significant amount of liquid or gaseous fuel that, after uncontrolled releases, may result in an accident. Depending on the type of the release, it is possible to identify some typical scenarios, outlined in Figure 2.14.

The sources of the releases may be different, depending on the plants, equipment, machines, open or closed tanks placed inside a building or outside. The context gives its

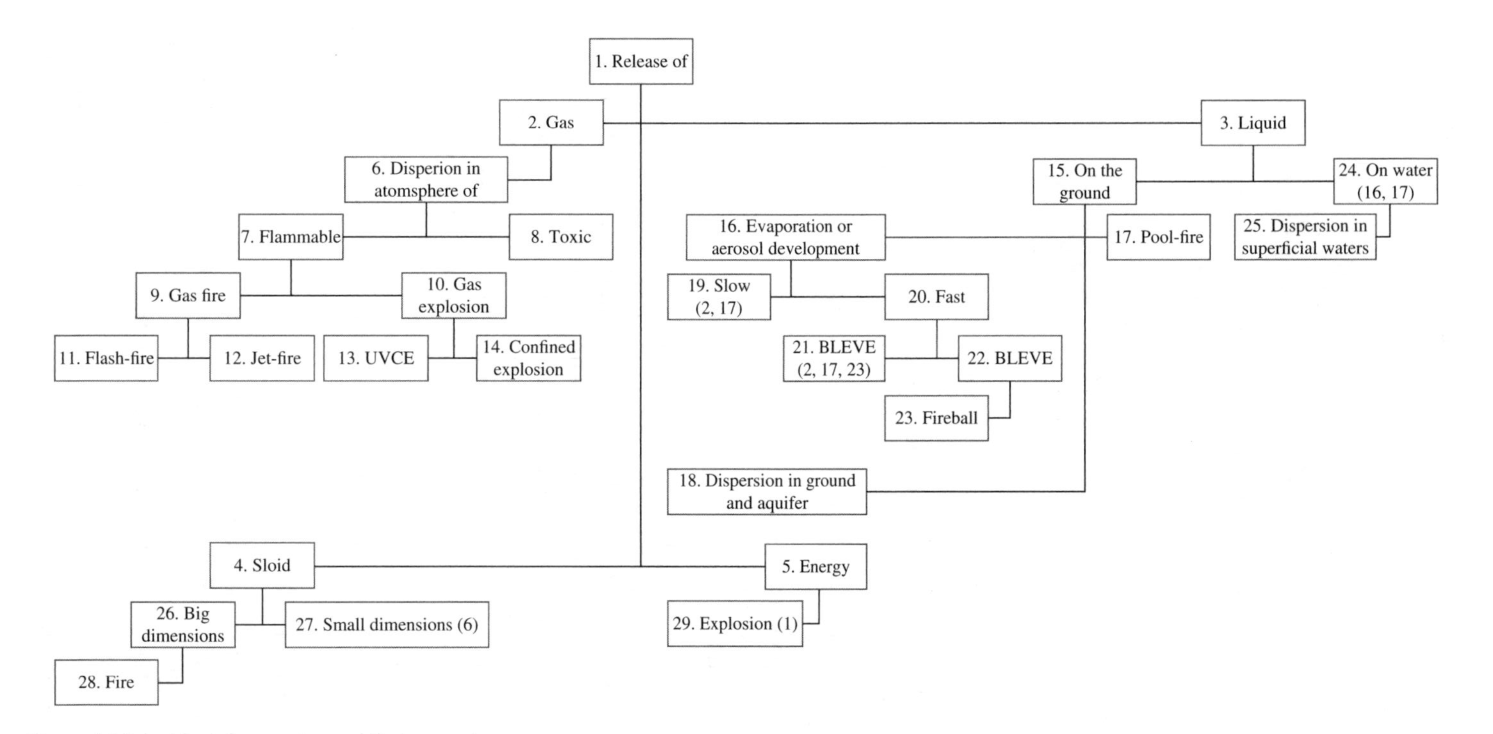

Figure 2.14 Incidental scenarios and their genesis.

contribution in defining the produced effects. Toxic, flammable or energy releases may be the consequence of the incident or the cause that generated them.

Having in mind the information provided by Figure 2.14, it is soundly accepted the following distinction for fires:

- **Flash Fire**. It is a sudden blaze with a limited duration of few seconds. It is caused by the ignition of solids, vapors or gases. A rapid and subsonic flame front is its main feature. Figure 2.15 shows an example.
- **Jet Fire.** It occurs when the mixture of oxidiser and gaseous fuel is ignited in one or more directions. The Jet Fire has a significant diffusion and power. The release of a gaseous substance, from a tank in pressure or piping, is its main cause. It may have different shapes (horizontal, vertical or inclined jet) depending on the local conditions, such as the presence of wind and the geometrical configuration.
  There are several parameters to evaluate the radiated power. Some of them are:
  - Quantity of fuel taking part in the reaction;
  - distance covered by the jet; and
  - the distance of the defeat (i.e. the crack) from the source of ignition.

  Those parameters are then correlated to other conditions like the dimension of the crack and the inner pressure of the tank. Several parameters contribute to model the Jet Fire in a very complicated way. The interested reader can find further information on [23]. Figure 2.16 shows: on the left, the modelled jet fire for a fire investigation [29], further discussed as case study at Chapter 6; on the right, an example of jet fire.
- **Pool Fire**. It is typical of cylindrical tanks. Generally, it is due to the spill of a flammable or combustible liquid. The horizontal dimension of the flames is comparable with the spill dimension while the height is almost double. It may be confined or not, depending on whether the spill occurs in a tank or on the unconfined ground. When the fuel is spilt on water, the pool fire may produce, under specific conditions, the so-called "boil over". It is the boiling of the underlying water with the indirect involvement of a significant quantity of fuel. This sudden phenomenon produces an increase of one order of magnitude in the combustion rate. To evaluate the radiant power emitted by the pool fire, it is necessary to know:
  - The geometry of the flame;
  - the features of the flame;

Figure 2.15 An example of Flash Fire. Source: Frame from [28].

- the combustion rate; and
- the radiation and the geometrical related factors.

Typically, the flame is modelled as a vertical cylinder, even if a more accurate model, taking care about its inclination, provides better results, especially in windy conditions. For more details, see [23]. Figure 2.17 shows an example.

- **Fireballs**. They can be the consequence of two events:
  - The collapse of an LPG tank, with the consequent vaporisation and ignition of the containment; and
  - the ignition of a cloud of gaseous fuel (rarer. Generally it generates a flash fire).

In both the two cases, the cloud of fuel burns with a diffusive flame, as shown in Figure 2.18.

Modelling a fireball allows an estimation of the thermal radiation produced. To do so, it is necessary to evaluate:

- Involved mass of fuel;
- diameter and duration; and
- radiation and geometrical related factors.

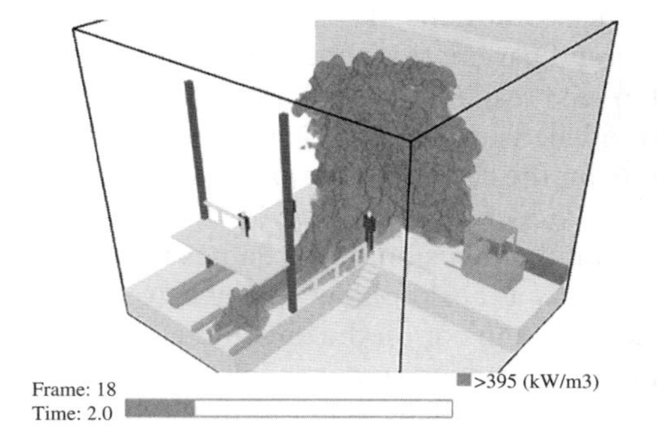

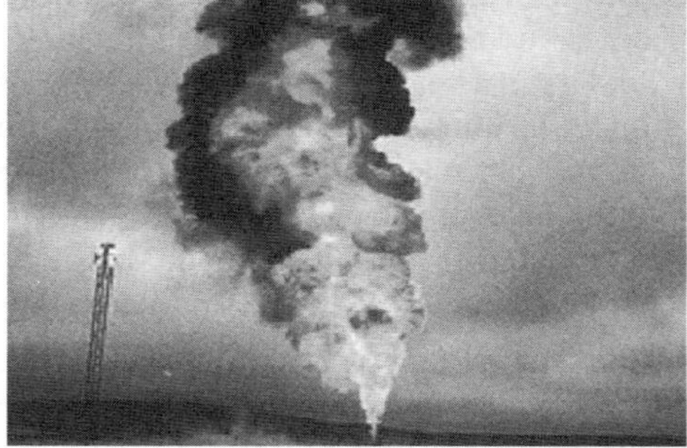

Figure 2.16 On the left, a modelled jet fire for a fire investigation Source: [29]. On the right, an example of a jet fire Source: [30].

Figure 2.17 Example of Pool Fire.

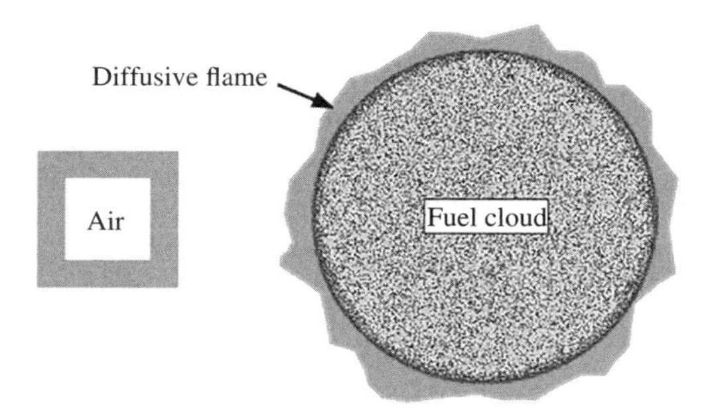

Figure 2.18 Schematic representation of a fireball in the stationary stage.

Figure 2.19 A Vapor Cloud Explosion test. Source: Reprinted with permission from [31].

Three of the models widely adopted to calculate the emitted power per unitary surface of the fireball (thus an estimation of the radiation on the surroundings) are the following [23]:

- Point source model. It considers that all the energy is produced from one single point, i.e. the centre of the fireball;
- solid flame model. It estimates the power radiated by assuming that the flame is equivalent to a grey body; and
- flame emissivity. It evaluates the emitted power starting from some peculiar parameters like temperature, dimension, and composition of the flame.

Regarding explosions, the following incidental scenarios can be observed:

- **Unconfined Vapor Cloud Explosion (UVCE).** It is an unconfined explosion of vapor in the atmosphere, even if partial confinement is usually the real context because of natural obstacles, the presence of buildings or simple openings that deeply influence its dynamics. This kind of explosion is worldly considered among the most dangerous in the chemical industry sector. It is generated after the release of a significant amount of vapors in the atmosphere. The ignition and the subsequent combustion create a flame front and an expansion of the combustion products, causing a pressure wave that sometimes may evolve in a shock wave. The flame front may accelerate significantly in case of congestion (i.e. in presence of many small obstacles like pipings, trees) to supersonic value such that locally a detonation may occur. One of the most sadly famous examples is the one in Flixborough, which occurred in 1974, where 40 tonnes of cyclohexane were released from a reactor, generating a cloud of half a million of cubic meters whose ignition caused 28 victims, the destruction of the plant and an economic loss of 150 billion US$.

- **Confined or partially confined explosion**. In this kind of explosion, energy is released inside a containment structure, like a tank, a reactor, a room or building. If the explosion, involving piping or a vessel, is generated by a flammable gaseous mixture, it is possible to have deflagration or detonation. Instead, it is not clear whether detonation may be generated from a dust explosion, in the context of an industrial plant. Confined explosions produce pressure waves that may cause, in an interconnected system, the so-called "pressure piling". The phenomenon is caused by an increase in both temperature and pressure inside a tank, determining a similar growth in the connected system, thus generating further increases. To avoid this complex behavior and isolate the various systems, it may be useful to install rapid depressurization valves.
- **Boiling Liquid Expanding Vapor Explosion (BLEVE)**. It occurs when a tank, containing a liquid under pressure, collapses suddenly causing the rapid depressurization and the subsequent evaporation of the fluid, resulting in an extremely dangerous explosion (Figure 2.20 and 2.21). The most frequent cause is the engulfment in flames of a tank. Flames warm the upper part of a tank (the one above the liquid level), while

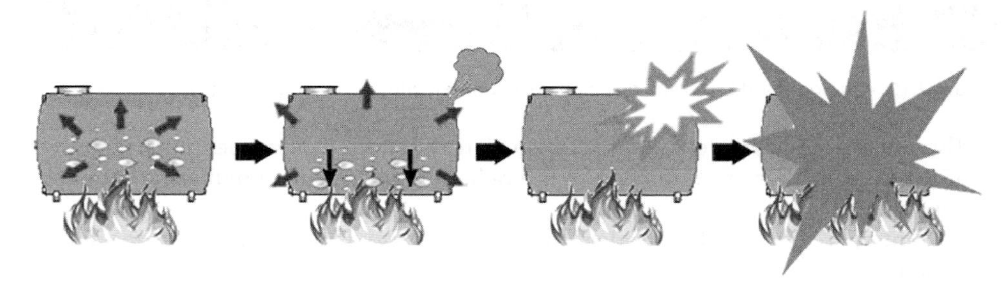

Figure 2.20 Sequence events to BLEVE.

Figure 2.21 Example of BLEVE. Source: Frame from [32].

the lower part remains at a lower temperature, because heat is transferred to the liquid inside that changes its phase. This may result in a decrease of the mechanical resistance of the metal above the liquid level and a parallel increase of the inner pressure because of the increase of the vapor pressure. Therefore, from a structural point of view, actions increase and strength decreases. When the crack appears, the subsequent prompt evaporation, due to the sudden depressurization, causes a catastrophic rupture. A famous example is the incident of Mexico City, which occurred on 19 November 1984, which caused about 500 victims and more than 7000 injured people. On that occasion, the rupture of a piping containing LPG was the reason for a release of a flammable cloud that was ignited by the torch of the plant. The Vapor Cloud Explosion (VCE) and the jet fire that followed were the two contributors of the first BLEVE of an LPG spherical tank. Fifteen further BLEVEs followed, causing the complete destruction of the plant.

2.2 Near Misses

A **near miss**, also called "near hit", is an incident that had no adverse consequences but only the potentiality to generate a loss. Some authors summarise this concept saying that the main difference between an accident and a near miss is then luck or chance [4]. We can additionally say that while an incident is an actual loss or harm, a near miss (according to the definition given by [33]) is an occurrence in which an accident (i.e., property damage, environmental impact, or human loss) or an operational interruption could have plausibly resulted if circumstances had been slightly different. With this given definition, a near miss could be confused with a potential incident. Indeed, the two concepts are intimately different. Let us use the same example reported in the previous Paragraph to illustrate the difference more clearly. It was established that, by definition, a wrench falling from the top of a three-storey scaffold and stopping at an intermediate layer is a potential accident. We have already said that if the wrench hits a worker at ground level, then it is an accident. However, what happens if the wrench continues its fall without stopping at an intermediate layer and hits the ground causing harm to nobody, since nobody is present under the trajectory of the wrench? No one is hurt simply because nobody was present, but all the safeguards put in place failed. In this case, a near miss occurred. In terms of a Fault Tree Analysis (FTA) (this technique is discussed in Chapter 5), a near miss is characterised by one or more negative inputs to an AND gate [8]. Typical examples of near misses, related to the process industry, include:

- Processes running outside safety operational limits without a severe direct consequence;
- shut down system unnecessarily being activated; or
- the release of toxic chemicals, but nobody is affected.

A near miss is different from an **undesired circumstance**: with this definition, a set of conditions or circumstances that have the potential to cause injury or ill health is intended. The conceptual differences among accident, near miss, and undesired circumstance are shown in the example of Figure 2.22.

It is important to deal with near misses, since they represent unique occasions for learning from experience without any severe consequences for human safety,

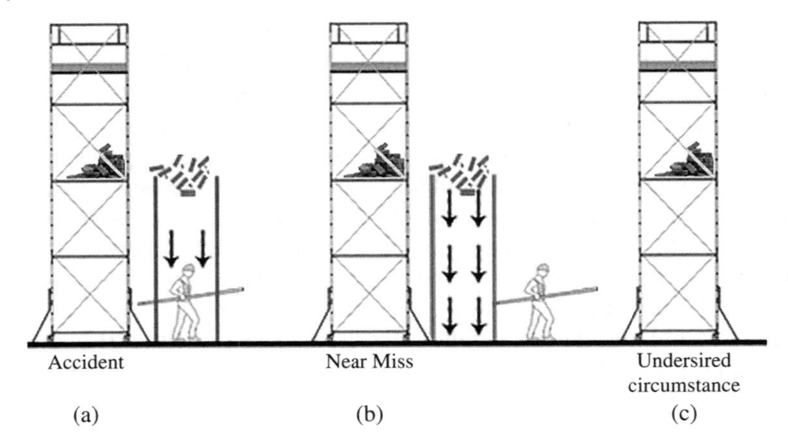

Figure 2.22 Differences between accident (a), near miss (b), and undesired circumstance (c). Source: Adapted from [3]. Courtesy of Health and Safety Executive.

environmental impact or business loss. They are free lessons to learn, with the goal of preventing future accidents and then to avoid future harm and loss, including major expenses for the companies dealing with an incident. They are opportunities to prevent expensive accidents, identify systematic problems in the safety management and to face them. This is why they are investigated as deeply as the incidents are, being a robust performance indicator about the global safety and reliability of the system.

How to understand a near miss and treat it is part of Chapter 6.

2.3 Process Safety

In the process industries, the continuous efforts in preventing an industrial accident or a near miss are known as **Process Safety**. This definition covers mainly the personal safety (which is primarily faced by occupational safety and health) but also those efforts to prevent environmental harm or economic loss. It is intended as a complex challenge to avoid explosions, fires and accidental releases of potentially dangerous chemical substances. Applying process safety and establishing the right paths to follow is not an easy task. It requires strong capabilities and sound knowledge of the specific process one intends to analyse. Process safety is part of the strategy to ensure the **Operational Excellence** (OE) of a company. It is a complex mix of regulations, procedures, internal standards but also the habits, culture and knowledge necessary to push the overall company profile on to an excellent level of performance. The indicators used include those about profitability (thus efficiency, quality and quantity) and safety (thus prevention from health-related risks, environmental risks and economic loss-related risks). In some companies, reducing the risk profile as much as possible, and ensuring a high level of operational excellence are implemented through specific strategies, whose objective is to reach the "**Goal Zero**". The expression "Goal Zero" means the asymptotic result of zero incidents, no environmental harm and zero economic losses during all the lifecycle of the company. It is an asymptotic result since empirically we attend, year by year, industrial accidents causing losses on varying scales of severity. During recent decades, a great deal has been done to reduce the frequency of the occurrence of accidents. Nowadays a quasi-horizontal levellling off of the cost-benefit curve related to the prevention of industrial accidents has been reached. This means that, compared to the past, a bigger effort in terms of resources (mainly budget and time) is necessary to reach the same reduction in the frequency of occurrence of accidents. This happens because more complex problems need to be faced, requiring high-cost solutions that use high-technologies, engineered only after many years of high-costly research and the development of projects.

The strategy adopted by industries to deal with process safety has evolved over time [34]. The oldest one was a standards-based strategy. For a long time, experience-based standards defined the process safety and loss prevention efforts of many companies. Standards include both internal company rules and external technical standards, like the ones promoted by ANSI, API, ASME, and NFPA. However, the experience-based dataset at the basis of standards development is not big enough to develop effective strategies, since accidents are rare events. It is a prescriptive approach. Public attention on industrial incidents forced national governments to create laws and regulations to ensure a minimum level of accident prevention to protect workers, communities and the environment. The Process Safety Management (PSM) standard by OSHA and the Seveso Directive in Europe are an example. Companies were therefore forced to be compliant with the new set of national and international regulations: it is compliance-based process safety management. However, this approach defines only the minimum requirements and, in some cases, is not sufficient to manage the risks properly. Moreover, those facilities having a quantity of hazardous materials below the threshold are not asked to be compliant, even if they remain risky activities. Thus, this kind of approach may not be the best strategy for some companies. A further step has been done with the continuous improvement-based process safety management.

2.3.1 Management of Safety

In recent decades, the quality management programs changed substantially, continuously improving themselves. Many companies applied the same emphasis to process safety management, being aware that a stationary strategy would have been dangerous, badly affecting performance and global competitiveness, not reaching the full satisfaction of the society's safety expectations. This is the moment where companies raised the bar, in a proactive approach about how to learn from experience. But this kind of strategy actually also fails if performance indicators, named lagging indicators, are low-frequency and high-consequence past events. This leads to the currently most used risk-based process safety management. In this approach, companies continue to be compliant, adherent to well-established technical standards, valorise their experience applying the lessons learnt, and continue using the lagging indicators. What makes the difference in a risk-based process safety management is the use of leading indicators. In other words, independently of any loss events, risk information is used to predict the performance, in a full prevention-based approach. The adoption of this strategy requires an accurate understanding of the risks, a proper selection of the performance indicators, an adequate discipline to monitor them, a developed organization integrity to review performance, and a powerful and flexible management system, capable of applying the corrections suggested by the predictive metrics. It is a performance-based approach, in opposition to the prescriptive one. A risk-based process safety management simply answers the following questions:

- What can be wrong? (Hazard identification);
- how bad could it be? (Consequence evaluation); and
- how often might it happen? (Frequency evaluation).

Understanding hazards and risks, managing them, and learning from experience are three important pillars of a risk-based process safety, but a fourth one needs to be introduced: commitment to process safety. Commitment to process safety, as reminded by [34] and [12], means:

- Developing and sustaining a Process Safety Culture;
- identifying, understanding and being compliant with standards;
- constantly improving the managerial skills and competencies; and
- ensuring an adequate workforce and stakeholders involvement.

It is difficult to define uniquely what Process Safety Culture is. A useful definition is given by [34]: "the combination of a group of values and behaviors that determine the manner in which process safety is managed". The best way to ensure a successful Process Safety Culture is to apply the requirements of conventional safety culture, since the two topics share many concepts. The topic is extensive, thus the interested reader is invited to consult the references cited in this Paragraph together with the suggested further readings, to study in depth this pillar of Process Safety. For the scope of this book, it is sufficient to present the key principles that should be addressed when developing, evaluating, or improving a management system and its Process Safety Culture:

- Maintain a dependable practice. It means that the implementation of good practice is ensured over time. It generally requires strong leadership, establishing process safety as a core value, written procedures and documents, establishing high-performance goals;
- develop a sound culture and implement it. It means that the organization should maintain a sense of vulnerability and be modest with respect to its capability of managing risk. An open and effective communication must be ensured, together with a constant training for both groups and individuals. A questioning and learning environment, mutual trust, and a prompt response to process safety issues are also other important elements; and
- guide and monitor the culture, continuously monitoring the performances.

Learning from case histories is a good option to gain an immediate outline about process safety and how it reflects on both daily activities and unwanted incidents. [35] and [12] show many examples of chemical process safety incidents. The attention to Process Safety is quite recent, as Figure 2.23 shows. Similarly, the evolution of the safety culture is shown in Figure 2.24. Today, industrial accident analysis, strongly based on the Process Safety concepts, has widely accepted the Swiss Cheese Model by Reason, whose details are discussed in Chapter 5.

Dealing with process safety also requires a multidisciplinary approach: mechanics, physics, chemistry, metallurgy, industrial process engineering, and thermodynamics are only some of the main topics that a process safety engineer handles on a daily basis. The required approach is an important link with the attitude and the skills that a forensic

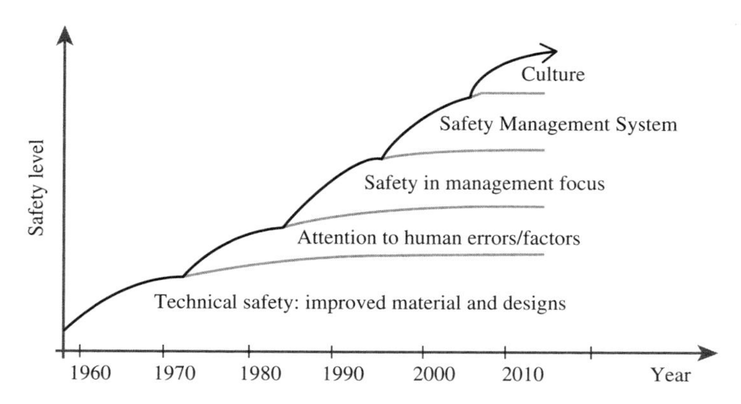

Figure 2.23 Contributing factors in improving loss prevention performance in the process industry. Source: Adapted from [1]. Reproduced with permission.

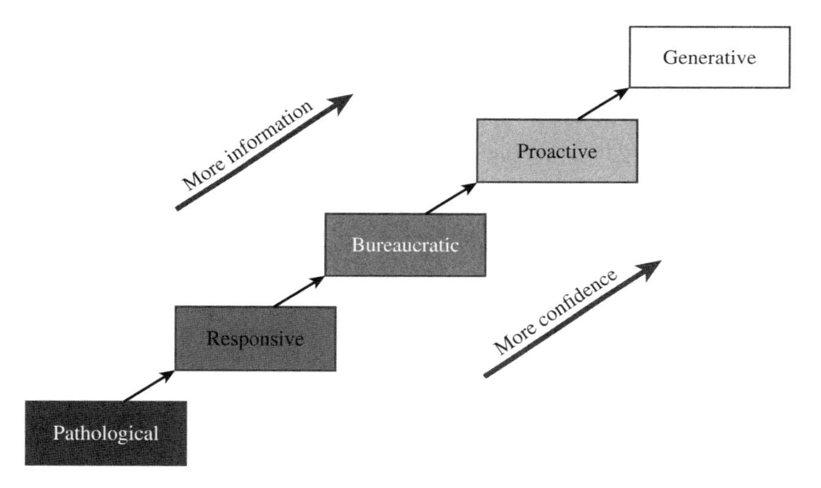

Figure 2.24 The evolution of safety culture. Source: Adapted from [36]. Reproduced with permission of Fiorentin.

engineer spends when dealing with industrial accidents. This point of contact is crucial since it shows a first connection between the context of the process industry and the forensic discipline. Following this observation, some of the main important arguments related to the "process safety" are now presented. The approach used is by choice a smooth path.

Plants and units of an industrial company are equipped with the Basic Process Control System (BPCS). It consists of all instrumentation, including the Distributed Control System (DCS), for process measurement, display, and regulation installed to support normal process operations. The DCS does not perform any safety instrumented functions with a claimed Safety Integrity Level (SIL) ≥ 1 (this concept will be discussed in depth in Chapter 5, where the Layer Of Protection Analysis is introduced). The DCS is a computer-based control system which divides process control functions (display, control, communication and data storage) into discrete subsystems interconnected by communication channels (data highways). The SIL indicates the degree of risk reduction allocated to a Safety Instrumented Function (SIF): they range from 1 (lowest integrity) to 3 (highest integrity). A SIF is an instrumented function with a specified Safety Integrity Level (SIL) necessary to achieve Functional Safety, i.e. part of the overall safety related to the process and the Basic Process Control System, which depends on the correct functioning of the Safety Instrumented System (SIS) and other protection layers. Finally, an SIS is an instrumented system used to implement one or more Safety Instrumented Functions. An SIS is composed of any combination of sensors, Logic Solvers, and Final Elements. The Logic Solver is that portion of either a BPCS or SIS that performs one or more predefined functions as a result of the condition of the input data. The logic solver may be pneumatic, hydraulic, electrical, electronic or programmable electronic. Sensors and Final Elements are not part of the logic solver. Indeed, the Final Element is the part of an SIS that implements the required physical action to achieve a safe state. Simple examples are on-off type valves and motor control starters. An automated instrumentation system or subsystem that performs a discrete action in response to Process Variables (i.e. a measured characteristic of a process such as pressure, temperature, flow, level or concentration) or physical conditions outside a prescribed limit is named Interlock. The affected device shall stay in the safe state until the condition which caused the action is corrected. An interlock may be designated to prevent hazards related to safety, environmental, asset protection/mechanical integrity or product quality excursion and may protect against one or more hazards (i.e. a chemical or physical characteristic that has the potential to harm people, property, or the environment).

A simple way to indicate the general flow of plant processes is the Block Flow Diagram (BFD). It is a schematic representation of the flow process through blocks. An example is shown in Figure 2.25. To have a first reference to the associated equipment, the Process Flow Diagram, also known as Process Flow Sheet (PFS), is consulted. It is a diagram commonly used in the process engineering that displays the relationships among major equipment of a plant facility. Figure 2.26 shows an example. Major information, like pipings and designations, are shown in the Piping and Instrumentation Diagram (P&ID), a detailed diagram where the piping and all the items in the process flow, together with the instrumentation and control devices, are shown. An example is in Figure 2.27.

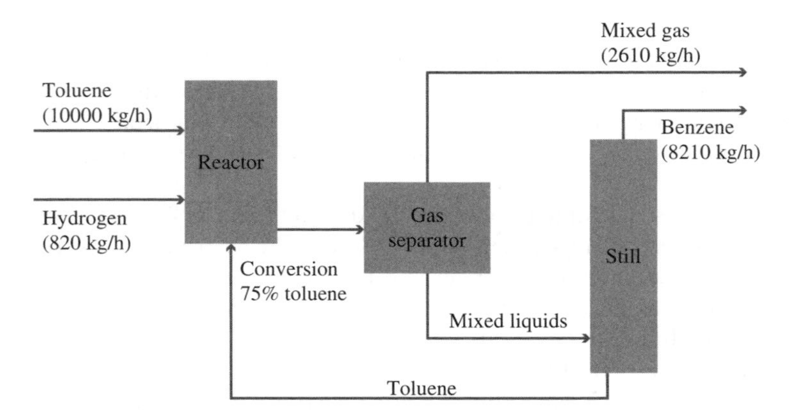

Figure 2.25 Example of BFD for the production of benzene by the HydroDeAlkylation of toluene (HDA).

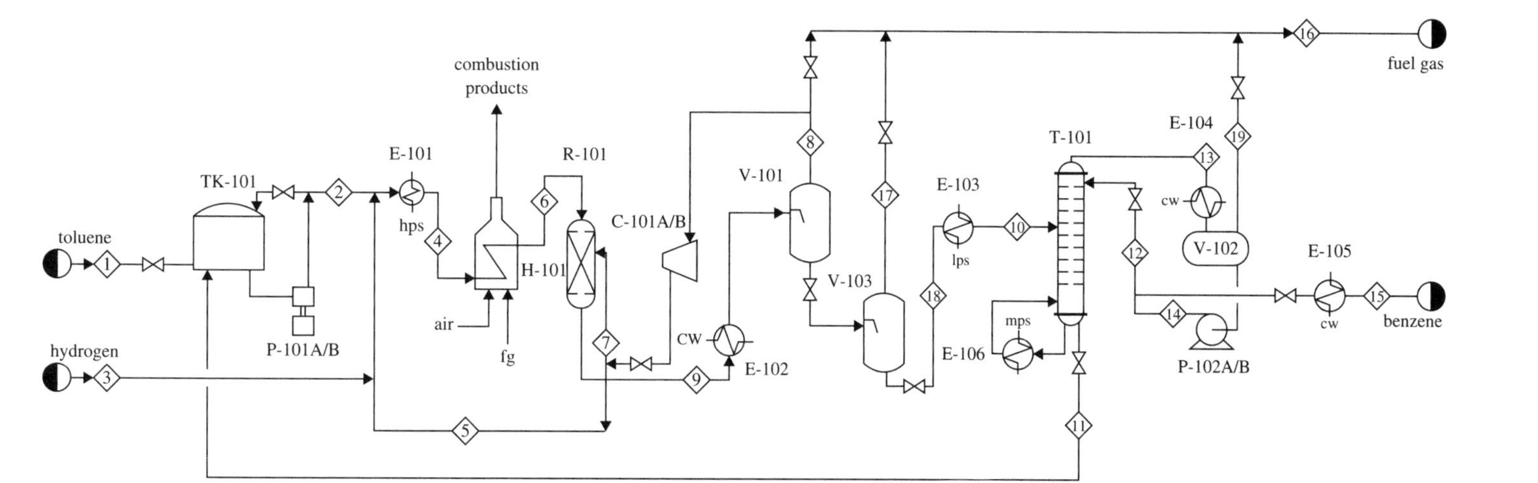

Figure 2.26 Example of PFS for the manufacture of benzene by Had.

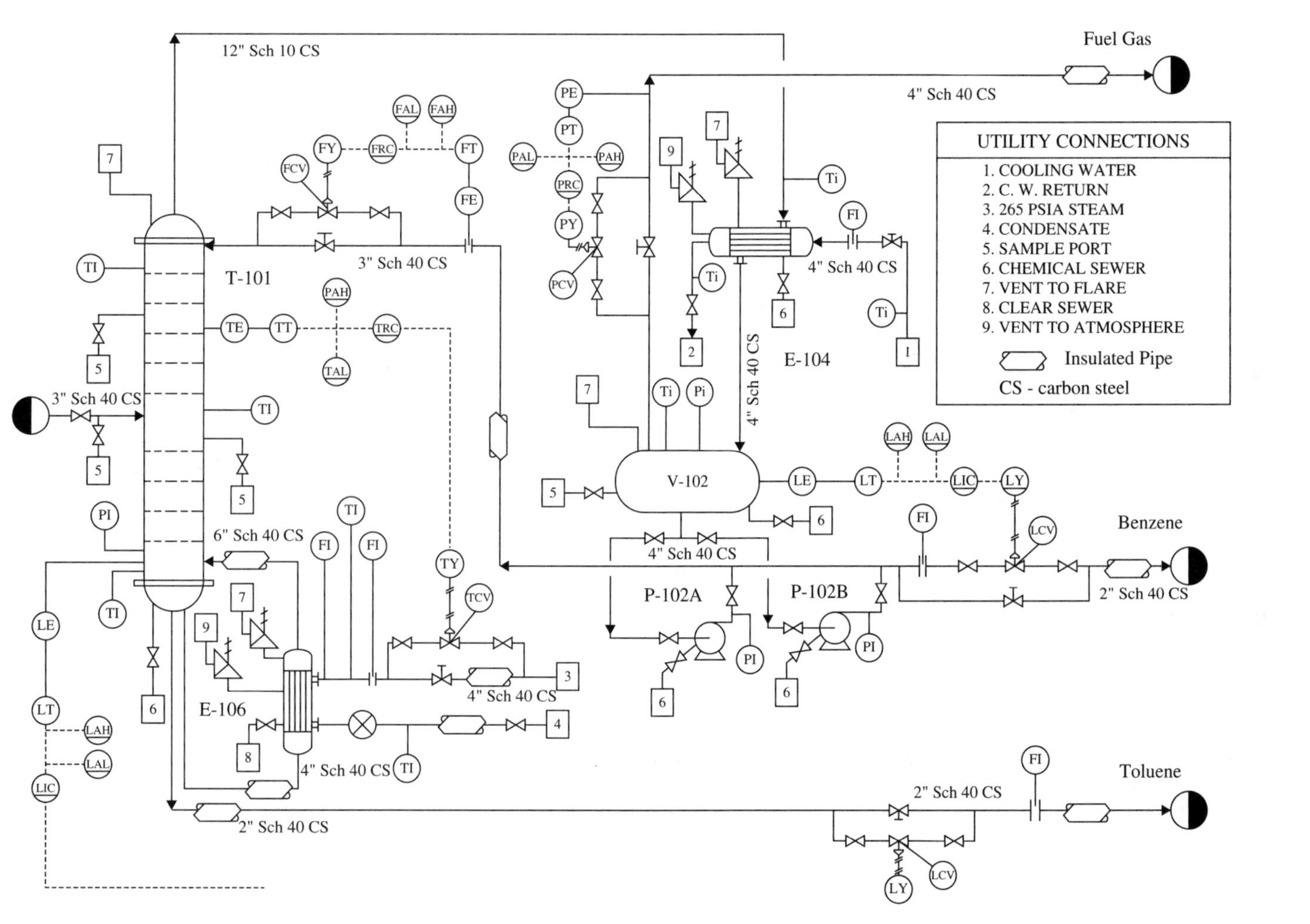

Figure 2.27 Example of P&ID for the production of benzene by Had.

Process Hazard Analysis (PHA) is one of the available tools to identify, analyse and control the industrial hazards. Its results can be seen as an organised effort to face the consequences associated with deviations in process and operations, equipment or in handling the hazardous chemicals. A Process Hazard evaluation uses Risk Assessment (RA) techniques to determine the magnitude and frequency of consequences, assessing whether adequate safeguards are in place and developing recommendations whether additional safeguards are required. A safeguard is defined as any device, system or action which would likely interrupt the chain of events following an Initiating Event (IE). Finally, an IE is an event or deviation that results in a sequence of events that could lead to an undesired consequence.

The interested reader can find a rich literature about Process Safety, in the "Further reading" section of this Chapter.

2.4 The Importance of Accidents

When an accident occurs, the emergency machine (set up by local institutions and/or by the same company) intervenes to stop the cause of the accidental event – or at least to halt the propagation, thus the domino effect of the accident towards adjacent areas, plants or units – and also to provide emergency assistance where needed. Once this step is concluded, what remains is only debris and significant loss, including injuries to workers, leaks causing probable pollution and a destroyed plant, thus enormous economic damage. The occurrence of an accident then becomes an opportunity to learn some lessons, although at payment of a high cost. These lessons start with analysis about what went wrong. They allow a better understanding of unsolved questions – or unknown aspects – of which the impact on the process safety is considerable, as witnessed by the occurrence of the accident itself. To avoid the recurrence of the accidents, an accident investigation is carried out with the objective to find the so-called root causes that led to the event.

The Learning From Experience (LFE) process captures learning opportunities from the incident, audits or other events, which drive enhancements to the Operational Excellence Management System documentation of the company and requests for actions through directives, advisories or LFE alerts. At this point in the book, it is important that the reader should have understood that an investigation can enhance learning only if:

- It is fact-finding, not fault-finding;
- it must get to the root causes; and
- it must be reported, shared and retained.

The importance of incident analysis relies on three simple principles, depicted in Figure 2.28. However, the final improvement is the scope of a manager, as Figure 2.29 reminds us. Indeed, the incident investigation is also a fundamental act for third parties and institutions, whose real interest is not the actual improvement of safety levels, but to protect private and public interests, respectively.

To immediately understand the real importance of accidents, some widely recognised accidents are described next, highlighting their key characteristics in terms of lessons learned and measures taken to avoid any recurrence. The incident investigation workflow is discussed in depth in Chapter 4; however, Figure 2.30 shows a brief

Figure 2.28 Principles of incident analysis.

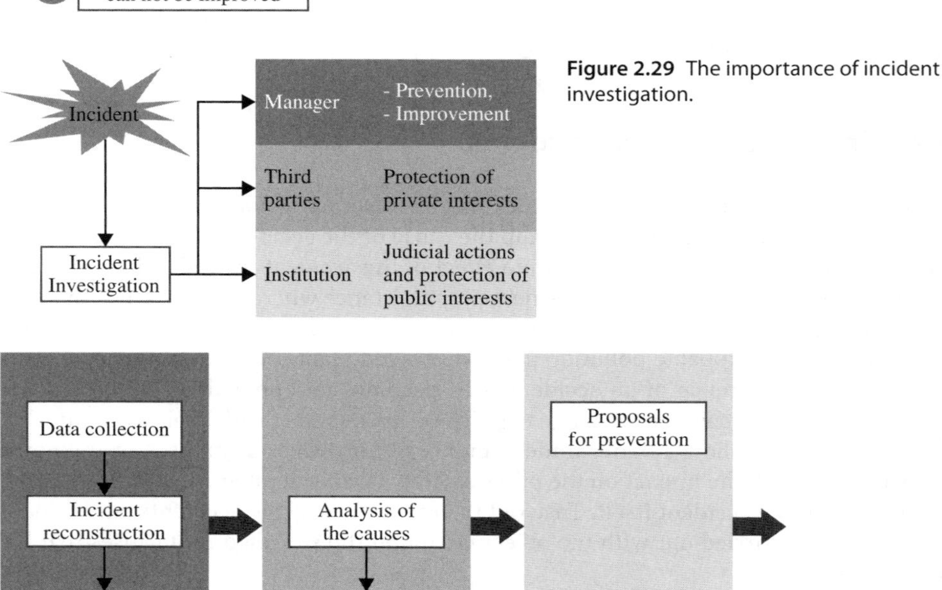

Figure 2.29 The importance of incident investigation.

Figure 2.30 Steps of incident analysis.

summary, in order to provide a very basic knowledge to the reader to better consider the following paragraphs about some real case studies.

2.4.1 Seveso disaster

The Seveso incident is among the most tragic of known industrial incidents. The level of the recorded consequences was so significant and shocking as to radically change the approach to process safety within the industrial community, promoting the birth of specific European regulations, currently known as "Seveso directives".

On Saturday 10 July 1976, in a chemical plant near the town of Seveso, near Milan, a bursting disc on a chemical batch reactor ruptured. The plant produced

2,4,5-trichlorophenol (TCP) (a product used to make herbicide) from 1,2,4, 5-tetrachlorobenzene and caustic soda, in the presence of ethylene glycol. Dioxin (2,3,7, 8-tetrachlorodibenzodioxin or TCDD) is not normally formed (except in negligible amounts). But that day the reactor got too hot and a runaway reaction occurred.

The reason why the reactor became too hot is now explained. Italian Law required the plant to shut down for the weekend, regardless of the possibility of being in the middle of a batch. This is what happened at the weekend of the incident, when the reaction mixture was at 158°C (the temperature at which the exothermic uncontrolled reaction was believed possible is 230°C). Regardless of the chemical concepts related to that specific process, it is now known that the exothermic reaction can start at 180°C, proceeding slowly at this stage. The reactor was heated by an external steam coil, where the steam was provided at a temperature of 190°C. Because of the interruption to operation for the weekend, the turbine, from which exhausted steam was taken to feed the steam coil, was on a reduced load. The consequence was a temperature of the exhaust steam of about 300°C.

Obviously, the temperature of the liquid inside the reactor could not be greater than its boiling point at the operating pressure (about 160°C). This caused the following temperatures at the Seveso reactor (Figure 2.31):

- Below the liquid level: gradient of temperature with 300°C outside the reactor wall and 160°C inside; and
- above the liquid level: the wall was at 300°C.

Once the steam was isolated, the part of the wall reactor below the liquid level was in touch with the cooler liquid inside, while the part above the liquid level remained hotter. Thanks to radiation, the heat was transferred from the upper wall to the surface of the liquid. In 15 minutes the temperature of the liquid rose to 180–190°C, triggering a slow exothermic reaction. A runaway reaction occurred then after about seven hours.

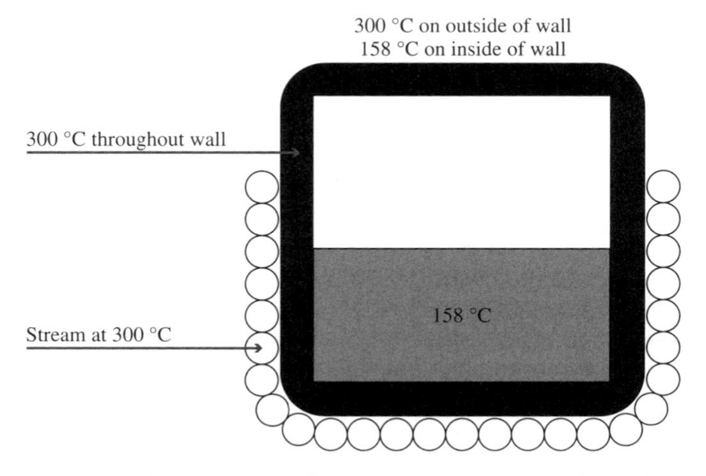

Figure 2.31 Temperatures at the Seveso reactor. Source: Adapted from [35]. Reproduced with permission.

The runaway would not have occurred if:

- Italian laws left sufficient autonomy to the company management to complete the batch, having the weekend provision;
- the batch was not stopped at an unusual stage; and
- a HAZOP (a hazard identification technique further discussed in this book) had been conducted and a "more temperature" deviation had been analysed.

As a consequence of the runaway, dioxin formed and pressure rose resulting in the bursting of the rupture disc. The plant did not have a catchpot where to collect the discharge of the reactor, therefore about 6 tons, including about 1 kg of dioxin, were released over the surrounding area [37]. The emission was limited thanks to a foreman who opened the cooling water supply to the reactor coils, alerted by the noise of the vent. The dense white cloud dispersed in the atmosphere, reaching considerable altitude. The release lasted for twenty minutes about, and during the next days the lack of communication between the company and the authorities, called to manage this type of situation, leaded to much confusion. No human victims of dioxin were registered, even if many people fell ill (250 people developed the skin disease chloracne and about 450 suffered bursts by caustic soda). Moreover, 26 pregnant had abortions attributable to the toxic substance and thousands of animals, living in the contaminated area (17 km^2 contaminated and about 4 km^2 declared uninhabitable – see Figure 2.32), died while others were slaughtered to avoid the ingress of dioxin inside the food chain.

The incident investigation showed a series of failing in technical measures [38]:

- Operating procedures. The production process was interrupted, but without any agitation or cooling. This caused the reactants to continue their reaction. Also, according to the plant procedures, the charge was acidified after the distillation, reversing the original sequence prescribed by the original method of distillation patent;
- Relief/Vent systems. The rupture of the bursting disc was set at an excessive pressure (3.5 bar). With a lower pressure set, venting would have occurred at a less hazardous temperature;
- Control Systems. The overall system for monitoring fundamental parameters and providing automatic control was inadequate;
- Reaction/Product Testing. The awareness of the company about the thermal stability of the reaction was only partial;

Figure 2.32 A photograph of the signs used to forbid access into the infected areas in Seveso.

- Design Codes – Plant. There was no individual protection layer to destroy, or at least collect, the toxic material once vented;
- Secondary Containment. No secondary receiver was installed to recover the toxic materials, in contrast to the bursting disc manufacturer recommendations; and
- Emergency Response/Spill Control. Information about the hazards associated with the toxic materials was not provided by the company. Moreover, the communication between the company and the local authorities to manage the emergency was poor.

Following this incident, the European authorities promoted the Seveso directive to prevent similar accidents. As reminded by [1], the Seveso Directive III came into authority in 2012 and its objective is the control of major-accident hazards involving dangerous substances. With respect to Seveso II, Seveso III introduces the Safety Management System (SMS), as a new tool to be included in a Major Accident Prevention Policy. In particular, the SMS should deal with the following issues:

- Organization and personnel (responsibilities and training);
- hazard identification;
- operational control (adoption of procedures);
- Management Of Change (MOC);
- emergency planning;
- monitoring performance, with process safety performance indicators (see Paragraph 2.5); and
- audit and review.

Therefore, the safety report has to demonstrate that SMS has been put into effect and measures taken to have an acceptable risk. The effectiveness of the Seveso Directive has also been evaluated and a positive contribution to improve the safety levels has been recorded, without affecting the competitiveness of European companies.

Despite the severity of the incident, the real cause of the accident still remains partially remains undetected and different mechanism hypotheses were proposed. Because of these doubts, the Seveso incident can be seen as a "black swan" incident [39], having the following three peculiarities: it was not expected; it had an extremely high impact; it has been explained and predictable after its occurrence. The interested reader may find additional information in [39], where the incident is further analysed with three different methods.

In Figure 2.33, a simplified conceptual Bow-Tie of Seveso incident is shown. The Bow-Tie technique is discussed in Chapter 5. At this stage, it is sufficient to say that it is a method to graphically collect the causes, including the failed preventing safeguards, on the left side; the main event at the centre; the failed protecting safeguards on the right side together with the resulting consequences.

2.4.2 Bhopal Disaster

Another impressive disaster, well known by the experts of this field, is the Bhopal disaster. Its lessons learned encouraged many people to rethink the basic concepts of process safety. One of these outcomes is this book [40].

On 3 December 1984, during the early hours of the morning, a relief valve on a storage tank lifted. Its containment was a highly toxic substance (Methyl-IsoCyanate or MIC, an

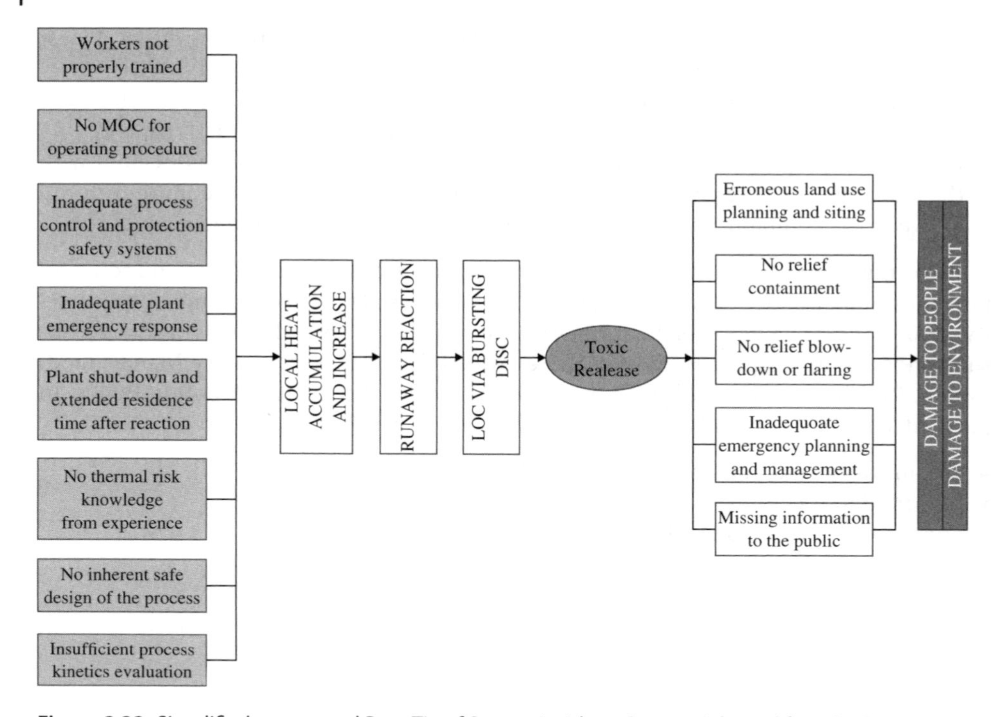

Figure 2.33 Simplified conceptual Bow-Tie of Seveso incident. Source: Adapted from [39]. Reproduced with permission.

intermediate product in the manufacturing process of an insecticide). Probably because of sabotage, the MIC was contaminated with water and a runaway reaction consequentially occurred. It caused a significant increase in temperature and pressure that lifted the relief valve of the storage tank. All the prevention and protection systems failed to work: the refrigeration system (to cool the storage tank) was shut down, the scrubbing system (to adsorb toxic vapors) was not immediately available, and the flare system (to burn any vapor residual from the scrubbing system) was out of use [37].

The resulting cloud of 25 tons of MIC moved towards the nearby houses (Figure 2.34). In the night of the day before, an operator actually observed that the pressure inside the storage tank was abnormally high, even if within the working range. Simultaneously, likewise in the night of 2 December, MIC releases were reported both near the Vent Gas Scrubber (VGS) and in the process area. Rumbling and screeching noise were heard; and attempts were made to set the VGS on, but it was not in operational mode. In a short period, more than 2000 people died and tens of thousands were injured. The emergency services were in difficulty attempting to manage and provide help to such a high number of people. The hospitals were unaware of the toxic substance involved and its risks. The exact number of victims and injured people is actually unknown, since deaths continued for years because of the toxic release. Bhopal incident is undoubtedly the worst recorded within the process industry.

The incident investigation showed a series of failings in technical measures [41]:

- The flare system, a critical element for the plant's protection, was out of commission for almost three months before the incident;

Figure 2.34 The chemical plant in Bhopal after the incident.

- hazards related to a runaway reaction within a storage tank were not deeply studied;
- the access of unwanted material (water) caused the exothermic reaction with the process fluid, but the exact point of ingress is uncertain (even if few modifications may have given their contribution);
- decommissioning of the refrigeration system significantly contributed to the accident. Indeed, the absence of this system caused the temperature to rise above the design temperature of 0°C; and
- the emergency plan was ineffective. There was unawareness about which medical action to take.

The importance of this incident, as usual, is in the lessons learnt. Firstly, the incident highlights the important concept that "what you don't have, cannot leak". It is a concept shared by the Flixborough incident, next discussed. Indeed, it was not required to stock the leaked material, since it is only an intermediate product, not a final product or raw material. The necessity to store MIC was old, when it was imported; later it was manufactured on site, but the logistics of this situation was unmodified. Avoiding the storage of intermediate materials resulted, at the end of 1985, in a reduction of 74% of 36 toxic chemicals inventoried by the company. This strategy requires a review of the plant design and the whole process, but it does not in any way mean the need to renounce productivity levels, as a number of alternative solutions shows [35].

Similarly, the high magnitude of the Bhopal incident is also attributable to the plant's location in respect to the adjacent small town. The simple concept here is that a person who is not there, cannot be killed. The town of Bhopal was too much close to the hazardous plant. Moreover, the types of buildings (shantytown) and its poor urbanistic management are an additional factor where many efforts should be dedicated to avoid their growth. In a shantytown, houses cannot be considered equal to the permanent buildings. For them, the closed windows may ensure a certain time interval during which people can be considered safe, before evacuation; the same cannot be said for a shantytown.

Some doubts still remain about the ingress of water, but when you are looking for the root causes, rather than the immediate causes, the path by which the water entered the MIC storage tank does not matter anymore. In addition, according to what is currently known, no HAZOP was carried out. Probably it would have been the powerful tool to avoid any source of water nearby the MIC storage tank, having considered the hazards related to its proximity.

Another important lesson learned is the necessity to keep protective equipment in working order and size it correctly, as we have already listed the deficiencies detected. It could be stated that the refrigeration, scrubbing and flare systems were not designed to face a runaway reaction of that size. However, their contribution in reducing the magnitude of the incident is not negligible at all. The incident highlights the limitations of hazard assessment techniques too. Indeed, if we would have carried out a Fault Tree Analysis (FTA) to estimate the Probability of Failure on Demand (PFD) of the refrigeration, scrubbing, and flare systems we would not have considered the combined probability of their contemporary switching off. The key concept is that hazard identification becomes useless if the initial hypotheses (e.g. the safeguards are switched on) are no longer true. The Bhopal incident also reminds us that procedures are likewise affected by a "corrosion" phenomenon, similarly to the steelwork, that can reduce its effectiveness. It is a sort of temporal weakening, often increased by the managers who lose interest.

The Bhopal plant was a joint venture between a US and an Indian company. In this case, it is essential to have cleared which are the responsibilities of each company, in order to avoid the general feeling that a multinational company and its partners in joint ventures, because of the international nature of their business, are not fully responsible for what happened in their plants.

Another relevant question that arose after the incident concerned the training in loss prevention. The question concerns the adequacy of the training in loss prevention for those people in charge of the plant, including its designers. The necessity of such a step, as reminded by the Bhopal incident, relies on several aspects:

- Loss prevention is an integral part of the plant design, not a later issue;
- some issues apparently out of the scope of the chemical engineer's activities are actually quantifiable if considered within the loss prevention knowledge; and
- universities do not generally provide sufficient knowledge about loss prevention, therefore there is the necessity to carry out internal training sessions after the degree.

However, the lack of commitment to safety and the poor management are key points of the Bhopal incident: the errors that were made (disconnection of safety equipment) are so basic that inexperience cannot be considered as the root cause for them. The acknowledge of the principles of loss prevention, reversely of what happened in this case, would certainly avoid the occurrence of such a sad event.

The necessity to have a coordinated emergency plan between the company and the local authorities is another issue, showing the necessity for a better management of those services, as well as encouraging the local legislator in promoting regulations to favour the cooperation, the identification of hazards and their possible consequences.

The Bhopal incident has been an occasion to rethink about the safety measures too. The ones put in place at the Bhopal plant (like the scrubber and the flare system) were too near the top event. This means that if they fail (as they actually did), there is nothing to fall back on. To prevent the recurrence of a similar incident, safety measures need to be taken starting from the bottom of the chain, that is to say from the root causes which are apparently far too close to the top event like managerial decisions.

2.4.3 Flixborough Disaster

The Flixborough incident is a milestone in the history of the chemical industry. It occurred on Saturday, 1 June 1974 in a chemical plant in Flixborough (UK), manufacturing nylon. It was a severe explosion, causing 28 victims and 36 injured people. The magnitude of the event could be higher if it had occurred on a weekday: indeed, the main office block was not occupied on Saturday. The consequence of the explosion also went outside the boundaries of the site, reporting 53 injured people and several damages to the properties in the surrounding area. To explain the reasons that led to the occurrence of the incident, a step backwards has to be done. On 27 March 1974, a vertical crack was found in reactor no. 5, causing its containment, cyclohexane, to leak. The subsequent investigation showed the presence of a critical problem with that reactor, so serious that the corrective action taken was to remove it and install a bypass to connect reactor no. 4 and no. 6, allowing the plant to continue its production (Figure 2.35). During the afternoon of 1 June 1974, the 20-inch bypass system ruptured, probably because of a fire on the adjacent 8-inch pipe. Consequentially, a significant quantity of cyclohexane leaked and formed a flammable mixture which then found a source of ignition. At 16:53, an enormous VCE (Vapor Cloud Explosion) occurred, causing significant damages and starting several other fires on the site. The serious structural damages in the control room caused the death of 18 people. After ten days, some fires were still active, impeding the rescue activities.

The incident investigation showed a series of failings in technical measures [42]:

- The integrity of the bypass line was evaluated on the basis of limited calculations;
- no drawings of the proposed modification were produced;
- the installed pipework modification was not tested to pressure;
- the position of occupied buildings did not consider the consequence of a major instantaneous disaster;
- the structural design of the control room was not robust enough to resist against major events; and
- the incident happened during the startup phase, in a stressful context with a high number of critical decisions to be made.

The UK government, as usually happens for those major events, set up not only an investigation to understand the immediate causes, but also a special committee to

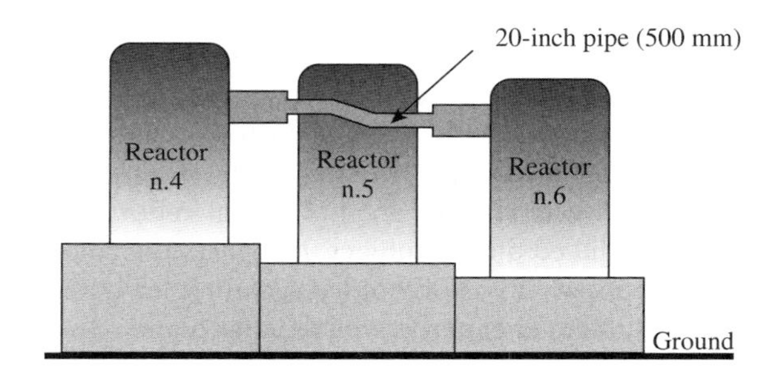

Figure 2.35 Arrangement of reactors and temporary bypass. Source: Adapted from [37]. Reproduced with permission.

Figure 2.36 The chemical plant in Flixborough after the incident.

discover the deeper causes of the explosion [37], in order to avoid the recurrence of similar major incidents.

The main reaction in the Flixborough plant, during the normal process, was the oxidisation of cyclohexane with air to a ketone/alcohol mixture. In order to avoid the production of unwanted products, the reaction was slow, so a high quantity of cyclohexane was stored in the plant (about 400 tons). The reaction took place in six reactors, each containing 20 tons.

The bypass pipe was not straight, but it had two bends to allow the overflowing of the liquid from one reactor to another (for the same reason, reactors were placed at a different height). The bypass pipe was 20 inches diameter while the bellows were 28 inches. The design, construction and installation of the bypass were carried out by two men charged with these tasks. They were men of great experience, but not professionally qualified to design a bypass pipeline and its supports. Two months after their installation, a slight increase in pressure occurred, causing the temporary pipe to twist. However, the increased pressure was too low to activate the relief valve. There are still different hypotheses about the triggering event that produced the small increase in pressure [37], like the stop in the agitation of reactor no. 4 causing a layer of water to vaporise during the heating startup phase. The resulting bending moment tore the bellows, creating two 28-inch holes. The cyclohexane, at a pressure of 10 bar, was at the temperature of 150°C, above its boiling point (81°C). Consequentially, a massive leak occurred as the pressure was reduced. About 30–50 tons leaked in 50 seconds, then a source of ignition (probably a furnace) was reached and the explosion occurred (Figure 2.36).

This incident has great lessons to provide. The most important one concerns the reduction of inventory. The smaller the inventory, the less the leakage. This observation, so simple, was ignored in the official report and in the majority of the papers that were produced on the subject. It repeats the concept already expressed in the previous Paragraph: what you do not have, cannot leak. Reducing inventory is a design approach that results in cheaper as well as safer plants. They are cheaper because less protective equipment is needed, with lower direct (purchase) and indirect (test and maintenance) costs. Flixborough incident shows how the inventory reduction is desirable, especially for flashing flammable or toxic liquids. It is an approach guiding towards inherently safer plants, even if it consumes more time during the early stages of the design, in parallel with HAZOP sessions involving Health Safety and Environment experts since the conceptual stage. Several papers demonstrate that at Flixborough, the inventory reduction was possible. The interested reader may find additional information on [37].

The control of the process and the plant modifications are other key lessons. No changes should be made without the formal authorization of a professionally qualified manager who can help in identifying all the possible consequences of the proposed modification. Moreover, an inspection should be carried out once the modification is complete, to check its technical compliance with state of the art. In addition, a checklist should also be provided to help the identification of the possible consequences and a training program needed to educate people to the Management Of Change (MOC).

Flixborough shows us also the importance of qualifications. Undoubtedly, having practical experience is highly desirable but it is not sufficient to ensure a proper design of the bypass pipeline. Obviously, this book does not intend to find culprits, but also to understand how to prevent the recurrence of similar events. Having this concept always in mind, this lesson learnt can be summarised as an invitation to look at your own capabilities and check if they cover the task you have been charged for.

Also, the preference for robust equipment clearly emerges from the lessons learnt from this incident: flexible hoses and bellows are inevitably a weak link and they should be avoided when hazardous materials are used.

Some considerations about the plant layout and locations can be also made. Even if it is highly probable that a leakage of the size of the one occurred at Flixborough ignites, there is a chance to prevent it if the plant layout is designed so as to minimise the chance of reaching a source of ignition. Moreover, other aspects of design concern also the structural design of the control room and, in general, of the buildings occupied by people. Indeed, the risk level of an event depends also on its magnitude and the number of people potentially affected by the incident is a key parameter in this context. Basically, following the principles that "what you do not have, cannot leak", the best solution is to not have people within the boundaries of a potential incident. No people, no victims. Also, emergency operations should be performed remotely, in a safe location. This concept also applies outside the boundaries of the plant site. It therefore becomes important to provide proper tools to govern the territory nearby a chemical plant handling hazardous materials, in cooperation with the local authorities having jurisdiction (as Seveso Directive imposes). It is a preventive strategy and it should be preferred. If, for space availability or control necessity, it is a requirement to have people nearby the potential incidental area, then a robust structural design needs to be pursued. It is a protective strategy.

After the Flixborough events, a number of papers were written about the behavior of large leaks, under various wind and weather conditions. Only a few papers were written on the reasons for large leaks and the actions required to prevent them. This happened because large leaks, being also occasional, were considered inevitable. We know that this is not true. Having observed how large leaks are caused by pipe failures, the most effective action to prevent them is to follow the piping design strictly, and to test and maintain them after construction.

Just like the Bhopal plant, the Flixborough plant was also jointly owned. It becomes therefore crucial to clarify the responsibilities of the joint venture's members, regarding safety, design and operations, to avoid a misleading tasks assignment.

The plant was then rebuilt, with a different process to manufacture cyclohexanol: it was made by the hydrogenation of phenol instead of oxidizing the cyclohexane. The process, however, is as hazardous as the previous one and it was not carried out in Flixborough but elsewhere: the risks were only exported, not really diminished. The rebuilt plant was then closed after a few years for commercial reasons.

2.4.4 Deepwater Horizon Drilling Rig Explosion

The Deepwater Horizon platform disaster was an explosion which occurred in the Gulf of Mexico on 20 April 2010. The incident is taken as a key example in [43], to present a seven-steps strategy for effective problem-solving and the Apollo RCA™ method, next discussed in this book. Also [1] is taken as a reference to introduce this incident. The Deepwater Horizon was a semisubmersible, huge, and very advanced drilling rig. It was nine years old and substituted a different rig, the Marianas, that started drilling the Macondo well but was then damaged during a hurricane. The sea floor was at 1500 m depth and the intent of the project was to drill up to 5500 m under the sea floor, using progressively smaller diameter casing strings.

On 20 April 2010, 11 workers died and 17 were seriously injured by an explosion on the Deepwater Horizon (Figure 2.37). The offshore drilling rig was located approximately 50 miles off the coast of Louisiana and it burned continuously for two days, eventually sinking. The incident is the largest oil spill in US history.

Different authorities and investigators, including the U.S. Chemical Safety Board (CSB), carried out their examinations to discover the technical, organizational, and

Figure 2.37 The Deepwater Horizon drilling rig on fire.

regulatory factors that contributed to the accident. They also suggested the corrective actions necessary to prevent a similar occurrence.

Drilling an offshore well means the creation of a pathway between the drilling rig and the reservoir under the seafloor. The bore is drilled through layers of rocks that can trap water, crude oil, and natural gas under pressure. An unplanned flow of these fluids into the wellbore may happen: in the industry, this phenomenon is known as "kick". A kick can be very dangerous, since it can lead to a blowout, that is an uncontrolled release of flammable oil and gas. A blowout may bring severe consequences, as oil and gas reaching the drilling rig can ignite resulting in a fire or explosion. In order to prevent kicks, drillers pump a dense slurry, named drilling mud, to create a barrier between oil and gas (undersea) and the piping leading to the rig. If this barrier is not sufficient to contrast the blowout or it is removed, then the safety is entrusted to the Blow-Out Preventer (BOP), a critical equipment located on the seafloor.

The BOP is an essential device to control the well and to prevent a disaster. It is both electrically and hydraulically powered. The BOP and the rig are connected through a large diameter pipe named riser. When a kick occurs, the BOP prevents the travel of oil and gas through the riser up to the drill. This prevention system is done by sealing the annular space (the area around the drill pipe). To do so, the crew can manually close pipe rams using rubber devices known as annular preventers, which are donut-shaped. If they fail, the last protection layer is a pair of sharp metal blades designed to cut the drill pipe and seal the well. This protection can be activated either manually or automatically. On the evening of 20 April 2010, at 8:45 p.m., a kick occurred. The undetected oil and gas ingress passed above the BOP and travelled up the riser towards the rig. The drilling mud, pushed by the rising oil and gas, suddenly blew out onto the rig at about 9:40 p.m. Being aware of this occurrence, the crew members closed the upper annular preventer in BOP, but it did not work as intended and flammable oil and gas continued to flow into the riser towards the rig. Then, the pipe ram was also closed, properly sealing the well. Unfortunately, this was only a temporary fix, since the oil and gas already above the pipe ram continued to flow towards the Deepwater Horizon. At 9:49 p.m. the flowed materials ignited, resulting in a violent explosion. The pressure in the annular space above the pipe Ram immediately decreased because of the oil and gas escaped from the riser. At the same time, the pressure in the drill pipe increased. So, the drill pipe was closed at the top but the flow of oil and gas kept continuing from the reservoir. According to the investigation carried out by CSB [44], this difference in pressure caused the buckling of the drill pipe, bending it and limiting the barrier effectiveness of the shear ram blades. The explosion and the subsequent loss of energy and hydraulic power activated an automatic system on the BOP, known as AMF deadman. This system closes the blind shear ram and cuts the drill pipe. But electricity, hydraulic pressure and communications from the rig have been lost. The AMF deadman worked with two redundant control systems on the BOP, named the yellow pod and the blue pod. The two pods are independent; thus, the reliability of the system is supposed to be increased. They were comprised of identical embodied computer systems and solenoid valves which controlled fundamental BOP functions, including closing the blind shear ram. On 20 April 2010, the electrical supply was lost. The two control pods could rely on backup 27- and 9-volt batteries: the 9-volt batteries supplied power to computers that would activate the solenoid valves, powered by the 27-volt batteries. The collected evidence showed that the blue pod was mis-wired, causing the drainage of the 27-volt batteries and resulting, in the end, in the impossibility to operate with the blind shear ram. Also, the yellow pod was mis-wired. Each solenoid valve was controlled by two coils, designed to work in concert. In the mis-wired solenoid valve, the two coils acted in contrast to each other, leaving the valve immobile. A third failure allowed the yellow pod to operate: one of the 9-volt batteries had failed, consequentially the computer system was not capable of giving the command to energise the mis-wired coil. Thanks to this failure, opposite forces on the solenoid valve were not generated and the working coil succeeded in opening the solenoid valve. This opening started the closure of the blind shear ram, which should have cut the drill pipe and sealed the well. But, because of the bent condition of the buckled pipe inside the BOP, it was only partially cut. Once the last barrier failed to prevent the blowout, the massive oil spill could not be stopped and the destruction of the rig prevented. According to the CSB investigation, the effective compression was the phenomenon that caused the buckling of the pipe. It happens because of invisible irregularities of the pipe that, even if it appears perfectly straight to the naked eye, curve the pipe. As a consequence of these irregularities, one side of the pipe is slightly longer and offers a wider surface respective to the opposite side. Now, with the limited difference in pressure, these geometric irregularities are negligible, but with the large difference experienced in this case, the geometric diversity eventually causes different forces and a resulting buckling effect. According to CSB, similar pressure conditions could be experienced by many others existing drilling rigs, significantly reducing the effectiveness of the BOP barrier. The spillage following the Deepwater Horizon accident lasted 87 days, with 5 million barrels of oil spilt in the Gulf of Mexico: regardless of the impact on humans (victims and injured people), it resulted in one of the worst environmental disasters of history. Figures 2.38 to 2.40 show the application of the Apollo RC™ method to this incident, using RealityCharting® [43]. At this stage, the reader is not asked to fully understand the RCA methodology, but having a first look at the trees obtained is a useful introduction into the investigation methods, fully described in Chapter 5.

2.4.5 San Juanico Disaster

This incident is due to a gas leak, resulting in a VCE and a series of BLEVEs. On 19 November 1984, at 5:35 a.m. a series of catastrophic explosions occurred at the LPG Terminal in San Juanico, Mexico City. This incident is among the worst in the history of the chemical industry, counting about 500 victims and the total destruction of the terminal.

Before the incident, the plant was being filled with LPG from a 400 km distant refinery, which since the previous day was almost empty: 2 spheres and 48 cylindrical vessels were filled to 90% and 4 other spheres were filled to 50%. Both in the control room and at the pipeline pumping station, a drop in pressure was noticed. A pipe between a sphere and some cylinders ruptured, resulting in a continuous release of LPG. For 5–10 minutes, while no operator was able to identify the cause of the pressure decrease, the gas cloud continued to grow, reaching the estimated dimension of 200 x 150 x 2 m. Then it moved to a flare stack and consequentially it ignited. The resulting explosion (as well as the subsequent ones) was also recorded by the seismograph at the University of Mexico. Different ground fires occurred and only at a late stage was the emergency shut down button pressed. The first BLEVE occurred after 15 minutes from the initial release. A series of other BLEVEs, caused by violent vessels explosions, occurred for the next hour

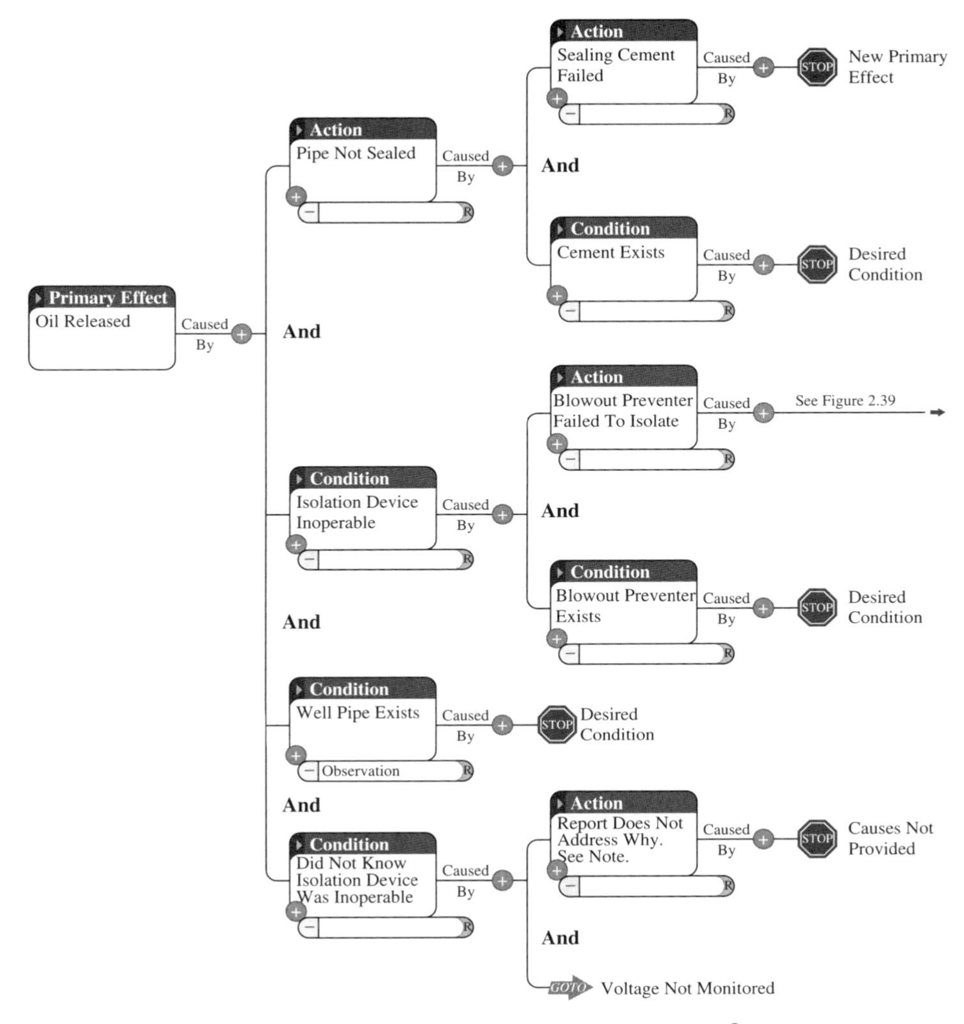

Figure 2.38 Application of the Apollo RCA™ Method using RealityCharting® to the Deepwater Horizon incident. Source: Reproduced with permission from [43].

and a half (Figure 2.41). Because of the numerous explosions and fires, the engulfing LPG continued to increase the devastating effects of the incident, setting alight the covered surfaces.

The incident investigation showed a series of failing in technical measures [45]:

- Plant Layout/Isolation. The destruction of the terminal is attributable to a failure in the overall basis of safety, including the plant layout and the emergency isolation features;
- Active/Passive Fire Protection. The firefighting system did not activate when the first explosion occurred. Moreover, the spray systems produced inadequate results;
- Leak/Gas Detection. No gas detection system was installed. It is likely that the emergency isolation was initiated too late. A more effective gas detection and emergency

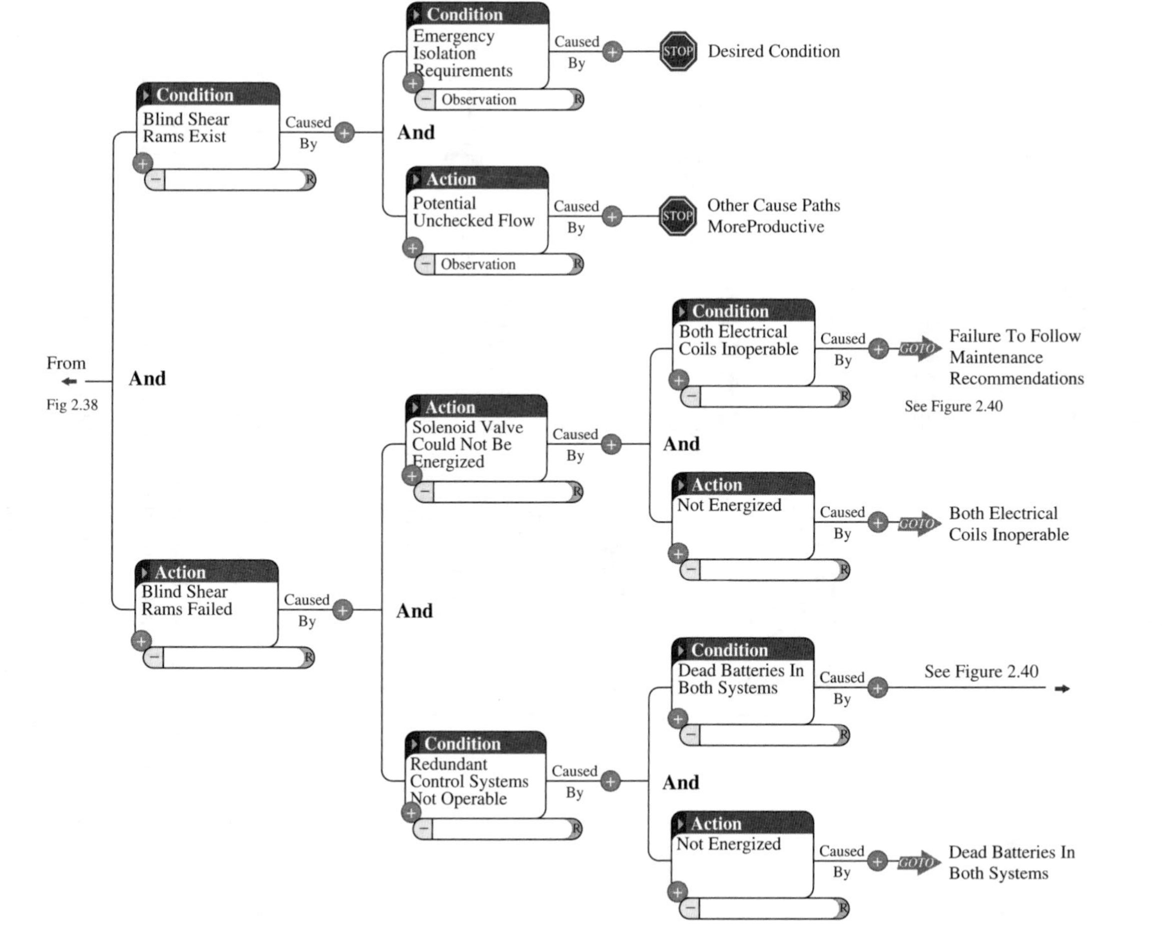

Figure 2.39 Application of the Apollo RCA™ Method using RealityCharting® to the Deepwater Horizon incident. Used by permission. Taken from [43].

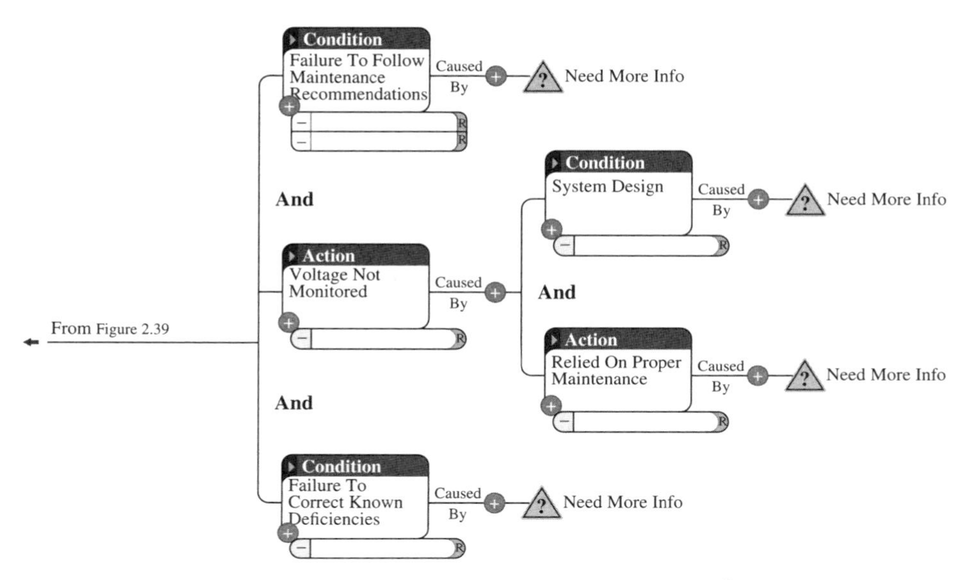

Figure 2.40 Application of the Apollo RCA™ Method using RealityCharting® to the Deepwater Horizon incident. Source: Reproduced with permission from [43].

Figure 2.41 Some LPG spherical tanks during the San Juanico disaster.

isolation system could have avoided, or at least minimised, the consequences of the incident; and

- Emergency Response/Spill Control. The emergency services were hindered by the traffic chaos (residents tried to escape the area). The site emergency plan was ineffective and access to emergency vehicles was not guaranteed.

2.4.6 Buncefield Disaster

The Buncefield disaster is one of the major incidents in the process industry. This paragraph has been written taking inspiration from [34, 46].

In 2005, the Buncefield oil storage depot experienced an unconfined vapor cloud explosion as never seen before, resulting in severe economic losses and, fortunately, in no victims. The storage and transfer depot was a tank farm 40 km northwest of London. In December 2005, three different operating sites were at the depot; all of them were "top-tier" sites under the Control of Major Accident Hazards Regulations 1999 (COMAH). The fuel was transported using three different pipelines. Fuels were transported in batches and were separated into dedicated tanks according to their grades. Road tanks loaded the fuel to transport it from the depot to the final destination. The storage depot also served the Heathrow and Gatwick airports. Moreover, the depot was above a major aquifer, providing drinking water.

On Saturday, 10 December 2005, a batch of gasoline was transferred through the pipeline into Tank 912 (25 meters of diameter and 14.3 of straight side height), with a flow rate of 550 m^3/h (at this rate, a car tank is emptied in 3-4 minutes). The tank was equipped with an Automatic Tank Gauging system (ATG) which measured the level in the tank and displayed it on a screen in the control room. In the early hours of Sunday, 11 December 2005, at about 3.00 a.m., the display showed a constant value, that is to say it stopped registering the rising level, while the tank continued to fill. As a consequence, the three ATG alarms, set at different levels, could not operate.

However, the tank was also fitted with an Independent High-Level Switch (IHLS) (Figure 2.42), which was intended to stop the filling process automatically when the level reached the high-level, also producing a soundable alarm. But the IHLS failed and, starting from approx. 5.30 a.m., the tank overfilled and the fuel started to spill out of the vents in the tank roof.

Closed circuit TV shows that a white cloud suddenly formed, reaching a diameter of about 360 m. including a car park and Tank 12, containing aviation kerosene. In 25 minutes, the cloud covered an area roughly 500 meters by 40 meters to a depth of 2–4 meters. The cloud was noticed by tanker drivers who alerted employees on site. The fire alarm button was pressed at 6.00 a.m., and the firewater pump was activated. Almost immediately, at 6.01 a.m., a vapor cloud explosion occurred, probably ignited by a spark from the firewater pump starting. The initial blast was recorded as 2.4 on the Richter scale. Two follow-up explosions occurred next.

There were no victims, and more than 40 injured. The resulting fire engulfed 23 fuel tanks (some of which are shown in Figure 2.43) and burnt for almost five days, affecting an area of about 150.000 m^2. The smoke cloud was visible in satellite photos. Moreover, fuel and firefighting chemicals flowed from leaking bunds both on and off-site, causing a significant environmental, social, and economic loss. Liquids also flowed down onto the M1 motorway, which was temporarily closed. About 2000 people had been evacuated

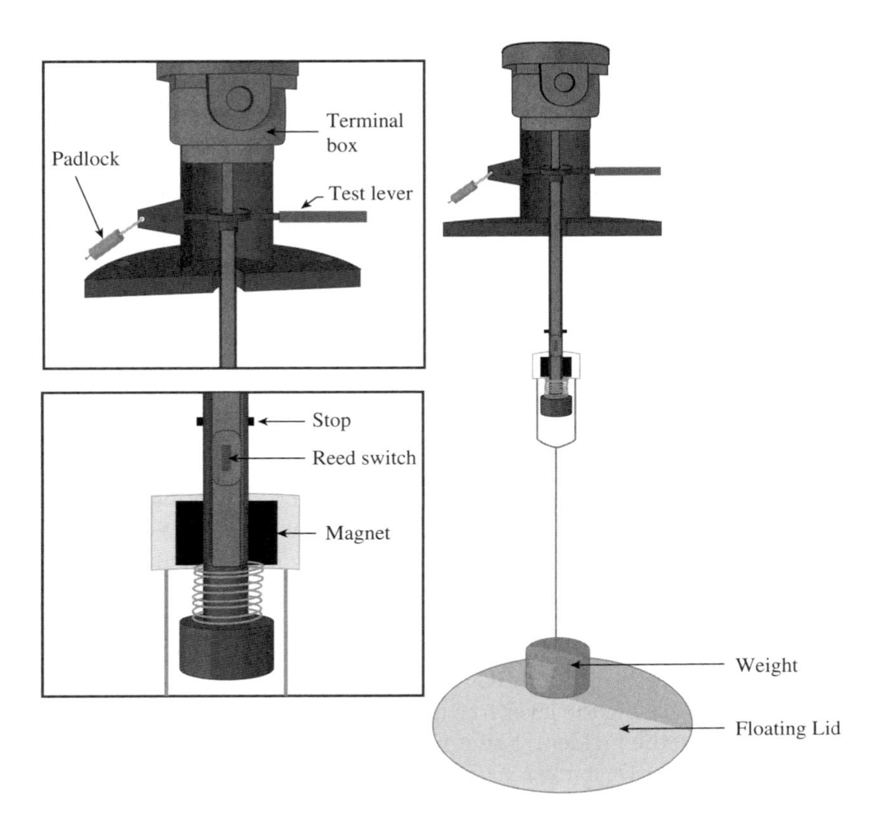

Figure 2.42 The IHLS. Source: [46]. Reproduced with permission.

Figure 2.43 The site after the incident. Source: © Chiltern Air Support. Taken from [46]. Reproduced with permission.

from their homes, and 180 firefighters were present, using 20 vehicles and 25 pumps. It took 32 hours to extinguish the main blaze. The quantifiable costs approached US$1.6 million. The consequences on humans could be worst if the event would not have occurred on Sunday morning.

In order to investigate the vaporization phenomena and the resulting vapor cloud intensity, a full-scale model (1:8) of the tank was constructed. Tests confirmed the increased vaporization from splashing of the fuel on the wind girder, boosting the vapor cloud formation.

The immediate causes of the incident are two:

- The IHLS and
- the ATG system.

The IHLS was designed to be tested using a lever. There were three positions for the lever: the horizontal position was the normal operating position, allowing the switch to work as intended. A padlock should be used to secure the lever in the horizontal position. To test the IHLS, the lever is raised to the upper position, activating the alarm circuit even if the floating lid is not high enough to activate it. Once the test is completed the lever would return to the horizontal position and secured with the padlock. However, the level switch can also be used to detect low levels of fuel in a tank; therefore, the check lever could be lowered too. Unfortunately, lowering the lever has no effect on the switch that is intended to operate in the high-level mode. It is evident how the padlock played a critical role in safety issue concerning the IHLS. The IHLS fitted on Tank 912 was found with no padlock, leaving the lever free to fall in the lower position.

The ATG had stuck before the incident, and it was not the first time: it occurred 14 times between 31 August and 11 December 2005. When it happened, supervisors solved the problem by "stowing".

This incident also teaches about ergonomics: there was only one visual display screen for the ATG system on a number of tanks: this means that the operator could only monitor the status of one tank at a time.

Going deeper in the incident investigation, the underlying management failures concern:

- The control of incoming fuel. Indeed, flow rate suddenly increased from 550 m^3/h to 900 m^3/h, without the knowledge of supervisors;
- the increase in throughput, since the adjacent terminal closed in 2002 and its throughput was absorbed into the terminal that suffered the incident;
- the tank filling procedures. As previously discussed, it was possible to see the status of only one tank at a time. Moreover, supervisors often relied on alarms to control the filling process. Therefore, "*when situations arise requiring staff to work outside the normal operating envelope they should be recorded and reviewed by management*";
- the pressure of work, since fuel deliveries were unpredictable. To overcome this, supervisors developed their own systems, like a small alarm clock into the control room to track the filling procedure and that tanks were getting close to their capacity. Moreover, working patterns did not help: supervisors worked 12 hours shifts for a total of 84 hours of working in seven days. "*Management has a duty to monitor working pressures on staff and take actions to keep workloads to acceptable levels so far as reasonably practicable*";

- the inadequate fault logging, being absent any system to monitor key safety parameters. *"Management should have in place systems to monitor the reliability of safety-critical equipment"*; and
- contractor control systems. *"For high-hazard risks duty holders should have formal arrangements that specify the roles of all parties involved to ensure so far as is reasonably practicable that the highest standards are provided for safety critical equipment"*.

Regarding the loss of secondary containment (i.e. the bund surrounding a tank or a group of tanks), the main causes were:

- Bund joints, that do not retain liquid if they do not contain waterstops (preformed strips of durable impermeable material embedded in the concrete during the construction, providing a liquid-tight seal during a range of joint movements);
- tie bar holes, introduced to hold in place the formwork used to cast the concrete. They are penetrating through the bund, plugged and grouted; and
- pipe penetrating through walls could no longer retain liquids when, for instance, a catastrophic failure of the walls at pipe penetrations happens or for the loss of seal between pipes and walls (Figure 2.44).

At Buncefield, the tertiary containment, i.e. the means by which liquids can be contained/controlled within the site boundary, was virtually not in place. Indeed, the containment outside the bunding was designed for rainwater, not for large-scale releases.

In conclusion, the management systems were inadequate because:

- Risk assessments did not consider the implications of more than one tank being on fire. They also failed considering that bunds do not fail structurally or their capacity is never exceeded;
- changes during the design and construction of bunds were not reviewed;
- bunds failures were not treated as "near misses"

Figure 2.44 Pipe penetrations for the loss of seal between pipes and walls. Source: [46]. Reproduced with permission.

- there was no periodic review on the bunds' characteristics;
- the safety critical parts list was not provided: it was an example of the poor focus on major hazard systems and plant; and
- the SMS focused too closely on occupational safety and lacked any depth about the control of major hazards, including the loss of primary containment.

The report in [46] reminds that *"good process safety management does not happen by chance and requires constant active engagement. Safety management systems [...] should specifically focus on major hazard risks and ensure that appropriate process safety indicators are used and maintained"*.

The HSE final report listed 25 recommendations for design and operation of fuel storage sites (to increase the defence provided by the primary containment – i.e. the tank - , and to improve secondary and tertiary containment too), 32 recommendations on emergency preparedness, and 21 recommendations about land use planning and control of societal risk. The 25 recommendations for design and operation include also some broader strategic objectives relating to sector leadership and safety culture.

In conclusion, the process safety controls on safety critical operations were not maintained to the highest standard, senior managers did not apply effective control, and effective auditing systems were not in place. Moreover, it clearly emerged that high standards expected of operators of safety critical equipment apply equally to all those involved in the supply of that equipment.

Figure 2.45 shows the diagram developed by company Governors BV (NL) for the analysis, through RCA, of the Buncefield explosion. In the presented RCA diagram, the immediate, underlying, and root causes of the incidental event are depicted.

2.5 Performance Indicators

According to [9], performance is "the degree to which you get what you expect from a person, a machine, or a process". Systems and individuals rarely over-perform; indeed, an incident, a near miss or a failure occur when the desired outcomes are not reached. Therefore, they can be seen as "deviations from an expected outcome". A process improvement or a system enhancement are also deviations from the expected result, even if they are good things. For clear reasons, in this book the interest is for those deviations that decrease the projected performance level.

In order to audit the overall level of safety of a process industry, some performance indicators should constantly be monitored. Indeed, most of the industrial accidents are the natural and predictable consequence of a potential situation where some performance indicators have been misunderstood or undervalued.

A distinction between "lagging" indicators (related to actual loss events) and "leading" indicators (related to the precursor events) is done. The Lagging Process Safety Performance Indicator measures the number of process incidents. The term "lagging" stands for "happened"; therefore, the Lagging Indicators are related to events which have already happened (incidents), with their consequences and the suggested corrective actions. Process incidental events are classified and communicated according to the internal procedures of the company. On the other hand, the Leading Process Safety Performance Indicator measures the effectiveness of the process safety management

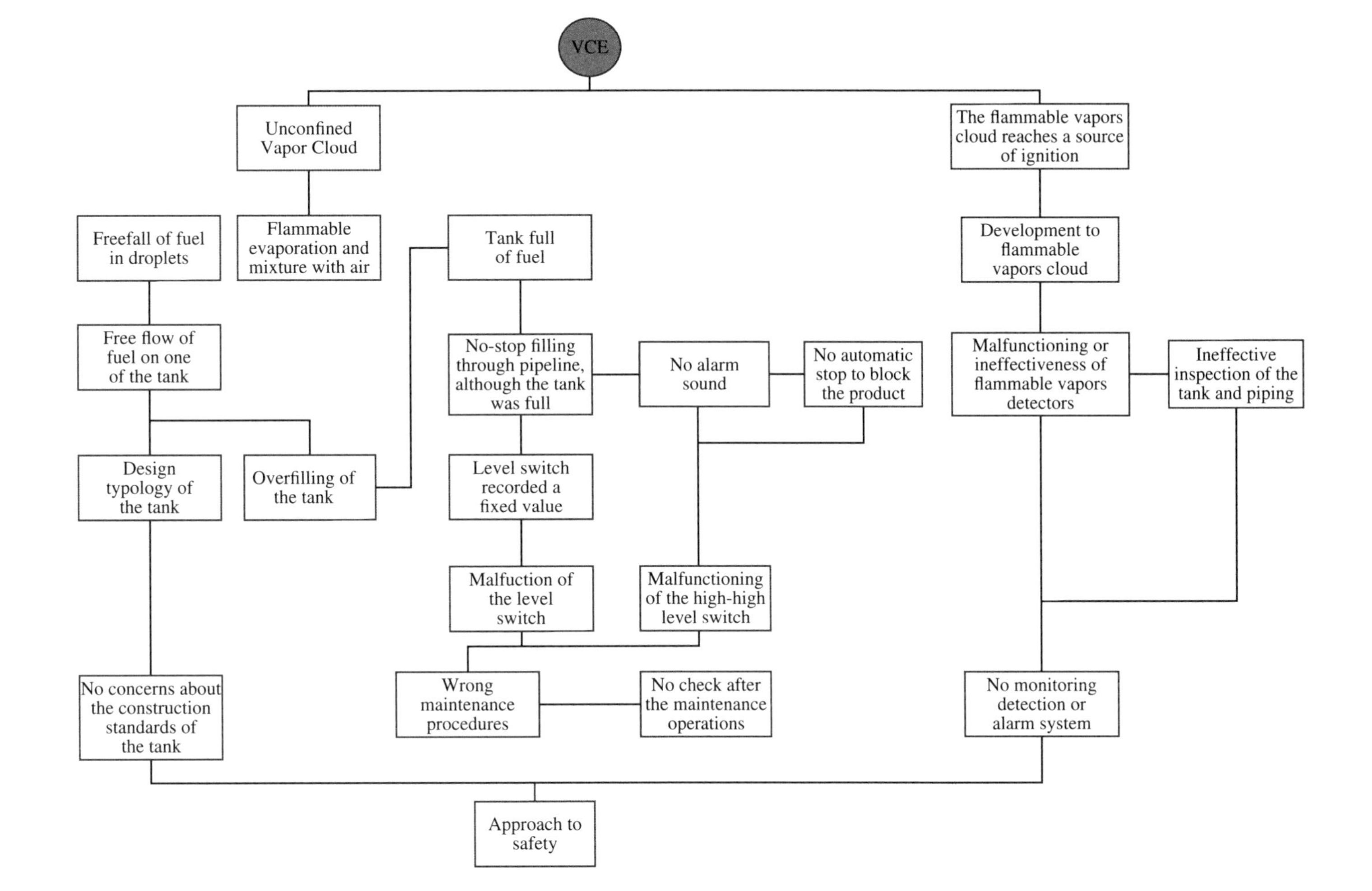

Figure 2.45 RCA of the Bouncefield explosion developed by company Governors BV (NL). Source: Adapted from [20]. Reproduced with permission.

activities. The term "leading" stands for "precursor"; so, it refers to activities, factors, and parameters related to events that did not happen. Therefore, the Leading Indicators allow the identification of preventive measures for incidents.

Some lagging indicators are:

- Major incident counts;
- monetary losses;
- injury/illness rates; and
- process safety incident rates.

On the other hand, the leading indicators include:

- Near misses;
- abnormal situations (like overpressure relief events, safety alarm or shutdown system actuation, flammable gas detector trips);
- unsafe acts and conditions; and
- other PSM element metrics.

Safety performance measurement and monitoring require the definition of some safety performance metrics, like the incident-based metrics and the related statistical measures cited in [23]. Examples of performance indicators are:

- Number of relevant safety recommendations, still open on the site. It is the number of recommendations identified during the PHA, safety audits, incident investigation audits, near-miss investigation audit, or similar activities. These recommendations have been evaluated as a priority, because they are related to high risks, and require an immediate action (within 1 year). The expectation on the evolution of this indicator is a decrease of the number of open recommendation in time, until zero. The number and velocity of resolution of the indicator must be constantly monitored. On a regular basis, the number of still open recommendations has to be notified;
- number of scheduled inspections and maintenance activities identified on safety critical equipment, with a delay of 60 days to be realised and without formal approval by the Direction. The indicator should distinguish the different equipment typologies;
- reaching of the Safe Operating Limit. The indicator includes: emergency automatic depressurizations, activations of safety valves and rupture discs, activation of safety interlocks, activation of cooling system (for exothermic reactions);
- training of plant operators. The indicator is expressed in hours/month for a single operator;
- Percent Evaluation Sheet. The indicator expresses, in percent, the execution of safety tours respect to the goal number;
- percentage of involvement in emergency training. The indicator expresses the annual frequency calculated as the number of emergency training executed by a single operator respect to the total expected number;
- number of spurious activation of SIF;
- number of bypasses in use over a process safeguard;
- number of bypasses in use for more than 30 days without implementing change (MOC);
- Permit to work. The percent of permits to work correctly closed;
- number of the "Pre Start-up Safety Review" not completed before the startup;

- number of implemented management changes without adherence with internal procedures; and
- number of simulated emergency tests carried out over one year.

They are a measure about how the Health, Safety and Environment company's policy, which is established from the corporate management and affects up to the line level, is effective in preventing and protecting from an incident.

The development of a performance indicator is discussed in [47], which identifies ten steps:

- Identify and record the business outcome;
- identify the process flow and record the process outcome;
- identify and record the process purpose;
- identify and record the most important outputs of the process;
- identify the critical stages of the process and the dimensions of process performance; develop and record the measurements for each dimension;
- develop and record goals for each measure. Goals must be specific, measurable, achievable, relevant and time-sensitive; and
- define and record the levels of success, indicating if the results have been achieved.

The incident analysis reveals a reoccurrence of some organisational factors playing an important role in triggering/causing/developing the incident itself. This identification process requires the previous definition of a classification meter for those factors. This approach may be a bit difficult, since it is not easy to establish, *a priori,* which are the organisational phenomena that may result, in the end, to an incident. Indeed, the linking between those factors and the undesired outcomes is not direct: they are often interlinked, and a latent period of time is very often observed between the occurrence of the factors and the one of the incident. This means that it is not possible to establish, *a priori,* a chronological order and therefore it is impossible to use a cause-consequence approach. However, these organisational factors can be considered as "indicators" which are symptoms of a future incident. This approach requires attention: the proposed subsets must not be too generic, to avoid a useless classification, nor too specific, to avoid the peculiarities of the particular incident analysis having an influence on the purpose of the classification. A selection of five recurrent factors is now discussed, taking inspiration from [48]. Some of them may be observed simultaneously, sometimes with mutual enhancement.

The first one is the weakness of the organisational safety culture. Anticipating what presented in Paragraph 6.5, and quoting [48], the organisational safety culture is intended as a set of factors *"put in place or favoured by a business, which concur to achieving the latter's production objectives thanks to the safe functioning of its operation processes."* At the base of this culture there is a sound structure, made of procedures, behaviors, best practise, design criteria, and so on. Therefore, the safety culture is not the mere sum of the conduct of each actor (who acts prudently and rigorously), but it is reached with a holistic approach. Some of the indicators related to this factor include: managerial deficiencies in safety instructions, the absence of risk analysis, inappropriate training, and "practices" in conflict with regulations.

A further factor is a complex and inappropriate organisation. The introduction of protection systems in process industries resulted in a safer complex system. This complexity is at the origin of the system's failure, when the organisation (i.e. the system of relations between the different actors) is inadequate. This means that having a large structured organisation may have the consequence of negatively affecting the decision-making process. The indicators include: coordination problems, lack of each owning responsibility, excessive tasks definition, and poor planning.

The limits of operational feedback are another organizational recurrent factor. Operational feedback is crucial to ensure safety enhancement. A superficial incident analyses, not taking into account the unfavourable organisational factors related to safety or giving to much space to formalism, is an example of possible "indicator", as well as censorship for some aspects of the analysis.

The unfavourable conditions for a safe environment may also be generated from production pressures, together with uncontrolled financial constraints. For instance, a culture that pushes over production imperatives or a financial approach to safety, considering risks as adjustment variables, are examples of "indicators" for these factors.

The last organisational recurrent factor is the failure of the control organisations. Generally, in order to guarantee an acceptable safety level, risky industries manage their performance with internal audits and monitoring measures. What is questioned is the reliability of these self-controls, their effectiveness. From this point of view, some "indicators" are: the presence of a conflict of interest between controllers and controlled, the lack of independence in the company, and its tendency to make obstacles for internal formal safety audits and analyses.

In Paragraph 2.3, the "Goal Zero" policy was presented, adopted from many industries to prevent, control and limit the incidents. The above-mentioned performance indicators intend to take a picture of the overall safety level of the company at a certain time. They should be monitored to follow how incident rates develop in time, taking into consideration the final goal of the adopted safety policy: having no incident. This desire is actually asymptomatic and frankly unrealistic, therefore it becomes crucial to ask yourself if zero accidents is the right goal. It is evident that no one wants to meet an incident in his/her career, but "zero" is a standard of absolute perfection, while the organisation (made of workers and managers, thus humans) is not absolutely perfect and, unfortunately, failures will occur. Perfection leaves no room for human error, which will inevitably occur. Moreover, it should be noted that asking workers to have zero incidents does not explain to them how to have zero incidents [9]. In other words, the adoption of a "zero accidents" goal does not make a safety program. The learning process ensuring a safety improvement is consumed by trial and error. This is why the "Zero" is not the right goal: pushing down the numbers to zero, trying to understand when and where the next accident will occur, is the real goal. And in order to foresee the next incident, performance indicators assume a key role.

2.6 The Role of 'Uncertainty' and 'Risk'

In everyday language the terms "Hazard" and "Risk" are used as synonyms, but they are not. It is therefore fundamental to give their different definitions. A hazard is an action that has the potential to cause harm to human health or the environment or economic loss. A risk, instead, is a measure to express the probability that a certain event appears with a specific magnitude (that is to say with its level of severity). Being a combination of these two factors, it can be expressed in formula as follow:

$$\text{Risk} = \text{Event Likelihood} \times \text{Event Magnitude} \tag{2.2}$$

The hazardous nature, in the industrial context, generally refers to used substances, including toxic and flammable ones. Taking inspiration from [12], process hazards include: chemical reactivity hazards, transportation of chemicals, static electricity, material properties concerning fires and explosions development, and many others. Several factors can lead to a hazard: equipment failures, human factors, operational or managerial problems. As already stated since the beginning of the book, the natural phenomena (like earthquakes or hurricanes) are excluded from the list. Making a parallel with the entropy law, the investigated systems tend to increase their "disturbance", causing an accident to occur. This happens until a sufficient amount of energy is added to the system: in our case, this "energy" is the risk analysis and its management [49]. Performing a risk management means:,

- To identify the hazards involved with the specific chemical plant, including a prevision of the incident scenarios;
- to analyse the risks related to the identified scenarios, including the evaluation of the consequences of the scenarios; and
- to develop the Safety Management System (SMS), in order to prevent the identified scenarios and/or to mitigate their consequences.

Examples of some methods related to the probability of occurrence (like the Fault Tree Analysis or the Event Tree Analysis) are described in Chapter 5, where the broad topic of the human factor effects is also discussed. Some of the hazard identification techniques (deeper described in Chapter 3) are:

- "What-if" Analysis. It is the simplest technique, based on the repetitive question "What will happen if" a certain component or procedure related to the examined process does not work properly;
- Hazard and Operability Analysis (HAZOP). It is a structured method to identify the hazards related to a process, analysing all the possible deviations from the normal operating conditions; and
- Failure Modes and Effects Analysis (FMEA). It is used to identify potential failures in equipment or system design, analysing the effects on their selves.

Qualitative techniques are used by the experienced team to evaluate the hazards of an existing technology, with documented long past experience. This book does not intend to discuss such qualitative approaches because differently from the above-mentioned methods, they do not generally allow the conceptual link between risk assessment and incident investigation.

In Chapter 1, the transition between two ways of understanding the world has been briefly mentioned referring to [50]. On the one hand, there is the deterministic approach, son of the scientific method where everything is rigorously obtained from logic processes in a complete and satisfactory way. It is the method of the time reversibility where the objects – intended as physical reality – are put in the centre of the reasoning. On the other hand, there is the probabilistic approach, son of the complex theory where certainties drop in favour of a likelihood of occurrence, of a limited knowledge because relations among objects are now preferred rather than their physical reality. As a consequence, the concept of "**uncertainty**" strongly imposes itself and leads to another important related concept: the one about "risk". The **risk** can be defined as a measure of economic loss, human injury, or environmental damage or reputation regarding both the incident likelihood and magnitude of the loss, injury, or damage. To establish the likelihood, a frequency assessment is performed, while the definition of the magnitude requires consequence assessment. The consequence is the ultimate result of an initiating event, deviations or multiple deviations, intended as a change in a state beyond specified limits, conditions or status (whose boundaries are monitored by the performance indicators).

It is interesting to note how the perception of risks may be different from how they actually are [35]. Most people fear the trivial risks and underestimate the significant dangers of the everyday life. Probably, this happens because risk perception is driven by emotions, being the human response guided by survival. To have an idea, according to the United States Department of Labor (Bureau of Labor Statistics), it is safer to work in a US chemical plant than at a grocery store. Indeed, the chemical industry established excellent safety records in the last decade. It can be impressive to know how dangerous is to be a timber cutter, a fisher, or structural metal workers. In conclusion, the approach to industrial risks requires an open mind, free from prejudices. This is especially required to a forensic engineer, in order to carry out a correct investigation.

The risk acceptability is a criterion intimately connected to both the company policies and the compliance with the national laws and the technical standards recognised worldwide. Thus, different companies may accept or not the same risk, depending on their own managerial choices, even if a minimum level of risk acceptability comes from the compliance with standards.

The **risk matrix** is a great tool to have a graphical visualization of risks and their combination of magnitude and frequency (Figure 2.46). It is particularly used when a

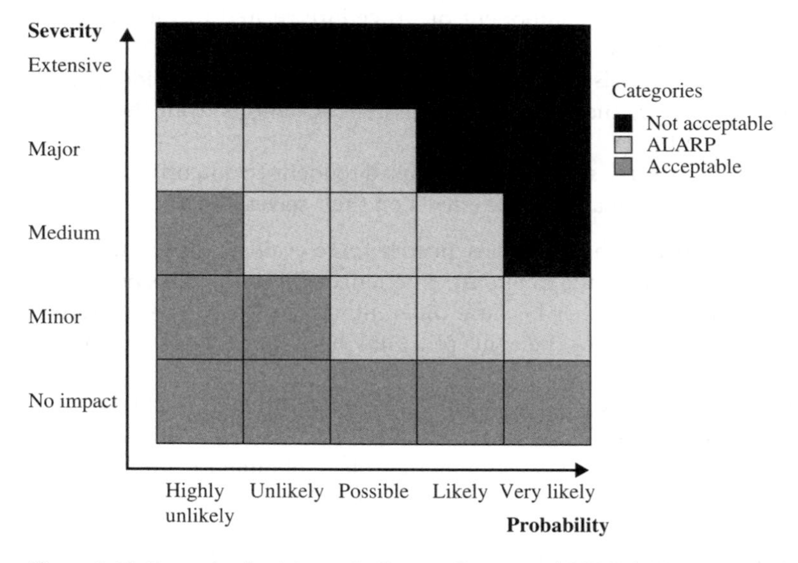

Figure 2.46 Example of a risk matrix. Source: Courtesy of CGE Risk Management Solutions (NL)).

semi-quantitative risk analysis is performed. This type of analysis uses a numerical approach, which is typical of full quantitative risk assessment (discussed in Chapter 5), together with simplifying and conservative assumptions regarding the consequence severity assessment, the frequency assessment of the initiating events, and the effectiveness of safeguards. The results of a semi-quantitative risk analysis are generally expressed in orders of magnitude. However, a risk matrix is also generally used with a **qualitative** risk analysis, like in the example of Figure 2.46. Here, both the probability and the severity are expressed in qualitative terms, that need to be evaluated from an experienced team to assign the proper risk level, given by the combination of a determined class of severity with a specific one for probability. Instead, in a semi-quantitative risk analysis, probability is usually expressed in occasion per year (yr^{-1}) while the consequences are identified through a progressive level from 1 (the less severe) to 5 (the most severe), depending on the severity of the foreseen consequence. In the example, the black regions define the most severe risks. A risk in this region often requires the immediate stop of the industrial process, being absolutely not acceptable. The light grey area of the matrix usually identifies a particular region where the risk *could* be accepted. Risks in this region require an "As Low As Reasonably Practicable" (**ALARP**) study. Briefly, it is a cost-benefit analysis of the potential intervention required to mitigate the risk to the acceptable region. Since the mitigation may require an economic effort which is not justified by the reduction of the risk category, a risk falling into the ALARP region may be accepted as such: managers will take the accountability of this justified cost-based choice. When performing an ALARP study, two key questions must be addressed:

- Which alternatives are available for eliminating, reducing or managing the risk; and
- which factors determine the practicability of each risk mitigation alternative.

Finally, the dark grey region regards the acceptable risks, so no further mitigation or ALARP study is required.

Risk mitigation is possible thanks to the **Individual Protection Layers** (IPLs). They are instrumented safety functions, or mechanical devices, or administrative controls that guarantee whether a reduction of the frequency of occurrence or a decrease in the level of severity of the event.

References

1 Pasman, H. (2015) *Risk analysis and control for industrial processes.* 1e. Oxford: Elsevier Butterworth-Heineman.

2 Beveridge, A. (2012) *Forensic investigation of explosions.* 2e. Boca Raton: CRC Press.

3 Health and Safety Executive. (2004) *HSG245: Investigating accidents and incidents: a workbook for employers, unions, safety representatives and safety professionals.* 1e. Health and Safety Executive.

4 Oakley, J. (2012) *Accident investigation techniques.* 1e. Des Plaines, Ill.: American Society of Safety Engineer.

5 Bird, F., Germain, G., and Clark, D. (1985) *Practical loss control leadership.* DNV GL-Business Assurance.

6 OSH Academy (2010) *Effective Accident Investigation.* 1e. Portland: Geigle Communication.

7 CCPS (Center for Chemical Process Safety) (2008) *Incidents that define process safety*. Hoboken: John Wiley and Son.
8 Sutton, I. (2010) *Process Risk and Reliability Management*. Burlington: William Andrew, In.
9 Conklin, T. (2012) *Pre-Accident Investigations*. 1e. Farnham: Ashgate Publishing Ltd.
10 ESReDA Working Group on Accident Investigation. (2009) *Guidelines for Safety Investigations of Accidents*. 1st ed. European Safety and Reliability and Data Association.
11 Crowl, D. and Louvar, J. (2011) *Chemical process safety: Fundamentals with Applications*. 3e. Upper Saddle River, NJ: Prentice Hal.
12 CCPS (Center for Chemical Process Safety). (2016) *Introduction to process safety for undergraduates and engineers*. 1e. Hoboken: John Wiley & Sons.
13 Stauffer, E. and NicDaéid. (2013) Chemistry of Fire. Encyclopedia of Forensic Sciences. pp. 161–166.
14 Chiaia, B., Marmo, L., Fiorentini, L.,et al. (2017) *Incendio della motonave Norman Atlantic: indagini multidisciplinari in incidente probatorio*. IF CRASC 17. Milan: Dario Flaccovio Editore pp. 129–140.
15 Marmo, L., Piccinini, N., Russo, G. Et al. (2013) "Multiple tank explosion in edible oil refinery plant: a case study." In: *Chemical Engineering and Technology*, vol. 36 n. 7, pp. 1131–1137.
16 Marmo, L., Danzi, E. Tognotti, L. et al. (2017) Fire and Explosion risk in Biodiesel production plants: a case study paper presented at the HAZARDS 27, Birmingham (UK), 10–12 May 2017.
17 Quintiere, J. (2016) *Principles of fire behavior*. 2e. Boca Raton: CRC Press.
18 Babrauskas, V. (2003) *Ignition handbook*. Issaquah, USA: Fire Science Publisher.
19 Crippin, J. (2013) Types of Fires. *Encyclopedia of Forensic Sciences. pp.*406–408.
20 Fiorentini, L and Marmo L. (2011) *La valutazione dei rischi di incendio*. Rome: EPC Editore.
21 Heskestad, G. (1991) *Sprinkler/hot layer interaction*. Gaithersburg, MD: National Institute of Standards and Technology, Building and Fire Research Laboratory.
22 Doyle, S. (2013) Explosions. *Encyclopedia of Forensic Sciences. pp.*443–448.
23 Mannan, S. and Lees, F. (2012) *Lee's loss prevention in the process industries*. 4e. Boston: Butterworth-Heinemann.
24 Yallop, J. and Kind, S. (1980) *Explosion investigation*. Harrogate: The Forensic Science Society.
25 National Institute of Justice (U.S.). (2000) *Technical Working Group for Bombing Scene Investigation*. A guide for explosion and bombing scene investigation. U.S. Dept. of Justice, Office of Justice Programs, National Institute of Justice.
26 Eckhoff, R. (2003) *Dust explosions in the process industries*. Amsterdam: Gulf Professional Pub.
27 Augenti, N, and Chiaia, B. (2011) *Ingegneria forense. Metodologie, protocolli, casi studio*. Palermo: D. Flaccovio.
28 YouTube. (2015) Flash Fire caused by a dropped object hitting a pipe [Internet]. [cited 13 August 2017]. Available from: https://www.youtube.com/watch?v=ph2S5vNIFTY

29 Marmo, L., Piccinini, N., and Fiorentini, L. (2013) Missing safety measures led to the jet fire and seven deaths at a steel plant in Turin. Dynamics and lessons learned. *Journal of Loss Prevention in the Process Industries,* 26(1):215–224.

30 Jet Fire [Internet]. (2017) [cited 15 August 2017]. Available from: http://www.hrdp-idrm.in/e5783/e17327/e27015/e27713/

31 Bakerrisk. (2017) Vapor Cloud Explosion Test [Internet]. [cited 15 August 2017]. Available from: http://www.bakerrisk.com/services/specialised-testing-and-rd/vce-testing/

32 YouTube. (2017) BIGGEST Explosion Compilation 2015 – BEST EXPLOSIONS EVER! [Internet]. [cited 15 September 2017]. Available from: https://www.youtube.com/watch?v=zhBxfNm3rJE

33 CCPS (Center for Chemical Process Safety). (2003) *Guidelines for investigating chemical process incidents.* 2e. New York: American Institute of Chemical Engineers.

34 CCPS (Center for Chemical Process Safety). (2011) *Guidelines for Risk Based Process Safety.* New York: Wiley.

35 Sanders, R. (2015) *Chemical Process Safety: Learning From Case Histories.* 4e. Oxford: Elsevier Science.

36 Fiorentini, L. and Heimplaetzer. P. (2017) *Analisi del rischio di incendio nell'ambito delle stazioni ferroviarie (Fire risk assessment in railway stations).* Bergamo, Italy: SafetyExpo EdItaly.

37 Kletz, T. (2001) Learning from accidents. 3e. Oxford: Gulf Professional Publishing.

38 Health and Safety Executive. Icmesa chemical company, Seveso, Italy. 10th July 1976 [Internet]. Hse.gov.u. Undated [cited 15 September 2017]. Available from: http://www.hse.gov.uk/comah/sragtech/caseseveso76.htm

39 Fabiano, B., Vianello, C., Reverberi, A. et al. (2017) A perspective on Seveso accident based on cause-consequences analysis by three different methods. *Journal of Loss Prevention in the Process Industries.*

40 Bloch, K. (2016) *Rethinking Bhopal: A Definitive Guide to Investigating, Preventing, and Learning From Industrial Disasters.* Amsterdam: Elsevier.

41 Health and Safety Executive. Union Carbide India Ltd, Bhopal, India. 3rd December 1984 [Internet]. Hse.gov.uk. 2006 [cited 15 September 2017]. Available from: http://www.hse.gov.uk/comah/sragtech/caseuncarbide84.htm

42 Health and Safety Executive. Flixborough (Nypro UK) Explosion 1 June 1974 [Internet]. Hse.gov.uk. [cited 15 September 2017]. Available from: http://www.hse.gov.uk/comah/sragtech/caseflixboroug74.htm

43 Gano D. (2011) *RealityCharting®– Seven Steps to Effective Problem-Solving and Strategies for Personal Success.* 2e. Richland: Apollonian Publications.

44 U.S. Chemical Safety Board (2014) Macondo Blowout and Explosion – Investigations | the U.S. Chemical Safety Board [Internet]. Csb.gov. 2014 [cited 15 September 2017]. Available from: http://www.csb.gov/macondo-blowout-and-explosion/

45 Health and Safety Executive. PEMEX LPG Terminal, Mexico City, Mexico. 19th November 1984 [Internet]. Hse.gov.uk. [cited 15 September 2017]. Available from: http://www.hse.gov.uk/comah/sragtech/casepemex84.htm

46 The COMAH Competent Authority. Buncefield: Why did it happen? The underlying causes of the explosion and fire at the Buncefield oil storage depot, Hemel Hempstead, *Hertfordshire on* 11 December 2005. 2011.

47 Forck, F. and Noakes-Fry, K. (2016) *Cause Analysis Manual*. 1e. Brookfield, Conn.: Rothstein Publishing.

48 Dien, Y., Llory, M., and Montmayeul, R. (2004) Organisational accidents investigation methodology and lessons learned. *Journal of Hazardous Materials,* 111(1–3):147–153.

49 Assael, M. and Kakosimos, K. (2010) *Fires, explosions, and toxic gas dispersions. Effects calculation and risk analysis*. Boca Raton, FL: CRC Press/Taylor & Francis.

50 Dekker, S., Cilliers, P., and Hofmeyr, J. (2011) The complexity of failure: Implications of complexity theory for safety investigations. *Safety Science,* 49(6):939–945.

Further reading

ABS Consulting (Vanden Heuvel, L., Lorenzo, D., Jackson, L. et al.) (2008) Root cause analysis handbook: a guide to efficient and effective incident investigation. 3e. Brookfield, Conn.: Rothstein Associates Inc.

CCPS (Center for Chemical Process Safety) (2009) *Inherently Safer Chemical Processes: A Life Cycle Approach*. 2e. Hoboken: John Wiley & Sons.

CCPS (Center for Chemical Process Safety). (2012) *Guidelines for Evaluating Process Plant Buildings for External Explosions, Fires, and Toxic Releases*. Hoboken: Wiley; 2012.

CCPS (Center for Chemical Process Safety). (2003) *Guidelines for Facility Siting and Layout*. New York: American Institute of Chemical Engineers.

CCPS (Center for Chemical Process Safety). (2009) *Guidelines for Process Safety Metrics*. Hoboken, NJ: John Wiley & Sons.

CCPS (Center for Chemical Process Safety) (2008) *Guidelines for chemical transportation safety, security, and risk management*. 2e. Hoboken, NJ: Wiley.

Health and Safety Executive. (2015) The control of major accident hazards regulations 2015. *3e*. UK: Health and Safety Executive.

Hopkins, A. (2012) *Disastrous decisions: The Human and Organisational Causes of the Gulf of Mexico Blowout*. North Ryde, NSW: CCH Australia.

Lees, F. and Ang, M. (1989) *Safety cases within the Control of Industrial Major Accident Hazards (CIMAH), Regulations 1984*. London: Butterworth Scientific.

Lentini, J.J. (2013) Fire Scene Inspection Methodology. *Encyclopedia of Forensic Sciences*. 392–395.

Martin, J. and Pepler R. (2013) Physics/Thermodynamics. *Encyclopedia of Forensic Sciences*. 167–172.

Mathis, T. and Galloway, S. (2013) *STEPS to safety culture excellence*. Hoboken, NJ: John Wiley & Sons.

3

What is Accident Investigation? What is Forensic Engineering? What is Risk Assessment? Who is the Forensic Engineer and what is his Role?

3.1 Investigation

Investigation of incidents and near misses, together with subsequent related activities, is one of the most valid methods to improve the safety and reliability of process plants and, by reflection, of the entire process industry. It has already been pointed out how other methods, like the hazards identification or the management of change, share the same objective of the investigation (safety improvement), but they are predictive methods. This characteristic implies the following limitations [1]:

- The analyses are speculative. Therefore, it is highly possible that all the plausible events are not identified;
- it is difficult to predict the real level of risk, because it is usually based on approximated likelihood and consequence assessment;
- it is difficult to identify multiple-cause events, as the accidents are; and
- it is difficult to predict human error and to take it into account in the risk assessment (this error concerns many incidents).

On the other hand, investigation of actual incidents provides useful information, even if hard to extract, cutting prejudice, ignorance, and misunderstandings that may affect the theoretical preventive analysis. Incident investigation is a core element of a Safety Management System (SMS) [2], whose main goal is to prevent the incident. Indeed, a fundamental assumption of incident investigations concerns the possibility to find, as the root cause, a malfunctioning in the SMS. In other words, it is always possible to find some aspects of the SMS that, if properly organised and applied, would have prevented the occurred incident. That malfunctioning can be related to a lack of planning, organization, actualization, or control. Taking inspiration from [3], when the management system for incidents is developed, or evaluated, or improved, it is essential to:

- Involve competent personnel, define an appropriate scope, implement the program consistently throughout the company, and monitor the effectiveness of incident investigation to maintain a dependable investigation practice;
- identify the potential incidents for investigation, monitoring all possible sources, and ensuring the reporting activities;
- adopt proper methods to investigate, collecting appropriate data, being rigorous, providing expertise and tools to the investigation personnel;
- report the incident investigation results, with a clear link between causes and recommendations, and developing recommendations;

Principles of Forensic Engineering Applied to Industrial Accidents, First Edition.
Luca Fiorentini and Luca Marmo.

- follow-up the results of investigations, resolving recommendations, sharing the findings externally and internally; and
- analyse data to identify a trend in the recurrence of similar incidents.

Performing an accident investigation requires the usage of an investigation method (some are discussed in Chapter 5). However, an effective accident investigation does not simply stop at the application of the selected method. Indeed, it also requires those personnel involved:

- To establish trust and confidence, thus a favourable environment to discuss the incident;
- to be prone in listening to what people say, and to base all findings on verifiable facts;
- to establish a clear cause-effect link, based on sound evidence, together with a timeline;
- to be assisted by technical experts when dealing with specialised issues;
- to understand the identified root causes; and
- to manage the accident investigation as an ordinary project should be (scheduling activities, budgeting).

The fundamental basis for investigation is scientific method. Its rigorous approach is the key to being able to carry out effective investigations, capable of looking beyond the widget, as already discussed in Chapter 1. But it is also systematic, thorough and intellectually honest. The accident investigation process consists of many activities. In the context of a simplified approach, it is possible to identify the three phases in Figure 3.1. The first one is collecting data. It is immediately carried out together with the analysis of the evidence, which may guide the collection of the evidence towards the new objectives, generating a new hypothesis or rejecting others. At the end of the investigation, once the findings have been discovered, there is the last phase of the recommendations development.

Incident investigation is a process that may be required for different purposes by different entities [4]. The main purpose of the investigation is to determine the cause of the accident, exploring both immediate and root causes, and to develop recommendations to avoid its recurrence. As has already been stated, its purpose is not to assign blame.

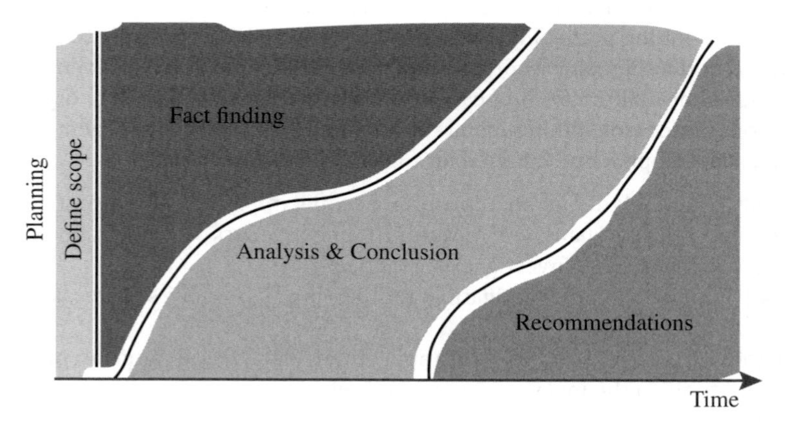

Figure 3.1 Phases in accident investigation.

However, there may be other goals in conducting an accident investigation, such as to check the compliance with law and standards or to solve issues about insurance liability for compensation [5]. For example, after an industrial accident, the administration of a State may suppose that a crime has been committed, so it decides to run an investigation. In this case, it is carried out to evaluate the basis for potential criminal prosecution, thus blame finding is legitimated. Very often, the investigation is commissioned by the same company which experienced the accident. This happens not only to comply with internal rules established at the corporate level but also to ensure the understanding of the offered lessons, thus preventing future reoccurrences. Investigating accidents is also a good way to demonstrate how positive is the attitude of the company to health and safety, regardless of the onset of an actual litigation. It is common to investigate why a part, component, material, procedure, or management system fails. In theory, the reasons why they have been successful before the occurrence of the incident should be investigated as well, but the reasons for successful results are generally taken for granted and the attention is focused only on the undesired outcomes [6]. Even if the main purpose of an investigation is clear, very frequently an in-house investigator or an external consultant is diverted to serve other ends, like blaming or exonerating certain people or things. Obviously, this method tends to introduce bias, since some positions are a priori defended or offended, by strengthening the speculative approach even before any evidence is collected.

A formal definition of investigation is given by [7], recalling [8]:

"An investigation can be defined as the management process by which underlying causes of undesirable events are uncovered and steps are taken to prevent similar occurrences."

Another interesting definition is provided by [9]:

"A structured process of uncovering the sequence of events that produced or had the potential to produce injury, death, or property damage to determine the causal factors and corrective actions."

This definition recalls the one of the root cause. According to the definition provided in [7], a root cause is a fundamental system-related reason why an accident occurred that identifies a correctable failure or failures in management systems. There is typically more than one root cause of every process safety incident.

Incident investigation groups a series of activities. It is a process for reporting, tracking, and investigating incidents, including a formal process for investigating them and their trending to identify recurring events [3].

Incident investigations usually start from the end of the story: once the fire, or the explosion, or the collapse, or the toxic release occurred, people ask how it happened. Starting from the chronological endpoint, the investigator begins his/her work, in a tentative way to determine who, what, when, where, why, and how it happened. Only when the event has been explained, its sequence reconstructed, and the main causes found, then the investigation has been solved. The investigative analysis is based on physical evidence and verifiable facts. The investigator then uses scientific principles and selected methodologies to collect, recognise, organise, and analyse evidence. These topics are discussed in depth in Chapter 4. At this stage, it is sufficient to understand that the incident investigation is structured like a pyramid [10] (Figure 3.2). The collected facts and physical evidence form the large base of the investigation pyramid. They are then the basis for the analysis, carried out in adherence to the scientific principles. Finally, the analysis is the base to support a small number of conclusions (the apex of the pyramid).

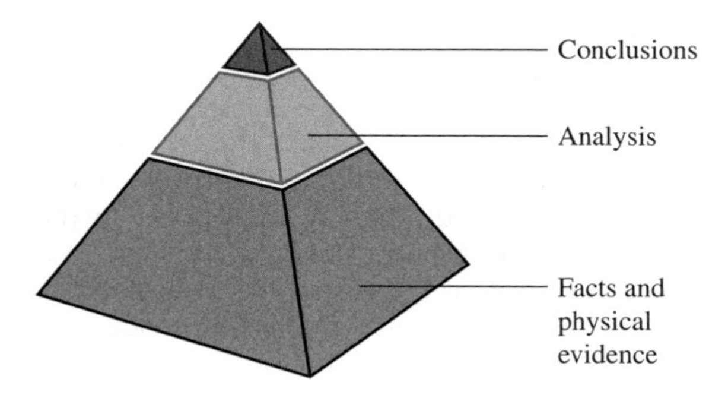

Figure 3.2 The Conclusion Pyramid. Source: Adapted from [10]. Reproduced with permission.

Conclusions should be self-evident as in Figure 3.3. Usually, this characteristic automatically complies when facts are logically arranged in chronological order and with clear cause-effect relations. Conclusions must not be based on other conclusions or hypotheses, otherwise the investigation pyramid collapses.

Even if it is possible to classify the accidental scenarios and to find similar peculiar consequences, each accident is a stand-alone case. The uniqueness must be sought in the progress of facts, which strongly depends on the context and the intrinsic features of the plant. The goal of the scientific investigation, in the industrial context, is to reconstruct the dynamics of the incidental event, finding all the causes and their interconnections, as well as underlining the lack of technical compliance regarding plants, procedures, and machinery.

Different guidelines identify the crucial aspects to be considered during the investigation; however, they rarely find a unique methodology to be followed, thus providing only general information. This is due because a technical investigation cannot be faced as the resolution of a scientific or mathematic general problem. Indeed a "problem" is a question where the aim is to find unknown data which are logically obtainable from the already-known ones. From this point of view, if a problem is well-posed then the solution stands in its definition, requiring only to be extracted from the person in charge to solve the problem through a quantifiable method. Conversely, writing a technical report has a high level of uncertainty (the consequence of introducing the complex theory in the investigation context has been already discussed). Uncertainty is given by:

- The peculiarity of each incident;
- the complexity of the problem;

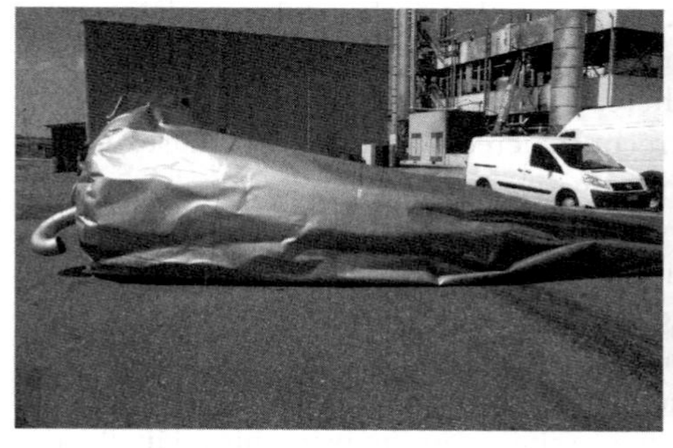

Figure 3.3 A damaged item under investigation.

- the lack of all the useful data, for the resolution from the beginning; and
- the subjectivity, given by the personal contribution of the technical consultant.

Therefore, the complex problem to be faced is defined step by step, proceeding with the learning process and developing on different levels at the same time. A further difficulty is implicit in the required equilibrium between elasticity and rigour. On the one hand, to prove the claimed assertions, it is necessary to comply with the scientific literature and the laws but, on the other hand, it is not suggested that an investigation be faced only from a pragmatic point of view. The reason is in the following citation:

"When you go looking for something specific, your chances of finding it are very bad. Because, of all the things in the worlds, you're only looking for one of them. When you go looking for anything at all, your chances of finding it are very good. Because, of all the things in the world, you're sure to find some of them".

Daryl Zero in the film "Zero Effect" (USA, 1998).

It has been already pointed out that older investigations were superficial, since they only identified obvious causes and developed poor recommendations. In the more modern layered approach, a deeper analysis is carried out and additional layers of recommendations are developed: immediate technical recommendations, recommendations to avoid the hazards, and recommendations to improve the management system.

To sum up, the research into the causes of an incident span over three different levels [11]:

- **Immediate cause**. It is the most obvious reason why an adverse event happens (e.g. the valve is in the incorrect position). A single adverse event may be correlated to several immediate causes identified;
- **underlying cause**. It is the less obvious reason found at the end of the investigation outcome and it concerns the system. Examples are: preliminary checks not carried out by supervisors; not robust risk assessment; too great production pressures, poor safety culture, and so on;
- **root cause**. It is the initiating event from which all other causes come. Root causes are generally related to management, planning or organisational failings.

Generally, recommendations are also developed over these three different levels, reflecting the distinction presented for the causes.

Having clarified which reasons to investigate, the information gained from an investigation and the benefits arising from it, one may question about which events should be investigated [11]. Indeed, the injuries suffered on the occasion cannot simply determine the level of investigation, since the potential consequences and the likelihood of recurrence should also guide in what in depth should be carried out the investigation. The severity and the immediacy of the risk involved determine also the urgency of an investigation. It suggested that adverse events are investigated as soon as possible, also to enjoy the best memory and motivation.

In this sense, an important question that may arise is: "How thoroughly should the accident be investigated?" [12]. Rasmussen, in [13], answered the question by identifying the so-called stop-rules. Reason, in [14], suggests that when the identified causes are no longer controllable, then the investigation stops. This rule of thumb actually identifies different stopping points for various parties. For example, companies should go back to their own management systems to develop effective preventing measures. Supervisory authorities, like national commissions of inquiries or permanent investigation boards, should look at regulatory systems to understand if legal weaknesses could contribute to the accident. Instead, the police and the prosecutors are generally interested in the outer layer to evaluate the basis of a potential crime. Insurance companies are focused on the liability for compensation, therefore their investigation stops at a further different level respect to the previously listed cases. Stopping at the root cause level is a recurring challenge. Increasing the depth of the analysis implies an increasing level of learning, which results in an increasing scope of the corrective actions. In simpler words, solving the management system issues is much more effective than repairing the failed equipment or blaming human error.

A common error is to consider an event as a root cause [5]. But events are not root causes; they are the consequences of the underlying causes. For example, an LOPC or a malfunctioning SIS are not root causes, but events. Similarly, a lack of knowledge or insufficient skill is not a root cause. It is therefore fundamental to push the investigation down to the root cause level, even if the stopping point could not be easily identified, otherwise ineffective recommendations will be produced. It is undoubtedly true that finding in-depth causes is a real challenge. Depending on the depth of the analysis, it is possible to develop recommendations also to prevent similar incidents, not only the very same ones [3]. One of the problems affecting the analysis of what can be called an "organizational incident" is also the socio-cultural environment surrounding the analyst, as discussed in [15].

Some common terms of the art are now described, taking suggestion from [6] and [4], to help the reader in the learning process and in a wider comprehension of the topic:

- Failure analysis. It is the determination of how a specific part, equipment, machinery, component has failed. It also concerns the design, the adopted materials, methods of production, and product usage;
- evidence. From a legal point of view, it is an information used and accepted by the court to resolve disputed issues of fact. The different sources of evidence are presented in Chapter 4. Fundamentally, there are two types of evidence: direct and circumstantial. The difference between them is that direct evidence proves to a certainty that a fact happened, while a circumstantial evidence brings a level of probability in its definition. Generally, direct evidence is accepted by the courts. Circumstantial evidence is taken into account only if it is not decreed as irrelevant, not obtained illegally, not a hearsay, and being proved by one logical step, at least;
- Root Cause Analysis (RCA). It connotes the determination of the managerial and human performance aspects of failure. It is discussed in depth in Chapter 5;
- forensic. This modifier connotes that something is related to the law, the courts, the debate, and so on;
- contributing cause. It is a factor that does not cause the event to occur, not triggering the incident sequence, but it significantly gives its contribution in increasing the magnitude of the event or the likelihood of its occurrence;
- causative factor. It is a pre-existing condition that increases the likelihood of the event. It can be:
  - Direct cause. It existed immediately before the occurrence of the event and directly allowed or promoted it; or
  - indirect cause. It is the same as a contributing cause;

- root cause. It is a type of direct cause. It is defined by some as the fundamental cause, that is to say, once removed or modified, it would have prevented the event from occurring (or recurring). This definition implies that only a single root cause exists: this was a conviction of the past, when the "one event-one cause" tenet was extremely appealing. However, even if some incidents may have a single root cause, the current definition of root cause establishes the simultaneous presence of other root causes. Indeed, an incident investigation rarely found a single root cause: more than one root cause typically exist. A cause that cannot be controlled by a person is not a root cause (e.g. lightning);
- apparent cause. It is also named "immediate cause". It is the cause found by a limited investigation. It usually concerns failures in equipment or human error, without considering the managerial context. An investigation stops at immediate cause when the problem is small or limited in scope and there is no risk in performing a limited inquiry. This is not the extent of this book, which intends to go deeper in the root cause analysis;
- programmatic cause. It is a deficiency in a managerial construct (like procedures and training) that increases the likelihood that human error will occur;
- reconstruction. It is the explanation of a failure, a crime, an incident, or, more generally, an event;
- Human Performance Evaluation Process (HPEP). It is a method to evaluate how people's behaviors and actions contribute to causing the incident. The human factor is discussed in Chapter 6;
- corrective action. At the end of the investigation, it is the developed recommendation to fix the problems or weaknesses that are identified in the root cause. How to develop recommendations is discussed in Chapter 6;
- extent of condition. It is the speculative effort to evaluate if similar incidents can occur elsewhere. Thus, the knowledge gained from the experienced incident is used to prevent further events; and
- falsification. It is a principle used when applying the scientific method to the incident investigation. It simply means that the working hypothesis must provide the predicted outputs (facts and collected data prove that the hypothesis is correct), but the hypothesis must not be proven incorrect (facts and collected data prove that the hypothesis is not incorrect). Falsification is important in incident investigation: it is not the quantity of evidence supporting a hypothesis that count, nor the authority of those people supporting the same hypothesis. What counts is the quality of the collected evidence and of those facts that falsify (or fail to falsify) a hypothesis. The value of falsification is dealt with in depth in [6].

Falsification is extremely important to avoid an unwanted bias during the incident investigation: the confirmation bias. Briefly it occurs when the investigator tends to enforce one hypothesis solely because "there cannot be another explanation", and the reconstruction of the event is carried out selecting only those pieces of evidence that may confirm the prejudice in the mind of the investigator, even unconsciously. Falsification is a strategy that tends to eliminate this bias, which is difficult to detect because investigators are usually unaware of being affected.

In order to prevent the recurrence of similar incidents, it is a requirement to [5]:

- Identify and understand the scenario (what happened and how it happened);

- identify the underlying and contributing causes (why it happened). Rejection of proposed hypotheses should be based on physical evidence;
- develop recommendations (identify preventive measures); and
- implement recommendations and share the lessons learnt.

There are some decisions to be made before an investigation begins [9]. They will be discussed further in this book and concern:

- The level of the investigation, that is to say how much detail the investigation should uncover;
- the decision about who will investigate. Usually, a team approach is encouraged;
- the decision about how much time will be dedicated to the investigation; and
- the determination about eventual additional resources will be needed, like experts, testing equipment, or software.

Reference [16] acts as a reminder of the necessity to have protocols for conducting investigations. They are required to identify the roles and the responsibilities of those people involved in the investigation, specifying the steps to be taken and establishing a shared terminology. Protocols also facilitate the sharing of information, resulting in a desirable tool to be established at the very early stage of the investigation process. Moreover, taking into account the possibility to have more than one investigative body, efforts should be addressed in coordination, to avoid the useless waste of resources like duplication.

The composition of the investigation team is discussed in the next Chapter. The multidisciplinary approach is likely to be used for an industrial accident [17]: chemical engineers contribute with the knowledge of the process and the chemical reactions that occurred before, during, and after the incident; mechanical engineers contribute with the technical knowledge about failed equipment, failure modes and causes; instrument and control engineers provide useful information about the monitored parameters, the SISs, the functionality of the BPCS; electrical engineers identify electrical components' failures also providing action items to restore or increment the electrical reliability. Members of the team, which should involve both management and workforce to have a detailed knowledge of the work activities involved, have to be familiar also with standards and legal requirements [11], while possessing the required investigative skills.

The proper investigative attitude is stressed also in [18]. The incident investigator shares the same logic of a criminal investigator, but the mindset is completely different especially when evaluating human behavior. This happens because a criminal investigator deals with wilful malicious acts and it was sufficiently stressed that these events are out of the scope of an incident investigation and of this book. Obviously, from an accident investigation it may result that the initiating event of the sequence resulting in the incident was a criminal act and the principles described in this book can be fully applied. If this is the outcome of the accident investigation, then it is time to turn the investigation over to the police and the human resources department. This is because the attitude of the accident investigator cannot be to look for the bad person or to assign blame, but it is to develop recommendations so that the organization may build the proper defence to prevent a further recurrence. However, it is possible to learn from error, but not from crime. There is not a single layer of protection strong enough to prevent a malicious act. Therefore, one of the pillars of the accident investigation (i.e. learning from experience [17]) collapses: this is why investigating malicious acts is out of the scope.

In this book, accident investigations are handled. However, having an event in order to learn how not to have the very same event is an approach that should be always coupled with prevention strategies, based on the organization's ability to learn [19]. These pre-incident investigations are based on constant activities in monitoring performance indicators, discussed in the previous Chapter. It is therefore required to look for high consequence activities, for small signals indicating a weakness, for error prone conditions, to listen to the workers and to whatever keeps you awake at night. This is why near misses are also investigated.

In conclusion, what makes a good investigation is the capability to identify root causes, from which organizations can learn how to prevent future failures [11]. This capability is ensured by following a structured approach, since unstructured approaches often lead to ineffective results [12].

3.2 Forensic Engineering

Recalling Noon in [10], forensic engineering is the application of engineering principles, knowledge, skills, and methodologies to answer questions of fact, usually associated with incidents, catastrophic events, and other types of failures, that may have legal ramifications. In short, the job of the forensic engineer is to answer the question "what caused this event to happen?", knowing only the end result and applying reverse engineering. The final result of the forensic engineer's job is the reconstruction of the incident, that is to say the full explanation about the incident that has been solved. To do so, failure analysis and root causes analysis are used. The reader should be now aware of the difference between the two definitions, even if they are sometimes used interchangeably. Familiarity with codes, standards, protocols and usual work practices is also required. There are also several guidelines promoted by different organizations that suggest how to conduct forensic investigations depending on the type of incident. The main duties of a forensic engineer include:

- To assess the conditions before the event;
- to assess the conditions during and after the event;
- to hypothesise how the pre-event conditions become the post-event conditions;
- to search for evidence that supports or falsifies the proposed hypotheses; and
- to apply the scientific method and the engineering knowledge to link facts and observations, thus reconstructing the incident.

Those activities are always conducted with an extensive and constant application of logic, which provides order and coherence to all the facts, principles, and methodologies used.

An incident investigation usually requires a multidisciplinary approach. This reflects also on the forensic engineer, who is not a specialist in a given engineering discipline. On the contrary, several scientific disciplines are involved in the solution of forensic engineering problems. When reconstructing the incident, a discipline may give its contribution to developing a further step in the overall reasoning, and so on. From this point of view, a skilled forensic engineer is usually an excellent engineering generalist.

The forensic world is both shrinking and expanding, as M.M. Houck said in the preface of [6]. The global scenarios are responsible for this double trend: on the one hand, forensic experts travel around the world no longer limiting their profession within a laboratory. On the other hand, the increased interest in the topic led to a growing knowledge in size, complexity, and depth. The role of forensic engineering in the recent investigation is doubtless predominant, especially when talking about industrial accidents. The main reason for such a success is based on the rigorously adopted approach, previously discussed, that allows treating forensic engineering as a discipline. In Europe, forensic engineering is a new discipline: it is sufficient to run a search on the internet to find how low the level of related contents is. The same conclusion cannot be considered for other disciplines; for instance, an extended set of information about legal medicine, like definitions, classifications, and scopes, is already available to the community. The scientific community widely recognises the legal medicine all over the world, while the forensic engineering has experienced a lower trend in Europe (not in the US and in the Anglo-Saxon world, where significant funds have been provided to support the investments on this field).

Talking about forensic engineering implies the application of the techniques, i.e. the principles and the methodologies typical of engineering, aimed at the resolution of complex problems, dangerous events claimed during a judicial proceeding, thanks to the role of the technical consultant (who serves the judge or the prosecutor or one of the parties). Forensic engineering consists of a complex match between engineering, intended in all its different sectors (including industrial, structural, chemical, mechanical, electronic, and electric engineering), and law, in trying to use a comprehensible language to support who is in charge to make judgment. Obviously, such a transversal topic is complex and vast. The intent of this book is to provide an organic approach to this immense complexity, in order to develop a technical investigation related to industrial accidents.

The activities of the forensic engineer are removed from the events that he/she is trying to reconstruct. Time passes in one direction, and some details, useful to have a complete reconstruction, are left behind. What remains are traces, that is to say only partial evidence is provided to the forensic engineer, who can enjoy only a necessarily incomplete and occasionally vague knowledge. Therefore, the goal of the forensic engineer is to connect those dots, trying to reconstruct the actual sequence of events that leads to the unwanted incident. Sometimes data could be part of a too large population or they could be too unwieldy to measure directly. In these cases, statistics and probabilities are the tools to obtain a robust and wide base of the investigation pyramid. Facts are put in sequence using logic and scientific rigours. If an event comes always after a specific fact, then this is often the proof that the prior caused the latter. However, this is not absolutely true: coincidence, correlation, and causation have different values and a forensic engineer has to look for causation only in his/her reconstruction activities. These concepts are discussed in depth in Chapter 5. Causation is not only a scientific requirement but also a legal necessity: indeed, the cause-effect relationships are the fundamental basis to establish not only the actual incident path but also the eventual legal accountabilities.

When an incident occurs, regardless of the risk-based process safety or any other prevention loss strategy, tragic losses of life, property, reputation, and treasure are experienced. It is therefore normal, especially for those people who are directly affected by such losses, to ask why it happened. Therefore, the forensic engineer covers a delicate role when carrying out the investigation, and he/she must not allow biases to affect the logic and the proven scientific principles, otherwise he/she will likely fail. Practice is undoubtedly the best way to become a good investigator. However, studying the available literature gives you a discount on the amount of time required to be proficient. This is the goal of this book, also allowing the reader to learn from others' experiences and mistakes.

Forensic engineering is a discipline. Possessing the basic scientific knowledge cannot be substituted by the mere application of an investigation method, since the methodology is simply the framework to organise and assess knowledge, evidence, and facts. The application of the scientific method to determine a root cause cannot be the mere shift of the laboratory-methodology to the context of actual investigation. Indeed, in a laboratory experiment, variables are studied whilst being free from other influences, without needing to consider the simultaneous presence of other elements and with pre-defined and controlled boundary conditions. This approach, in theory, could also be applied to industrial accidents, to reconstruct the real sequence of events. This means that, in theory, variables are changed and combined until the combination leading to the experimental duplication of the event is found, thus solving the investigation case. However, this approach has many problems. Firstly, each industrial incident is unique: varying and combining variables until they perfectly match the event being assessed is costly, time-expensive, logistically difficult, and risky for safety too. However, a large collection of facts and observational evidence can be seen as a substitute for the direct experimental data. This is generally true, but only the correct reconstruction hypothesis will fit them, being also scientifically rigorous. An example is the determination of an equation from a set of points on the Cartesian plane: the larger the number of points, better the fitting curve. Therefore, the application of the scientific method for the reconstruction of accidents and failures consists in:

- Proposing a first working hypothesis, based on first verified information;
- modifying the first working hypothesis as more information is collected, to fit the observations progressively gathered; and
- testing the working hypothesis to predict the presence of unobvious or overlooked evidence.

Finally, a working hypothesis is considered the real complete incident reconstruction if:

- It encompasses all the verified observations;
- it predicts (when possible) the existence of additional unknown evidence; and
- it is consistent with the scientific method (principles, knowledge, and methodology).

However, incidents sometimes destroy the evidence and observational gaps are not so uncommon. When few data are available, more than one hypothesis could fit the evidence gathered, preventing a unique solution. It is the consequence of accepting the complex theory in the incident investigation context: knowledge of facts cannot be fully possessed, time is not fully reversible and observational gaps have to be accepted. This is why the collection of the evidence is a fundamental part of the accident investigation workflow: only a wide basis of the investigation pyramid results in robust conclusions. In this sense, forensic engineering is like solving a picture puzzle: disjointed pieces may not provide much information, but when they are sorted methodically, they fit in a logical context and the overall picture starts to emerge. Continuing the similarity, when a great part of the puzzle has been solved, it will be easier to put in position the remaining pieces.

Engineering is often seen as an exact science, that is to say there always exists a unique exact solution to engineering problems. Forensic engineering does not share this peculiarity of traditional engineering [10]. Actually, it has the same "soft" attributes and uncertainties that affect other disciplines like sociology, economics, or psychology. No matter the engineers' qualifications or experiences: in the context of a court, what is important are the facts that the engineers provide to the judge and the jury. In this sense, a clear distinction with the role of the attorney emerges. Indeed, the attorney may have a stake in the outcome of a trial, since its compensation may be a benefit if he/she will win. It is not unusual that the attorney pressures the engineer to manufacture the technical report in order to gain a better position for his client. While the attorney is legitimated in defending his client, fully representing him in front of the judicial authority and doing the best to protect him, the role of the forensic engineer is quite different. Even if the commissioner may want to cover some facts, the forensic engineer does his best job when he informs the attorney about all the facts he has uncovered: this also results in providing the attorney with a full awareness about what goes wrong, allowing him to best prepare the case for the presentation in court. Moreover, accepting to cover facts or an extra remuneration to do so, can result in the suspension or the revocation of the engineer's license. This does not mean that the investigator is regarded as neutral to the investigation outputs [16]. Indeed, the position of the investigator towards the event and his role regarding the investigation results are two aspects that could impact the outcomes of the investigation. For example, the investigator could be part of the company where the event occurred and also part of the plant; or could not be part of the plant and attached to the corporate headquarters; or he could be from outside the company where the event occurred. Therefore, different positions of the investigator may result in different outcomes: investigators that are too close to the event may hide or disregard some root causes, because they may not have access beyond the organizational limit of the company to which they belong, thus with no possibility (or authority) to explore potential underlying causes. Or, depending on his position, the investigator may develop a corrective recommendation that he knows to be within the organizational boundaries available to him. Not only the position, but also the role of the investigator may affect the investigation outcomes. Indeed, an investigation is rarely launched by the investigators themselves, but it is often requested by someone having the authority to do it. Consequentially, the information gathered during the investigation could be filtered by the authority before the release of the investigative report. The higher the investigator's independence, the greater the amount of information released in the final report.

Forensic engineering can be specific or general in scope, depending upon the nature of the dispute [6]. In this sense, the distinction between a failure analysis and a root cause analysis has already been done. Indeed, while a **failure analysis** is carried out to determinate how a specific component, machine, or equipment has failed, a **root cause analysis** concerns the managerial or human performance aspects rather than the failure of a single part, being addressed in preventing the incident from recurring through a deep analysis about how to enhance procedures and managerial techniques. This is why it is usually adopted where there is a heavy emphasis on safety and quality, such as for industrial incidents.

3.3 Legal Aspects

A particular sensitivity to legal issues can help when the outcome of the investigation may trigger a litigation. This book does not intend to provide legal guidance, but only to present some aspects related to the topic. It should be kept in mind that the final goal, for both the investigation and the legal team, is to prevent similar incidents.

In the legal context, the role of the forensic disciplines is to assist with proof [20]. This is obviously also true for forensic engineering, where evidence is provided to help the trier of fact to determine, beyond a reasonable doubt, guilt in criminal proceedings or liability in civil proceedings. The role of the investigator is to testify, in deposition or in court, about the findings of his/her investigation. Generally, the investigator is asked to answer a set of questions posed by the judge, or the prosecutor, or the attorney for an involved party. This task is accomplished with the deposition of the incident investigation report. The judge, or the prosecutor, or the attorney for an involved party are generally interested in [6] and [10]: the investigator's qualification for the specific incident analysis; the assumptions at the base of the investigator's analysis; the reasonableness of the findings reached; the possible alternative incident reconstructions, initially not considered. The investigator is a customer for his attorney, who will pay his fee. Therefore, before the proceeding, experts are obliged by the court to be objective, refraining from being an advocate. The best rule is to remain professional and honest, discussing with the client both the favourable and the unfavourable findings of the analysis, prior to testifying.

Some legal aspects of forensic engineering include [20]:

- The chain of custody. There are often specific legislations, or guidelines, providing standard rules about how to collect, transport, handle, and store the gathered samples that might be used for legal purposes. Indeed, many of the problems related to this topic have been solved through administrative solutions, such as labelling, barcodes, or restricted access. Figure 3.4 shows the handling of an item under investigation;
- the admissibility of forensic science. In order to be considered in a criminal or civil proceeding, the forensic evidence must be obtained legally, properly following formal rules. Moreover, the weight of the evidence, that is the value assigned by the trier of fact, is also fundamental in establishing the relevance of the evidence. Admissibility standards and practices should be known by the investigator, in order to provide admissible evidence from the beginning of his activity, focusing only of what can be spent during the proceeding;
- the expert evidence at trial. An admitted evidence can be contested during the trial, for its probative value or weight of expert evidence. In many jurisdictions, the analyst who collected that evidence and performed the subsequent tests, finding the resulting conclusions, may be asked to testify when (part of) his work is contested. Direct examination, cross-examination, and re-examination are means to debate about the expert witness during a trial. The expert should pay much attention in the manner he expresses his opinions, to avoid misunderstanding and increasing controversy;
- the right of appeal review and postconviction. Some jurisdictions allow parties to appellate on the admissibility of expert evidence. The investigator should be prepared for this eventuality;

Figure 3.4 Handling of an item under investigation.

- the lay assessment of the forensic science. To fully understand the expert evidence, technical knowledge should be possessed by everyone in the court. Obviously, this is not the case. This is why communication aspects need to be considered;
- the plea bargains and interrogations. The objectivity of the forensic evidence is leverage in plea negotiations. A similar role is assumed in police interrogations, persuading a suspect that forensic science implies his guilt. However, in this book deliberative malicious acts are not considered, and therefore this aspect may be of less interest;
- the wrongful convictions. There are several cases where forensic science failed in the reconstruction of disputed issues. Therefore, from a legal point of view, the forensic evidence can be as wrong as another source of evidence, like testimonies, regardless the scrupulousness taken from the scientific principles;
- the expert witness immunity. For a long time, forensic scientists enjoyed the immunity from negligence. Recently this changed, and experts are exposed as being liable for mistakes caused by negligence, inadvertence, and incompetence.

The necessity to seek legal guidance when preparing documentation is stressed in [21]. If the incident has potential liability for the company, a prompt involvement of legal counsel is suggested. A frank communication between the attorney and his client is encouraged, being enforced by the existence of the attorney-client privilege. It will help to have an open communication between the incident investigation team and legal counsel, and between the legal counsel and management. Documents regarding the investigation or any legal advice should be protected from disclosure, especially when they are considered privileged information. Each member of the investigation team is highly recommended to treat every written document he/she will produce (including emails, notes, reports) as if they would become of public domain and shared with the press. This allows sure protection from unwanted disclosure, while the team member remains professional if those documents become evidence during the trial. However, the attorney-client privilege can be denied by the court under certain circumstances. In this case, investigation reports are asked to be shared, disclosing them. For this reason, it is suggested to use header and footer designations to identify those documents containing confidential information. When writing documents, inflammatory terms should be avoided, like "disaster", "catastrophe", "lethal", and so on. Similarly, technical reports must not use judgmental words such as "negligent", "deficient", or "intentional", refraining from assigning blame. They should be free from opinions too, like the ones regarding contract rights, obligations, or warranty issues. The investigator should also refrain from presenting vague conclusions, unsupported by facts. Investigating and documenting near misses as deeply as is done with incidents, is also good from the legal point of view, being an excellent way to demonstrate the commitment of the company towards the risk-based process safety.

Considering the liability issues and how they vary from country to country, it is important that any documentation generated during the investigation remains discoverable [12] regardless the barriers that might be put in place. Moreover, this documentation could be used to sway the public opinion and to assign blame and negligence: this is why it becomes important to adopt a method, together with the attorneys, to control such documents. Indeed, documents management is as important as their writing: the collection, recognition and organization of evidence and documentation are crucial steps for a proper investigation. Issues about documents management concern the necessity to develop header and footer to identify those documents that cannot be duplicated, or to ensure their chain of custody. The retention of the incident investigation reports is a controversial topic between layers, who tend to not retain them to limit both the legal costs and avoid an eventual increase in the company liability, and engineers, who tend to retain them to preserve the lessons learned from the incident. Internal procedures and, in some countries, technical standards provide guidance about the retention requirements and timing.

Any statement, action, and decision to be taken after the occurrence of an incident might significantly affect the credibility and the reputation of the company, which are fragile assets. Therefore, it is important to invest in a proper communication management, informing employees, contractors, neighbors, local authorities, and the public with an appropriate degree of details, weighting each single word to avoid misunderstanding about the incident investigation and, by reflection, about the company's commitment to process safety and its reputation. Credibility is a value that is difficult to create, requiring years, maybe decades, to be solid enough; but it can be destroyed very easily in a few moments. It is therefore essential to provide corporate communication protocols to the investigation team, together with additional training, if required.

Learning from experience may also have legal repercussion. Indeed, the legal liabilities are likely to increase for the company that does not apply the lessons learned from previous incidents, not necessarily occurred in the same plant or company, to avoid the recurrence of similar events. Legal issues related to post-investigation generally arise when the provided recommendation has a high cost to be implemented, and then it is not followed up. It is therefore important to discuss alternative recommendations before they are officially formalised.

In the litigious environment, sharing knowledge could also be problematic for logistic issues, especially for large companies, which have to consider practical issues for a proper knowledge sharing, like the turnover of personnel. However, all these challenges can be faced through an adequate communication management.

Very often, employees' interviews are an important source of information for an industrial incident investigation. It is therefore recommended that the investigation team members be trained to properly conduct interviews, and to inform the employees about how to sit the interview, in order to minimise personal liability.

Sometimes, when a punishable offence has been committed, the technical consultant may have the additional task of comparing the behavior of the investigated person, and the omitted (or infringed) cautions with respect to the specific laws of reference. The sources of law may be widely different from country to country. A forensic engineer possesses a solid knowledge about the sources of law of the country in which he/she operates; indeed, being a sound technician is not sufficient to also be a good forensic engineer.

Disagreements with reports regarding the same incident, which have been written by another qualified colleague, should be presented professionally. The focus of the criticism must not be personal attacks (e.g. concerning the academic qualifications or the private life): it should be based on the facts, the reconstructed timeline, highlighting gaps or different conclusions.

It may happen that the investigator provides different explanations for a single accident. This is not necessarily a disadvantage, because even if the investigator does not know exactly what happened, he/she knows what did not happen for sure. This consideration could be useful especially for the defendant (the party accused by the prosecutor). Indeed, the defendant does not need to fully reconstruct the dynamics of the incident, but it is sufficient to prove that he was not wrong and did not have any correlation with the occurred incident: that may be all that is needed.

Sometimes, depending on the regulations of the specific country, the necessity to investigate an incident is required by law. It is interesting to ask if this will prevent the accident's occurrence [22]. The answer depends on a number of factors. Firstly, the companies should be equipped properly to perform an investigation, checking the availability of enough resources to find the root causes. Secondly, the interpretation of the law promoted by the public opinion and mass media is more oriented in finding the blame rather than the actual causes that led to the incident. When dealing with industrial accidents, it is generally, and incorrectly, assumed that the lack of safety is attributable to managerial choices in putting profit before safety. With this spirit, the help provided by law seems to vanish and the attention is not properly focused on the corrective actions. Moreover, it is important to question about the sharing of incident reports among different companies, even anonymously: indeed, the work involved in the investigation is largely wasted if the information found is not broadcast, thus producing only a limited enhancement in the short-term scale.

In conclusion, taking inspiration from [12], some final considerations about the legal aspects are confounded. Firstly, it is essential that proper legal assistance, by the organization's attorney, is provided throughout the investigation process. Then it is fundamental to keep a technical focus when performing the investigation, not answering questions about legal responsibility (this job is the task of the legal counsel). The investigation team should have the proper credentials to carry out its job: this will help to defend the organization and its position during an eventual trial. It is also important to follow the requirements of regulations about incident investigation including the internal organization's procedures. High-quality standards should be maintained to guarantee the confidential information will remain so and to enhance the credibility. Concerning the witness statements, attention should be paid to avoid wordiness that may lead others to misunderstand the content of the testimonies or to question about their credibility. Clarity in writing is a desirable skill to avoid ambiguities in interviews and reports. After a witness statement is recorded on a document, it is suggested that the witness sign on each page. Audio or video recording the interview may help in gathering as much information as possible, but their use should be extremely prudent since it may make the witness nervous and less willing to share information. When there is a high probability of legal consequences after an incident, the interviews should be structured formally, and the witnesses informed of this condition. The witness may be reluctant to share some information; it becomes important to try to relax the witness under this typical stressful condition. Remember to respect him/her even if it will not be possible to obtain his/her collaboration. Throughout the investigation, maintaining a proper chain of custody is legally essential to document the formal transfer of evidence from one person to another, having the objective to prevent alteration. From a legal point of view, it is also necessary that all the interested parties take part to the unrepeatable tests that could be performed: the investigator must invite them when physical data could be permanently altered.

3.4 Ethic Issues

Finding the Truth is a hard task. The job is rendered harder by the high ethic stature that is required to solve such a complex problem. Talking about the responsibility about the conduction and the outcomes of an accident investigation, several professional investigative companies have developed a "Code of Conduct" for their investigators [16]. The Code must not be confused with a technical guide about which forensic engineering methodology must be used or how to technically perform the task. Instead, it is intended as a guide to the pursued ethical principles to maintain the moral authority, integrity, objectivity, logic, credibility and independence to conduct properly an investigation. For example, a code of conduct may focus on:

- Integrity. Every activity should be performed in accordance with the high standard of integrity required by the role;
- objectivity. Facts should be collected, analysed, described, and communicated with objectivity;
- logic. The reconstruction of the incident should be carried out on solid logic basis;
- prevention. Findings from facts and evidence analysis should be used to develop recommendations that will improve safety; and
- independence. The investigation team should be independent from any actor involved in the incident, like the company where the incident occurred, the national authorities, and so on.

Very often, especially when being the technical consultant of the judge (thus when you are the impartial third party), you may receive particular attention from the parties, who could behave especially generously and familiarly with you. Even if this behavior does not represent a law infringement, you have to operate a clear distinction between the human interactions and relations (based on gentleness and cordiality) and the professional ones. It is a hard task, but it is necessary to ensure that your evaluation is free of dangerous prejudgement (it does not matter if positively or negatively developed) prestructured with human interactions that are external to the professional field. It is a matter of ethics.

It is expected that these principles are followed not only by the investigation team members but also by all the stakeholders involved in the investigation; sanctions should be also provided for eventual infringements.

It is possible to ensure a broad consensus through:

- Rigorous adherence to the goals of the investigation, without assigning blame or liability;
- the investigation of near misses to learn corrective lessons;
- the adaption of the budget, time and personnel resources according to the complexity of the incident; and
- the development and follow-up recommendations to reduce the risk factors.

To sum up, the investigation should be carried out by professionals possessing a high standard of competence and knowledge, strongly oriented to the objectives of the investigation, impartial and trained at safety and risk management.

Ethics reflect also when dealing with the media [12], to answer specific questions. The release of the names of victims must be absolutely avoided if families have not been notified. Moreover, the interviewed person is not obliged to tell everything he/she knows about the incident: what is important is that what is mentioned is the truth, to avoid misunderstanding and to avoid the spreading of incorrect information. Speculations have to be categorically banned: there is no room for opinions, conjectures, hypotheses: only confirmed events and solid conclusions should be described. When interviewed by a journalist, you should be prepared to describe the incident investigation approach, including how you intend to determine the root causes and which are the efforts to be made avoid a repetition of the incident: demonstrating an organised approach is a good starting point to receive positive feedback from public opinion. Finally, it is important to not bring up old histories. Be focused on the current incident and disregard the older ones: there is no reason to remember old incidents and feed inflammation against the organization.

3.5 Insurance Aspects

There are many reasons to conduct an incident investigation. It can be required by the law, because the judicial authorities hypothesise its infringement, or by the company's internal procedures. Among these conditions, a technical investigation can be also required for insurance purposes, even in combination with the other reasons presented. The occurrence of an industrial accident may bring about catastrophic consequences like victims, injuries, environmental damages or economic losses. Regardless of the judicial aspects, which might overcome the administrative scope and reach the penal context, the consequences of an industrial accident may cause serious damage to the company that may reflect into lost of production, administrative fines, re-buying of damaged equipment, paying for medical and legal assistance, procurement of technical consultants to investigate, paying to restore the afflicted damages. An example is shown in Figure 3.5. For the reasons listed, industrial companies subscribe to insurance policies to protect them against the administrative and economic consequences of an incident. Therefore the reason to investigate may also come from the insurance company, who desires to know whom to contest with the insurance claim and, in the end, who is blame. The reader should note that, also in this case, the interest in who is to blame does not belong to the investigator but to a subject which is not directly involved in the investigation (i.e. it does not perform the investigation in person, even if experts from the insurance company usually attend the investigation). As it has been already repeated, the investigator does not find culprits but, with a different approach, looks for the truth. The insurance company, for clear reasons, is interested in understanding the exact dynamics of the incident to establish if the subscribed policy covers the client's damages. Also the company which experiences an incident has the same interest, regarding its suppliers. For examples, an investigation may show how, among the root causes of an incident, there is a set of damaged valves that were certified as safe by the supplier. In this case, the company will ask its supplier to pay for the incurred damages, and the investigation report will be the starting point of the negotiation, having previously discussed both the shared and not-shared responsibilities, to define their boundaries. If the two parties support different incident dynamics (with different root causes and thus different responsibilities), then the controversy is usually shifted, as it often happens,

Figure 3.5 Explosion of flour at the mill of Cordero di Fossano (CN). The damages caused involved many insurance-related consequences.

in front of the judicial authority. Obviously it may happen that, because of the penal relevance of the event, the entire discussion requires a judicial verdict since the beginning. It will be important for the company, in this case, to have an adequate insurance policy to cover, or at least minimise, the economic requests that will be advanced by the offended parties to restore their damaged rights. During a trial, two or more parties present their own investigation reports and the insurance company is generally a further party. The presence of multiple points of view explains why supporting its own standpoint is usually not sufficient. Indeed the technical consultant who helps its client in the forensic debate should have to know also how to apply the principle of falsification in order to falsify an adverse hypothesis advanced during the debate, in addition to support its own theory. The adherence to the ethical principles, when using falsification, is desirable to find the true event dynamics and responsibilities as soon as possible.

3.6 Accident Prevention and Risk Assessment

In Chapter 2.6, by presenting the concept of "Risk", an intimate relationship between Accident Investigation (AI) and Risk Assessment (RA) emerged. Particularly in the first steps of an accident investigation, several points of contact could be found between AI and RA, as reminded by [16]. When an accident happens, the immediate consideration is the following: "Has this incident been studied earlier in a Risk Assessment?". The idea is therefore to discover if, in some way, the accident could be prevented, or at least foreseen, thanks to previous studies. The second step is to check for eventual risk-reducing measures, always related to the incident involved, which would be presented during the risk assessment sessions. The accident investigation should then identify if those measures were implemented or not, creating the basis for further deeper considerations. Reasoning this way, it automatically transpires that personnel having conducted the risk assessment become an extremely important source of information during the incident analysis, trying to understand if it could be predicted, providing assistance to describe the involved systems, the management procedures, and the infrastructure. Another point of contact concerns the methodologies used in an accident investigation that should be the same adopted for risk assessment (e.g. to verify domino effects or combinations of pre-existing circumstances). Finally, once the accident investigation is concluded, its recommendations should include performing a specific risk assessment for the particular system or situation related to the happened incident.

The relationship between risk assessment and accident investigation is also revealed from the risk assessment point of view. Indeed, the outcomes from an accident investigation (that is to say both data and knowledge) should become input for future risk assessments, providing important information about possible causes, the magnitude of the consequences, the likelihood of the scenarios, presence of domino effects. Moreover, the findings of an accident investigation could suggest updating an already existing risk assessment, taking subsequent actions according to the safety and risk management system of the company. Undoubtedly, those findings become a tool to check if the hypotheses at the base of the risk assessment were correct or not. Methods which are typical of risk assessment can also be used to evaluate the effectiveness of implementing the recommendations which emerged from the accident investigation report. When an accident investigation is concluded, it could be of interest to carry out a sort of risk assessment of the accident investigation process itself. This is useful to enhance the future use of the accident investigation methodology, questioning the accuracy of the investigation, the correctness of its conclusions, the exploitation of all the available data sources, and the implementation of the advanced recommendations.

Risk assessment and accident investigation are both important tools for a company that intends to minimise and prevent its risks. It becomes crucial to exploit the cross-learning from these two disciplines, mutually sharing some methods, to increase their quality and reliability and, as a consequence, to save time and resources. In conclusion, learning from experience is the key concept to ensure a holistic result in linking an *a posteriori* and *a priori* approach. The link is strengthened by a policy of transition from prescriptive rules to performance-oriented goals, by a courageous vertical sharing of information, and by a horizontal integration among all the actors involved in such a complex environment. An actual application of the link between risk assessment and accident investigation is presented in Chapter 5, where a software solution to implement these concepts is also briefly mentioned.

In the following section, we will discuss in more depth the Hazard Identification and Risk Analysis (HIRA), intended as all the activities involved in identifying hazards and evaluating risks [17] in those facilities where people, assets and environment are constantly monitored to guarantee an acceptable level of risk. It should be noted that those activities are usually indicated with synonymous titles like "Process Hazard Analysis", "Process Hazard Review", "Process Safety Review", "Predictive Hazard Evaluation", "Process Risk Review", "Process Risk Survey", "Hazard Assessment", "Hazard and risk analysis" or "Hazard study". However, in many companies the term "Process Hazard Analysis" (PHA) is used to indicate the methodology to follow to comply, for example, with the OSHA PSM standard. The first step to management of risk is to identify hazards, to establish later if those risks are acceptable or not. The performed hazard and risk analyses aim at providing a correct perception of risk, otherwise a limited knowledge might lead to undesired consequences because of exceeding the tolerance risk level established for safety reason by the company. Whichever is the methodology used, there are three main risk questions at their base:

- What can go wrong? (Hazard identification);
- how bad could it be? (Establish the magnitude of the consequences); and
- how often might it happen? (Establish the likelihood of the consequences).

Several tools are available to answer these questions. Hazard identification or qualitative risk analysis include Hazard and Operability Analysis (HAZOP), what-if/checklist analysis and Failure Modes and Effects Analysis (FMEA). Failure Modes, Effects and Criticality Analysis (FMECA) and Layer Of Protection Analysis (LOPA) are used for semi-quantitative risk analysis. Instead, Event Tree Analysis (ETA) and Fault Tree Analysis (FTA) are adopted for full-quantitative risk analysis, which also encompasses consequence modelling through Computational Fluid Dynamics. Depending on the method used, the results of the HIRA are reported in different types of worksheet form. The objective of the analysis and the acceptance risk criteria determine the methodology to be used, spanning from the simplest qualitative to the most detailed quantitative analysis. A regular follow-up and updating of a HIRA conducted on existing plants/units should be performed; requirements of the OSHA PSM establish that a PHA must be updated every 5 years. In the following Paragraphs, a brief description of "What-if", HAZOP and FMEA analyses is presented, taking inspiration from [23].

3.6.1 "What-if" Analysis

This method allows the identification of hazards by recursively questioning "What will happen if..." a certain component/equipment/plant does not operate as it would have to. The method is applicable also to procedures governing the processes. A team of people having experience in the analysed plant/process/equipment is asked to brainstorm to find potential deviations from ordinary activities that could lead to undesired events. Talking about the objective, it is similar to the one of HAZOP and FMEA: the only difference is the poor structure that this method has got if compared with the others. Good results are obtained if the individuals taking part in the team are experienced in design and operation, having serviced in similar equipment or facilities. A knowledge of design standards, past errors, maintenance procedures, and difficulties is suggested to create a winning team. Team members may belong to different business areas (engineering, maintenance, manufacturing, R&D, and so on) depending on the specific topic. After the team is created, the subsequent step concerns the collection of the information: from this point of view, it is highly recommended to visit and walk through the plant's site. Documents like P&IDs, Process Flow Diagrams, administrative and operational procedures are all essential to the review team and should be promptly available. Moreover, it is clear that a good analysis cannot be performed without up-to-date documents. This is why a reliable documentation is generally required for these analyses, independently from the used method. "What-if" analysis enjoys a great flexibility: this means that it can be applied at every step of the analysed process or plant, exploiting the information related to the available knowledge. Requiring skilled personnel with specific knowledge of the process or plant, capable of foreseeing deviation from ordinary operations could be disadvantageous [23].

In order to have a comparison between "What-if" and HAZOP, the following scheme will be analysed with both the two methodologies to underline pros and cons for each. The example takes inspiration from [23]. Let us consider the flow diagram in Figure 3.6. It is a feed line of propane-butane separation column. Initially the mixture enters in vessel D-1; then it is pumped (through P-1) towards the column T-1. An FRC valve controls the flow rate and the heat exchanger E-1 pre-heats the mixture before entering in T-1.

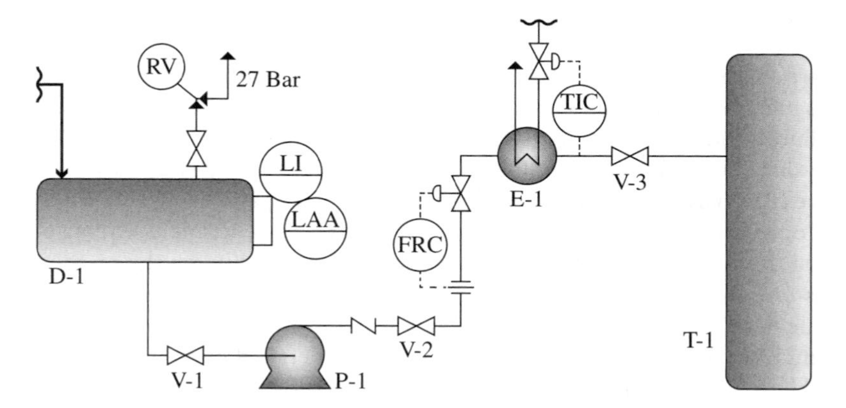

Figure 3.6 Feed line propane-butane separation column. Source: Adapted from [23]. Reproduced with permission.

In Figure 3.6, the following abbreviations are used:

- RV: Relief Valve;
- LI: Level Indicator;
- LLA: Low-Level Alarm;
- FRC: Flow Recorder Controller; and
- TIC: Temperature Indicator Controller.

In Table 3.1 there are two very easy examples of "what-if" scenarios.

3.6.2 Hazard and Operability Analysis (HAZOP) & Hazard Identification (HAZID)

The Hazard and Operability Analysis (HAZOP) was invented in the UK during the 1970s. It is a very structured technique, allowing identification of those hazards related to process deviations of parameters with respect to the normal range of activity. A HAZOP analysis can be used for every kind of process. The structured path to identify the possible deviations consists in the need to:

- Establish a list of key-words, intended as parameter's modifier. A typical set is in Table 3.2;
- establish a list of parameters, intended as those physical dimensions whose setting affects the process (like pressure, level, temperature, flow, composition, and so on);
- combine each key-word with all the parameters, to identify all possible deviations. Obviously, some resulting deviations might have no sense or might be not applicable in the specific context being analysed;
- determine all the possible causes for each deviation;
- determine all the possible consequences for each cause; and
- develop the necessary corrective actions to face (avoiding or mitigating) the hazardous scenarios being identified.

When performing a HAZOP analysis, a balanced team composition is crucial to obtain good results. HAZOP team requires:

- A Team Leader, to guide and help the team in reaching the objectives of the analysis. It is not necessary for the Team leader to know technically the specific process being investigated, since other members are required to bring that knowledge;
- a Scribe (eventual), to write down the results emerging from the brainstorming activities of the team;

Table 3.1 Example of "what-if" analysis [23].

Question "What-if"	Consequences	Recommendations
...the pump P-1 shuts down? ...valve V-1 is accidentally closed?	• Liquid level rises in D-1 • Feeding T-1 interrupted, causing operational upset	• RV will open if LI fails
...the FRC valve is leaking?	• Possible fire due to flammable mixture	• Schedule a more frequent maintenance • Substitute it with a double-seal system

Table 3.2 Guide words for HAZOP analysis.

Guide-word	Meaning
NO	Complete negation, Fully absence of
LESS	Quantitative decrease
MORE	Quantitative increase
REVERSE	Logical opposite
OTHER THAN	Complete substitution

- operator(s) experienced with the process being analysed and its standard and emergency procedures; and
- technical specialists, like instrumental, electronic, mechanical or plant operator(s), depending on the specific process/plant (this category may include technologist engineers).

It should be well kept in mind that the primary goal of a HAZOP analysis is hazard identification: this means that engineered solutions must not be found during a HAZOP session, thus avoiding waste of time. Clearly, if the corrective action is obvious, then the HAZOP team may recommend it; instead, when the solution is not reached so immediately, the task must be left to the engineering team.

Even if it is preferable that the HAZOP analysis is carried out at the earlier stages of the design, so as to positively influence it, on the other hand an already complete design is required to perform an exhaustive HAZOP. A compromise could be carrying out the HAZOP as a final check, once the detailed design is ready.

A HAZOP may also concern an existing facility and it is generally used to identify hazards due to plant modifications, or to propose modifications in order to reduce risks. Being a structured method, HAZOP is widely used in the process industry.

Taking inspiration from the example already discussed in Figure 3.6 [23] about the feed line propane-butane separation column, Table 3.3 presents the scenarios corresponding to the outcomes of the "what-if" analysis in the previous Paragraph, to make the difference immediately visible. The example concerns the parameter "flow" and, one more time, is restricted to the subset of causes found in the "what-if" analysis. A full HAZOP is more extended than that presented in Table 3.3.

A different tool that can be used in order to perform a Preliminary Hazard Analysis is the Hazard Identification (HAZID). The HAZOP is generally used late in the design phase; therefore, the identified safety and environmental issues can cause project delays or costly design changes. Instead, HAZID is a structured brainstorming (guideword-based) that is generally carried out during early design, so that hazards can be easily avoided or reduced. The objective of HAZID is to identify all hazards associated with a particular concept, design, operation or activity (as stated in ISO17776-2016). Typically, the structured brainstorming technique involves designers, project management, commissioning, and operation personnel. Like a HAZOP, HAZID is based on an inductive reasoning, so it is necessary that the analyst has a sufficient experience in "safety", to think as widely as possible in order to ensure that predictable

Table 3.3 Extract of example of HAZOP analysis.

Guide-word	Deviation	Causes	Consequences	Recommendations
NO	No flow	The pump P-1 shuts down (failure or power loss)	Liquid level rises in D-1 Feeding T-1 interrupted, causing operational upset	There is already a RV. Place a High-Level Alarm (HLA) on D-1
		The valve V-1 is accidentally closed by an operator	Pump overheats: possible mechanical damage of the seal, leakage and fire Feeding T-1 interrupted, causing operational upset	Point out the error in the operating procedures
LESS	Less flow	The FRC valve has a minor leak	Hydrocarbons in the air, possible fire	Schedule a more frequent maintenance Install a double-seal system

Source: Adapted from [23].

major accident hazards are not overlooked, including low-frequency events. To do so, the analyst focuses his/her attention on:

- The hazardous substances used as inputs, as intermediates, and as outputs;
- the chemical processes used;
- the equipment, components and materials being used;
- the plant layout;
- the environment surrounding the plant;
- the safety systems; and
- the inspection, control, and maintenance activities.

To conduct a HAZID analysis, it is suggested to subdivide the scope of the analysis into homogeneous areas or functional groups, as from the process schemes (Table 3.4).

Then, the hazards associated with the performed activities are taken into account (fires, explosions, toxicity, and so on). It is not necessary to list all the possible causes

Table 3.4 Subdivision of the analysed system into areas.

Area	Designation	Details	Flammable inventory	Toxic inventory	Comments	PDF
1	1st stage separator	1st separator	Hydrocarbons	–	–	#
2	Crude booster pumps	Crude booster pumps, process area	Hydrocarbons	–	–	#
3	…	…	…	…	…	…

for each incident; indeed, it is sufficient to identify a significant number of them to determine the probability of its occurrence. Pre-defined lists of hazards are generally used, as the ones provided in ISO 17776 related to the hazards that can be encountered in the petroleum and natural gas industries. The approach should be applied to each area and hazard guideword, asking the following questions:

- Is the guideword relevant?
- is there something similar that should be identified?
- what are the causes that could lead to a major accident?
- what are the credible potential consequences?
- what are the preventive and mitigating barriers already specified (or expected)?
- are there any additional barriers that could be proposed?
- are human barriers (if any) reasonable?
- is further (quantitative) analysis required to understand better the consequence of the hazard? and
- what recommendations can be made?

An example of a list that can be used during the preliminary analysis is shown in Table 3.5. Finally, the analyst identifies the consequences of each assumed event, considering the most conservative scenario. The consequences can be pre-defined too, as shown in Table 3.6, as well as the preventive barriers and the mitigating measures.

The results of a HAZID analysis are arranged using HAZID worksheets, showing clear linkages between hazardous events, hazards, underlying causes and control measures/safeguards (if any) as well as the corrective actions. An example of HAZID worksheet is shown in Table 3.7.

Table 3.5 Subdivision of the analysed system into areas.

Hazards	Assumed event
Hydrocarbons under pressure	Leakages
Toxic substances	Leakages
Lifting facilities	Falling parts
Transportation/Traffic	Collision
Utility facilities	Loss of function

Table 3.6 List of typical consequences.

Consequences
Pool fire
Jet fire
Toxic gas cloud formation
BLEVE
VCE
Others

Table 3.7 HAZID worksheet.

Node: **P&ID #:**					**Date:** **Revision**
No.	**Deviation**	**Cause**	**Consequences**	**Safeguards**	**Recommendation**
1	No flow	The pump P-1 shuts down (failure or power loss)	Liquid level rises in D-1 Feeding T-1 interrupted, causing operational upset	RV	There is already a RV. Place a High-Level Alarm (HLA) on D-1

3.6.3 Failure Modes and Effects Analysis (FMEA)

The Failure Modes and Effects Analysis (FMEA) is a technique used to evaluate how equipment may fail and what may be the effects of its failure. The analysis is usually used to foresee future improvements in the system design. When performing an FMEA, all the possible failures of equipment are considered independent, i.e. it is assumed that there is not a cross-influencing effect among them, except for the subsequent effects that a failure might determine. This method is particularly suitable to analyse hazards coming from mechanical equipment and electrical failures: therefore, it is in contrast with the HAZOP analysis, which intends to find hazards looking throughout the whole process, including its dynamics.

A Risk Priority Number (RPN) is defined to establish the priority in treating the identified risks. Without entering in the mathematical definition, the RPN is obtained by multiplying the severity of the consequences by the likelihood of occurrence over one year by the difficulty in the identification of the particular event [23].

The application of the method is effortless, being easy to understand and learn. However, it is crucial that the analyst has a sound knowledge of the components being analysed: failure modes and their effects on the whole system must be known deeply. This request highlights how the required approach is highly time-consuming and reveals why this method is generally used under specific requirements, leaving other methods (the HAZOP overall) to investigate widely the hazards in the process industry.

3.7 Technical Standards

Technical standards arose, in various sectors, more than fifty years ago. Initially, their purpose was to protect the internal market of individual countries by discouraging foreign entrepreneurs from competing with the local manufacturers. Then, with the gradual opening of borders and the free movement of goods, technical standards evolved, becoming a reference to harmonizing the various products in order to make them usable in all the countries. Currently, the technical standards are documents that define the dimensional characteristics, the performance, the quality and the safety issues related to a certain product, process or service, focusing their scope to all the stages of production or activity of the considered service, according to state of the art.

These rules also reserve attention to the aspects regarding the safety, the business organization and the environmental protection, thus favouring a protection against personnel, business, and environmental risks. Usually, their application is not mandatory but, considering the topics that they cover (i.e. areas of significant interest for the life of the workers, for the environment and the society), the states' administrations often recall them into the legislative documents, transforming them in mandatory prescriptions.

The standards authorities are different in each segment of interest and are structured at different levels, from the national one to the international context. The process that led to the creation of a technical standard is quite similar in each country and can be divided into four steps:

1. Set the study together with a preliminary public inquiry. After a request from the market, the consumers, the institutions or the official authority itself, a feasibility study is carried out by correlating the market needs with the regulative ones. If the outcome of these preliminary public inquiries is positive, then the second step is carried out;
2. draft the document. The draft of the standard is prepared by different working groups belonging to the technical office of the standard authority: these groups may be made up of social and economic parties as well as external experts. At this stage, the standard authority has a role of coordination and supervision, always remaining impartial;
3. public inquiry. The draft standard is made available to the market to collect observations, comments, and suggestions in order to obtain a consensus as broad as possible; and
4. publication. After the ratification, the rule is published and included in the technical standards' set. Consequentially it enters into force and it is available to everyone.

The activities of every company operating in the industrial sector have to be contextualised in a set of technical rules and regulations, which are written by organizations (like ASME or API), thus creating a written consensus standard. Even a sad event like an incident actually helps those organizations in improving shared standards, thus developing a safer environment. The timing required to transform some developed recommendations into real technical standards could be very long: however, the effort is necessary to reach the goal of unifying the process safety culture among the different industrial realities [1].

Being compliant with standards requires a system to identify and provide access to the codes applicable to process safety. When a company experiences difficulties in building such a system, then a further root cause can be likely considered: this is what happened on 20 February 2003 with the dust explosion of the manufacturing facility in Corbin, Kentucky [17]. By "standards" we refer to a wide set of practices, regulations and laws, also promoted by industry associations, and established at the different levels of government (regional, national, multinational). Knowing these rules, and ensuring the compliance with them, helps a company to operate in a safer way, being always up-to-date with the most advanced technical requests. Moreover, being such a good approach to process safety-demonstrable compliance helps in minimizing the legal liability in case of incidents. It is also the complex standard system that, indirectly, provides the evaluation criteria for audit program oriented to the process safety, in addition to a communication system about the company's compliance status among managers and personnel. There are a significant number of technical standards to be taken into account for a company operating in the industrial sector: this book does not intend to provide a deep knowledge about the specific contents of these standards. However, there is a subset of standards, belonging to the Process Safety Information element of the U.S. Occupational Safety and Health Administration (OSHA) PSM (Process Safety Management) and EPA (U.S. Environmental Protection Agency) RMP (Risk Management Program) regulations, that the interested reader can study in deep. A good starting point could be [17]. These standards are the shared way to face some engineering activities (e.g. about vessel design, relief valve dimensioning, software and hardware requirements to ensure the Functional Safety, and so on). Clearly, a company may be compliant with standards in different ways: developing internal codes, reviewing every project, or other methods.

Some of the most famous standards are OSHA PSM and EPA RMP in the U.S., Seveso Directive in Europe, and OOMAH regulations in the UK. There are several third-party standards: each company, with its safety process department, decides which to follow. For instance, the chemical engineers in a company might be asked to know the American Petroleum Institute (API) standards (e.g. the API 752, Management of Hazards Associated with Location of Process Plant Buildings), and the National Fire Protection Association (NFPA) fire standards (e.g. the NFPA 30, Flammable and Combustible Liquids Code). A mechanical engineer may need to know the API standards about construction and corrosion or the American Society of Mechanical Engineers (ASME) standards.

Finally, the duties of a control engineer also concern the Safety Instrumented System (SIS), already discussed in Paragraph 2.3. Two examples of standards related to this topic are:

- IEC 61508, Functional Safety of Electrical/Electronic/Programmable Electronic Safety-related Systems (E/E/PE, or E/E/PES), 2010; and
- IEC 61511, Functional Safety – Safety instrumented systems for the process industry sector, 2003.

To ensure the risk reduction that a SIS should guarantee, it is necessary to be familiar with safety manuals of the Final Elements, the Logic Solvers and the other components taking part in the specific SIF. Functional Safety is a topic very influent in the current investigation of industrial accidents. This is why the IEC 61508 and the IEC 61511 are now briefly presented in their details.

The technical standard EN/IEC 61508 concerns "Functional Safety of Electrical/Electronic/Programmable Electronic (E/E/PE) Safety-related Systems". It is divided into seven parts and provides a general approach to all the global Life-Cycle activities of E/E/PE safety-related systems (SIS). The standard EN/IEC 61508 establishes the performance objectives (SIL) that the Safety Instrumented Functions (SIF), activated by the E/E/PE safety-related systems (SIS), must satisfy and reach. The standard EN/IEC 61508 introduces the SIL (1, 2, 3, 4), with the aim to set, for each SIF acting on demand, the probability of failure on demand (PFD), because of the lack or erroneous answers on demand, within the preset time and conditions. In the case of SIF operating continuously, the four discrete values of SIF correspond to four ranges of hazardous failure frequency. In Table 3.8 PFD values concern the operative modality of the intervention of the SIF on demand while the PFH values are about the operative modality of the intervention of the SIF working continuously, in a Probability of Failure per Hour.

Giving these definitions by IEC 61508, the technical standard EN/IEC 61511 intends to explain in depth in "Functional Safety: Safety Instrumented Systems for the Process

Table 3.8 Relations between discrete values of SIL and continuous range of PFD and PFH.

SIL	PFD	PFH
4	$10^{-5} \leq PFD < 10^{-4}$	$10^{-9} \leq PFH < 10^{-8}$
3	$10^{-4} \leq PFD < 10^{-3}$	$10^{-8} \leq PFH < 10^{-7}$
2	$10^{-3} \leq PFD < 10^{-2}$	$10^{-7} \leq PFH < 10^{-6}$
1	$10^{-2} \leq PFD < 10^{-1}$	$10^{-6} \leq PFH < 10^{-5}$

Industry Sector". It is therefore specifically addressed at the industrial sector and establishes the hardware and software requirements to ensure the functional safety, according to the SIL-level determined by IEC 61508. It can be said that the SIS for the process industry sector (i.e. Oil and Gas, refinery, chemistry, non-nuclear energy plant) are governed by these two technical standards. On the one hand, there is the EN/IEC 61508, especially addressed to manufacturers and equipment suppliers (sensors, logic solvers, final elements, and so on); on the other hand, there is the EN/IEN 61511, calling the designers and all the SIS users to respect it. A key concept of those standards is the Functional Safety Life-Cycle. It is defined as the required activities to implement one or more SIF performed by Safety Instrumented System (SIS). The Functional Safety Life-Cycle is used as a base to meet conformity to the EN/IEC 61508 and/or EN/IEC 61511 technical standards. In particular, EN/IEC 61511 details the Safety Life-Cycle (SLC) activities, identifying eight different phases:

- Phase 1 – Hazard and Risk Analysis;
- Phase 2 – Allocation of Safety Functions to Protection Layers;
- Phase 3 – SIS Safety Requirements Specification;
- Phase 4 – SIS Design & Engineering;
- Phase 5 – SIS Installation, Commissioning & Validation;
- Phase 6 – SIS Operation & Maintenance;
- Phase 7 – SIS Modification; and
- Phase 8 – SIS Decommissioning.

A system should also be created to manage Functional Safety, perform the SIS verification and the SIS Functional Safety Assessment, Auditing & Revisions.

Talking about these technical standards, the concept of Individual Protection Layer (IPL) needs to be presented in detail (it was briefly mentioned previously, when talking about Process Safety). In order to reduce the risks connected to relevant incidents, all the measures aimed at the prevention and mitigation of those risks must be carried out. These measures are often named "protection layers" or "defence lines". The typical sequence of IPLs is here listed:

- Inherently safe process;
- BPCS;
- alarm system;
- Safety Instrumented System (SIS);
- safety valve/rupture disk;

- Fire and Gas system;
- external equipment of protection (bunker, containment basin, etc.);
- Emergency Internal System;
- Emergency External System.

The compliance with EN/IEC 61511 requires the establishment of the safety requirements for each IPL and to quantify its contribution in reducing the risk through a LOPA analysis (see Chapter 5 for details about LOPA).

The correct SIL allocation becomes therefore a crucial step. According to the requirements of EN/IEC 61511, part 1, a Process Hazard & Risk Analysis must be conducted to determine and assign the SIL to every Safety Instrumented Function (SIF), which is used as Individual Protection Layer (IPL). The standard EN/IEC 61511 requires that the SIL is assigned for each SIF, taking into account the goal of risk reduction, to be reached in accordance with the established criteria of classification and acceptability of the risk. The SIL assignment can be conducted following different methodologies, as written in the EN/IEC 61511 part 3 standard. Examples of these methodologies are: risk matrix, risk chart, LOPA, ETA, and FTA.

The American National Fire Protection Association (NFPA) standards are developed through a process approved by the American National Standards Institute (ANSI), which tends to establish consensus on fire and other safety issues. Among the numerous set of NFPA technical standards, there are two of interest for a forensic engineer: the NFPA 550 and the NFPA 921.

The NFPA 550 Standard [24] is a "Guide to the Fire Safety Concepts Tree". The authors of this book are among the members of the Technical Committee on Fire Risk Assessment Methods. The Standard intends to provide useful information about the structure of the Fire Safety Concepts Tree, showing its applications, limitations, and usages. A Fire Safety Concepts Tree is a tool to communicate fire safety and protection concepts. In particular it is structured so as to analyse the potential impact of fire safety strategies, identifying gaps, redundancies, and helping in making fire safety decisions. The structure of the Fire Safety Concepts Tree (Figure 3.7) shows the relationships of fire prevention and fire damage control strategies. The advantage of the Fire Safety Concepts Tree is its approach in considering the mutual influence of the single fire safety features, traditionally considered as independent of one another. The hierarchical relationships are drawn by logic gates "and" and "or". The Fire Safety Objectives are shown in the top box of the Tree that represents the goal to reach. The strategies for achieving the objectives are classified into two subsets: "Prevent Fire Ignition" and "Manage Fire Impact". The two branches concern different approaches to Fire Safety: the former is about prevention, that is the reduction of the likelihood of occurrence of a fire, while the latter is about protection, that is to say the efforts to minimise the magnitude of the hazard and limit its effects. For further details, a full reading of the 550 Standard is recommended, here only briefly cited. Having such a powerful tool to communicate and organise Fire Safety is crucial to ensure an effective implementation of the recommended technical and administrative barriers to prevent and protect against fire. Moreover, in case of an incident involving fire, consulting the Fire Safety Concepts Tree, if available, could be a step towards providing tangible advantages in the following investigation phase. An application of the Fire Safety Concepts Tree is shown in Chapter 7, where a refinery's pipe-way fire is presented as a case study.

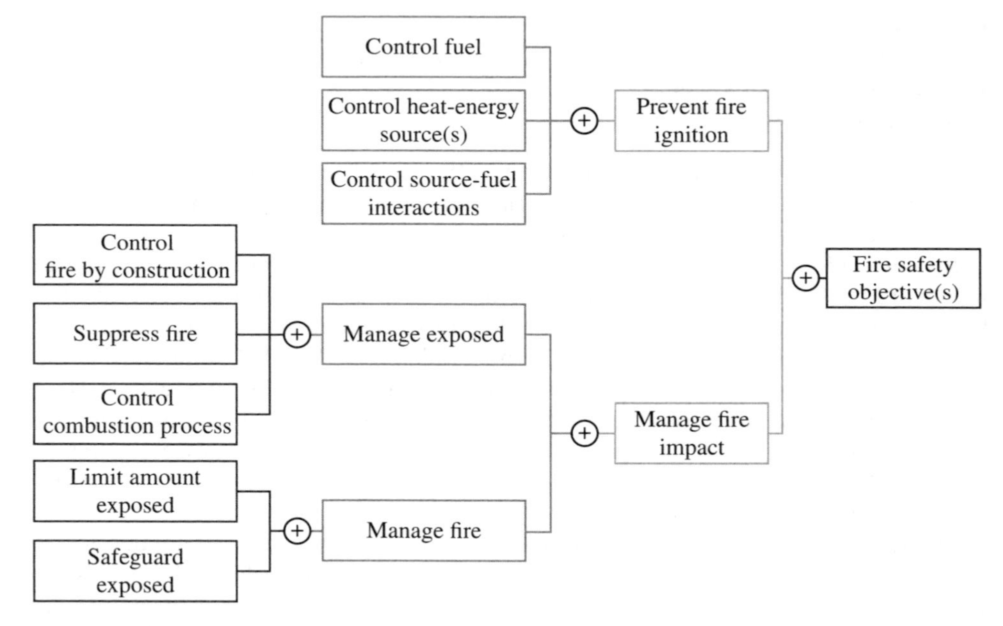

Figure 3.7 Top Gates of the Fire Safety Concepts Tree. Source: Adapted from [24]. Reproduced with permission.

Another important reference is the NFPA 921 Standard [25]. It is a Guide for Fire and Explosion Investigation and its reading is highly suggested to have a full idea about what a worldly recognised technical standard says about the topic. In particular the 921 Standard wants to assist the technicians in charge to investigate and analyze fire and explosion incidents, formulating a hypothesis about its cause, origin, and dynamics. The technical standard describes how to properly determine both the origin and the consequence of fire and explosion incidents, also for statistical reasons. Indeed, accurate statistics are the base for fire prevention codes, standards, and training. The 921 Standard explicitly recalls the Scientific Method, as the recommended systematic approach that provides an organizational and analytical process that is desirable and necessary for a successful fire investigation. The application of the Scientific Method requires the following steps (Figure 3.8):

- Recognise the need, i.e. identify the problem (be aware that it exists);
- define the problem, i.e. how the identified problem can be faced and solved;
- collect Data;
- analyse the Data;
- develop a Hypothesis, using an inductive reasoning from the empirically collected data;
- test the Hypothesis, using a deductive reasoning to compare the hypothesis to all known facts, also trying to falsify the hypothesis; and
- select final Hypothesis.

These steps will be discussed in the next Chapter, dedicated in detail to the forensic engineering workflow. It is crucial that the investigator will avoid presumption,

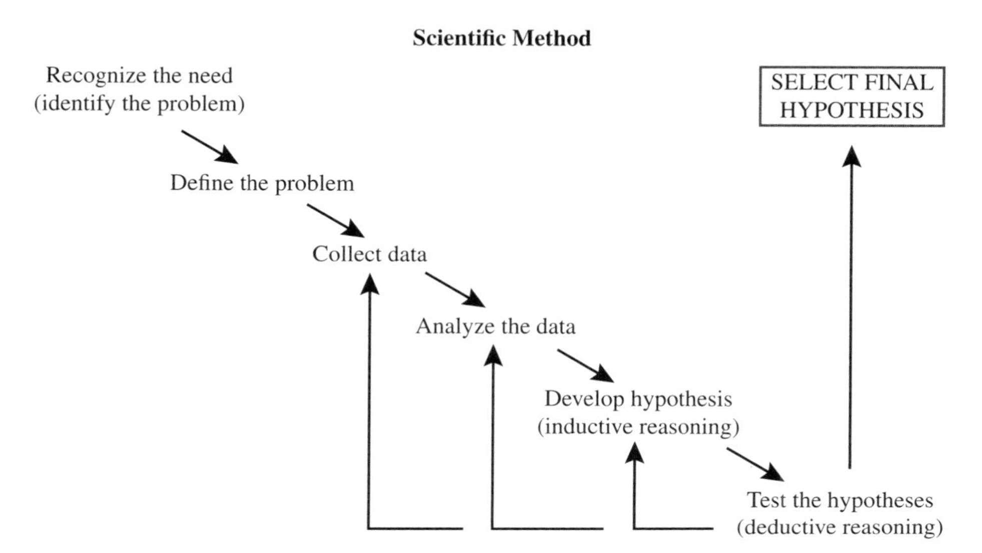

Figure 3.8 Use of the Scientific Method according to NFPA 921. Source: Adapted from [25]. Reproduced with permission.

expectation bias (i.e. avoid reaching premature conclusions without an exhaustive examination of all the data in logic and scientific manner), and confirmation bias (i.e. avoid relying only on partial data that support only a "preferred hypothesis", disregarding properly falsifying the alternative hypotheses). The NFPA 921 Standards also provides the reader with Basic Fire Science, to have a general view of the main relevant scientific concepts involved in a fire investigation. These concepts have also been briefly presented in this book, but we suggest reading the NFPA 921 to have a more concise knowledge about those topics. The Standard helps the investigator to recognise fire patterns, whose analysis is essential to reconstruct the entire fire dynamics, and shares also some knowledge about the building systems, the active fire protection systems, the relationship between electricity and fire, and the fire-related human behavior. It also provides particular information about motor vehicles fires, wildfires, marine fires and incendiary fires. The topics discussed, presented extensively in NFPA 921 Standard, suggest a direct reading of the Standard by everyone who is approaching to the forensic engineering for the first time, it not being possible to provide an exhaustive explanation in the context of this introductory volume.

In conclusion, the purpose of technical standards is to provide a guide, specifications and procedures to ensure the specific product or service will reach the required expected level, creating confidence in its outcomes. The forensic science standards ensure consistency of laboratory tests, consistency in procedures across laboratories, the definition of quality and reliability, the minimum requirement, judicial confidence in the output of a forensic science laboratory test. The only challenge in developing such standards is to not develop prescriptive recommendations with respect to a specific methodology. Indeed, technical standards are not designed to replace internal procedures or methods. Instead, they are designed to establish the expectations. Some considerations about the goal-oriented and the prescriptive regulations are discussed in [26]. Whichever is the considered national context or the standard authority, the "core" forensic standards cover the universal aspects of forensic science [27]. In particular:

- Collection standards focus on recognition, preservation, recording, collection packaging, transport, and storage;
- analysis standards focus on continuity, recording, sampling, analysis, comparison, and identification;
- interpretation standards focus on observations, results, calculations, interpretations, verifications, opinions, and conclusions; and
- reporting standards focus on format, methods, results, opinions, conclusions, limitations, and qualifications.

References

1 Sutton, I. (2010) *Process Risk and Reliability Management*. Burlington: William Andrew, Inc.

2 Agenzia per la protezione dell'ambiente e per i servizi tecnici (APAT). (2005) *Analisi post-incidentale nelle attività a rischio di incidente rilevante*. 1e. Roma: APA.

3 CCPS (Center for Chemical Process Safety). (2011) *Guidelines for Risk Based Process Safety*. New York: Wile.

4 Sklet, S. (2002) *Methods for accident investigation*. 1st ed. Trondheim: Norwegian University of Science and Technology

5 Mannan, S. and Lees, F. (2012) Lee's loss prevention in the process industries. 4e. Boston: Butterworth-Heinemann.

6 Noon R. (2009) *Scientific method*. Boca Raton, FL: CRC Pres

7 CCPS (Center for Chemical Process Safety) (2003) Guidelines for investigating chemical process incidents. 2e. *New York: American Institute of Chemical Engineers.*

8 CCPS (Center for Chemical Process Safety) (1989) Guidelines for technical management of chemical process safety. 1st ed. *New York: American Institute of Chemical Engineers.*

9 Oakley, J. (2012) Accident investigation techniques. 1e. Des Plaines, Ill.: American Society of Safety Engineers.

10 Noon, R. (2001) Forensic engineering investigation. 1e. Boca Raton, FL: CRC Press.

11 Health and Safety Executive. (2004) *HSG245: Investigating accidents and incidents: a workbook for employers, unions, safety representatives and safety professionals*. 1e. Health and Safety Executive.

12 ABS Consulting (Vanden Heuvel, L., Lorenzo, D., Jackson, L. et al.). (2008) *Root cause analysis handbook: a guide to efficient and effective incident investigation*. 3e. Brookfield, Conn.: Rothstein Associates Inc.,

13 Rasmussen, J. (1990) Human error and the problem of causality in analysis of incident. *Human Factors in Hazardous Situations. Oxford: Clarendon Press. pp.* 1–12.

14 Reason, J. (1997) *Managing the risks of organizational accidents*. Aldershot: Ashgate.

15 Dien, Y., Dechy, N., and Guillaume, E. (2012) Accident investigation: From searching direct causes to finding in-depth causes – Problem of analysis or/and of analyst? *Safety Science*, 50(6):1398–1407.

16 ESReDA Working Group on Accident Investigation (2009) Guidelines for Safety Investigations of Accidents. 1e. European Safety and Reliability and Data Association.
17 CCPS (Center for Chemical Process Safety) (2016) Introduction to process safety for undergraduates and engineers. 1st ed. *Hoboken: John Wiley & Sons.*
18 Forck, F. and Noakes-Fry, K. (2016) Cause Analysis Manual. 1e. Brookfield, US: Rothstein Publishing.
19 Conklin, T. (2012) Pre-Accident Investigations. 1e. Farnham: Ashgate Publishing Ltd.
20 Edmond, G. and Cole, S. (2013) Legal Aspects of Forensic Science. Encyclopedia of Forensic Sciences. Academic Press, pp. 466–470.
21 CCPS (Center for Chemical Process Safety). (2003) *Guidelines for investigating chemical process incidents.* 2nd ed. New York: American Institute of Chemical Engineers.
22 Kletz, T. (2002) Accident investigation – Missed opportunities. Hazards XVI: Analysing the Past, Planning the Future. Manchester: Institution of Chemical Engineers. pp. 3–8.
23 Assael, M and Kakosimos K. (2010) Fires, explosions, and toxic gas dispersions. Effects calculation and risk analysis. Boca Raton, FL: CRC Press/Taylor & Francis.
24 NFPA (National Fire Protection Association) (2017) NFPA 550: Guide to the Fire Safety Concepts Tree. NFPA (National Fire Protection Association).
25 NFPA (National Fire Protection Association) (2017) NFPA 921: Guide for Fire and Explosion Investigations. NFPA (National Fire Protection Association).
26 Pasman, H. (2015) Risk analysis and control for industrial processes. 1e. Oxford: Elsevier Butterworth-Heinemann.
27 Brandi, J., Wilson-Wilde, L. (2013) Standard Methods. *Encyclopedia of Forensic Sciences. pp.* 522–527.

Further Reading

Crowl, D. and Louvar, J. (2011) Chemical process safety: Fundamentals with Applications. 3e. Upper Saddle River, NJ: Prentice Hall.
Kirkcaldy, K. and Chauhan, D. (2012) Functional Safety in the Process Industry: a Handbook of Practical Guidance in the Application of IEC61511 and ANSI / ISA-84. 1e. Leipzig: Amazon Distribution GmbH.
Perlmutter, D. (2013) Forensic Chemical Engineering Investigation and Analysis. *Encyclopedia of Forensic Sciences. pp.* 477–482.

4

The Forensic Engineering Workflow

4.1 The Workflow

Providing a detailed forensic engineering workflow is frankly impossible, since the organization of the activities strictly depends on the particular incident. However, a general path can be described, highlighting the essential steps that define a complete investigation. The related bibliography is quite extensive and small differences can be found in the proposed paths.

According to [1], the investigation workflow can be divided into six steps, as shown in Figure 4.1. Notification is one of the preliminary processes: it is required to inform all the necessary personnel about the incident, to trigger the initial emergency activities and to alert also external authorities and the public [2]. Once the emergency response activities have been carried out, the initial investigation starts. This step aims to give immediate feedback to the appropriate people about what has happened and which are the immediate corrective actions to be put in place to restore safety and ensure that similar events will not recur. During the emergency response activities, data can be altered. However, their primary goal is to prevent further injuries, so loss or alteration of evidence should be taken into account, besides the efforts to avoid them. Only if the investigation is carried out in parallel with the emergency response activities, is it possible to preserve and collect data before they are lost or altered.

This step is followed by the team formation, which is separately discussed in the next Paragraph. The collection of evidence and their analysis is also discussed in this Chapter, together with the reporting activities, while the timeline development and the cause analysis are discussed in Chapter 5. Finally, Chapter 6 will discuss how to develop effective recommendations, concluding the investigation workflow.

A seven-step methodology is instead proposed by [3]. It marks the same path previously discussed, but it focuses the attention on slightly different sides of the investigation. These seven steps are:

- Scope the problem;
- investigate the factors;
- reconstruct the story;
- establish contributing factors;
- validate underlying factors;
- plan corrective actions; and
- report learnings.

Principles of Forensic Engineering Applied to Industrial Accidents, First Edition.
Luca Fiorentini and Luca Marmo.

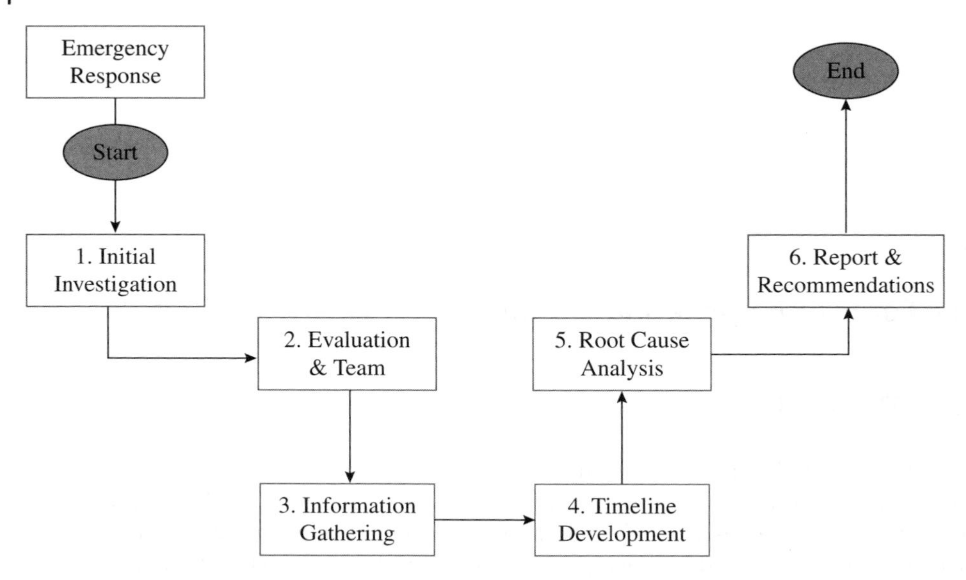

Figure 4.1 The forensic engineering workflow. Source: Adapted from [1]. Reproduced with permission.

In this path, what emerges is probably really the very first step in an incident investigation: recognizing that an incident has occurred. Defining the scope of the problem is the very first thing an investigator needs to do. Taking inspiration from [3], a powerful method to face this step is to write a concise statement about the reason for the investigation, as though it was a newspaper headline. Stating the problem requires stating both the affected object (person, place, or thing) and the defect or deviation. Obviously, at this stage, the stated deviation is likely to belong to the set of immediate causes. Indeed, root causes are revealed once the investigation is concluded. Once the problem is stated, the investigator can then proceed with the problem description, including information about time, location, involved equipment, personnel, and consequences, clearly differencing where the problem is and is not (difference mapping). A final extent of condition review can be done to establish if other activities, processes, organizations, programs, or businesses may experience the same unwanted event.

The six-step process described in [4] marks again the general path already described. In particular, it establishes, as the first step, the need to secure the incident scene. Indeed, investigators can be exposed to several hazards, because of the context in which they operate, like oily or unstable walking surfaces, sharp debris, uninsulated sources of energy (electrical, pneumatic, thermal, or hydraulic), usage of crane baskets to reach inaccessible areas, and so on. Each member of the investigation team has to be competent in the use of personal protective equipment and should be prepared for eventual emergencies occurring during the investigation itself. It is not safe to start if hazards have not been properly mitigated.

A more detailed workflow is instead described in [5]. The investigative activities are summarized in the flowchart shown in Figure 4.2, that can be used as a detailed guideline.

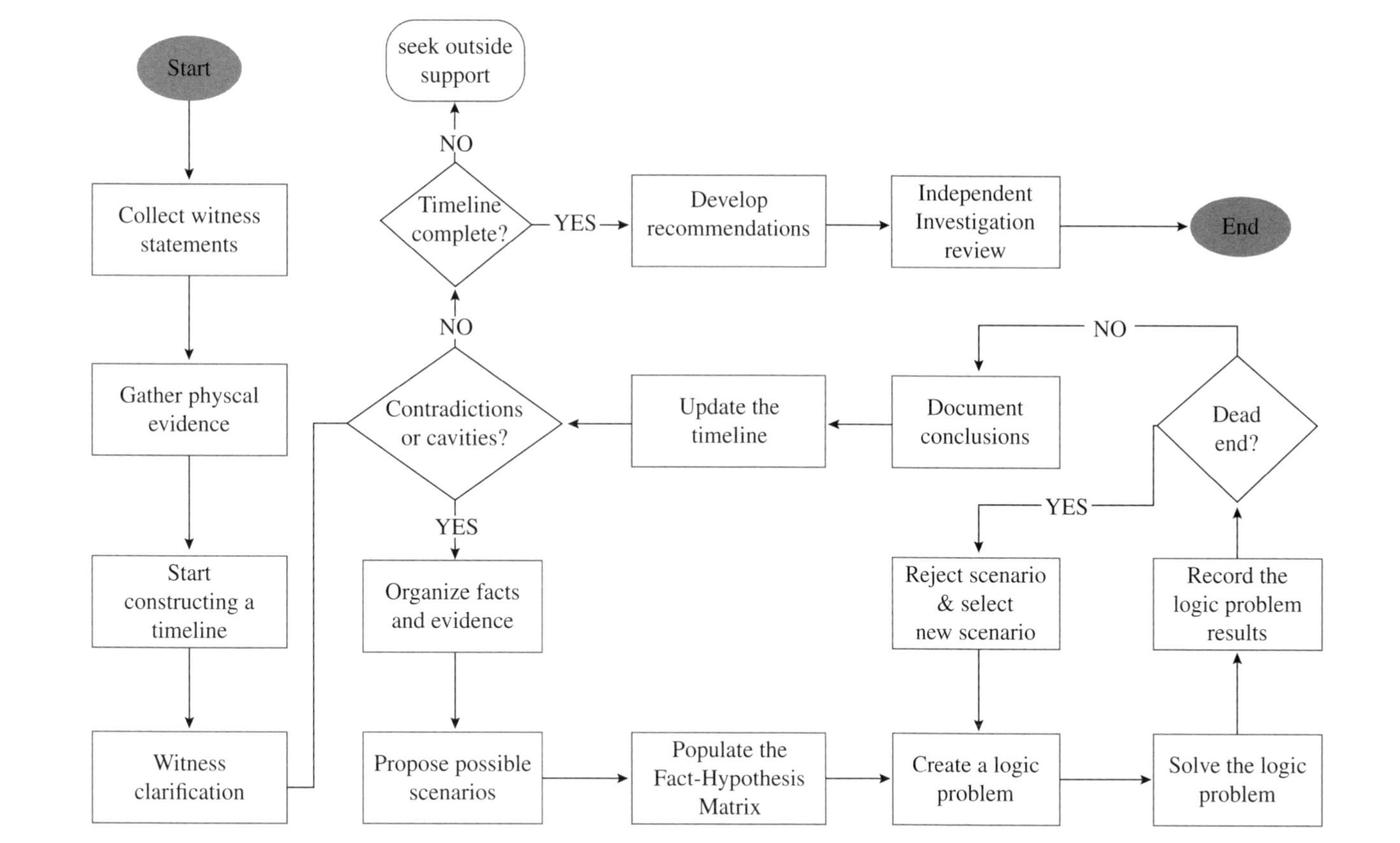

Figure 4.2 A detailed investigative workflow. Source: Adapted from [5]. Reproduced with permission.

Sharing the same philosophy, discussed also in [6], [7] proposes a set of 24 questions to take into account while moving inside the incident investigation workflow. They are useful to drive the investigation towards some specific aspects. They are:

- Gather information:
  - Where and when did the adverse event happen?
  - who was injured/suffered ill health or was otherwise involved with the adverse event?
  - how did the adverse event happen?
  - what activities were being carried out at the time?
  - was there anything unusual or different about the working conditions?
  - were there adequate safe working procedures and were they followed?
  - what injuries or ill health effects, if any, were caused?
  - if there was an injury, how did it occur and what caused it?
  - was the risk known? If so, why was not it controlled? If not, why not?
  - did the organisation and arrangement of the work influence the adverse event?
  - were maintenance and cleaning sufficient?
  - were the people involved competent and suitable?
  - did the workplace layout influence the adverse event?
  - did the nature or shape of the materials influence the adverse event?
  - did difficulties using the plant and equipment influence the adverse event?
  - was the safety equipment sufficient? and
  - did other conditions influence the adverse event?
- Analyze the information:
  - What were the immediate, underlying and root causes?
- Identify suitable risk control measures:
  - What risk control measures are needed/recommended?
  - do similar risks exist elsewhere? What and where? and
  - have similar adverse events happened before?
- The action plan and its implementation:
  - Which risk control measures should be implemented in the short- and long-term?
  - which risk assessments and safe working procedures need to be reviewed and updated?
  - have the details of the adverse event and the investigation findings been recorded and analysed? Are there any trends or common causes which suggest the need for further investigation? and what did the adverse event cost?

4.2 Team and Planning

After an industrial incident, the set of typical recurring circumstances (including traumatized employees, residual fires, destruction of large part of the plant, and no longer working utilities) requires a robust incident investigation management system, since there is only little opportunity to collect time-sensitive evidence (e.g. computer control historical data overwritten, outside scene exposed to atmospheric agents, oxidizing of chemical residues). Actually, this is one of the reasons to start the investigation as soon as possible. The other main reasons are the necessity to contrast the fading or changing of the witness memories, the will to avoid other similar incidents, and the fact that restart may depend on completing actions to prevent recurrence. The investigation timing can also be imposed by legal or corporate requirements (e.g. U.S. OSHA PSM requires the start within 48 hours). Obviously, starting as soon as possible implies some challenges: firstly, the team must be selected and assembled; it also may need to be trained and equipped; team members may need to travel to the site; authorities or others may block access; site may be unsafe to approach/enter.

In [6] is underlined the necessity to clarify the investigation needs before the investigation starts. Planning, training and preparedness to investigate require strong organization capabilities in choosing the right people, at the right times, in the right places, equipped with the right tools, following the right procedures [8]. In other words, a high-quality incident investigation program begins with the management's support. A proper management system for investigating process safety incidents encourages employees in reporting incidents and near misses; it also ensures the identification of root causes and effective preventive measures [9]. These objectives are written down in a policy document regarding incident reporting and investigation, which is periodically reviewed to ensure the achievement of the desired results. Obviously, the incident investigation management system includes a description of the team organization and its functions, and the team leader's responsibilities are made explicit. Flexibility in the team composition is a key factor that demonstrates a well-designed management system. Moreover, for major events (when outside investigators are preferable, as discussed next) corporate support should be provided giving the necessary tools to the teams, software included. Training of the team occurs in two stages. A first formal training is conducted prior to the event, to have a group of potential team members who know expectations, methods, definitions, and objectives of an incident investigation. Then, when the investigation is launched, training to refresh the team members is provided, focusing on the nature of the event.

The evaluation of the incident is the next step. Major incidents involving injuries, or releases of chemicals, are always investigated in depth, but many of them produce minor effects, or are actually near misses. However, minor events should be investigated as thoroughly as major incidents are, since the "quality" of the lesson learnt is often the same [1]. Indeed, the analysis of minor events can prevent the major ones.

Once the incident has been evaluated and a first estimate of the required efforts have been carried out, the core step is to put together the investigation team. It has already been pointed out that facing an accident investigation requires a complex and multidisciplinary approach. Even if it is always possible to work as a stand-alone professional, it is highly recommended to shape the job in a teamwork activity. Firstly, the convenience of this approach relies on the honest consideration that it is almost impossible for a single person to investigate, properly collect, analyze, and extract conclusions from a complex issue like an industrial accident (if so, consider the possibility that you are overestimating your capabilities). The seriousness of the incident determines the dimension of the team and its composition. Among the extensive bibliography, the necessity of a multidisciplinary team is also highlighted in [10] and [11]. Moreover, a team approach may be also required by regulatory authorities. Sometimes, for more complex investigations, more than one team are involved, changing their composition during the progress of the investigation itself. Obviously, people involved in the event, managers, and anyone having potential responsibilities on the incident are excluded from being team members.

It is unnecessary to say that an investigation cannot be assigned to neophytes, requiring experienced skilled experts.

There is not a preconfigured team composition: a specific incident requires a specific investigation team, depending on the type, the severity, and the complexity of the event being analyzed. However, for industrial accident investigation, a potential team composition typically consists of the following:

- Team leader;
- process operators;
- process engineers;
- process safety specialist;
- instrument technicians, inspection technicians, and maintenance technicians (if needed); and
- contractor representatives (if the incident involved contractors).

Senior management, even if not directly involved in the investigation, plays an important role, reviewing and commenting informally the progressive activities. Moreover, keeping senior management informed demonstrates the company commitment to the safety management system.

The accuracy of the incident investigation depends upon the capabilities of the team members. Regardless of the team size, the team approach is desirable because it enhances the quality of the investigation activities. Indeed, it ensures multiple technical perspectives when looking for findings, thanks to the different backgrounds and skills possessed by the members: this gives redundancy, helping to reach more robust conclusions and final recommendations. Actually, having different viewpoints enhances objectivity, eliminating subjective biases, like the confirmation one. Moreover, internal peer reviews provide constructive critique, having relevant knowledge of the investigated process, thus enhancing one more time the overall quality level of the investigation. Finally, the team approach also allows a subdivision of the tasks: indeed, the investigation activities may require a workload exceeding the capabilities of a single person. Parallel tasks are also easier to meet, being scheduled by a third party, as the team leader does. The incident investigation team for process incidents usually requires a cross-section of skills related to the specific process and nature of the event (fire, explosion, or toxic release). There is no specific number of members established a priori; however, the team size and composition are generally comparable to the PHA team [10].

The investigation team structure includes the incident owner. The line supervisor, responsible for the area in which the event occurred, usually covers this role. He/she is not directly involved in the investigation, but is made aware of its progress. The incident owner contributes to the creation of the right atmosphere to look for truth, without pursuing blame. He/she reports to the facility manager, who is interested in any significant findings of the investigation, since the report's recommendations are likely to be funded by his/her authorization. Like the incident owner, the facility manager is not an active team member, since it can be responsible for some of the uncovered events.

The investigation team leader is specifically trained in the investigation process. The leader is a sort of project manager, managing the schedule, the budget and the final report. He/she selects the team members and assigns responsibilities and tasks to them. He/she set up the logistics and is the intermediary with external entities, like national authorities, press, and the public. The leader is also responsible for coordination with the legal representatives. He/she ensures the chain of custody for evidence, preserving the potential one, especially during the first stages when the emergency activities might (voluntarily or not) delete them. He/she writes the final report, and calls on outside experts when extra expertise is required. The team leader develops a specific investigation plan (see Table 4.1), whose primary objectives are to identify the physical causes (process and chemistry), the root causes related to the PSM, and the recommendations to prevent further recurrence, ensuring also their implementation.

It is fundamental for the team leader to establish the terms of reference. They usually concern the objective of the investigation, the identification of the team members, its scope, the methodology to be used, its priority, the proposed timeline to deliver the final report, the estimate about required time and cost, and the required depth of the root cause analysis.

The area supervisor is often a crucial team member, because of the knowledge he/she possesses about people and technology on the plant. If his/her role may have contributed to the occurred event, it is suggested to invite a supervisor who was not on duty at the time of the event. A member of the HSE department is encouraged to join the investigation team, when the regulations affecting the plant's operations need to be taken into account. Process engineers and maintenance technicians can help in understanding what was going wrong with the process and which maintenance activities could be critical in determining what happened. Depending on the structure of the company and on the severity of the incident, other team members could be the process safety

Table 4.1 Possible checklist for developing an investigation plan.

Possible checklist for developing an investigation plan
• Establish priorities
• Rescue activities and medical treatment
• Secure the site and preserve the evidence
• Environmental issues
• Evidence gathering
• Plan for witness interviews
• Rebuild/restart
• Team leader selection
• Team member selection, training, and organization
• Initial visit and photography
• Plan for evidence recognition, collection and organization
• Establish a communication protocol
• Identify the required equipment and tools and plan for their procurement
• Plan for special or refresher training
• Schedule progress

Source: Adapted from [9].

management coordinator and the union representative. If the incident involves also equipment or services from contractors and vendors, representatives of those companies can be asked to join the investigation team. Useful information can be provided by the emergency response specialists: even if they are not part of the investigation team, their observations about the immediately post-incident scene can be valuable, especially those concerning that evidence that were moved or changed during the emergency phase. It is also suggested that one member of the team be designated as a full-time coordinate witness interviews, another one to manage photos, videos, and other types of evidence.

Before starting the investigation, an initial team meeting is usually carried out, to introduce team members and their skill sets, to establish the communication protocols, the evidence management, the chain of custody, and to assign tasks and responsibilities to the team members.

An investigation team member should possess both hard and soft skills, being objective, painstaking, using logical thinking, avoiding jumping to conclusions, not being haughty and showing empathy. Other team members can be involved in a part-time consulting role, depending on what the particular needs are, like a chemist, structural engineer, equipment specialist, process control engineer, environmental scientist, HR representative, and other specialists. However, job assignments should be flexible and modifiable, so to adapt when team members admit they require extra help, not having the specifically needed competence. Table 4.2 shows what the investigation team members should and should not do.

It is not a requirement to be an expert in the process being investigated, but a general understanding of the process technology should be possessed by each team member, together with the chemical and process vocabulary and a deep knowledge about the selected incident investigation methodology. Moreover, the investigation team is also responsible for the evidence processing. Therefore, a proper team structure may also include [12]:

- Evidence custodian;
- forensic specialist;
- logistics specialist;
- medical examiner;
- photographer;

Table 4.2 Investigation team members should and should not.

Investigation team members...	
...should	**...should not**
• Have an open and logical mind, with an independent perspective; • work well in a team; • have analysis, writing, and communication skills; and • be an expert on a particular side of the investigation.	• Have pre-formed opinions; • identify causes before the investigation starts; • be close to the incident, or emotionally involved; • have conflicting work assignments or priorities; and • impose schedule restraints that are not compatible with the investigation timing.

- procurement specialist;
- safety specialist (structural engineer);
- searchers/collectors; and
- sketch artist.

Being aware of personal limitations, it is fundamental to look for outside experts, if they are needed. Some factors influencing the selection of an expert are outlined in [13]. They also include the expert's qualifications, experience, education, training, and membership in professional associations. Outside investigators, free from the management involvement, are perceived as more independent [10], thus they are preferable. It is not a matter of objectivity, since an internal team member may be as objective as an outside investigator: indeed, credibility is a perception issue. Generally, an investigator from outside has developed a high level of expertise that internal personnel do not likely possess. In addition, outsiders will probably better manage confidential information, working in contact with the attorneys. Moreover, internal investigators cannot work full time on the investigation, like an outsider does, because they have to carry out their "principal" jobs. Finally, being external from the company can help to reach the real root causes, without any fair to discuss openly with the management about the findings and the subsequent action items. Clearly, those events that are perceived to offer only limited lessons can be investigated by an internal investigation team, also to reduce costs.

A prior preparation is fundamental, since the team members generally do not know the details of the situation until they arrive at the incident site. Preparing the right equipment and tools is a preliminary step for any investigation. Because of the unpredictable nature of an incident, it is necessary that investigation supplies and equipment are prepared and maintained to be promptly used when required. The nature of the equipment may vary depending on the incident scenario, however some common tools should be always taken. They include evidence collection kits (bags, tags, labels, and so on), first-aid kit, protective shoes, glasses, gloves, helmets, disposable suits, particle masks, flashlights, batteries, telephone, sketchbooks, pencils, camera, measuring equipment, food, and water. GPS equipment, chemical test kits, vapour detectors, and trace explosives detectors are specialized equipment that can be required, depending on the particular event.

The role of the investigation team in the incident investigation is also discussed in [10]. Typically, the team members are asked to answer questions concerning the use of a safe work procedure, the changed conditions that made the normal process unsafe, the availability of proper tools and materials, the equipment majorly involved in the incident, the availability of safety devices and their working conditions. Operators' and supervisors' log sheets and log books need to be looked at by the investigation team to obtain information about equipment problems, materials movements, and shifts, together with the maintenance logs to known completed, in progress, and planned activities. The investigation team inspects the items which may be involved in the incident, like pumps, vessels, piping, control valves, transmitters, and so on. It has also to prepare the list of questions to be asked to potential witnesses.

Finally, regular team meetings need to be defined, to resolve questions, update all the members on new information, share intermediate results, report on performed activities, and so on. The last phase of the team's activities is the presentation of the results (both findings and recommendations), usually in a written formal report.

4.3 Preliminary and Onsite Investigation (Collecting the Evidence)

The initial visit to the scene is one important activity of the investigation. Its goal is to obtain an overview of the incident scene, before attention is focused on details and the collection of the evidence disturbs the original scene. At this stage, it is important to:

- Check for potential safety hazards for the team members;
- look to the scene at a macro level, not focusing only on the details;
- take notes about what is damaged and what is not;
- take notes about what should be present and is not; and
- take notes about what should be absent and is not.

To ensure safety, the first responders must evaluate and mitigate the identified residual hazards, eventually establishing safety zones. Lifesaving activities are undoubtedly the priority; however, where rescue activities are taking place, care should be taken to minimize disturbance. Generally, the investigator will promote a briefing to ensure safety and security of the scene. Procedures are also established to protect the scene integrity. Only at this step, does the investigator carry out an initial walkthrough, where he/she has the possibility to identify evidence and eventual hazards [12]. The interested reader may find additional information about the initial site visit in [9].

There are different methods to initially document the incident scene [4]. Some options, to be used alternatively or in parallel, include: taking notes on personal observations, getting initial statements from witnesses, taking photos or video clips, and sketching the accident scene.

During the immediate response activities, the investigation team members must not perform any actions that could make the situation worse, strictly following the requirements regarding the safety issues, that is to say controlling hazards and isolating energy sources. Taking safety precautions during the gathering phase, like using the Personal Protection Equipment, is taken for granted, see Figure 4.3. Only when the emergency

Figure 4.3 During the preliminary and onsite investigation, remember to wear the PPE.

response activities have been completed, is it possible, for the authorized people, to access the incident site. It is suggested that more evidence be preserved than what is apparently necessary, because it will be difficult to collect this when the investigation is at an advanced stage; after all, unneeded items can be released later. The investigation team has a dedicated room where to store the personal protection equipment and the personal belongings, when team members are on the scene. In the ideal team room, there is also enough empty space on the walls, to develop charts, timelines, and logic trees.

The preliminary and onsite investigation has a fundamental role in the investigation process. With the term "evidence" we mean all the data, different in nature, that are collected at the incident site or are generically related to it, permitting the reconstruction of the dynamics of the event. Some evidence may instantaneously help when collected; others may require deeper analyses before providing useful information. The evidence becomes a proof only when it is put in the right position within the general context: when supported by scientific literature, specific laws, and final tests, it can be considered as the basis for the deductions. Generally, it is not necessary that all the collected evidence support a specific deduction. However, it is undoubtedly essential that none of them discredits the supported deduction.

A thorough investigation starts with the evidence preservation, to ensure that data are not changed, contaminated or lost. To prevent the risk of contamination, that affects every type of evidence and their examination, separation from an outside source should be provided. This approach will help also to demonstrate and maintain the integrity of the investigation, ensuring any conclusions gained from it [14]. Preservation of evidence is the most important step in the investigation, as also noted by [10]. In this perspective, clean-up activities that are not part of the first responder activities, should be allowed only once the physical evidence is collected. The necessity to create a Security Chain of Custody is especially true for the physical evidence and the documents, ensuring a scrupulous attention in their handling, avoiding any form of alterations, and ensuring the certain traceability of the proof. The chain refers to all the aspects related to the gathering stage, like:

- Collection or requisition of the evidence;
- custody;
- control;
- transfer;
- analysis; and
- positioning, arrangement.

For any of the above-listed steps, proper documentation must be provided, certifying the gathering conditions, the identities of the people who collected and handled the evidence, the safety and security precautions adopted during the handling, the custody, and the transfer procedures. The number of transfers and people who deal with the evidence should be limited. If the chain of custody breaks, then the evidence might be considered untrustworthy.

Other important tools are the witness recollection statements, to record facts recalled by the observers, the witnesses' interviews, and the Pareto analysis, to get more from the collected evidence [3], as next discussed.

Evidence collection is merely the research of various types of information that allows knowing, by its comprehension, the context in which the incident happened: the final aim is to support the theory concerning the incident dynamics and its causes. However, collecting and analyzing evidence is not so easy as it could appear. From this viewpoint, [15] provides a list of the major evidence-related issues. Indeed, some evidence must be put in chronological order by means of logic, inferring statements; instead, other evidence, having date and time stamps, can be used to test the reconstructed sequence in the timeline, being anchor points. Other evidence may be red herrings: even if they are unimportant, it is necessary to properly investigate them, in order to achieve a robust conclusion (remember the falsification principle). Some crucial evidence may remain covered, never being discovered, or destroyed by the incident itself. Other evidence may be misinterpreted: typically, this happens when the investigator does not have sufficient expertise. In addition, some evidence could be deliberately placed to drive the investigator toward an incorrect solution. Finally, in the case of a fire, the damages are mainly related to the effect of the heat while mechanical damages usually arise after an explosion. This suggests that the different incidental scenarios, having their own peculiarities, may generate a major number of evidence in a defined category rather than others.

In order to have an extensive, complete and rigorous collection, an important factor must be taken into account: the time. Immediately after the occurrence of an industrial accident, time is an urgent need. The priority is to rescue the injured people, if any, providing them with the first aid, together with the restoration of the safety conditions of the site. It is self-evident how these operations may alter the scene: as already stated, this effect has to be always considered during the next phase of analysis. Moreover, some evidence is particularly sensitive to the flow of time which may cause the loss, the alteration or the rupture of the proof. Some examples of "losses" are a sign that disappears, a witness that forgets something, an interruption in the energy supply that causes the loss of data memorized on temporary data storage. The alterations or distortions happen when an object is moved, a witness does not remember correctly, digital data is over-written. An example of evidence "ruptured" by time is an influenced witness, whose testimony is no longer trustworthy. The higher the fragility of proof with time, the higher the priority of its collection. It is not generally possible to provide a list of priorities, since it really depends on the specific investigation. However, among all the data that can be collected, those with a high sensibility are the paper ones, the electronic files, the materials experiencing a rapid decomposition, the objects with fractures that can oxidise (so the ones that are relevant from a metallurgic point of view). The position of things and people is undoubtedly the most fragile data, having also an enormous potential: indeed, almost every evidence becomes significant when the information about its position is known. Given the priority of collection, it must be noted that the gathering sequence is generally unimportant, being commutative.

The evidence is the cornerstone of the whole investigation. Their collection requires a proper approach to maintain a vision as wider as possible, without preconceptions. This approach is useful to not exclude *a priori* some hypotheses and respect, at the same time, the steps and the schedule defined by laws, or internal procedures.

In conclusion, talking about the collection of the evidence, it is possible to define three keywords:

- Rapidity, because data are usually sensitive to the flow of time;

- accuracy, in creating the collection, respecting standards and laws to certify and guarantee the evidence; and
- traceability, of whatever is collected, through a security chain of custody.

4.3.1 Sampling

The collection of the physical evidence takes place in two steps. The first one concerns the identification of what is found on the incident site. This step is often mandatory when conducted within the boundaries of a criminal trial. The objects that need to be next taken by seizure are selected by the attorney, also taking into account the recommendations of his/her consultant. The second step, much deeper, concerns the selection of samples, starting from the evidence that is considered substantial, to be further analyzed in specific and accredited laboratories.

Samples are taken to be generally tested, in order to have a quantitative measure of their chemical and physical features, when these parameters could be of interest for the investigation. The topic is ample and providing a concise knowledge is not within the goals of this book.

It is highly important that the operation is reported and specific technical standards are respected. For example, some of them are:

- ISO 11648:2003 (Part 1 and 2). Statistical aspect of sampling from bulk materials;
- ISO 10715:2001. Natural gas - sampling guidelines;
- ISO 3171:2001. Petroleum Liquids - Automatic Pipeline Sampling; and
- ISO 4257:2001. Liquefied petroleum gases -- Method of sampling.

Regardless the nature of the sample, the sampling process consists of five steps:

- Selection of the sample;
- collection of the sample;
- packaging of the sample;
- affixing a seal at the packaging; and
- transport to the analysis laboratory.

4.3.1.1 Selection of the Sample

A unique criterion does not exist for how to select a sample, since it depends on the specific case and on what the investigator is looking for. Sometimes there is the necessity to define the main features related to a big set of data. In this case, the selection of a sample is generally performed by selecting that part that may be representative of the whole evidence, with its properties and peculiarities. Only if this condition is respected, then the evidence can be considered as proof during the eventual trial. In other cases, there is the necessity to find a needle in a haystack. This could happen when sampling fire residuals, where the samples are taken from several points of the site, not having a particular criterion for their selection but only preferring those materials exposed to the smoke.

For a fire investigation, some criteria to be taken into account are the following:

- The volatility of the flammable liquids. Indeed, they generally tend to easily evaporate, leaving only small concentrations in the combustion residuals. It is necessary to consider that they are often a mixture of liquids, with different volatilities; therefore, the relative composition may also change;

- the features of the substrate, particularly its surface. The probability of survival for flammable liquids or their traces depends on it: an affine substrate will trap the accelerant more easily; and
- the localization of the sample. It is referred not only to its physical position within the boundaries of the incident site, but also, and above all, respect to the fire. Pieces that were directly exposed to the fire, where the highest temperatures have been reached, are likely to trap few traces of fire accelerants.

In fire investigation, it is essential to have one sample to be used as reference: it is named the "white" sample. It is taken from those positions where the presence of accelerants may be excluded with certainty. This allows finding eventual substances already present in the materials, thus having a list of the materials that cannot be related to the incident.

4.3.1.2 Collection of the Sample

The extraction process of the sample from the evidence depends on the nature of the evidence itself. It is fundamental to take photos (Figure 4.4), maps, and notes that allow keeping in memory the position and the features of the sample before it is collected. It is a common procedure to perform a triple collection for each sample or, at least, to collect a sufficient amount of material allowing its subdivision into three equal parts: one for the analysis by the party, one for the counterparty, and another one for the eventuality of future controls by the judicial authority.

In the fire investigation, it is generally distinguished between direct and indirect sampling. In the direct sampling, a portion of the substrate is collected, like a piece of cloth or wood, and laboratory tests are performed to search fire accelerators. The dimensions are critical: it is not generally true that the more you get, the better it is. Indeed, it may result very complicated to move and store a sample of large dimensions. During this operation, much attention should be paid to the risk of contamination. This risk can be minimized using:

- Disposable gloves;
- new tools for each sampling; and
- tools not altering the chemical composition and the physical features of the sample.

Figure 4.4 Collection of some portions of metal sheet from the processing tape and their subsequent enumeration, ThyssenKrupp investigation.

The indirect sampling uses a "collection" substrate, which adsorbs the substance of interest. The nature of the substrate depends on the nature of the substance which is going to be collected (gas, vapour, liquid or dust) and its typology. For instance, active carbon is used, as a substrate, to sample organic solvents vapours. If it is necessary to capture polar substances, like the amines, vials with silica gel are adopted. If vapours and corpuscular units are contemporary present (like in the Polycyclic Aromatic Hydrocarbons – PAH), a particular collection substrate named "double train" is generally used. It is composed of a membrane and a vial containing an adsorbent substance. Generally, the membranes are used to capture dust while liquids are sampled by absorption using tampons or sponges.

4.3.1.3 Packaging of the Sample

Once the sample has been collected, it is inserted into a container to be protected during the storage and the transportation to the laboratory analysis. In particular, the aim is to preserve it from three types of alterations:

- Loss;
- contamination; and
- degradation.

Several factors may alter a sample, including the material of the container, the presence of humidity, the solar exposition, or the presence of a heat source or of oxygen.

There are different typologies of packaging. Some containers used for sampling, their main features, pros, and cons are in Table 4.3. Even if the final choice is strictly related to what is put inside, the characteristics of the ideal container are:

- Easily available;
- not expensive;

Table 4.3 Some containers for sampling, their main features, pros, and cons.

Typology	Features	Pros	Cons
Metallic can	Friction closure cap Can be covered internally Different dimensions (1 or 4 litres are the most common)	Not expensive Robust Easy to stock, open and close Real physical barrier against losses and contaminations	Not suitable for samples with particular shapes Corrosion risk, if the sample is humid
Glass can	Metallic cap Different dimensions	Transparent Non-contaminable (pay attention to the cap)	Brittle
Plastic bag	Different types of closures systems (by pressure, zip, heat) Different materials (like nylon and PVC)	Transparent Flexible Occupy little space when empty Does not break them by falling	Can be teared up and punctured It suffers from thermal dilatations and presence of solvents or oxidizing substances It is permeable to many substances

- easy to transport and store;
- easy to be sealed;
- not contaminated; and
- resistant to damages, ruptures, and cuts.

At this stage, in order to reduce the possibility of contamination, it may be convenient to use similar containers (bought in stock) and analyse one of them, still empty, to know the eventual already-existent contamination, even before their usage. In addition, in order to control pollution due to the transportation stage, an empty container is usually analysed. Informally, it is defined "white" and is adopted as a reference to interpret the results of the analysis.

In Table 4.3, there is only a mention of the plastic containers since they belong to a category that is much extended for typologies and purposes.

Generally, a good practice is not filling the container completely: as a rule of thumb, the contents should not exceed 70% of the container capacity.

The packaging operation ends with the tagging of the container, by a numeration system to identify each sample uniquely (see Figure 4.5).

4.3.1.4 Sealing the Packaging

Affixing a seal to the packaging is the last step before the transportation of the sample to the laboratory for tests. It is a key stage since it certifies that the container has never been opened, thus contaminated, or altered.

Also, these operations are recorded by photos and a special form is also used to record:

- Identification number of the sample;
- date of the collection;
- person in charge for the collection;
- type of container and its closure;
- contents and typology of the sample; and
- indications about how to prevent deteriorations.

4.4 Sources and Type of Evidence to be Considered

The type of evidence that should be collected depends on the particular incident. However, a general discussion about the potential sources of evidence can be done.

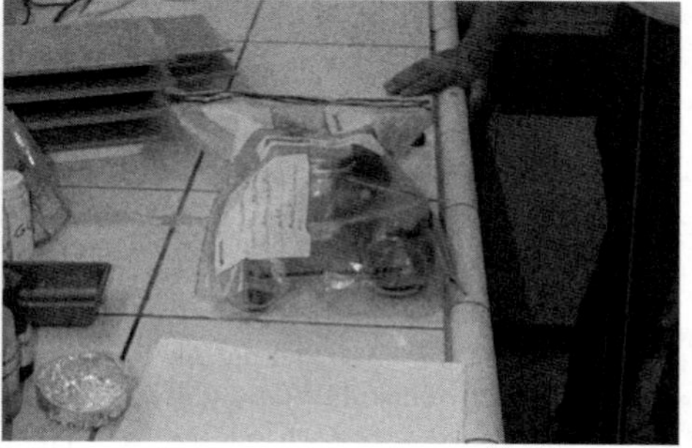

Figure 4.5 Samples in glass cans and in plastic bags with zipping closure.

Firstly, whichever is the evidence being considered, its position is usually documented, as well as the use of photographs taken from different angles to collect any potentially useful information more accurately [10]. Clearly, the referred position describes the objects "as-found" (including the position of valves and switches), as is also reminded by [16]: eventually, this evidence needs to be properly discussed and analysed before stating a conclusion. Useful information can be derived from the PHA studies, providing potential incident scenarios. This information is obviously coupled with the required PFDs, P&IDs, records, and logbooks, whose analysis provides a complete understanding of the operating conditions of the process. Also, maintenance inspections, engineering drawings and MOC documentations are a valuable source of information. Evidence can be found also beyond the area in which the incident occurred.

According to [9], there are five types of evidence (Table 4.4):

- People, providing testimonies or written statements;
- physical, like mechanical parts, equipment, materials, finished products, but also damages and details related to the site conditions (like structural deformations, incisions, burns, and so on);
- electronic, such as data recorded by control systems, video recordings, and emails, which are often a powerful source of information;
- position, of both people and physical objects; and
- paper, like alarm logs, written procedures, inspections registers, blueprints, and training records.

According to [1], evidence can be also classified differently in: interviews, documentation, field information, instrument records, and testing/lab analysis. The different classification, from a different perspective, does not change the general considerations that

Table 4.4 Checklists to evidence examination.

People-related data	Identification of personnel directly or indirectly involved, including:	• Operators • Maintenance personnel • Process technicians • Laboratories personnel • Emergency teams • People involved in the start-up of a plant • Security personnel • Designers
Electronic data	Record of control systems and data gathering	
	Records from the video surveillance systems	
Chemical and physical data	Observations about positions and conditions of:	• General equipment and items (tanks, valves, gaskets, flanges, connections) • Mobile working equipment and PPE • Samples of materials (raw materials, final products, residues, wastes) • Missing parts

Table 4.4 (Continued)

Position data	Position of involved equipment, like valves, controllers, switches, and safety devices	
	Extension and position of damages:	• Flame and heat traces; • Melting, weakening, mechanical deterioration of materials • Smoke traces • Traces of fragments impact, squirts, and so on • Position and sequence of debris layering • Direction of projection of fragments • Distances and corresponding masses • Fragments status • Position of people and witnesses
Paper data	Documents about early and final design, site plans	
	Documents about instrumentations, electrical schemes, set points of alarms	
	Recorded past incidents and near misses, training manuals	
	Operative procedures, checklists, manuals, shift logs, work permits, maintenance and control registers, quality control log books, MSDS	

Source: Data elaborated from [9] adapted from CCPS 2003.

are going to be discussed. What is important is that information is cross-checked by more than one source, when possible.

Data should be collected following a scale of priority, depending on their fragility (Table 4.5). A prescribed priority does not exist, since it depends on the specific incident. Generally, time-sensitive data should be collected first of all. This includes data stored in software files on computers with a limited battery backup, papers on the control room, to ensure they are not lost or destroyed, materials undergoing chemical decomposition or, more generally, a chemical reaction (like the oxidation of fracture surfaces). Missing information is generally due to equipment being returned to a safe condition, people losing memory quickly, people remembering incorrect facts and fixing on them.

The investigator knows that the data collected could not reflect the real condition after the incident, because emergency response activities might have altered them. For example, for safety issues, the position of some valves could be changed during emergency activities. The investigator will document the as-found position, but a comparison with data collected by the emergency team is mandatory.

The expert investigator, if required by the peculiar incident, can be asked to carry out a tool-mark examination. It is a discipline especially used for crime scenes, where malicious acts are involved, to determine the type of tool that may have been involved. Being typically out of the scope of this book, the interested reader can find additional information in [17]. Similarly, the analysis of explosives, both pre-blast and post-blast, is generally carried out when a malicious act is suspected. However, since an explosion can also be the consequence of an industrial incident, familiarity with the principles

Table 4.5 Forms of data fragility.

Data Source	Form of fragility		
	Loss	**Distortion**	**Breakage**
People/Position	Forgotten Overlooked Unrecorded	Remembered wrong Rationalized Misrepresented Misunderstood	Transferred Influenced Personal conflicts
Physical/Position	Taken Misplaced Cleaned up Destroyed	Moved Altered Disfigured Supplemented	Dispersed Taken apart
Paper	Overlooked Misplaced Taken	Altered Disfigured Misinterpreted	Incomplete Scattered
Electronic	Overwritten RAM lost in power outage Destroyed	Data averaged and individual samples overwritten	Incomplete

Source: Data taken from [9] adapted from CCPS 2003.

of the post-blast analysis is recommended. The interested reader can find additional information in [18].

The five sources of evidence are now treated in detail, adding a further one: photography. Photographic documentation should be included in the permanent scene record [12]. In particular, the investigator should take overall views of the scene, minimizing the presence of personnel in photos/videos, before taking photos of relevant details. They will be a valuable source for the future proceeding of the investigation. During the investigation, the different sources of evidence are intrinsically connected each other, like the colours of a painting: all contribute in giving shape to the drawing, no one excluded.

4.4.1 People

People being interviewed are both those directly and indirectly involved in the event. The former were present during the event and had seen, or listened to, something, providing useful direct information. The latter witnessed the event or its consequences, such as workers on a previous shift or safety professionals. Indirect witnesses were not directly involved in the incident, but they are still capable of providing information and reconstructing the conditions that are probably at the origin of the event. Anyone having information relating to the incident should be considered a potential witness. Valuable information can also be provided by recently retired or transferred employees. Also, people who visit the process plant on a routine basis can provide useful information about eventual unusual conditions. People participating in the emergency activities, including firefighters, have to be interviewed, since they may have disturbed, altered, or destroyed data for emergency reasons, and can provide information about the original status of equipment and about the post-incident scenarios (secondary fires or explosions). The Fire Brigade members, who are among the first responders, possess peculiar skills allowing a basic reconstruction of the event, by pointing out the most damaged areas on which their intervention was mainly focused. The investigator should pay particular attention to those areas. Interviewing the Fire Brigade is a unique source of information because data are provided by skilled people, who know and understands the physics related to an incident, so they can autonomously individualize the critical points and communicate them in the best way. In addition, their limited psychological involvement helps for a better and more fruitful interview.

Information from witnesses may also concern operating practices not formalized in written procedures, process response to various conditions, small changes in process variables, the reliability of the control instrumentations, history of past problems and related actions.

In any case, in order to use the testimonies during the trial, it is fundamental that they are collected following the procedures established by the local laws. They may vary from country to country, however they generally share the same values of traceability, accuracy, and rapidity. Compliance with specific procedures does not have to turn the interview into an interrogation, since the technical consultant is looking for causes, not culprits. This aspect should be always taken into account since the interviewees could be emotionally vulnerable (they were at the scene, may have seen the injuring of a colleague or their lives at risk).

The ideal interviewer, in addition to the general team member attributes, takes care of the rapport and transmit trust, possesses technical skills, is able to recognize critical factors, is objective, is able in taking effective notes, possesses strong communication and listening skills. In addition, some human characteristics need to be taken into account when dealing with potential witnesses. Human observations are far from being objective: they tend to ignore data that are assumed not to be influential, and to memorize (sometimes exaggerating) data that are assumed to be important. There is no voluntary behaviour in telling a false story; instead, the witnesses try to tell the story as best as they remember it. Witnesses are not video recorders or computers: they do not have a complete view of the event, and discrepancies are likely to be present when collecting more testimonies, because of different perspectives. In addition, human memory fades rapidly: a rule of thumb is that people forget 50-80% of the details within 24 hours [1]. Another characteristic is that memory does not always recall events in the right chronological order. This is the reason why witnesses are often asked to retell the story, helping them to remember additional details. Moreover, humans have a capability and tendency to see and hear what we expect to see and hear [10]. The direct witnesses generally report something coming from the visual memory, which is not so reliable as it may appear. Sherlock Holmes justified his deduction, reproaching Dr. Watson by saying to him: "You look, but don't see". The majority of people are like Dr. Watson. Our brain, in order to recognize information as quickly as possible, performs a selection, automatically ignoring what is not necessary because it expects this ignored information remains the same in time and space. It is a sort of simplification process and people are often unaware of it. Sometimes, this procedure overlooks details that are not entirely useless. In addition, the brain also adopts some automatic corrections that allow understanding the whole meaning of a world only with the first and the last letter in the right position. *Acirocdng to a Presofsor of the Cbidmarge Uinivsrety, tihs is baeusce the hmaun mnid deos not raed ervey sngile lteter but the wrod as a wlhoe.* Memory in general, and the visual one, in particular, is something personal and incomplete, so every single fact is unconsciously filtered during the acquisition process and the subsequent comprehension. This may bring in different reconstructions that are not perfectly aligned.

The emotional aspect, the peculiarity of the human mind and other dynamics that may be present, are all elements that the technical consultant has to take into account when he/she evaluates the testimonies.

When proceeding, the interviewer should:

- Be helpful and empathetic, but also severe, if necessary: at this moment the technical consultant is in the position to demand cooperation;
- be as much neutral as possible, not shining with amazement, aversion, annoyance or satisfaction for a statement, to not influence the witness;
- never express a judgment, even when there is awareness that the testimony is not completely true;
- not persevere on a particular aspect, to avoid the testimony focusing only on that aspect;
- encourage the interviewee to repeat the same statement more times and on different occasions, allowing them to add details continuously and to give them a chronological order; and
- never forget the influence on the interviewee.

Fear of punishment (also for colleagues) may influence the witness, who could decide to not tell, or purposely modify, the story. To overcome this possibility, it is necessary to establish a proper investigation environment, focusing, one more time, on finding the causes, not who is guilty. From this perspective, an investigator cannot promise the avoidance of potential punishments, but he/she can clarify the main purpose of the evidence gathering: to reconstruct the dynamics of the incident, not to lay blame.

The quality of the interviews is essential to obtain those narrative links connecting information from other sources of evidence [1]. It is recommended to have no less than two interviewers: one asks questions, and the other writes down the answers and provides a supplementary perspective on the interview. Open questions should be preferred to closed questions, both to avoid preconfigured answers (forcing the witnesses toward two preconfigured solutions like "yes" or "no"), and to avoid enforcing the erroneous feeling that you are looking for culprits, since closed questions recall interrogations rather than interviews (and a "no" answer can be perceived as an assignation of blame). For the same reasons, questions should not be too much specific and should not contain legal terms.

A list of potential witnesses includes the following:

- On-shift operators;
- off-shift operators;
- maintenance personnel (company and contract) assigned to the area;
- process engineers;
- operations management;
- maintenance management;
- chemistry and other laboratory personnel;
- warehouse personnel;
- procurement personnel;
- first responders/emergency response personnel;

- quality control personnel;
- research scientists;
- personnel involved in initial startup of the system;
- manufacturer's representatives;
- personnel previously involved in operation/maintenance of the system;
- personnel involved in previous incidents associated with the process;
- janitorial, delivery, and other service personnel;
- relevant off-site personnel and visitors;
- original design/installation contractors or engineering group; and
- security force (roaming guards or sentries).

Sometimes, the first step in gathering verbal information from people is asking them to write a "first report" [10], within the first 24 hours after the event. They are not guided when writing this report, in order to not be influenced.

In the organization of the interviews, some advantages may come from:

- Providing the same questions to the witness, in order to obtain comparable versions; and
- scheduling the interviews so that the witnesses may not influence each other.

When planning the interviews, an additional time between two interviews should be considered, in order to properly manage the collected information in a timeline, compare it with other information, and organize the notes efficiently. To sum up, in order to optimize results:

- Interviews should be prompt;
- the information provided should not be selectively accepted, that is to say the investigator should not have identified a preferred scenario;
- investigation team preliminary findings should not be disclosed with interviewees;
- interviewers remain neutral, avoiding suggesting a desired response;
- interviewers do not make a promise during the interview;
- statements are always written down and signed;
- room configuration should avoid a face-to-face opposition between interviewer and interviewee, because the witness can feel uncomfortable. A ninety-degree orientation is preferred; and
- attention should also be paid to non-verbal communications, like eye contact and body posture.

Also [4] provides a similar guideline about how to conduct effective interviews. Once all the remarkable testimonies have been collected, actual evidence is found and eventual contradictions are evaluated.

4.4.1.1 Conducting the Interview

The interview is a simple conversation during which testimonies are collected. It requires a well-defined structure, in order to comply with all the peculiarities of this typology of evidence. According to [10], the interviews can be divided into four parts:

- Opening, where the initial rapport and trust are established. Objectives and introductory information are provided;
- witness statement, in the form of uninterrupted narrative;

- interactive dialogue, where specific questions are asked to verify and clarify previous statements; and
- closing, where the interviewer makes a summary and asks for other facts of interest and potential witnesses.

The first stage is the meeting, where the investigator tries to establish a relationship of mutual trust and collaboration, crucial for finding as much information as possible. It is suggested that to start with, the final goal of the investigation should be put forward, setting it out and making some simple questions or neutral statements, not involving the witness personally: this will break the ice. In the next stage, the witness is left free to expose the facts, without any interruption and without steering the conversation towards a particular aspect. It is fundamental to respect the time and the ways adopted by the witness to recall the facts to his/her mind, the modality of their presentation, the silences. Therefore, the second stage is mainly dedicated to listening. The third step is the actual interview, with questions and answers. It is suggested that a set of "standard" questions be prepared, in order to make comparisons. Usually, witnesses tend to generalize, using expressions like "always", "everything", "all", "they" that need to be corrected, asking him/her to specify better. Moreover, different people have different definitions of the same word, so it becomes crucial to ask what they mean clearly. An example, in the Norman Atlantic case, is the expression used by the members of the crew of "Richter truck" which was used as a synonym of "Refrigerated truck".

Generally, the areas of interest for questioning deal with:

- Time of occurrence of the events;
- position of people, objects, indicators;
- direction of development of the phenomenon, towards which a witness was looking;
- past and recent changes in plants, operative procedures, solved and reoccurred faults; and
- opinions, to give a complete picture.

Conclusions are the final stage of the interview: it is good practice to ask for the confirmation of the notes taken during the conversation. In this way, the interviewee is given a chance to correct potential mistakes or to be more accurate on some details previously overlooked. At the end of the interview, the witness is invited to contact the technical consultant if he/she feels the need to add something respect to what declared.

Before proceeding with another witness, it may be useful to evaluate the level of interest and reliability of the received information, highlighting the most significant and eventual discordances which may have emerged with respect to other evidence, already gathered. It is also recommended that information is collected from people before expressing any judgment: indeed, only the analysis of all the testimonies may highlight potential inconsistencies. In this case, it could be necessary to interview a witness again, looking for new elements related to a specific aspect, or to pay more attention in the evidence collection on a specific area of the site.

In [1], the actions to conduct a successful interview are discussed. The interviewer should:

- Clarify the purpose of the interview from the beginning, and its structure;
- respect the interviewees, treating them equally, and start by asking general questions;

- provide personal information about the investigator, to establish a friendly environment;
- record names, dates, times, and every provided description;
- conduct the interview at the site of the event, if possible; this allows the interviewee to point directly to any equipment involved and to walk through the events;
- do not hurry and take the required time;
- interview before witnesses talk to each other, influencing mutually;
- ask for definitions of unknown terms;
- avoid acronyms;
- use simple language;
- ask for feedback about the correctness of the information provided;
- ask for suggested things to consider for the investigation;
- ask if there is anyone else to talk with, about the incident;
- do not drive the interview towards a preconceived hypothesis;
- keep notes in a timeline format;
- do not interview groups of people;
- use two interviewers, if possible;
- avoid questions related to the feelings;
- ignore attempts to blame others;
- recognize that information is filtered by the person's experience;
- recognize that events could not be presented in chronological order;
- encourage making sketches;
- distinguish observations from opinions;
- respect the witness, especially if involved in the incident, and his/her feelings; and
- ask interviewees to sign their statements, especially if litigation is probable.

A similar guideline can be provided to the employees for legal interviews. In particular, they should keep a log of which documents have been examined by the investigator, and should be instructed to not give any documents without the superior's approval. Moreover, employees should be refreshed about the training and certification they have received, and about their rights to representation. The interviewee should answer only to the specific question, avoiding adding information that is not required which could make the situation worse.

4.4.2 Paper Documentation

The collection of the paper documentation during the investigation is especially useful to create the knowledge *pre*-incident. It is important to carry out it rapidly, in order to prevent the risk of falsification; indeed, it may be appropriate to ask for the acquisition of paper documentation through seizure operated and controlled by the competent authority. Priority is given to those documents that are stored on the sites involved in the incident, in order to avoid their further deterioration. Cleaning or decontamination of them may be required. It is then important to define a chain of custody as soon as possible, to identify, record, and safeguard all the original documents. Copies should be taken, if necessary.

It is difficult to provide a complete list of potential paper evidence, since it is quite extensive. Just to give the reader an idea about how wide is the category, it is sufficient to mention process data records, operating procedures, checklists, manuals, shift logs, work permits, maintenance and inspection records, batch sheets, process and instrumentation drawings, detailed instrument and electrical drawings, design calculations, alarms and set points for trips, safe operating limits, PHA, Material Safety Data Sheets (MSDSs), material balances, corrosion data, interlock drawings, MOC records, training manuals, meteorological records, dispersion calculations, phone logs. All these provide evidence about management policy and programs, engineering, hazard analysis results, purchasing, operations and maintenance, and personnel's development. Batch sheets, operator logs, and similar paper data may be extremely useful when the investigator wants to analyze reactive chemistry thoroughly, since they can provide information about materials used, mix composition, sequence, rate, and volume of additions.

In particular, paper documentation concerning industrial accidents may be classified into the following three categories (the sub-lists are only for exemplifying):

- Safety documentation:
  - risk assessments;
  - specific risk assessment (like fire, explosive atmospheres, and chemical risk);
  - report of inspection, carried out for different reasons from authorities, third agencies, fire brigade agency, and so on; and
  - compulsory documentation about safety, imposed by the local laws.
- Company documentation:
  - organizational chart;
  - documents about delegates and appointment (including the first aid responders, fire team, cooler team, and so on);
  - documents defining roles, tasks, authorities (providing information about hierarchy and functionality);
  - contracts with suppliers (including the cleaning company) and related documents;
  - registers about the training of workers;
  - documents related to the company management, like the safety management system;
  - records of past incidents, promoted recommendations, and follow-up documents; and
  - emergency plans;
- Production and design documents:
  - architectural drawings;
  - electrical schemes;
  - instrumental schemes;
  - driving patterns;
  - design documents;
  - description of normal and anomalous chemical reactions;
  - alarms and interlocks thresholds;
  - material safety data sheets (MSDs); and
  - documents about the logistics, the transportation of materials, products and wastes inside and outside the boundaries of the site (such as registers and delivery notes).

Typically, the major difficulty in collecting paper data is to find the necessary documentation and, once found, to collect the required information within the paper document. The extraction of data from paper evidence can be time-consuming, so it is suggested that proper priority be assigned to their collection, even if are generally not affected by alteration over time (with the exception of those documents exposed to fires, explosions, chemical release, or weather, which may require a decontamination and can deteriorate). Documents reporting events correlated with time could use a different time scale and have small or large time shifts. The comparison of one or more events reported in two or more documents can help to adjust the time scale and properly assign the events in the timeline.

Paper data collection may be so extensive to require a single person to manage exclusively this side of the investigation.

4.4.3 Digital Documentation and Electronic Data

In the context of a technical investigation following an industrial incident, with digital data we mean information transferred using IT. Instead, when the investigation is about IT-related crimes, so the technology is itself the subject of the offence, then it is better to talk of Computer Forensics, to specify an investigation dealing with the IT world and requiring specific skills which are beyond the purposes of this book.

Examples of electronic data include the databases written by the control and alarm systems, the volatile control instrumentation records (DCS data), the PLC (Programmable Logic Controller) set points, the security camera tapes, and the emails, which are generally a valuable source of information.

In order to preserve data and information, the physical interdiction to the PCs is not generally sufficient. Indeed, it is also recommended that they should be electronically insulated. To do so, the easiest way is to turn off the PCs, even if, operating this way, it will not be possible to monitor specific parameters. The approach is more or less severe depending on what you are looking for.

Electronic data should be collected taking into account their potential volatility (Table 4.6). For example, a loss of electric power forces the evidence to be accessible only for a limited period of time, according to the battery backup capacities.

It is important to verify the time of the CPU of the machine that is considered as a source of the data; otherwise, it may prove complex to be able to locate events properly in the timescale. This aspect was of primary importance in the Norman Atlantic investigation, where the necessity to define a "time zero" clearly emerged from the first stages of the investigation, in order to have a comparable set of data.

Table 4.6 Digital evidence and their volatility.

Evidence	Volatility
CPU registers and cache	Nano-seconds
Main memory	Nano-seconds
State of a network	Milli-seconds
In operation processes	Seconds
HD	Minutes/Years
Backup memories	Months/Years
CD-R, DVD	10 Years

In the context of an industrial incident investigation, it may be necessary to carry out the following actions:

- Examine the PCs, usually to find and print emails (it could be convenient that all those present sign them);
- acquire files on a tamper proof medium (if possible and useful, recovering the deleted files);
- copy the hard disk (if necessary, evaluate a bit by bit copy to ensure that data are not altered during the process);
- seize the HD; or
- seize the whole hardware.

Digital documentation is collected and treated with high priority, since it is fundamental to preserve conformity with the original, in order for this to be used as proof during an eventual trial. For this type of evidence, it is essential to have a traceability chain, from its collection to its usage during the trial, in order to certify that data do not undergo changes. The collection process of digital data is shown in Figure 4.6.

4.4.3.1 An Example About the Value of Digital Evidence

Normally the fire engineering approach to the fire design is based on a known geometry of a building, and this provides the evolution of each fire event identified in a fire scenario.

This case study represents a different point of view, because the fire scenario is a well-known fire, occurred in a manufacturing warehouse, and some fire engineering quantitative tools were used to try to learn about the event.

The aim of the investigation was to understand the pattern and the spread of the fire, as well as to compare its growth dynamic with a conceivable fire scenario, on a 3600 square meters plant surface, 8 meters height, a manufacturing warehouse, operating in the expanded polymers treated with flame retardant additives, for application on the air conditioning and heat pumps

Some definite and indisputable elements were available, as the sequence of activation of smoke alarm based on a point sensors plant, the sequence and the exact time of the emergency calls to the fire station, and the external video record of the fire, recorded by pedestrian witness and uploaded on a web streaming video service, and the melted metals (Aluminium and Copper) inside the compartment, with a temperature reference for the fire zone. In this video the exact time of arrival of the fire team is unequivocally fixed by the recorded radio call, as per the requested procedure, because a radio signal clock is associated with each track.

The first element, the smoke sensors activation sequence shown in Figure 4.7, returns a smoke spreading scenario with 80 sensors that have been activated in less than three minutes, with a final open loop event, resulting by the first melting of a sensor in a different area of first activation.

The second element is the time of the calls to the emergency service, witnessing a fire with high flames over the roof and a huge, very dark plumes elevating over the building. The time of the first call to emergency number 115 followed less than two minutes from the first smoke alarm signal inside the building.

On this basis, some backward considerations were developed, with a “reverse Fire Engineering” approach.

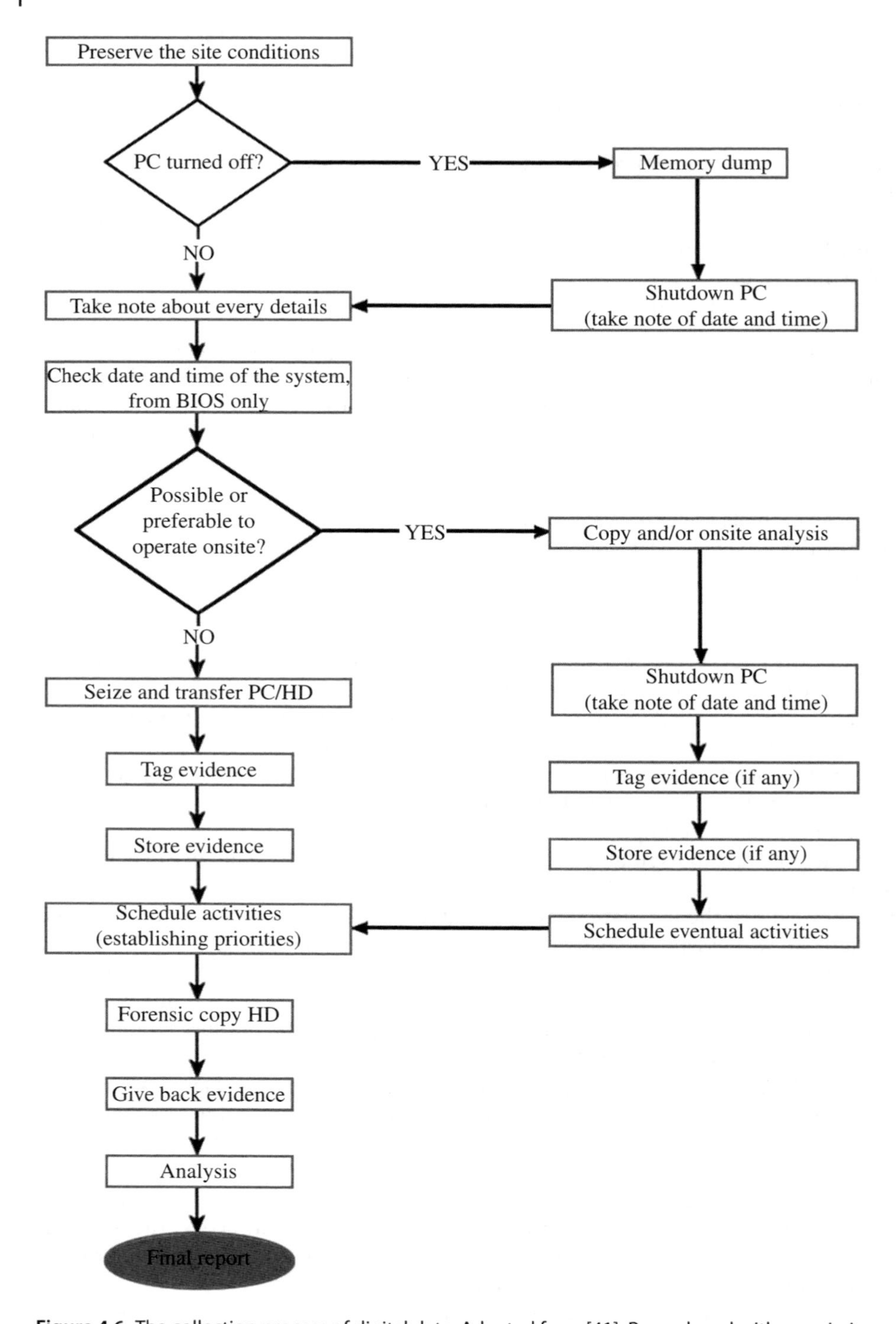

Figure 4.6 The collection process of digital data. Adapted from [41]. Reproduced with permission.

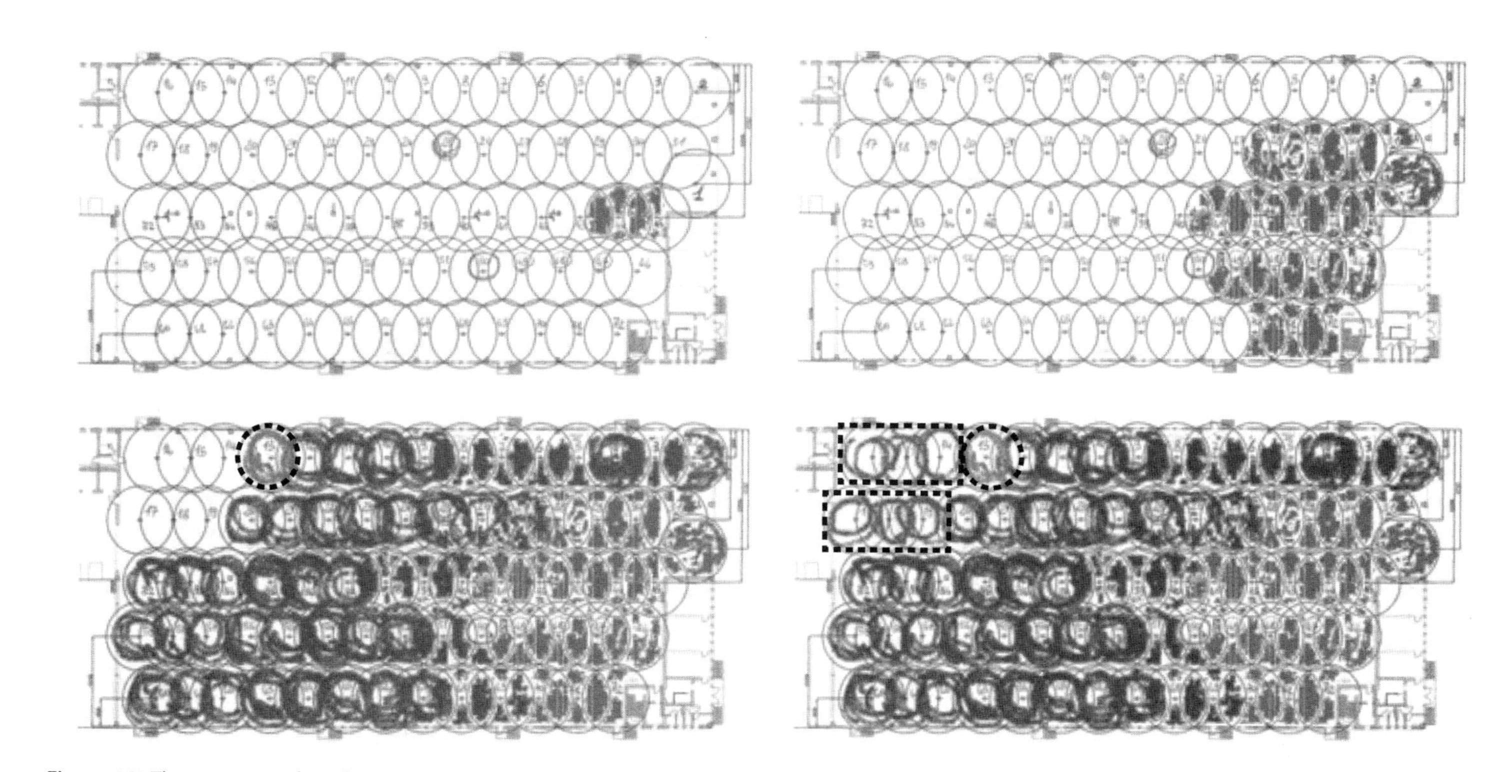

Figure 4.7 The sequence of smoke sensors activation. In grey the first group, in dark grey the following 60 seconds, in dashed circle the first open loop and in dashed circle and dashed rectangles the residual activation, all in less than 180 seconds

By means of the NUREG spreadsheet, with the Herkestad correlation for the fire plume, with a flame 12 meters high, we can simply evaluate an energy release of 12 MW on a 27 square meters of equivalent pool fire (the thermoplastic behaviour of the expanded LDPE can justify this assumption).

From Figure 4.8, as the fire immediately spread over the entire length of the building, we could suppose various equivalent pool fires, on the side of the combustible deposit. Just considering the length of 90m, we obtain 15 pool fires, with a heat release rate estimated at 825 MW.

By comparing the declared values at the time of the project approval, the medium fire load inside the warehouse overtakes the declared data by at least an order of magnitude.

Moreover, some consideration must be conducted to consider a scientific explication to the event. Without other consideration on the ignition source, the cause for the spread of the fire over the whole length of the deposit could be the direct spread of the flame, thermal radiation between rolls and coils and the convective heat transfer from hot gases in the fire smoke and the fire load.

For the first path of growth, the literature clearly identifies a magnitude order to direct flame propagation of fire of centimetres/second, with regard to a solid fire load, here polymers fulfilled with flame retardants.

The second path of propagation, this could be assessed by simple radiation heat transfer, as described in the Lawson and Quintiere Method (NFPA Handbook of Fire Engineering). Considering an initial fire of 1 MW (i.e. a Stuffed Chair) with regard to a hemispheric envelope, at 4 meters the radiant flow will be close to 1 kW/m^2, the same as the sun's radiation.

The third way to evaluate the fire spread is to consider heat transfer from hot smoke gases, but the volume of the complex was 28800 cubic meters (3600 square meter surface and 8 meters high) with smoke evacuation system that immediately activated, after the automatic smoke alarm.

In order to activate all smoke alarms in a time period of less than 180 seconds we will need a magnitude order of 10 cubic meters of smoke at each second, in a single source of emission, compatible with an accidental, involuntary single point of ignition. On the other hand, the first radiation of hot gases stratified on the ceiling will be emitted

Figure 4.8 The wall collapse a few minutes after the arrival of the fire brigade unit.

at 6 meters from the high placed coils and rolls, and the evidence of the compartment found by the inquiry was that it didn't reach a flash over, but only a localized very hot fire.

In fact, on the opposite side of the fire, close to the operating machines, the granular polyethylene bags survived without be ignited nor melted, as well as some cardboard stacks.

Furthermore, a pile of rolls placed outside the warehouse, on the rear platform roof, invested by the hot smoke flow from the fire raging inside, along the full length of the wall collapsed after few minutes, as in Figure 4.8.

The pile, survived without any ignition but only the upper rolls dissolved as per the heat transfer (Figure 4.9).

The above considerations, with some simple quantitative evaluation of fire engineering, leave with any valid doubt about an accidental fire ignition scenario, because the order of magnitude of the growth of the fire, raging with an extraordinary power compared with a solid fire load, overcomes any foreseeable behaviour, even though there were expanded polymer rolls stored.

The discordance for orders of magnitude between the size of fire growth reconstructed by aligning smoke alarm records, emergency calls, fire brigade communications, video records and post fire scene evidences and the fire engineering and transport phenomena evaluation for any accidental scenario is driving us to an arson technical verdict, adopting a deductive approach known in philosophy as Occam's razor.

4.4.4 Physical Evidence

By physical evidence, you mean all those objects, generally but not exclusively found at the incident site, that may be a useful source of information to understand what happened. Physical data does not only come from the process system, but also from auxiliary and adjacent systems, such as the control, safety or support systems. Examples of physical objects that may need to be investigated are tanks, valves, gaskets, flanges, relief system devices (including the condition of the rupture disk), explosion fragments, data recorders, sensors, switches, as well as residual materials (mainly liquids and solids).

Figure 4.9 Rolls of expanded LDPE with flame retardant included invested from heat.

Moreover, physical evidence also includes damage affecting plants, structures and objects in general: fractures, corroded parts, dents, colour stains and any modifications to their original status are considered physical evidence. From this perspective, further examples of physical data to observe are:

- Fractures, distortions, surface defects/marks, and other types of damage;
- items suspected of internal failure or yielding;
- misaligned or miss-assembled parts;
- control devices in the wrong position;
- incorrect components;
- raw materials;
- parts, products, and chemical samples; and
- portable equipment (including tools, containers and vehicles).

It was already noted that for every kind of evidence, a proper security chain of custody must be defined: this is especially important for physical evidence.

In general, regardless of the type of physical evidence, collection is carried out as follows:

- Identification, by tag using a numbering system defined a priori (Figure 4.10);
- acknowledgment, by photography (Figure 4.8);
- conservation in an adequate location, depending on the dimension of the evidence, its critical features, and the time required for its custody (with the imperative goal to not alter it at all); and
- appointment of the person in charge of its custody.

Key items are photographed and tagged before they are moved. A good rule of thumb about what to collect is: "too much is better than too little" [9]. The reason is that having all the necessary information may require the investigator to come back to the incident site and carry out a further collection of evidence, but the passage of time could alter or remove them.

4.4.5 Position Data

Position data are associated with people and objects. They help in understanding several issues, like the origin of a fire, the position of the highest pressure, the point of view of a collected testimony, the barrier which failed first. Position evidence to be collected include: the as-found position of valves, controls, and switches, liquid levels, the position of fire marks, materials, debris, soot, the position of people, witness movement, and so on. Maps can be used to document the location of people, equipment and components. Position data are a valuable source of information for explosions, where fragments and debris locations can provide useful information to reconstruct the dynamics of the incident.

All this information is documented with an extensive photography report, to crystallize the evidence, especially those having legal value. As usual, it is recommended to take photos as soon as possible after the occurrence of the event, before post-event activities (like cleanup or rebuilding) that can alter the original conditions.

Having a list of data to be collected can help in organizing the evidence gathering phase. The list can include the following:

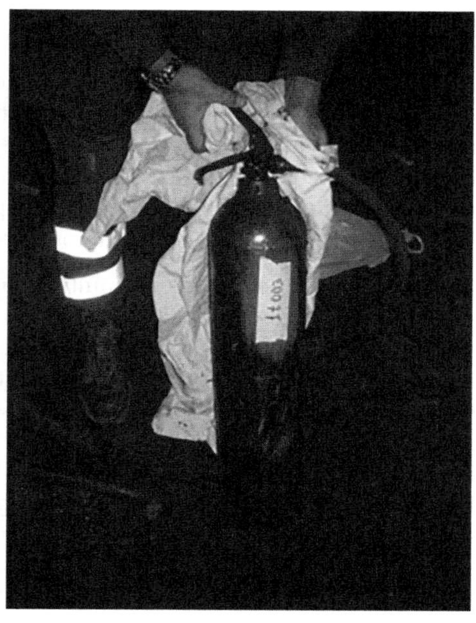

Figure 4.10 Identification of fire extinguishers by tags (on the left) and acknowledgement by photography (on the right), ThyssenKrupp investigation.

- As found position of every valve related to the occurrence;
- as found position of controls and switches;
- position of relief devices;
- tank levels;
- location of flame and scorch marks;
- position and sequence of layers of materials and debris;
- direction of glass pieces;
- locations of parts removed from the process as part of maintenance;
- locations of personnel involved in the maintenance and operation of the process;
- locations of witnesses;
- location of equipment that should be present that is missing;
- smoke traces;
- location or position of chemicals in the process;
- melting patterns; and
- impact marks.

4.4.6 Photographs

Photographs are a sort of window on the incident site. They become memories, details, positions, the starting point for analyses, supporting considerations. Photos mark the evolution of the investigation, from the first site visit to the subsequent inspections, even if carried out after months or years. In order to be considered as proofs during an eventual trial, it is fundamental that photographs are representative, identifiable and easy to consult. This means that they have to show clearly their subject, position, and context; moreover, all the pictures should be catalogued, in order to have a complete overview, particularly in details and time.

Normally, photographic equipment is not designed to be used in hazardous conditions: once the possible residual hazards have been identified, it is important to take necessary precautions (e.g. avoid any potential spark in potentially flammable vapour conditions, use an impermeable camera in wet conditions, a robust camera to be protected from dust or accidental falls and slips). A good investigator should consider the following practices concerning photography:

- Photos should be taken from multiple orientations, positions, and distances;
- an object of known size should be placed in the photo;
- potential shadows due the flash should be considered;
- a spare battery should be always brought;
- each image must be recorded in a register;
- autofocus devices should be avoided or, at least, the investigator should be aware of their limits; and
- distribution of copies needs to be managed.

There are two main steps:

- Collecting the photographs from the incident site, during the inspection; and
- cataloging the collected photos, after the inspection.

This procedure is followed for each inspection, increasing little by little the number of pictures in the investigator's collection.

4.4.6.1 The Collection of the Photographs

At this stage photos are taken and identified, in order to simplify the following step about cataloguing. Since the first inspection, it is suggested to set date and time on the digital camera. The investigator will take note of the ID-number of the first useful picture, to define the beginning of the collection.

It may be helpful to use different lens, like:

- Quadrangular lens, to reveal the spatial correlation among various objects;
- telephoto lens, to take photos of distant objects; and
- macro lens, when the objects are particularly close.

A good investigator knows that what is photographed is not exactly what our eyes see because a lot depends on the lighting conditions; this is the reason why it is convenient to take several photos of the same object from different points of view.

In order to have a representative picture, objects with a known dimension are usually included in the photo, to have a geometric reference and to highlight specific details. For this purpose, the investigator can also use his/her hand (Figure 4.11). For a more precise reference, it could be useful to utilize a straight graduated ruler (Figure 4.12).

The position of the subject in the picture respect to its initial (even only hypothetical) collocation is valuable information, being of interest for the investigation mission: cameras equipped with GPS are a powerful tool to be considered in the investigator's equipment. An alternative solution is to measure the distance between the interesting evidence and a chosen reference through a flexible meter. The information is then recorded on the sitemap, where the adopted reference is clearly identified.

Figure 4.11 Detail of a small imperfection on the edge of a metal sheet, ThyssenKrupp investigation

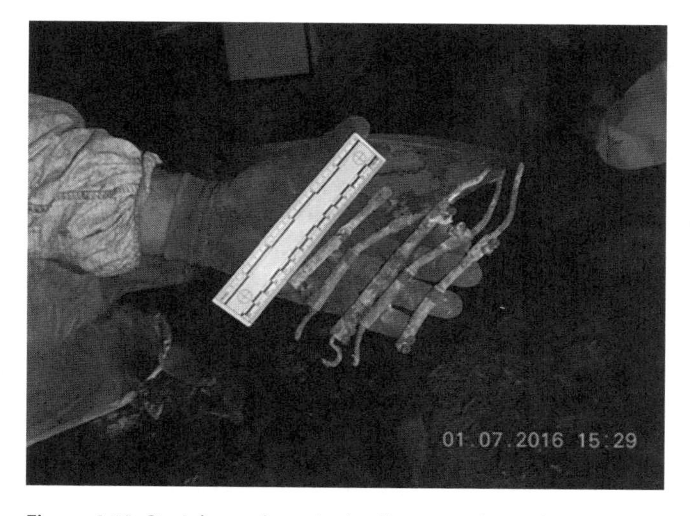

Figure 4.12 Straight graduated ruler, Norman Atlantic fire investigation.

Taking notes on the subject of the photos can be beneficial to the further steps of the investigation, but only if it is a fast and easy operation. A paper sheet hanged to the neck, using a folder, is generally a good idea: this solution allows having free hands to take photos, move objects and, of course, writing. Depending on the crime scene, its complexity, and the site typology, different forms may be used. An example is shown in Table 4.7. The adopted form must be personalized according to the specific scope.

The collection of photos should be carried out following the usual approach, from the general to the particular: firstly, shooting pictures about the context, then about the single machinery, and finally about the details of the machinery (such as a specific damage, a hose, or an oil leak). The subject of the photograph is identified by simply placing a cross in the corresponding box of the form, using the "note" field for some useful detail:

Table 4.7 Example of form to use for the collection of pictures.

ID	O	M	dM	C	dC	E	Note

Legend:
ID = Photo identification number
O = Overview picture
M = Machinery
dM = Detail of the machinery
C = Circuit
dC = Details of the circuit
E = Picture of the single element

the whole operation requires very little time. It could be equally beneficial to take note of the position from which the photograph is taken. It usually happens that the investigator finds important details while he/she is performing other activities. In this case, the form allows to have a memory, also highlighting at the end of the inspection, the necessity for additional photos, by looking the column that has a fewer number of crossed boxes.

4.4.6.2 Photograph Cataloguing

In order to understand the importance of the photograph cataloguing, it is interesting to provide some numbers about the investigation on the ThyssenKrupp plant in Turin: 12 inspections, spread over two years, and 468 photos. Among these, only 65, the most meaningful, were attached to the technical report. Similarly, more than 4000 photographs were taken for the investigation on Norman Atlantic, mainly because of the unrepeatability of the conducted inspections. With this numbers in mind, it is clear that a cataloguing procedure requires a powerful tool to manage the huge amount of data. There are different software programs on the market, both freeware and commercial, helping to catalogue the pictures (and also documents and different types of files). The idea at the base of their logic is to provide a set of metadata for each photograph. Metadata are a sort of identity card of the picture itself. They preserve information about the characteristics of the image like its dimensions, resolution but also the name, address, and other information about both who took the photo and who is cataloguing it.

An example of how this information is shown is in Figures 4.13, 4.14, and 4.15.

Photographs may be classified by tags, like colors, categories, stars (to prioritize their importance), and so on. It is also possible to give a name to the different labels. Using the keywords related to the pictures, it is also possible to group the images according to a particular criterion, filtering them by keywords or other selectable parameters like date, or position. Keywords and tags can be added, removed, or edited during the investigation process, having an always updated catalogue. It is convenient to perform these activities after each inspection.

This approach allows the creation of an ordinated album where it is possible to run a search for highlighting, for example, all the pictures about the fire extinguishers, regardless the date, or only the ones depicting ruptured hoses. In this way, the consultancy becomes fast and extremely efficient.

In conclusion, in an incident investigation, the organization of the evidence photography is probably one of the most important issues to solve. Photography (and video) is widely used during an investigation, to document the "as-found" position of physical evidence, their damage pattern, layering, status, and so on [10]. Photos are also used to present the investigation report and distribute the lesson learned during training sessions. If some items or conditions change because the investigation needs change,

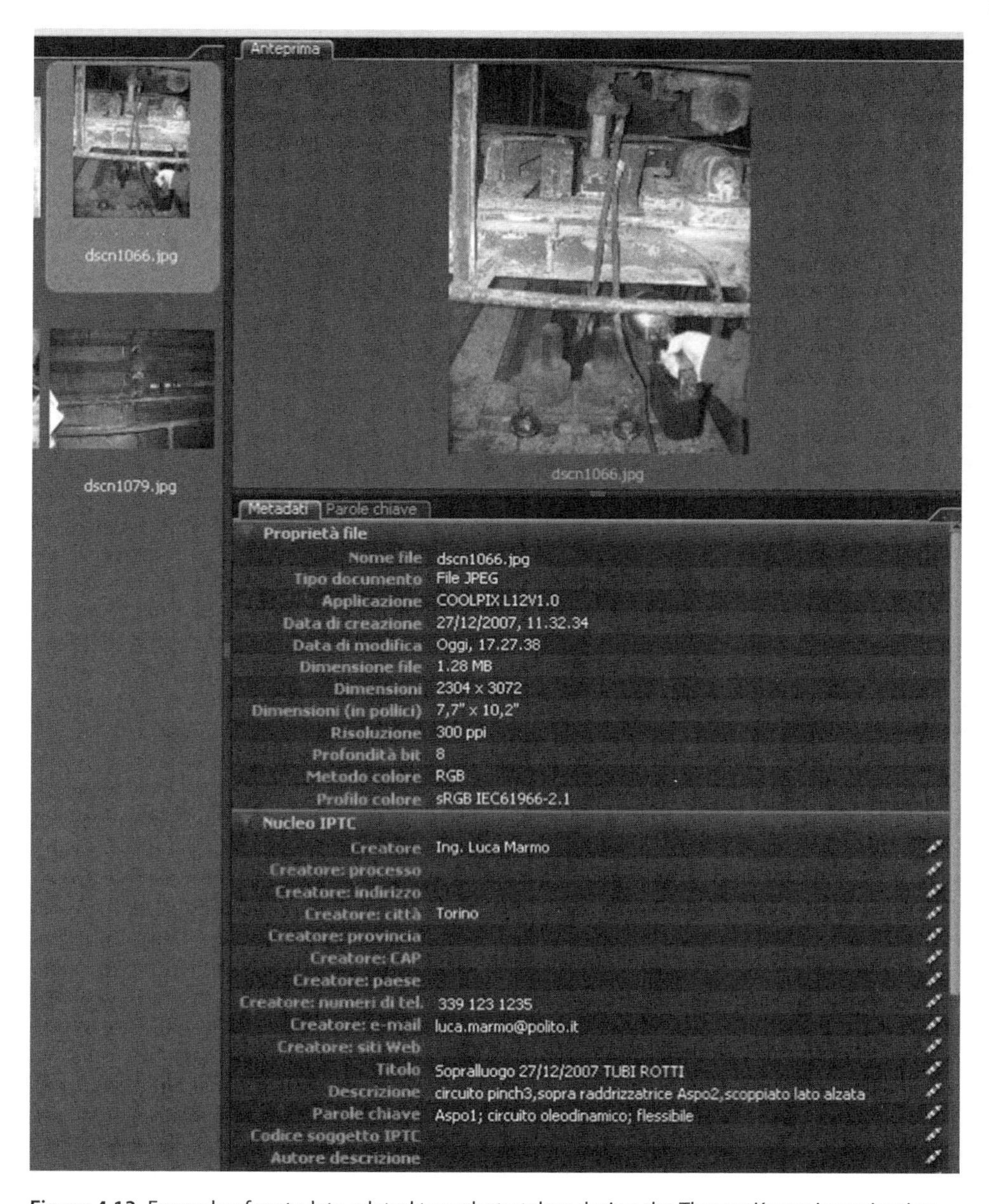

Figure 4.13 Example of metadata related to a photo taken during the ThyssenKrupp investigation.

photographs are taken to freeze promptly the scene, usually from multiple perspectives and distances. Having such a number of photos, it is important to organize them efficiently, using a formal log to indicate, at least, the date, the time, the photographer, and the contents of the image. A record of distribution should be maintained too, if copies are shared. In any case, for digital images, master copies and backup copies should be maintained in a controlled way.

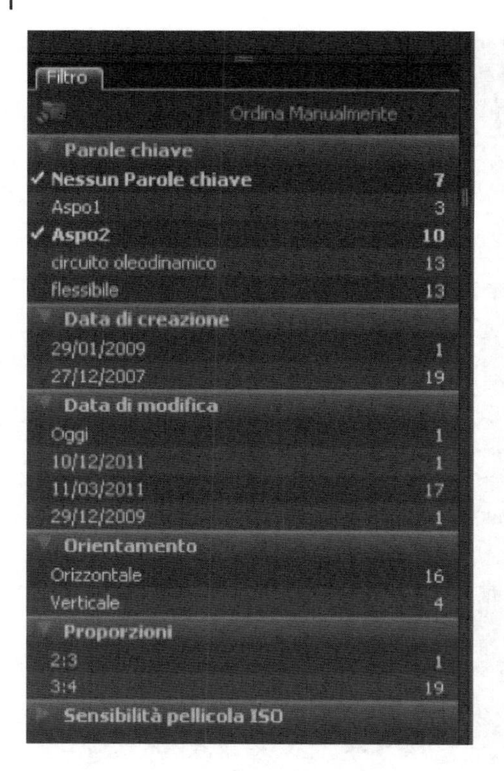

Figure 4.14 Example of keywords for filtering the picture of a collection.

4.5 Recognise the Evidence

Considering the wide range of possible sources of evidence, it becomes crucial to establish priorities and avoid useless waste of resources. One method to organize data and determine the few factors that really affect the problem is the Pareto Analysis [3] (Figure 4.16). The method uses a bar chart of failures, generally ordered by frequency of failure. Thanks to the Pareto chart, it is possible to have a visual tool to establish priority, focusing on areas where largest probabilities exist. Therefore, it helps in defining the investigative areas that should be improved the most.

Regardless of the sources and type of evidence, the investigator firstly asks: "What data should I collect or not collect?" [2]. It is extremely difficult to answer, because a decision needs to be taken before data are gathered. The Pareto analysis is a first solution, where the 80/20 rule applies, that is to say the 80% of the incidents are related to the 20% of the recurrent causes. Predicting which parameters will help is not an easy task. However, some of them can be chosen, reasoning. For example, facility age is likely to be a parameter affecting the reliability-related failures, since older facilities experience different types of failures respect to the newer one. Severe weather conditions can influence some type of incident, so they are a parameter to track. If the incident involves personnel injuries, individual clothing is included among the data to collect, together with the elapsed time between training on the task involved in the incident and its occurrence.

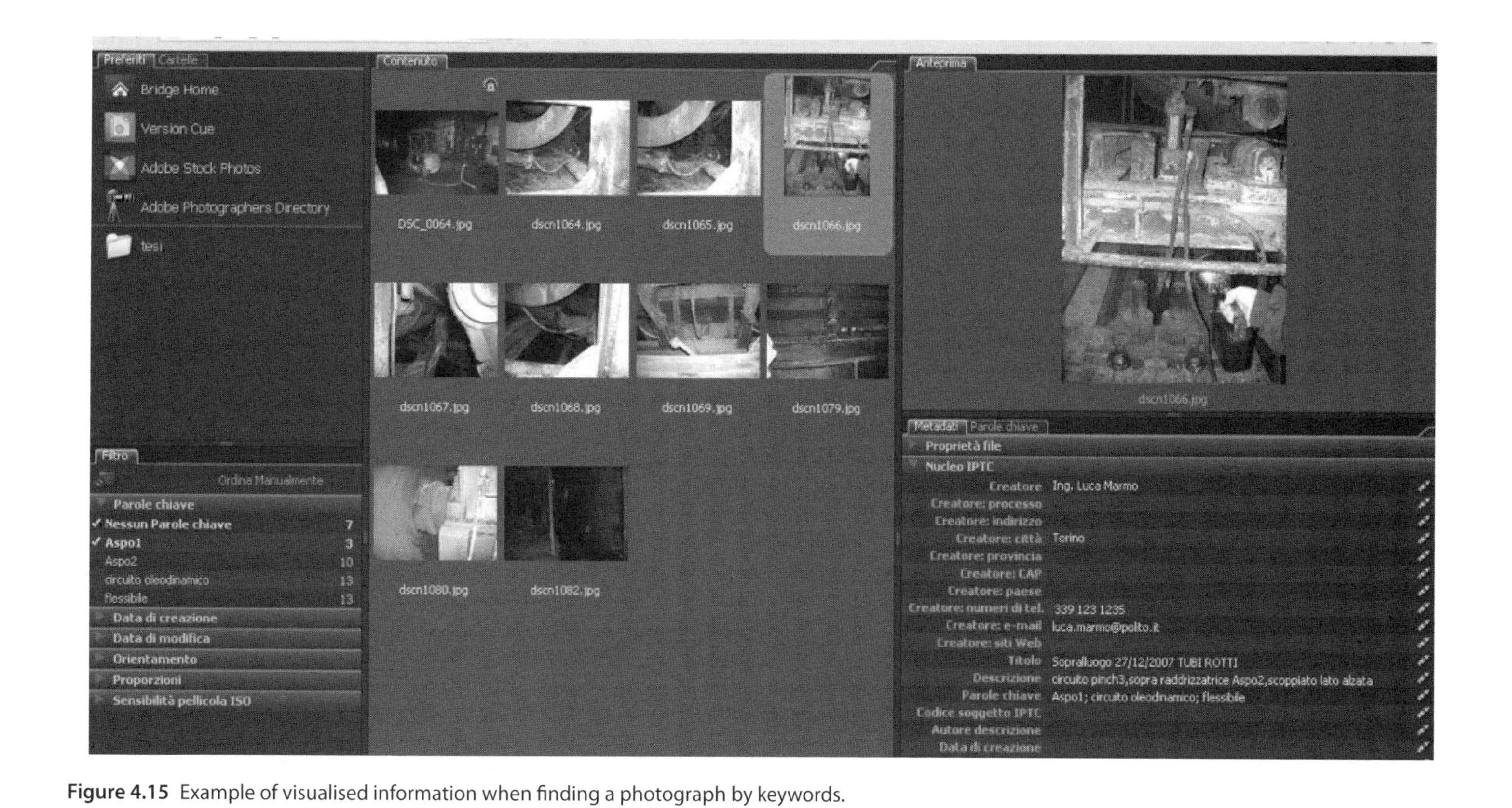

Figure 4.15 Example of visualised information when finding a photograph by keywords.

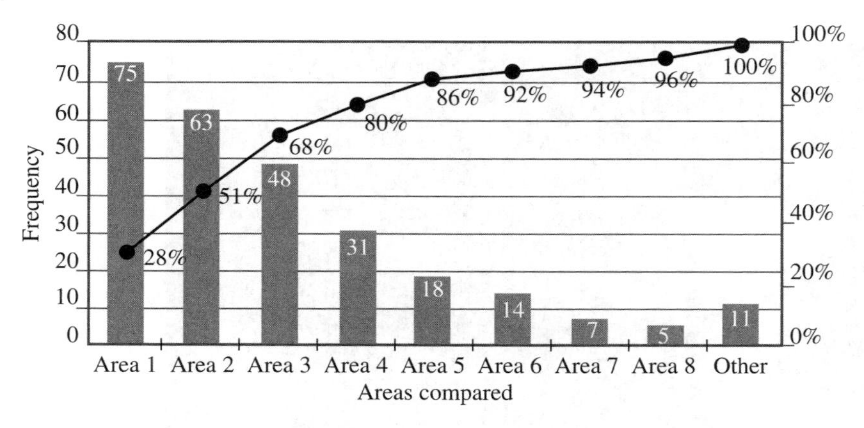

Figure 4.16 Example of Pareto Chart. Adapted from [3]. Reproduced with permission.

The following activities can help in defining the data to collect:

- Determine the type of decisions based on the evidence analysis;
- identify the necessary trends to make these decisions;
- determine the necessary data to identify these trends;
- determine if the required data can be collected;
- determine if other teams may have already collected the required data (such as the quality control, operations, or maintenance groups);
- determine if the required data can be obtained from already collected evidence;
- determine how to manage the evidence gathering and the storage system (chain of custody); and
- identify who will analyze data and the frequency of such analyses.

An effective evidence collection and preservation is ensured by paying proper attention to the organization and composition of the evidence processing team [12], whose members have been already listed in Paragraph 4.2. An evidence management is typically required for major chemical process incidents [10].

Throughout the evidence collection process, the investigator should constantly review the scene, in order to adapt to eventually occurred changes. Reviewing consists in reevaluating the boundaries of the scene, the safety issues, the technical requirements, the evidence storage locations, and the administrative and legal aspects, ensuring that time-sensitive evidence is preserved and collected. Controlling contamination is an essential requirement to protect not only the integrity of the incident scene, but also the forensic value of evidence and the safety of people working in the area. This is why the evidence storage location is typically placed outside the incident scene, clean protective equipment is used, and evidence is also properly packaged.

The evidence processing consists of the following six activities:

- Identify;
- collect;
- preserve;
- inventory;
- package; and
- transport.

In order to maximize the evidence process, the investigator should:

- Prepare an evidence recovery log, taking note of item number, description, location found, collector's name, markings, packaging method, and eventual comments;
- identify evidence by assigning personnel to specific search areas, establishing a search pattern and procedures, documenting event consequences (like structural damages and thermal effects), examining equipment, items and structures, documenting the location of injured and victims;
- collect the evidence, including suspected components, fragments, residues and other trace evidence;
- ensure that evidence is photographed, packaged, preserved, labelled, recorded and secured;
- place evidence from different areas in separate containers;
- label evidence, also identifying possible hazards; and
- arrange for the evidence transportation.

All physical items should be documented in place through photography or other methods. It is important that this documentation is produced before the evidence is moved or disturbed, taking note of the location, the orientation, and the time of collection. Assigning unique evidence numbers to each physical evidence is a common practice [10]. If the investigation requires disassembling equipment, the activity should be documented with photos and/or videos, tracking both evidence entering in the chain of custody by the investigation team and items that are not of interest. Being a modification of the original status, this operation may require the formal approval of litigants before it is touched. If litigation is likely to happen, long-term solutions should be provided to store physical evidence. Attention should be paid when distributing copies of documents, in order to avoid generating confusion and sharing reserved information.

One "tool" to use to recognize potential evidence is the thermal degradation [19]. It describes chemical decomposition due to the heat. Thermal degradation must not be confused with thermal decomposition or pyrolysis, even if these terms are often used interchangeably in fire investigation. Indeed, pyrolysis is the thermal degradation of solid fuels. It can be seen as the equivalent of evaporation for liquid fuels: both pyrolysis and evaporation describe the preliminary stage that allows a flaming combustion. Materials also undergo thermal degradation in inert atmospheres, such as nitrogen. The phenomenon can be monitored using thermogravimetric instruments (analyzing how mass evolves over time, changing the temperature) while pyrolysis products can be detected using special techniques, such as gas-chromatography combined with mass-spectrometry. However, an expert investigator knows that the interpretation of these analyses is complicated, because interfering products are usually present in the fire debris samples, affecting the analysis. The interested reader can find additional information on the topic on [19] and [20].

4.5.1 Short Case Studies

In the following paragraph some significant cases relevant to the subject matter are proposed. For each, a succinct reconstruction of the event has been provided together with the list of the substantial evidence which led to its solution. One or several deductions derive from each piece of evidence, sometimes even antithetical to each other; a compatibility rating of deductions is therefore also reported. Compatibility must be seen as referred to other deductions to identify a comprehensive and mutually compatible set [21].

4.5.1.1 Explosion of Flour at the Mill of Cordero in Fossano

The Cordero Mill suffered a disastrous event due to the explosion of wheat flour, which caused five victims [22, 23].

At the time of the explosion, followed by the partial collapse of the building and by fire, part of the excess flour was being re-pumped into a tanker, and the excess material was transferred into silos by a pneumatic conveyor.

Investigations discovered that the grounding connection of the tank was missing and that the silos containing the flour were not equipped with an explosion venting system.

Investigations were conducted on several occasions, first by the advisor to the Public Prosecutor and by independent ones, and secondly by one of the Authors, who was participating as an expert of the judge.

The gathering of evidence, performed in the field and using the photographic material, made it possible to outline the dynamics and the violence of the explosion through an examination of the state of the places, equipment, their position and mapping of the projection of fragments. The cause and location of the ignition were identified by cross-checking the objective findings with the testimonies relating to the events immediately preceding the accident and the testimonies regarding its dynamics. In particular, the ignition of the explosion was located in the re-pump line which had clear signs of internal overpressure (Figure 4.17). Finally, the medical reports of five victims, even in the absence of reliable localisation of four of them, were found to be compatible with the hypothesized incidental dynamics. In brief, the operation of flour unloading from the cistern created an accumulation of electrostatic charge that created an ignition in the re-pump pipe of the silos flour. From here, the explosion spread to the flour silos and the entire structure causing damage to the walls and the collapse of part of the building (Figures 4.18 and 4.19).

Table 4.8 shows a summary of the main evidence, coupled with a set of deductions fully compatible with each other to define with reasonable certainty the causes and dynamics of the event.

Figure 4.17 Evidence: overpressure damage to a flours repump duct flange.

Figure 4.18 Building (south side) with noticeable damage from excess pressure.

Figure 4.19 Building (north side) with widespread collapse primarily from static collapse.

4.5.1.2 Explosion at the Pettinatura Italiana Plant

The explosion occurred at the textile factory "Pettinatura Italiana" located in Vigliano Biellese, led to the death of three persons, the burning of another five, three of whom very seriously, in addition to conspicuous damage to a part of the structure [24]. This is an example of the dust explosion risk provoked by flocculent materials [25, 26, 27, 28].

Table 4.8 Summary of the evidence and deductions.

Evidence	Deductive reasoning	Compatibility
Damage to the structures and walls mainly from overpressure in the south wing, mainly from static collapse in the north wing	• Ia) Primary and secondary explosions chain • Ib) Increased quantity of flour-available in the south wing	• Total • Total
Projection of fragments over 100 m south side, to approximately 40 m north side	• II) A high quantity of flour available in its wing	• Total
Flour silos made of wood completely destroyed	• IIIa) Destroyed by fire	• Incompatible with IV (the magnitude of the damage involves the involvement of a large quantity of flour, such as that dispersed in a silo where this is repumped)
	• IIIb) Destroyed by internal explosion -	• Total
No damage from overpressure in the remaining flour silos, all steel	• IV) No explosion occurred in the steel flour silos	• Total
No obvious damage from internal overpressure in the equipment found, much damage due to the collapses	• V) No explosion was found within the equipment	• The trigger within an appliances incompatible with VI
Damage from internal overpressure in the flours repump pipe	• VI) Explosion was discovered within the flours repump pipe	• Total
Architecture of the repump pipe not defined	• VII) The repumped flour destination silos is not identified	• Total (without prejudging IIIb)
No. 1 victim died instantly as a result of trauma and burns. Defined position	• VIII) Violent shockwave into the place occupied by the victim	• Total
No. 4 victims died later, 1 as a result of lung trauma and 1 due to burns. Uncertain positions	• IXa) shock wave into the place occupied by the victim • IXb) Diffuse flame front	• Total
Testimony regarding prolonged hissing followed by a loud roar	• X) Primary explosion inside a pipe or equipment + secondary explosions	• Total
Testimonials regarding the operations of loading of the flour into a tanker and the discharge of a small excess quantity immediately after loading without tank earthing connection	• XI) Accumulation of electrostatic charge in the flour, insufficient charging relaxation time	• Total
Explosive properties of the flour (MIE, LEL)	• XII) Compatible with the hypothesis of electrostatic trigger in the repump pipe	• Total

The explosion was found to be a secondary explosion of dust produced by the process of removing burrs at the carding stage.

"Pettinatura Italiana" has carried out the activities on behalf of other businesses of washing, carding and combing the greasy wool for over a century; it is by far the most important Italian company in this sector. The production cycle also included the collection of different types of wastes (for example burrs) or waste of wool (for example's blouses). In particular, the burrs are lumps or wisps of wool containing vegetable parts of the original fleece of the sheep; these were extracted using carding machines and conveyed with a pneumatic system into suitable collection boxes located on the ground floor (Figure 4.22), in a part of the plant that was unmanned except for sporadic interventions of maintenance or unloading of these burrs. There were two boxes for each carding machine: one at the loading phase and the other at the cleaning phase. They were medium sized chambers (approximately 2.6 × 5.5 × 3.6 m) positioned in a line, fitted on the top (approximately 1.7 m starting from the ceiling) with three sides of tight mesh netting to allow the exit of pneumatic conveying air from the box being loaded to the adjacent ones (empty) or into the access corridor.

The blast occurred on the ground floor in the cell area but also involved much of the first floor in the area of the washing line cards (Figures 4.20, 4.21, and 4.23).

It should also be remembered that on the first floor, just before the explosion, around one carding machine, smoke was noticed coming from the subcard. This fact alerted the department manager and two technicians who, proceeding from two different staircases, descended to the ground floor to perform an inspection. While this inspection was in progress, the blast occurred.

Table 4.9 shows the list of investigative activities.

Table 4.10 describes the aetiology of the accident which is the result of four events which occurred in succession: that it was a typical case of domino effect.

Figure 4.20 Explosion of wool burrs, state of places.

Figure 4.21 Explosion of wool burrs, state of the places, card rooms.

Figure 4.22 Explosion of wool burrs, burrs storage boxes.

4.5.1.3 Explosion of the Boiler of the SISAS Plant of Pioltello

The explosion of the boiler of the SISAS plant of Pioltello originated from the spontaneous combustion of a mixture of air and unburned gases (CrO, CH, etc.) that had accidentally formed inside the boiler during a restart following a stop of 8 minutes. No external trigger was required as large areas of the boiler were undoubtedly at high temperatures, as confirmed by the temperature records found.

The identification of the causes that generated such a significant formation of combustion gases was made possible by a detailed examination of all traces of the recorders found in the control room.

Figure 4.24 reports only the traces of the fuel flow rates (CH_4) and the combustion air. In fact, at 4.45 p.m. the boiler was stopped by the operators; after approximately

Figure 4.23 Explosion of wool burrs, state of places, burrs collection boxes corridor with visible in the foreground signs of material fragment projection on the white bin.

Table 4.9 Summary of technical assessments, explosion of wool burrs at Pettinatura Italiana.

N.	Technical assessments
1	Examination of the places
2	Collection of testimonies of the first attendees
3	Results on the search for possible gas leaks (CH, thermal plant, fermentation gas)
4	Evaluation of the type of damage
5	Collection of testimonies of injured people
6	Analysis of the production cycle and, in particular, of the mode of the loading of the burr collection boxes
7	Sampling and analysis of the findings in order to search for accelerants (petrol, diesel, solvents, etc.)
8	Analysis of projection of combustible wool wisps
9	Sampling and physical-chemical characterisation of dusts
10	Evidence of flammability and of explosivity of dusts
11	Localisation of the trigger
12	Assessment on the possibility of explosion of unburned gases (CO, hydrocarbons and fumes)
13	Estimation of the quantity of fuel involved in the explosion

8 minutes of being stopped (the flow rate of methane was zero while that of the air was maintained at a constant value) there was a restart of the boiler supplying power with a sharp increase of the flow rate of methane. As the airflow had previously been kept constant, the conditions were generated to create in the boiler an atmosphere rich in unburned gases, with probably switch off of the burners. The boiler was devoid of a blocking system for air/fuel ratio errors. Apart from the very difficult context not

Table 4.10 Sequence of events that led to the explosion.

N.	Event
1	A fairly contained initial fire, at a ceiling close to a burrs collection box, whose flames were not seen by anyone. The smoke escaping from the subcard alerted the technicians on the floor above who proceeded to perform inspections
2	A modest initial explosion (perceived as a "bang" by witnesses) that occurred inside the box next to the fire and caused little material damage in addition to burns to the two technicians performing the inspection
3	The fires caused by the flame front of the explosion and, in particular, that of the wooden structures of the nearest boxes. These fires were extinguished by the intervention of fire-fighters
4	The evolution of the previous explosion with the contribution of other fuel (300+500 kg of dust adhering to the burrs collection box or contained in the sleeve filter bags of the air conditioning system) that generated both an increase in pressure with consequent damage to the roofs of the warehouses and a conspicuous flame front also visible outside the plant and cause of the fatal burns to persons who on the first floor were not far from the cards

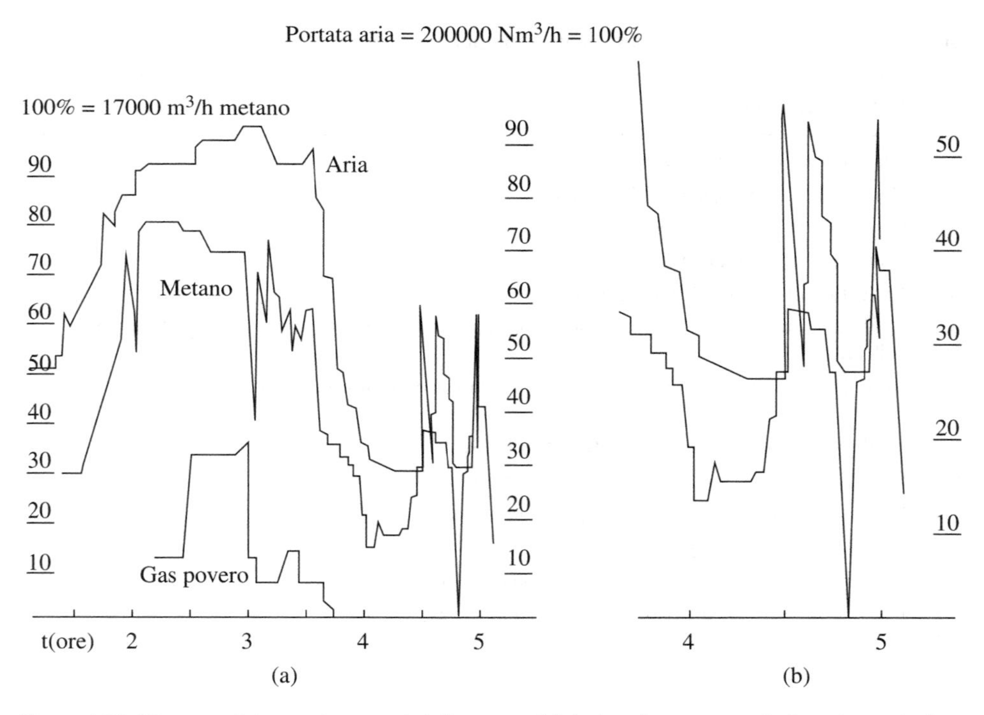

Figure 4.24 Diagram of the methane and air flow rates (a) during the moments before the explosion and (b) enlarged detail.

Table 4.11 Summary of the evidence and deductions.

Evidence	Deduction	Compatibility
Typical structural damage to the boiler due to overpressure generated inside	• I) Explosion inside the boiler	• Total
The explosion happened after 8 minutes of shutdown, during a manual restart phase with two burners out of the six available	• II) inside the boiler were high temperature areas able to act as a trigger	• Total
Blocking of automatic photocell intervention relay due to the insertion of wedges	• III) Impossible to detect the switching off of one or several burners	• Total
An examination of the traces of process variables shows that, minutes after the restart, given a constant air flow, the flow rate of the methane was consistently increased	• IV) Formation of a mixture of unburnt gases able to explode	• Total
The absence of an automatic protection due to incorrect air/fuel ratio	• V) Inability to identify and stop the unburnt gases formation process in case of incorrect ratio	• Total

reported here, in which the operators had to work that night, the triggering event of the explosion was undoubtedly a human error in the handling of the boiler air/fuel ratios.

Table 4.11 shows a summary about evidence and deductions.

4.5.1.4 Explosion of the Steam Generator of the Plant Enichem Synthesis at Villadossola

The explosion of the steam generator at the plant of enichem synthesis at villadossola resulted in the deaths of two employees and burns to six others. The accident occurred after a shutdown of approximately 30 minutes during a manual restart. A detailed examination of the terminal traces of process variable recorders resulted in being able to attribute the explosion to extended operation (approximately 4 minutes) of the boiler, in conditions of insufficient combustion air. This lack of air, certainly detectable by the air/fuel ratio differential pressure switch alarm, was ascribed to the erroneous operational behaviour of the stoker who, later questioned, admitted their mistake.

Table 4.12 shows a summary about evidence and deductions.

4.5.1.5 Aluminium Dust Explosion at Nicomax in Verbania

Aluminum dust explosion are very common in the process industry [29, 30, 31, 32, 33], due to the properties of aluminium dust [34]. The aluminium dust explosion at Nicomax in Verbania, described in [29, 30, 31], involved a company that straightened aluminium artefacts in a building measuring 450 m^2 wide and 5.5 m high. Sixteen semi-automatic straightening machines produced aluminium alloy powder, which was captured by a capture system created using 16 collectors connected to one main one. The plant consisted of a cyclone, followed by a bag filter with a system of periodic cleaning using

Table 4.12 Summary of the evidence and deductions

Evidence	Deduction	Compatibility
"Book"-type opening of the sides of the boiler for a significant overpressure generated inside	• I) Explosion inside the boiler	• Total
Correct functioning of the vent doors of the explosion located on the flue discharge pipe at the base of the chimney	• II) Relevant power of the explosion	• Total
The blast occurred after approximately 30 minutes of shutdown, during manual restart	• III) Inside the boiler were high temperature areas able to act as a trigger	• Total
Correct functioning of the components of the power system of the CH_4, including: • Flame sensor; and • control valves (pneumatic) of the fuel flow rate	• IV) Flammable mixture formation not attributable to a failure of components in service	• Total
An examination of the traces of process variables highlights subsequent increments of the air and methane in sequence. After a few repetitions there is an opposite intervention, with an increase first in fuel and then in air	• Va) Switch off due to exceeding of the fuel in the combustion chamber	• Total
	• Vb) Formation of a flammable mixture due to control error	• Total
Presence of alarm due to air and fuel ratio	• VI) No alarm detection	• Total
Absence of automatic protection systems for incorrect air/fuel ratio	• VII) No intervention of protection	• Total

counterflow compressed air. The cyclone and filter were located outside the building. A fan was located downstream of the bag filter. At the time of the explosion, the plant was working under normal operating conditions, and no special operation was in progress.

The explosion caused major structural damage. The dust extractor system was destroyed, with severe damage to the wall structures and the projection of fragments and missiles over a long distance (over 60 m), with the almost complete breaking of the glass in the windows of nearby buildings.

At the time of the explosion, there were ten workers, all inside the building: there were no casualties, three workers suffered minor injuries or burns. The investigation showed that the ignition occurred in the casing of a grinder (Figure 4.25) due to the breaking of the sanding belt. This was wrapped around one of the pulleys, knocking against the casing of the machine. The shocks produced mechanical sparks and caused the re-dispersion of the dust accumulated in the casing. This provided the fuel for an initial modest explosion, whose blaze burned the processing operator and caused the further production of sparks and burning of sanding belt fragments, which were aspirated up into the reduction system (Figure 4.26). These fragments ignited the aluminium powder at the bottom of the cyclone, where the presence of an explosive atmosphere is almost constant: this gave rise to a second explosion; the flame front funnelled towards the filter, igniting the abundant dust and causing the third, violent explosion.

Figure 4.25 Abatement system, detail of exploded fragment.

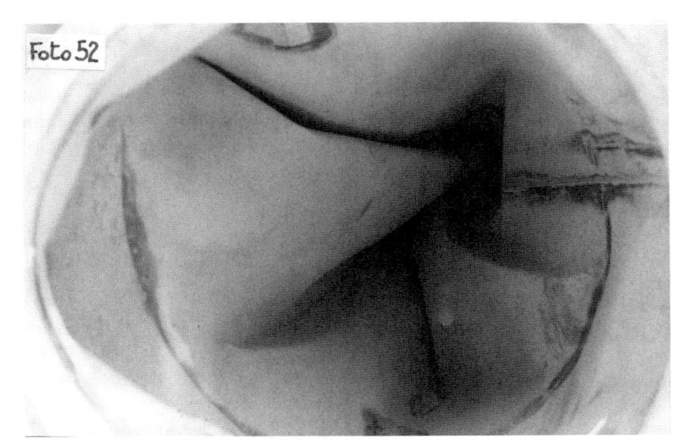

Figure 4.26 Reduction system, detail of the flue discharge pipe inside the cyclone.

Moreover, the increase in pressure generated inside the cyclone during the second explosion spread along the intake pipes, inside the warehouse, disrupting the machine casings and damaging the same ducts (Figure 4.27 and 4.28).

Table 4.13 shows a summary about evidence and deductions.

Figure 4.27 State of places and damage to the abatement system.

Figure 4.28 Remains of the bag filter.

4.6 Organize the Evidence

The available information is like pieces of a puzzle found scattered on the floor. The similarity, also presented in [15], gives an immediate understanding of how important it is to organize the evidence. Evidence is generally grouped according to a shared characteristic, or typology, or simply because of their position in the whole picture, just like we do when arranging the puzzle pieces before and during our attempts to solve it. Indeed,

Table 4.13 Summary of the evidence and deductions.

Evidence	Deduction	Compatibility
The breaking of a belt witnessed by an operator, immediately prior to the sequence of explosions described in 2	• Ia) Cause of mechanical sparks that are an effective ignition source	• Total
	• Ib) Random contemporaneity with the accident	• The assumption in itself is not strictly incompatible but of low probability
After an initial modest "bang", there was a second and then a third massive explosion very close together, the latter of which was the more violent one	• IIa) Explosion in the casing of a grinder	• Total
	• IIb) Separate explosions in the cyclone and in the bag filter	• Total
Damage to the cyclone of the abatement system from internal overpressure	• III) Explosion inside the cyclone	• Total
Destruction of the bag filter with signs of internal overpressure	• IV) Explosion inside the bag filter	• Total
Damage from overpressure at the collectors of the extraction system	• V) Propagation of explosion in the system collectors	• Total
Projection of missiles and fragments of the reduction systems over a long distance	• VI) Significant overpressure inside the reduction systems	• Total
Composition of the melting alloys used in the construction of mainly aluminium-based artefacts	• VII) The dust generated from sanding operations can be explosive.	• Total
Powder morphology	• VIII) The dusts are explosive	• Total

without an initial systematic organization of the evidence, the number of tentative to fit the "puzzle pieces" of our investigation suddenly increases, wasting more time than the required to properly organize data at the beginning.

It is therefore suggested to disregard those evidence that do not belong to the current scene investigation, to work in group to create puzzle pieces' islands that progressively grow and merge each other, to classify evidence for their relevance, belonging to equipment, location of collection, and any other criteria which can help in an efficient organization, depending on the peculiar incident and specific evidence. When all the pieces of the puzzle are fitted, then the image if fully recreated; similarly, when the collected evidence (and their analyses) fit together, then the incident investigation is solved. Even when a piece of the puzzle is still missing, it is still possible to see the entire picture. The same is valid for incident investigation: if one piece of evidence is missing but all the other ones fit together, the investigator can equally enjoy the global view of the incident reconstruction.

Evidence collection can be optimized using a standard data collection form. An electronic database facilitates the data management, leaving free the investigator's attention on the content of the evidence, performing the required analyses, not on their management. Organizing the evidence is also essential to ensure information security and chain of custody. Taking inspiration from [1], an example of Chain of Custody form is shown in Figure 4.29.

It contains information about the initial submittal, the description of the evidence submitted, the records of transfers, information about its disposal, and supplemental information. Every type of records should be retained until the final report is issued. Once the final report is issued, some government agency may require the destruction of any interim documents. Establishing a chain of custody also means that a protocol should be used before removing any evidence, in order to have a standard procedure about how to transfer, analyse, and test physical evidence. Obviously, the extraction of parts of interest needs to be controlled, accurate, and precise, in order to ensure its integrity. Depending on the evidence to be collected and on the surrounding conditions, the removal process may require simple tools (like a simple screwdriver), or more massive solutions (using cranes or forklifts). However, before the removal process, as already noted, the "as-found" condition must be photographed or video-recorded, as well as the removal process itself.

Uncovered findings, as well as the final promoted recommendations, can be recorded in an incident register, for example using an intranet-based database within the company. In this case, it is fundamental to manage the access to the register. On the one hand, it is preferable for this to be shared widely among the employees, to better share the lessons' value. On the other hand, the eventual confidential material should be protected from uncontrolled sharing.

4.7 Conducting the Investigation and the Analysis

The objectives of the investigation are defined in the Terms of Reference, as discussed previously. They mainly establish answers as to what, how, and why the incident happened. To do so, facts and evidence are collected: it is now that the actual investigation is just commencing. In order to conduct and manage the investigation process, [6] suggests some basic requirements to follow. Firstly, the investigation needs to be prepared, with those preliminary activities already discussed that are materialized in the Terms of Reference document. Moreover, the investigation must be conducted scrupulously, performing a comprehensive site overview, and paying attention to volatile evidence. The work process must be efficient and controlled, applying a "stop rule" for evidence gathering, adopting a structured search strategy, and including stakeholders in the investigation activities, like unrepeatable accesses, or evidence collection, or analysis. What is important is to maintain a transparent decision-making process during the investigation, together with high ethical standards, as diffusively discussed. Both known facts and unknown information, requiring further collection or analysis, need to be structured. In addition, as already discussed, in order to conduct the investigation, the investigator should possess some basic competencies. Mainly, he/she must be familiar with a

<table>
<tr><td colspan="6">SECTION A – INITIAL SUBMITTAL</td></tr>
<tr><td colspan="5">Name and Title of Submitter:</td><td>Date Submitted:</td></tr>
<tr><td colspan="5">Company:</td><td></td></tr>
<tr><td colspan="6">Address:</td></tr>
<tr><td colspan="2">Telephone number:</td><td colspan="2">Mobile number:</td><td colspan="2">E-mail address:</td></tr>
<tr><td colspan="6">SECTION B – DESCRIPTION AND LOCATION OF ITEMS SUBMITTED</td></tr>
<tr><td colspan="2">Sampling Site:</td><td colspan="4">Site Address:</td></tr>
<tr><td>Collected By:</td><td colspan="2">Date Collected:</td><td colspan="3">Company:</td></tr>
<tr><td colspan="6">Description of each item, including number of containers, identification number(s) and a physical description.</td></tr>
<tr><td colspan="6"></td></tr>
<tr><td colspan="6"></td></tr>
<tr><td colspan="6"></td></tr>
<tr><td colspan="6"></td></tr>
<tr><td colspan="6"></td></tr>
<tr><td colspan="6"></td></tr>
<tr><td colspan="6">Submitter Comments:</td></tr>
<tr><td colspan="6"></td></tr>
<tr><td colspan="6"></td></tr>
<tr><td colspan="6">SECTION C – TRANSFERS*</td></tr>
<tr><td>Relinquished By (Submitter)</td><td>Organization</td><td>Date/Time</td><td>Received by</td><td>Organization</td><td>Date/Time</td></tr>
<tr><td>1.</td><td></td><td></td><td></td><td></td><td></td></tr>
<tr><td>2.</td><td></td><td></td><td></td><td></td><td></td></tr>
<tr><td>. . .</td><td></td><td></td><td></td><td></td><td></td></tr>
<tr><td colspan="6">SECTION D – DISPOSAL</td></tr>
<tr><td colspan="3">Disposition Site:</td><td colspan="3">Method of Disposition</td></tr>
<tr><td colspan="5">Performed by:</td><td>Date:</td></tr>
<tr><td colspan="5">Witnessed by:</td><td>Date:</td></tr>
<tr><td colspan="6">SECTION E – SUPPLEMENTAL INFORMATION</td></tr>
<tr><td colspan="6"></td></tr>
<tr><td colspan="6"></td></tr>
</table>

Figure 4.29 Sample Chain of custody form. Taken from [1].

broad set of disciplines, being also able to pursue multiple lines of investigation at the same time.

Once data are gathered, the task is now to convert them into useful information [4]. The investigator does not simply identify what was present or absent before the incident. Actually, the analysis is conducted to determine how behaviours, conditions, and the underlying system fragilities contributed to the incident. Performing an analysis means to divide the incident into its individual events, and then look for those conditions that contributed firstly to every single event, and then to the whole incident. To do so, basic assumptions about what caused or contributed to the incident need to be made.

The analysis stage requires the achievement of two major goals:

- To validate what happened and how it happened; and
- to answer why it happened.

The former implies a study to assess the plausibility of the advanced hypotheses, generated on the basis of the evidence collected; the latter requires the identification of the root causes.

At this stage, the intermediate product of the investigation may need a consensus, to reach an acceptable explanation of the event being investigated. Once the most probable scenario is identified, the analysis of root causes can start.

The analysis is the core of the investigation process: it is between the fact-finding phase and the development of recommendations. The evidence analysis is a distinct phase from the evidence gathering, even if they may overlap. The evidence analysis is an iterative process: it may identify the need for additional specific information, restarting the evidence gathering [9]. The analysis gives structure to what the investigator knows and does not. It has not prescriptive rules, relying on informed judgement under uncertainty. The evidence analysis requires performing a cross-check among the different collected data, which must not present incompatibility or temporal inconsistencies.

Investigation tools are provided to follow a structured approach, with the advantageous consequences that come from their adoption (described in the next Chapter). Basically, there are two types of model: the accident model, to structure the sequence of the events, allocating causal factors in the timeline, and the systems model, to link the causal factors to the systems (i.e. the management, the culture, and the context) in which the incident has occurred. In any case, the analytical reasoning must be always based on actual findings, not on opinions, in order to avoid fallacies. The types of causal factors have been already presented; in particular they are the immediate causes, the contributing causes, and the root causes. Cause and effect trees, timelines, and causal factor charts are examples of tools to perform the data analysis. They are discussed in the next Chapter.

There are three analytical approaches to reach the conclusions in an incident investigation: the deductive, inductive, or morphological approach [35]. As better explained in the next Chapter, the deductive approach involves reasoning from the general to the particular while the inductive reasoning is from the particular to the general. The morphological approach, instead, is majorly focused on the factors that influence most the safety: in this approach, the investigator is primarily focused on known hazard sources.

The goal of the data analysis is to identify causal factors first, and the root causes then. Data analysis also concerns about the organization and the judgement of the data collected, formulating a hypothesis about how the incident occurred. Analysing data consists of the following three steps [2]:

- Summarise facts emerged from the gathering activities (separating facts from supposition);
- develop a hypothetic scenario, based on deductive and/or inductive reasoning, to identify the causes of the incident; and
- verify the accuracy and completeness of the developed hypothesis.

When a proposed scenario is selected, the falsification principle should be used to verify how robust it is. The iteration process implies an evolution of the original hypothesis [15]. It can be frustrating for personnel having deadlines to receive information that is likely to evolve, but it is essential that everyone is aware of this intrinsic nature of the investigation process, to avoid jumping to an erroneous conclusion before the investigation is definitely concluded. Once a hypothetic scenario is identified, the evidence that should be observable - if the scenario was real - is specified. They are then compared with the evidence actually collected. Therefore, the evidence is not only the starting point of the study, but also a constant reference throughout all the investigation process, looking for confirmations to the hypothesis that is gradually taking shape during the investigation or alternative suggestions.

The root causes are uncovered through the root causes analysis. The topic is discussed in depth in the next Chapter. One of the major problems encountered with RCA is fixation [1]. People usually tend to view the surrounding world with a personal perspective, based on their own experiences and opinions. In simple words, the root cause analysis may suffer from the prejudice of the investigator. However, fixation is a concept that goes beyond the forensic engineering. One example is the following: imagine you are an engineer in the 90s and you have to design a lighting system for a spacecraft to be used on the Moon. Probably you are immediately oriented to the standard light bulbs, and you "engineer" the solution to install light bulbs. However, in this process you have probably forgotten why light bulbs have bulbs: to protect the tungsten from oxygen. But there is no oxygen on the Moon, so your solution is probably far from being the optimal one. In other words, you suffer from fixation, considering light bulbs as the "obvious" solution to provide lights, on the base of your daily experience.

One of the tools to reconstruct the story is the task analysis [3]. It is a technique used to understand a work activity related to the incident. It generally consists of two different steps: a first phase, known as "table-top" phase, in which the investigator, using written documents, obtains a guideline for the task completion, in order to determine skill and knowledge deficiencies; a second phase, known as "walk-through" phase, in which the investigator observes the task and takes notes about any difficulties in its performance. The task analysis is useful to have a clear understanding of how the task is normally performed, helping the investigator to understand what went wrong.

Logic diagrams are often used during the analysis stage, to help the investigator finding the root causes. They are useful also to point the investigators on what need specific actions, like additional evidence [16]. They are discussed in the next Chapter.

Finding all the root causes is undoubtedly a major challenge [10]: indeed, a common mistake is to stop the investigation before they are all found. It is not unusual to deal with an investigation team trained to find "the" root cause. This approach may lead to a poor investigation, where ineffective recommendations are developed. Indeed, industrial incidents, especially the most severe ones, rarely are ascribable to one single cause; they are often the result of more causes that give their contribution, in different ways, to the occurrence and the development of the event. Each root cause is associated with a risk level (combination of likelihood and magnitude); from this point of view, it makes sense to say that some root causes are more significant than others. But it is not possible to conclude that finding the riskiest one is equivalent to solving the incident investigation: all the root causes need to be uncovered, otherwise it will be difficult to correct those weaknesses in the management system that allowed the incident to occur.

In order to ensure an accurate representation of the scene for the permanent record, the investigator should review all documentation before releasing the scene [12], ensuring that all investigative steps are documented. It is also important to take photos of the post-investigation scene, after the evidence collection process has been completed. The investigator will then release the scene, communicating the known safety and health-related issues to the receiving authority. It is also important to submit reports to the appropriate national databases.

Finally, the investigation methods used to find the root causes are different. As discussed in Chapter 5, when the Bow-Tie method is presented, incident investigation and risk assessment are actually two sides of the same coin. This is why the two disciplines share some methods and it explains why this book also treats elements of risk analysis.

4.7.1 Method of the Conic Spiral

Dealing with a technical investigation has nothing to do with the resolution of a scientific-mathematic problem. Indeed, all the data and the formulas that are useful for the resolution of a scientific problem are available before the resolution process starts. Therefore, the investigative method of the "conic spiral" is the cognitive tool to face such a complex challenge. Its shape recalls the idea of the investigative evolution, from the general to the particular.

As shown in Figure 4.30, the investigation develops along one direction, typically representing the time-coordinate. It is possible to divide the conic spiral into three top stages:

- The initial stage. It corresponds with the evidence collection, consuming the biggest part of the investigation process;

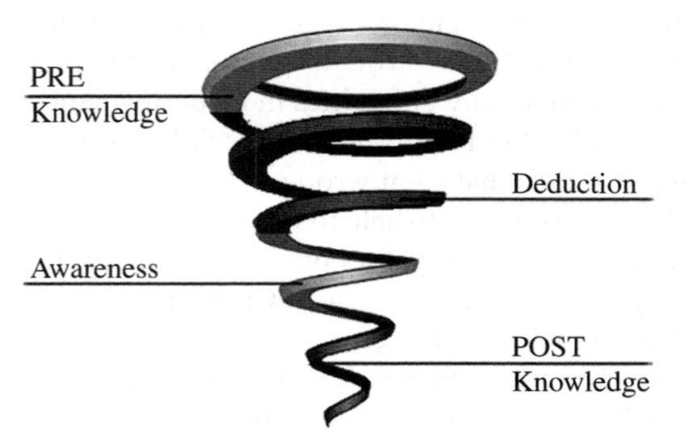

Figure 4.30 Front view of the conic spiral.

- the evidence analysis; and
- the cause analysis and the revealing of the findings.

Even if the analysis is logically subsequent to the collection, this does not imply that the two phases are distinct. It would be incorrect to test the significant evidence only at the end of the collection process. This is why these two steps should be considered as a unique greater step.

While the investigation proceeds, the spiral takes shape. The number of the spirals depends on the time consumed by the single phases: this is related to both the complexity of the incident and the questions formulated by the attorney or the parties that the investigator has to answer. Moreover, a clear separation between two adjacent phases does not exist: indeed, the structure of the spiral may also show some backlinks because the results of the analysis could focus the attention towards some particular aspects that were not identified from the beginning.

In the conic spiral, the different stages are covered following the logic-cognitive path that gradually drives to the centre of the spiral, corresponding to the full comprehension of the dynamics of the incident.

In order to reconstruct the incident, the investigator attempts to find those elements that develop both the *pre*-incident knowledge (for instance the knowledge of the site or the chemical process) and the *post*-incident knowledge (like the plant condition and the caused damages). The major difficulty of an investigation relies on the simultaneous development of these two sides of the knowledge, trying to find all the plausible hypotheses.

The awareness of the actual dynamics of the incident increases moving towards the centre of the spiral, with the evolution of the investigation. However, it is also not a linear path, because of the interactions with the parties. Indeed, the technical investigator (especially if working for the judicial authority) must take into account the observations and the objections promoted by the technical consultant of the other parties, being aware that some of them may be voluntarily misleading and the outcomes of tests are therefore necessary to confirm or question the hypothesized scenario.

If the convergence is not reached, then the investigation goes back to those points of the spiral where alternative hypotheses have been proposed, even if they were initially considered less probable. Therefore, the trajectory of the investigation path is modified, looking for new elements: it often happens that backlinks are used to find additional details.

In the final spirals, the investigator tries to give a structure to the collected evidence, giving support to the actual scenario. Using specific techniques, the investigator finds the causes at the origin of the incident, the conditions that favoured its evolution, the mode of occurrence and, in the end, if required, the responsibilities. At this level of awareness, the centre of the spiral has been reached and the investigation is concluded.

4.7.2 Evidence Analysis

What remains after an industrial accident reveals information about itself through the morphology of a fracture, the colour of a particular spot, or the smut distribution. The evidence analysis provides an objective confirmation or denial about the scenario proposed by the investigation team [10], always based on the scientific principles.

This Paragraph is a simple introduction to some of the techniques used for the evidence analysis. The technical standard NFPA 921 [36] is undoubtedly an important reference to read.

The evidence analysis intends to give voice to a potential direct testimony. It is important to follow the standards, which define the guidelines to obtain certified results, thus having data that can be used in the legal context. In the context of this book, it is not possible to provide clear and unique regulations, on the one hand to maintain a general discussion and on the other hand because some countries do not have standards with these attributes. What is essential for the investigator is to be adherent to the regulations, to define terminology, operative methods, equipment to use, procedures to be followed during a test in a laboratory. For example, the ISO/IEC 17025 defines the "General requirements for the competence of testing and calibration laboratories": having results from a certified laboratory ensures that the proofs cannot be disputed during the trial (obviously, at this stage, having a representative sample is taken for granted)

There are mainly five categories of evidence analysis [37]:

- Dimensional analysis;
- not destructive tests;
- chemical analysis;
- mechanical tests; and
- digital data analysis.

Dimensional analyses are measurements conducted to determine, for example, the length of a fracture, the extension of a corroded part, or the dimension of a thermic trace. The original dimensions of the item generally vary after an incident, because of possible deformations. Technical standards may define the instructions to check the equipment used for dimensional measurements, depending on the specific instrument (such as the calliper, the micrometre, or the comparator).

Chemical analysis is generally conducted to identify and quantify elements, ions, functional groups, composites. The analysis of explosives by mass spectrometry, or infrared spectrometry are laboratory tests, usually performed to investigate explosions. The interested reader can study them in deep in [38].

Non-destructive tests include exams, tests, and surveys conducted with methods that do not alter the material and do not require the destruction of the samples (sometimes it is not even required to collect a sample).

Some of the most significant non-destructive tests are the following:

- Visual exam. The visual exam is always carried out, unlike other tests, and it can be "macro" if it is sufficient an inspection with the naked eye, or "micro" if an instrument is required (like a microscope). This test allows finding several superficial characteristics of the considered material, like defects, fractures and so on;
- leakage test. it is performed to find a leakage in a tank or a similar object, due to defects passing. typically, they are found thanks to analytical surveys or through foaming liquid;
- test with penetrating liquid. these tests are performed to reveal superficial defects on sufficiently smooth and not porous surfaces;
- test with magnetic particles. also known as magnetoscopy, this technique allows to find superficial defects (even just under the surface), on ferromagnetic materials, with a sensibility of 4–5 mm;

- test with current by induction current or eddy's. these tests are used to determine, in both ferromagnetic and not ferromagnetic materials, the superficial discontinuities and sub-superficial ones;
- ultrasounds test; these tests, usually performed on metallic materials, allow revealing superficial and sub-superficial defects, using ultrasounds; and
- acoustic emission test. they are mainly indicated to find defects on fibreglass and other composite materials.

Conversely to the tests previously listed, mechanical tests deeply modify the characteristic of the analyzed element, making it not usable for the future. These tests are performed to support the fracture analysis, to determine the original specifications of the product and to verify if some modifications have been experienced or not. Some of these tests can be simulated: a sample, made of the same material of the particular component, is analyzed, obtaining information that is comparable to that coming from the specific evidence.

It is not possible to recommend a specific protocol for this type of tests, because it depends strictly on the specific case, in order to guarantee comparable, reliable and not contestable results.

Some of the most common mechanical tests are:

- Traction test. During this test, generally carried out at the room temperature, a sample is put in traction by a mechanical or hydraulic test machine. Traction resistance, yielding point, residual plastic deformation after rupture, striction coefficient, and other parameters are evaluated;
- hardness test. hardness is the resistance offered by a material to the penetration of an object. the shape of the penetrator, the applied load and the measurement modalities may be different depending on the specific methodology being used: brinell, vickers or rockwell;
- compression test. the name is self-explicative;
- resilience test. resilience is the capability of the material to resist to impacts. the test consists of breaking in a single shot, through a pendulum (charpy's pendulum) in free fall, a carved sample positioned on two bearings. the result is expressed as the energy (joule) absorbed by the impact to fracture the specimen; and
- bending test. it consists of loading a beam (whichever is the shape of its section) with a concentrated load placed at the centre of two symmetric bearings. through this test, the elastic parameters of the materials are investigated.

4.8 Reporting and Communication

An effective incident investigation takes into account not only the technical components, but also the human ones [1]. Indeed, a good investigator is also able to understand human behaviours and thinking, using effective communication when talking with the wide range of people involved in the investigation process.

For instance, the investigator encourages the front-line technicians to be open, since they often feel guilty when their colleagues have been injured, even if they often do not understand what caused the incident. The technicians' narrative flow should not be interrupted by questions or judgements about what happened. A proper communicative power has to be spent also with mid-level managers. Indeed, the developed recommendations usually concern changes in the facility's management system, but a manager could prefer to spend the managed resources on other goals. A proper communicative way, which empathizes the reasons why the action items need to be implemented, is therefore necessary. Typically, the major obstacle in accepting the investigation findings comes from the senior managers, who are resistant to their implications. A good investigator knows how to communicate with senior managers, who generally have strong personalities, indicating eventual systemic changes, and preventing defensive positions by the interlocutors.

Communication during an incident investigation is not a linear process, mainly because of the complexity of the relations among the several involved actors and its multiple dimensions [6]. In particular, communication is crucial when:

- Notifying the event;
- internally communicating with the investigation team members;
- externally communicating with stakeholders; and
- recommendations are developed and implemented.

Communication facilities should be set up immediately, at the incident location or nearby, to provide emails, telephone, and a meeting room. Anyone in the group must comply with the basic rule to not withhold information. Indeed, information should be shared and discussed also with the authorized stakeholder representatives who are expected to do the same with the investigation team. Daily meetings should be carried out to discuss the information obtained on a daily basis and consequentially plan the activities. Obviously, it is important to be aware of potential biases when collecting information, such as biases coming from witnesses. Models and tools (like infographic, or charts) help to communicate and share information, but the cooperation between investigators is probably the key factor to have a successful investigation.

What is needed is a report that clearly explains what, how, and why the incident happened [15, 39]. According to [1], four stages are required to issue the report:

- Writing it;
- presenting it;
- follow-up; and
- legal issues.

Several formats exist to report the outcomes of the investigation. The simplest one is fully narrative, where events are described in chronological order, like a diary. However, this format works well only with simple investigations: if there are many events to consider, together with large and complex evidence analyses, then the narrative report could be ineffective to communicate the cause-effect links, or the tests results, or the reconstruction of events from witnesses' interviews. As an alternative, the technical report can be structured like an academic paper, with equations, graphs, references, and footnotes. But this type of format could be readable only to experienced professionals, while a big part of people who will ultimately read it, may find it difficult to read.

Therefore, in order to determine the format to use, it is crucial to know who will read the report, who is the "audience". Examples of audience include:

- Claim adjusters. He/she will read the report to determine if the insurance company is jointly liable to pay the claim, according to the insurance policy, or if subrogation potential is suspected;
- law enforcement agencies. their interest is to understand if a crime has been committed. the technical report could be the basis for further investigations specifically aimed in establishing eventual criminal negligence or violation of the law;
- attorneys. both the plaintiff and the defendant party will read the report carefully, word by word. often, they will speculate over a word into a phrase, assigning it a different meaning than what intended by the investigator, trying to obtain a legal advantage;
- technical experts. this category possesses similar skills and knowledge as the writer of the report. the experts of the parties typically search for technical errors or omissions, and try to challenge every single conclusion of the report that could cause negative consequences to their client. sometimes, the review is shamelessly far from any ethics, trying to show that applied methods, scientific principles, or technical standards are – a priori – incorrect;
- the author. the investigator is often asked to testify about the investigation after several years from its conclusion, completely forgetting the contents of the report;
- the judge. if the litigation actually occurs, the judge decides if the report can be accepted as evidence into the trial. considering the limited technical knowledge of the judge, the report should be understandable, avoiding as much as possible equations or statistical data, if not strictly required;
- managers. the report can be used to understand what was wrong in the company business, also suggesting a corrective measure to prevent similar incidents. depending on the specific case, the findings of the report might cause the layoff of a person or they might be the base for solid investments; and
- other professionals. professionals from other industries may be interested in understanding the dynamics of the incident to prevent similar ones and enhance safety in their company.

To satisfy this variegate audience, it is suggested that a format should be used which is consistent with the conclusion pyramid, based on the argumentation style used by the Roman Senate members. This format allows the audience to decide which details level the reading is pushed at, selecting only the interested Paragraphs. The report's sections are the following, taking inspiration from [15, 39], and [4]:

- Report identifiers. this section contains the basic information: title and date of the report, author and client's name and address, and other information like the file number and the date of the incident;
- purpose. in this section, the objectives of the investigation are clarified. typically, a single statement is sufficient to describe the goals of the investigation. all the other sections of the report are written to satisfy the mission statement, otherwise they can be deleted. the conclusion of the report should explicitly recall the mission statement and provide the achieved answer;
- background information. in this section, a brief explanation of the work done by the investigator is presented, avoiding any analysis, conclusions, or opinions. general information about the incident is provided, not adding anything persuasive. facts are presented aseptically, including also the main actors, and a concise timeline of the main events can be provided. recalling [16] and [10], this section explains who, when,

how, and what happened. information can be provided pursuing a narrative approach, so that anyone unfamiliar with the incident can understand what happened;

- findings and observations. this section contains a detailed list of all the findings and observations related to the investigation. it includes the hazardous conditions and the weaknesses uncovered. just like the previous section, facts are described purely, not including any analyses or opinions. facts are arranged properly, from the general to the particular, helping the reader in a full comprehension, from immediate causes to root causes;
- analysis. in this section, the relations among facts are explained by the investigator. each fact is analyzed and explained to the reader. simple calculations can be added inline, while extensive data and bigger calculations are generally moved in the appendix. recalling [16] and [10], this section explains why the incident happened;
- conclusion. in the last section, the conclusion is stated in few sentences, answering the mission statement. the conclusion statement uses the indicative mode: there is no space to equivocation;
- remarks. it is an administrative section, where it is required to take care of something. for instance, the investigator may tell where the collected evidence is now stored, or which safety precautions should be taken when handling them. it also contains the "extent of condition", that is to say some considerations about safety issues related to the inspected equipment, process, or procedures can be extended to a similar context. recommendations are here developed, basing on the analysis of evidence, to prevent similar incidents;
- appendix. detailed calculations or extensive data are written here, to ensure a readable report; and
- attachments. this section includes those relevant items that cannot be inserted into the body of the report, such as photos, laboratory reports, or excerpts of regulations, to not compromise readability.

Finally, usually after the "Conclusion", the report is dated and signed by the main author and the other experts who take part in its writing. Some jurisdictions may require that only a licensed professional engineer can sign the report. For long and complex reports, an executive summary and a table of content should be provided too.

Typically, but not always, the team leader is responsible for writing the reports. Writing a report, it is suggested to:

- Be brief where possible;
- stick to the facts, when presenting data;
- state if a finding is an opinion or factual;
- determine the causes, for each finding;
- pay special attention to careful wording when writing recommendations;
- avoid emotional effects, such as by using superlatives; and
- avoid anticipating or mixing conclusions when presenting the findings.

In a report, whichever is the format being used, the following general information should be provided [3]:

- The anticipated consequences (what was expected);
- the real consequences (what actually happened);
- the potential consequences (what could happen);

- cause and effect relations;
- failed technical elements;
- organizational inappropriate actions; and
- failed barriers.

The importance of reporting is in two thoughts by Hendrick and Benner [40], also recalled in [35], observing how:

- Investigations are remembered by means of their reports; and
- a poor report will waste the best investigation.

Indeed, the objectivity and accuracy, pursued during the investigation phases, must also be transferred in the report [4]. The way findings are shaped will contribute to determining the subsequent corrective actions. Indeed, poorly designed reports might fail in preventing similar incident; usually this happens because causes analysis stops at the immediate causes, ignoring the root causes.

General managers are interested in the executive summary, where general lessons learned are communicated. The organizations that need to plan effective actions to implement the preventing measures are interested in the body of the report. Instead, the attachments are generally read-only from the technical experts and the regulatory bodies having a stake in the incident. Grade cards and scoresheets are used to evaluate the incident report, regarding the resolution criteria and the problem identification.

All employees should be informed as to what happened and injured ones should receive a copy of the final report from the management. The follow-up stage is rarely under the control of the investigation team; however, team members can provide useful information to best manage this step. Key information about recommendations needs to be maintained. They include the incident ID and date, the description of the findings and their owners, the description of the recommendations and their owners, their status, the target date, and the date of the last update.

The technical report should be sensitive to the legal implications deriving from the incident's losses, such as injured employees, victims, business interruption, or environmental damages. Therefore, the potential need to work within attorney-client privilege should be considered and attention should be paid to secure confidential information and notes.

Depending on the severity of the incident, an interim report may be appropriate [10]. This is frequent for process safety incident, whose uncovering may require years. In these eventualities, interim reports are generally provided at the end of each single investigation stage, sometimes facing only a specific subtopic [37]. The interim reports should be flexible, adapting themselves to new information [9].

Generally, a single report is provided at the end of the investigation, but it is not infrequent to write more than one version [37], differing the language (descriptive or technical), the degree of details of the analysis, the completeness (excerpts can be provided to stakeholders for specific purpose, excluding those parts related to patented processes, public security, judicial deeds, and so on).

As noted by [2], it is suggested to start writing the report since the beginning of the investigation. This approach will drive the efforts related to the collection of evidence towards an effective investigation. Having the report reviewed is also important, since grammar errors may call into question the technical accuracy of the investigation.

If more than one possible scenario is identified, any contradictory information needs to be clearly explained. Facts, conclusions, hypothesis, and recommendations should be clearly identifiable in the report as unambiguously distinguished. Further tips are:

- Write the report to address the needs of the audience;
- avoid unneeded information, and use additional information as needed;
- generally, names are not used. it is sufficient to identify a person by his/her position and role;
- manage reports as controlled documents (record it, mark it, include date and revision number, control proprietary data, destroy any drafts before the final report is issued); and
- follow technical writing guidelines (use past tense, avoid jargon, minimize acronyms and abbreviations).

Almost all companies have an internal intranet to share and search different types of documents, including the investigation report. This powerful tool allows the tracking of access activity, identifying who opened and read the report, and helps in the action items management. In this situation, it is also possible to add hyperlinks to the report, adding value to the intranet platform and the digital contents.

In conclusion, the investigation outcomes are shared according to who is the audience. The investigation report is generally too detailed to share the learnings with most interested persons. This is why an investigation summary can be used for a broader dissemination, such as to:

- Communicate to management;
- use in safety or security meetings;
- train new personnel; and
- share lessons learned with sister plants.

References

1 Sutton, I. (2010) *Process Risk and Reliability Management*. Burlington: William Andrew, Inc.

2 ABS Consulting (Vanden Heuvel, L., Lorenzo, D., Jackson, L. et al.) (2008) *Root cause analysis handbook: a guide to efficient and effective incident investigation*. 3e. Brookfield, Conn.: Rothstein Associates Inc.

3 Forck, F. and Noakes-Fry, K. *Cause Analysis Manual*. 1e. Brookfield, Conn.: Rothstein Publishing.

4 OSH Academy (2010) *Effective Accident Investigation*. 1e. Portland: Geigle Communications.

5 Bloch K. (2016) *Rethinking Bhopal. A definitive Guide to Investigating, Preventing and Learning from Industrial Disasters*. 1e. Amsterdam: Elsevier.

6 ESReDA Working Group on Accident Investigation (2009) *Guidelines for Safety Investigations of Accidents*. 1e. European Safety and Reliability and Data Association.

7 Health and Safety Executive (2004) *HSG245: Investigating accidents and incidents: a workbook for employers, unions, safety representatives and safety professionals*. 1e. Health and Safety Executive.

8 Nertney (1987) *Process Operational Readiness and Operational Readiness Follow-on.* Idaho Falls: U.S. Department of Energy.
9 CCPS (Center for Chemical Process Safety) (2003) Guidelines for investigating chemical process incidents. 2e. *New York: American Institute of Chemical Engineers.*
10 Mannan, S. and Lees, F. (2012) Lee's loss prevention in the process industries. 4e. Boston: Butterworth-Heinemann.
11 Craven, A.D. (1982) Fire and explosion investigations on chemical plants and oil refineries, in *Safety and Accident Investigations in Chemical Operations*, 2ed. H.H. Fawcett and W.S. Wood (eds). New York: Wiley.
12 National Institute of Justice (U.S.) (2000) *Technical Working Group for Bombing Scene Investigation.* A guide for explosion and bombing scene investigation. U.S. Dept. of Justice, Office of Justice Programs, National Institute of Justice.
13 Henderson, C. and Lenz, K. (2013) Expert Witness Qualifications and Testimony. *Encyclopedia of Forensic Sciences pp.*459–461.
14 Millen P. (2013) Contamination. *Encyclopedia of Forensic Sciences pp.*337–340.
15 Noon, R. (2009) *Scientific method.* Boca Raton, FL: CRC Press.
16 Mannan, S. (2014) *Lees' process safety essentials.* 1e. Kidlington, Oxford, U.K.: Butterworth-Heinemann.
17 Nichols, R. (2013) Tools. *Encyclopedia of Forensic Sciences. pp.*60-68.
18 Tamiri, T. and Zitrin, S. (2013) Explosives: Analysis. *Encyclopedia of Forensic Sciences. pp.*64–84.
19 Grimwood, K. (2013) Thermal Degradation. *Encyclopedia of Forensic Sciences. pp.*173–176.
20 Stauffer, E. and NicDaéid, N. Interpretation of Fire Debris Analysis. *Encyclopedia of Forensic Sciences.* pp.183–194.
21 Augenti, N. and Chiaia, B. (2011) *Ingegneria forense. Metodologie, protocolli, casi studio.* Palermo: D. Flaccovio.
22 Panzavolta, P. and Marmo, L. (2010) Analisi sulle cause dell'esplosione presso il Molino Cordero di Fossano. *Tecnica Molitoria,* 61:592–606.
23 Marmo, L. and Demichela, M. (2012) "Forensic reconstruction of the explosion that occurred at the Cordero flour mill, Cuneo, Italy. *Chemical Engineering Transactions,* 26, pp. 633–638.
24 Piccinini, N. (2008) Dust explosion in a wool factory: Origin, dynamics and consequences. *Fire Safety Journal,* 43(3):189–204.
25 Marmo, L. (2010) Case study of a nylon fibre explosion: An example of explosion risk in a textile plant. *Journal of Loss Prevention in the Process Industries,* 23(1):106–111.
26 Iarossi, I., Amyotte, P.R., Khan, F.I. et al. (2013) Explosibility of Polyamide and Polyester Fibers. *Journal of Loss Prevention in the Process Industries,* 26 (6): 1627–1633.
27 Amyotte P., Khan F., Boilard S. et al. (2012) "Explosibjxity of nontraditional dusts: Experimental and modeling challenges." In: *23rd Institution of Chemical Engineers Symposium on Hazards 2012, HAZARDS 2012, Southport; United Kingdom, 12 November 2012 through 15 November 2012.* pp. 83–90
28 Marmo, L., Sanchirico, R., Di Benedetto, A. et al. (2018), Study of the explosible properties of textile dusts, *Journal of Loss Prevention in the Process Industries,* 54, pp. 110–122

29 Marmo, L., Cavallero, D., and Debernardi, M. (2004) Aluminium dust explosion risk analysis in metal workings. *Journal of Loss Prevention in the Process Industries*, 2004;17(6):449–465.
30 Debemardi, M., Lembo, F., Marmo, L. et al. (2001) *Esplosioni da polveri nei processi di finitura di manufatti in alluminio e leghe nella realtà produttiva ASL 14 vco: analisi del rischio e misure di prevenzione.* Centro stampa Regione Piemonte. 2001;1–127.
31 Cavallero, D., Debernardi, M., Marmo, L., and Piccinini, N. (2004) Two Aluminium Powder Explosion that Occurred in Superficial Finishing Plants. International Conference on probabilistic safety assessment and management PSAM7.
32 Marmo, L., Piccinini, N., and Danzi E. (2015) Small magnitude explosion of Aluminium powder in an abatement plant: A Telling Case. *Process Safety and Environmental Protection*, 221–230.
33 Marmo, L., Riccio, D., and Danzi, E. (2017) Esplosibility of metallic waste dusts. *Process Safety and Environmental Protection*, 107, pp 69–80
34 Marmo, L. and Danzi, E. (2018) Metal waste dusts from mechanical workings – Explosibility parameters investigation. *Chemical Engineering Transactions*, 67
35 Sklet, S. (2002) *Methods for accident investigation.* 1e. Trondheim: Norwegian University of Science and Technology.
36 NFPA (National Fire Protection Association) (2017) NFPA 921: Guide for Fire and Explosion Investigations. *NFPA (National Fire Protection Association).*
37 Agenzia per la protezione dell'ambiente e per i servizi tecnici (APAT) (2005) *Analisi post-incidentale nelle attività a rischio di incidente rilevante.* 1st ed. Roma: APAT.
38 Beveridge, A. (2012) *Forensic investigation of explosions.* 2e. Boca Raton: CRC Press.
39 Noon, R. (2001) *Forensic engineering investigation.* 1e. Boca Raton, FL: CRC Press.
40 Hendrick, K. and Benner, L. (1987) *Investigating accidents with STEP.* 1e. New York: M. Dekker.
41 Vinardi, F. Computer Forensic: Metodologie di indagine in ambito tecnico-giudiziario.

Further Reading

NFPA (National Fire Protection Association) (1997) NFPA 902: Fire Reporting Field Incident Guide. *NFPA (National Fire Protection Association).*
NFPA (National Fire Protection Association) (1998) NFPA 906: Guide for Fire Incident Field Notes. *NFPA (National Fire Protection Association).*

5

Investigation Methods

5.1 Causes and Causal Mechanism Analysis

Once evidence has been collected and laboratory tests performed, a number of methods have been developed to carry out the analysis stage, in order to give a structure to hypotheses and evidence. Each method has its pros and cons, as will be discussed in this Chapter. Following the modern approach to incident investigation (see Chapter 1), the methods leading to discovering root causes are discussed next, providing only a hint about those historical methods that were ineffective in reaching such a deep level of knowledge. The available investigation tools, sorted by increasing structure, are [1]:

- Informal, One-on-One. It is the traditional informal interview, usually performed by the front-line supervisor. It is the poorest tool;
- brainstorming. The experience and judgment of the team are used to find credible causes;
- timeline. It is a chronological list of events;
- sequence diagram. The chronological data are depicted in a graph, allowing to show parallel events and conditions;
- causal Factor Identification. These tools (like the Barrier Analysis) are used to identify those negative conditions, actions, or events that contributed to the incident;
- checklists. Causal factors are reviewed against investigative checklist to determine why that factor actually existed;
- pre-defined trees. Ready-made trees are used to identify possible causal factors, discarding those branches that are not relevant to the investigated incident; and
- logic Trees. They use a multiple-cause, system-oriented approach to reveal the root causes affecting the PSM. They are the most structured tools.

Talking about the causes and causal mechanism analysis, one of the most known tools is the cause and effect diagram, created by Prof. Ishikawa in the 1960s. In this technique, also known as "Fishbone diagram" (because it looks like a fish skeleton), all the possible causes of a problem are discussed, using a diagram-based approach. It requires four major steps to follow [2]:

- Identify the problem;
- work out the major factors involved;
- identify possible causes; and
- analyze the diagram.

Principles of Forensic Engineering Applied to Industrial Accidents, First Edition.
Luca Fiorentini and Luca Marmo.

The Ishikawa diagram provides a list of possible causes to identify the real root causes of the incident. In this way, thanks to the one-shot identification, the investigator is helped in a better understanding, without recursively solving single smaller parts of the incidents. The possible causes are collected through brainstorming and they are graphically identified on the graph, reflecting a sort of mind map. The method leaves the investigator thinking more thoroughly about the root causes, thus leading a robust solution. Indeed, the investigator is driven in considering all the possible causes, not only the most obvious one. Causes are categorised to immediately identify the correct source.

The first step in drawing a Fishbone diagram is to identify the problem. A rectangle is drawn on the right side of the sheet, and the problem is written inside the box. A short brainstorming session may be required to define the scope. A straight arrow from left to right is the "spine" of the fish (Figure 5.1).

The second step is to identify and categorise causes. The method can be applied in any industrial sector; the categorization of the causes will depend on the specific context. For example, in the manufacturing industry, the "5M" categorization, promoted by Toyota, is usually adopted:

- Machine (technology);
- method (process);
- material (including raw material, consumables, and information);
- manpower/Mindpower (physical work and brain work); and
- measurement (data generated from the process, inspection).

The following "Ms" can be added to the previous list:

- Milieu (Mother Nature/Environment);
- management; and
- maintenance.

Different categorizations have also been proposed for the marketing industry, with the 7Ps: Product, Price, Place, Promotion, People, Positioning, Packaging. In the service industry, factors are categorised by 5Ss: Surroundings, Suppliers, Systems, Skills, and Safety. For each possible factor, a line on the fish spine is drawn as shown in Figure 5.2.

The third step is to brainstorm the possible causes. For each category, the investigator asks himself why the incident happened, and the question is posed recursively. The possible causes identified are written as shown in Figure 5.3. It is possible to continue

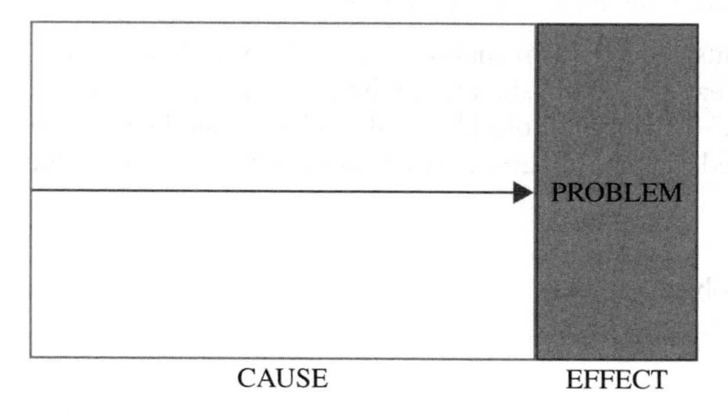

Figure 5.1 Fishbone diagram. Step 1: Identify the problem.

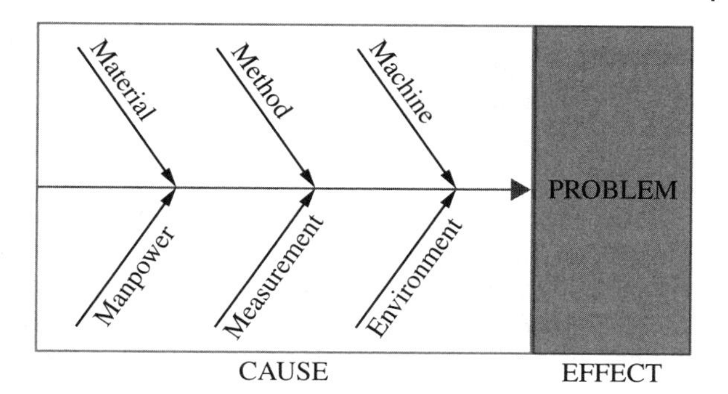

Figure 5.2 Fishbone diagram. Step 2: categorise the causes.

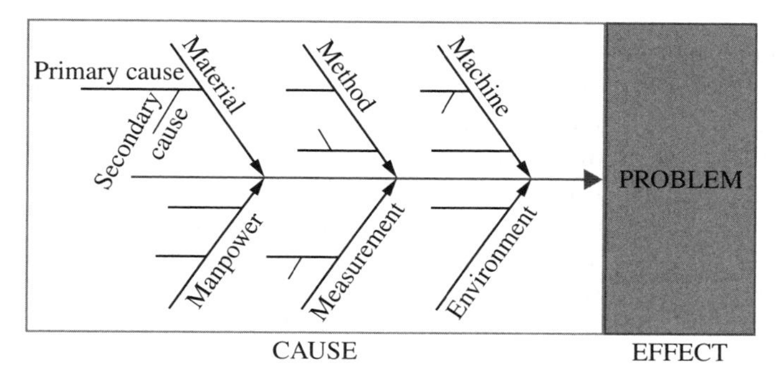

Figure 5.3 Fishbone diagram. Step 3: identify possible causes.

adding sub-branches, until a satisfactory result is reached. If too many causes are identified, it is possible to split the diagram into two or more parts, in order to guarantee readability.

Finally, the last step requires analyzing the resultant diagram. The investigator can see all the possible causes of the incident and can discuss them with the team members, investigating further to identify the root causes of the problem.

While developing a Fishbone diagram, it is important to identify the problem clearly, and to involve team members having experience with the identified problem. Some of the advantages of this technique are:

- It is very easy to understand, being a visual tool;
- it helps to identify the root causes and the bottlenecks;
- it prioritises further analysis; and
- it helps to take corrective action.

However, there are also some limitations:

- All causes look equally important: the investigation team members have to identify the root causes;
- the identification of less relevant causes may be wasting;
- it is more based on opinion rather than evidence; this is why it is necessary to have experienced members, and causes should be selected pursuing a "democratic" approach; and
- the discussion, if not properly managed, may deviate from its objective.

An effective and guided example of how to develop cause-and-effect diagrams is in [3]. The basic idea to construct them is very similar to the one governing the construction of a Fishbone diagram: identify the problem, ask why, identify the first-level causes, ask why, identify the second-level causes, ask why, and so on, until the root causes are revealed.

Basically, a cause-and-effect diagram is similar to an FMEA diagram but is constructed in reverse. The FMEA is used to know the ways in which a particular item may fail, also considering the consequence of each failure mode and its contributing factors and causes. FMEA is therefore used to anticipate the possible ways in which a product, component, or process may fail, in order to develop the corrective actions to prevent those failures. Usually, an example of reverse FMEA is provided in vendor manuals, as a troubleshooting guide.

A cause-and-effect diagram can be combined with a timeline into a single graph, known as "event and causal factor diagram". The events are linearly put on a sequence like a timeline (see next Paragraph); in addition, causal factors and conditions are connected to the events, in order to show the cause-effect links. By convention, validated events are represented by rectangles with a solid line, while hypothesised events, not yet validated, are in dotted lines. Pre-existing conditions, to be validated, are drawn with dotted curves, while a significant event, like a failure or another bad event, is drawn with a diamond shape. Ellipses represent causal factors; root causes are denoted by a vertical line on the right side of the ellipse. The final event, i.e. the incident, is shown by a circle, while brackets denote the time intervals. An example is shown in Figure 5.4, adapted from [3].

Events and causal factors charting are particularly useful in identifying multiple causes [4]. Some advantages of their usage are:

- They show the links among the immediate causes and the less apparent conditions;
- they identify information gaps, thus driving the gathering of evidence;
- they consider the possibility of multiple causes;

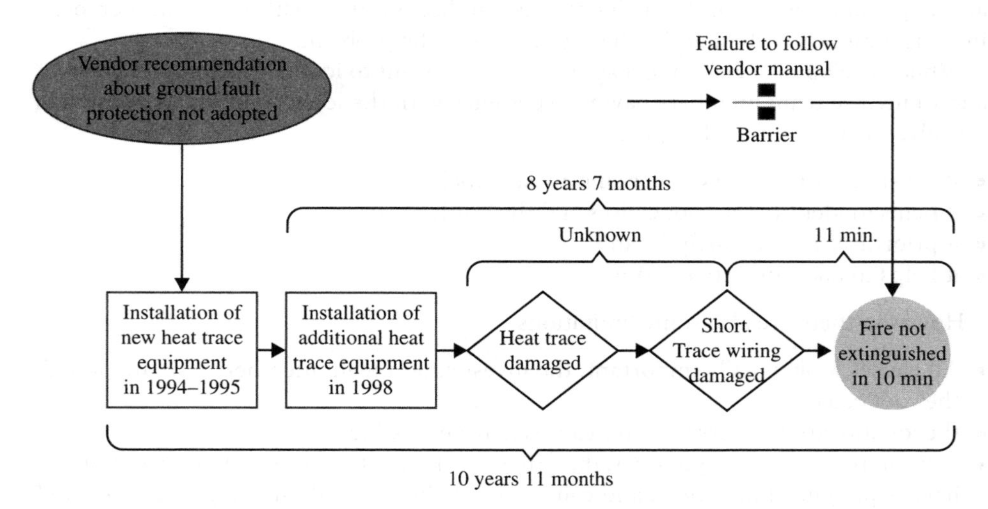

Figure 5.4 Example of event and causal factor diagram. Source: Adapted from [3]. Reproduced with permission.

- they clearly present the information about the incident, and can be used as a guide to writing the report; and
- they provide an immediate and effective visual feedback summarizing the key information.

The causal mechanism analysis can be also carried out through domino theories [5]. The first domino theory of accidents was developed by Heinrich in 1931 [6]; however, many others were developed later. The basic idea is that a first event starts the accident sequence, pushing on adjacent dominos and eventually causing the last domino to fall, that represent the accident. In this model, Heinrich identified five types of action (Figure 5.5):

- Ancestry and social environment;
- fault/person;
- unsafe act;
- unsafe condition; and
- injury.

Some examples of unsafe acts and unsafe condition are listed in Table 5.1, to immediately show their conceptual differences: acts are human-generated events, conditions are pre-existing factors.

An evolution of this domino theory has been developed by Bird and Germain [7], who promoted the Loss Causation Model. This model is based on different five dominos (Figure 5.6):

- Lack of control. It includes the failure to comply with standards;
- basic causes. They are the personal and job factors that trigger the accident sequence;
- immediate causes;
- incident; and
- loss.

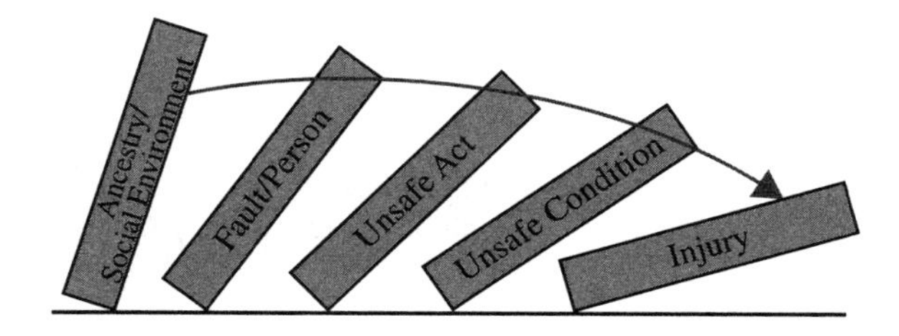

Figure 5.5 Domino theory by Heinrich (1931) [6].

Table 5.1 Examples of unsafe acts and conditions.

Unsafe acts	Unsafe conditions
Improper loading	Inadequate barriers
Improper lifting	Defective tools
Failure to secure	Fire and explosion hazards
Removing safety devices	Inadequate ventilation
Operating without authority	Inadequate protective equipment

Source: Adapted from [5].

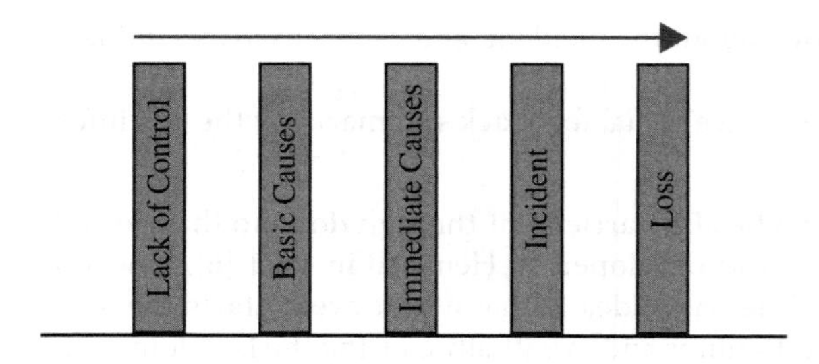

Figure 5.6 Loss Causation Model by Bird [7].

In this method, the meaning of basic causes and immediate causes is still under debate [5]. This is the reason why it is not further discussed in this introductory book.

Thanks to the domino representation, it is visually clear how dealing with the immediate cause (domino B, in Figure 5.7) will only prevent its occurrence while dealing with root causes (domino A, in Figure 5.7) can prevent the entire sequence of adverse events.

The process leading to determine the causal factors of an accident, starting from the discussed diagrams, is shown in Figure 5.8. This process is a preliminary step to determine the root causes of the accident. A deductive reasoning is required to reconstruct the events that lead to the accident.

Obviously, the investigator must ensure that adequate details are contained in the diagrams he/she has developed. Then he/she looks to the first event preceding the incident

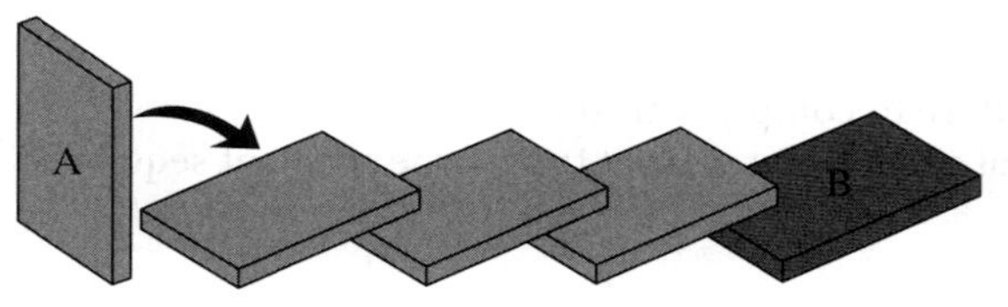

Figure 5.7 Sequence of dominos. Source: Adapted from [8]. Reproduced with permission.

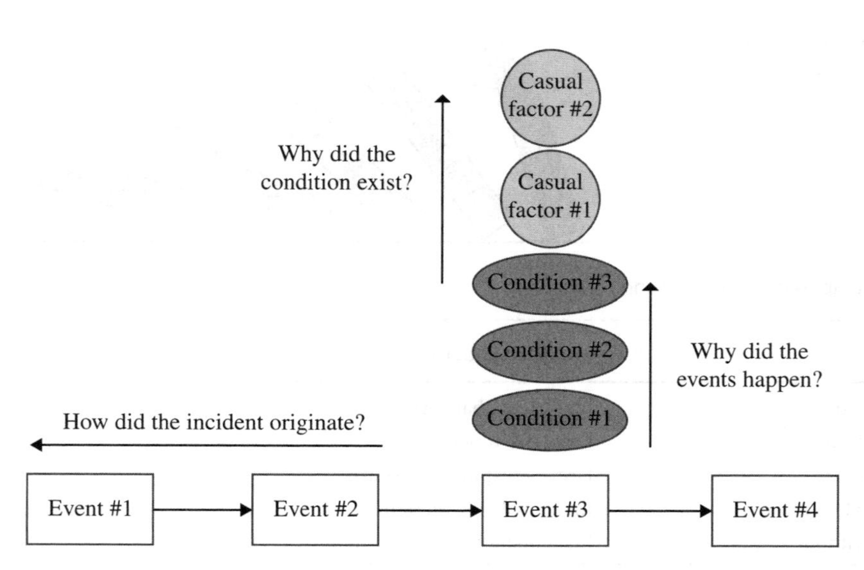

Figure 5.8 Events and causal factors analysis. Source: Adapted from [40]. Reproduced with permission.

and asks if the hypothetic removal of that event would avoid the accident to occur. If the answer is no, then the process continues to next event, going backwards from the incident. If the answer is yes, it is crucial to distinguish if that event represents a normal activity having expected outcomes or not. Indeed, if the event produced an expected outcome, then it is not significant, otherwise it is a significant event, deserving further root cause analysis.

It is interesting to compare the different nature of causality in human and technical systems [9]. These two types of causalities are shown in Figure 5.9. Human systems require a systemic approach, are hardly quantitative, non-deterministic, dynamic and circular (they do not have a linear causality); on the contrary, the technical systems enjoy a linear causality, are deterministic, quantitative, and they require an analytical approach be understood.

Indeed, it is difficult to predict behaviours in human and social systems, since feedback loops affect them, resulting in complexity and unpredictability. A specific Paragraph of this book is dedicated to the human factor.

It has been assumed that causes for effects can be always found [10]. As already discussed in Chapter 1, this is absolutely true in the Newtonian vision of the world, where cause and effect are seen in symmetry: they are both definitive, equal but opposite. This is generally true, when the incident analysis is restricted to the physical and materialistic world, where everything can be seen in the three-dimensional Euclidean space. The challenge is to apply the derived scientific method also for those systems that have not geometrical dimensions, like the human interactions and the social environment. Complexity arises for these reasons, pushing the incident investigation towards new objectives, which have been already discussed: going beyond the widget!

The representation with logic trees is among the most diffused, because of its high capability to incorporate useful information and to share them in a structured way [11]. They start with a known event, also known as top event, being the event under investigation. The causes at the origin of the top event are investigated, using AND and OR logic combinations (Figure 5.10). The symbols used in logic trees representing these combinations are called "gates". Above the gate is the effect, below the gate are the causes. Obviously, a single event can be seen as the effect of the causes below and, at the same time, as the cause of the effect above: in other words, ramification is possible, and the tree can develop on multiple levels (Figure 5.11).

In risk assessment, logic trees are developed to identify potential contributing factors that may lead to an incident; instead, in incident investigation, if a branch seems to be

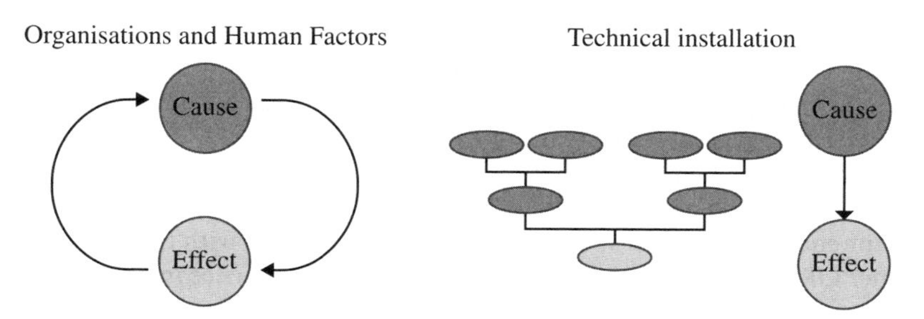

Figure 5.9 The different nature of human and technical systems. Source: Adapted from [9]. Reproduced with permission.

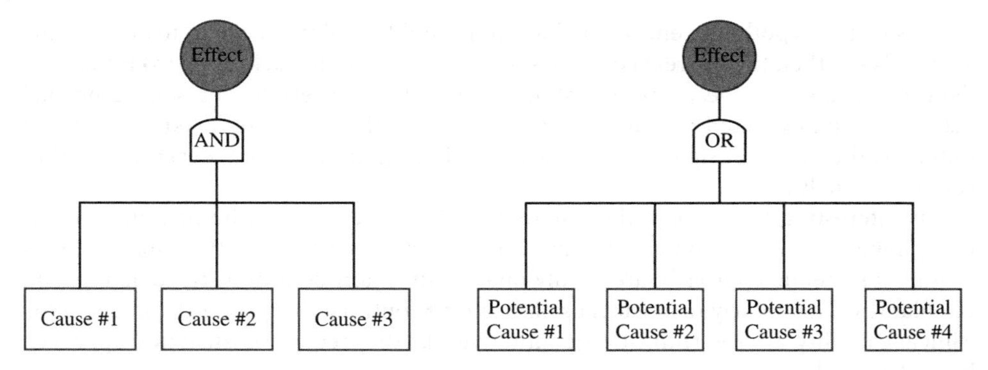

Figure 5.10 AND and OR combinations in logic trees.

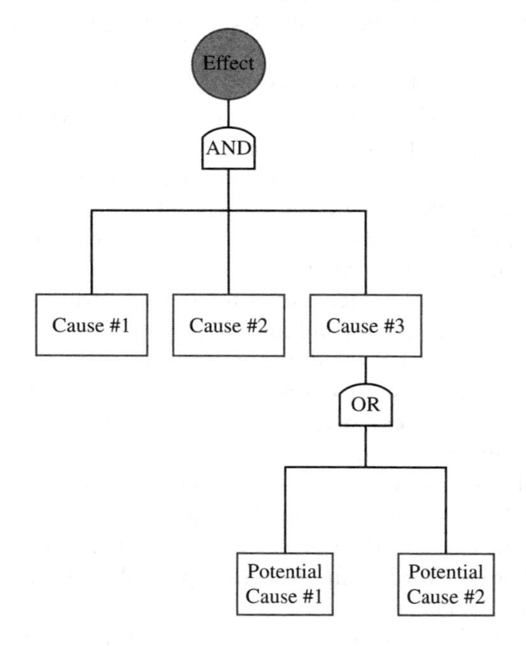

Figure 5.11 Multiple levels logic tree.

not credible, then it is not further developed, avoiding useless wasting of time. It has been already stressed out how a single event may have multiple causes at its origin. The fire triangle is a good example: in this case, one event requires three causes. Logic trees are therefore a powerful tool to investigate multi-causes events.

The "AND" gates are used when a combination of causes is required to have the effect above the gate. The AND combination may refer to multiple elements, or multiple pathways, or redundant equipment failures, or an initial event combined with a failed safeguard. Instead, the "OR" gates are used when one or more possible causes may generate the effect above the gate. The OR combination may refer to the failure of one or more multiple elements, or component failures, or inadvertent activation of safeguards.

The procedure to create a logic tree is the following (Figure 5.12):

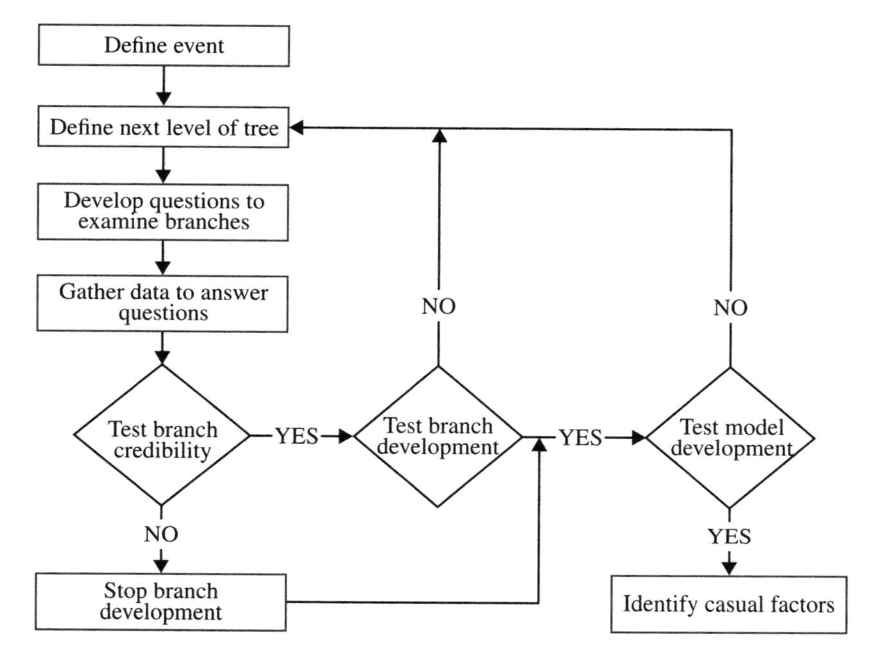

Figure 5.12 Procedure to create a logic tree.

- Define the top event. This definition will define the scope of the investigation;
- define the next level of the tree, using AND/OR gates. This step should be small enough to ensure all the possibilities are explored; indeed, a bigger logical step may overlook important information. The AND/OR gate logic should be tested;
- develop questions to assess the credibility of branches. The investigator asks what data can prove or disprove the constructed branches, highlighting eventual missing data or inconsistencies with gathering evidence, or the agreement between the hypothesised branch and the collected proofs;
- gather data to answer questions;
- determine if the branch is credible. If yes, go to the next step; otherwise, stop the branch development.
- determine if the branch is sufficiently developed. If yes, go to the next step; otherwise, define a further next level of the tree;
- determine if the tree is sufficiently developed. If yes, go to the next step; otherwise, define a further next level of the tree; and
- identify causal factors.

In an incident investigation, it is crucial to establish the right time sequence of an incident to reconstruct the real dynamics in a proper chronology. To complete the discussion of the present Paragraph, it is important to distinguish between the concepts of coincidence, correlation, and causation [3]. It can happen that two events occur closely in time: it becomes fundamental to establish if their chronology also reveals a cause-and-effect relationship or not. In other words, if event A occurred just before event B, can we conclude that A caused B? The answer is no, since an apparently ordered time sequence (i.e. a coincidence) does not automatically involve a cause-and-effect evidence. For instance, if you eat pizza the day before you sit an exam and pass it, this does not mean that the success you have is related to the pizza you ate. Coincidence is a random effect involving independent events. In the example, eating a pizza and passing an exam are events that occur independently. Very often, because of the improbability of the coincidence, people are prone to think that a cause-and-effect relationship must be present, because, according to them, the two events happening in the sequence are too much improbable to manifest by coincidence. This argumentation is only good for sophistries, not for forensic engineers. Even according to the law of large numbers, coincidence may exist: extremely low probable events can occur if we consider a large number of possibilities to occur (i.e. if the set of events, whichever is their final state, is big enough). If something is not just a coincidence, then a causal link must be found between the two events: in other words, it must be demonstrated that the first event triggers, encourages, sets up, makes the occurrence of the second event.

Therefore, the second concept needs to be introduced: correlation. A correlation exists when two events are linked with a demonstrable relationship. It implies repeatability of the chronological order and provides a useful tool to test the time and event sequence. Correlation is the first step to indicate the existence of a direct cause-and-effect link between two events. But it may be not sufficient. Indeed, correlation exists also when a common factor is shared between two events, regardless of a direct relationship between them. For example, the increase in the number of car accidents is regularly followed by a similar trend in the collapse of agricultural outbuildings. Obviously, this does not mean that the first event causes the second one. A correlation exists only because they share a common factor, which is the snow: snow makes the road wetter and slippery, increasing the car accidents, and increasing the structural loads on the agricultural outbuildings, which are not generally designed to resist in extreme conditions.

It is clear that coincidence and correlation should not be confused with causation, which is what an investigator looks for. The Latin *"post hoc ergo propter hoc"* fallacy synthesises this wrong approach to causation. Indeed, as discussed above, it is not true that if B comes after A, then A causes B. Once this concept is clear, we can proceed about how to organise data and evidence in timelines.

5.2 Time and Events Sequence

Cause and effect analysis is a very good technique to investigate the causal links that lead to an incident. However, it has one evident drawback: it does not provide information about the relative timing of the events.

A timeline is one of the most effective tools to organise and catalogue data. It is a chronological visual arrangement of the main events, data, and evidence associated with the incident being investigated [3]. It helps the team to see the events in chronological order and it is very useful not only for being an investigative tool, but also for its capability in graphically displaying the relationships among the different facts and the final incident. In addition to the events, which are active items like "pipe failed" or "the pump started up", a timeline may also include conditions. Conditions are passive items, like "the pump was running" or "the pipe was corroded" and represent pre-existing elements in the context of the incident (they are presented using the words "was" or "were"). Obviously, the timeline can also include failures and omissions, if relevant to the incident [1].

The development of a timeline covers the entire investigation, as data and information can be added throughout the investigation process to fill the gaps and solve the eventual inconsistencies.

In its simpler version, a timeline is a list of the events in columns, where it is easy to understand which came first, second, and so on. A tabular format is suggested to implement this type of timeline, as shown in Table 5.2.

More complicated formats of timelines are also available, to provide a higher number of information. For example, the Gantt chart format can be used. It enriches the previously discussed version providing the duration of every single event, which is therefore correlated in a general view together with all the events that occurred during the incident (note that a Gantt chart can be also used to plan the investigation activities). An example of the arrangement is shown in Table 5.3. Obviously, the suggested spreadsheets can be enriched with additional columns to specify additional information about the actors, the people or the equipment involved, and so on.

A very communicative way to arrange data in a timeline is shown in Figure 5.13. The example shows the first part of the timeline developed for the Norman Atlantic investigation, also discussed in Chapter 7.

Another way to arrange events in a time sequence consists in using a blackboard to draw the essential timeline and Post-it® notes to add information about the events. This alternative has the advantage of being extremely flexible (notes can be moved when needed) and has a considerable communicative impact. It is suggested that different colours of Post-it® notes are used for different types of data. Especially in the first stage of the timeline reconstruction, using software with a pre-defined approach could be limiting. It is preferable to use a simple and flexible format. Moreover, the level of detail should be maintained in a manageable way, avoiding adding everything that is known to the timeline. During the construction of the timeline, the timing of events and conditions may have different accuracy. For instance, data from BPCS have the accuracy of

Table 5.2 Example of spreadsheet event timeline.

Time	Remarks
03:24:02	Description of Event #1
03:24:09	Description of Event #2
03:24:44	Description of Event #3
03:25:58	Description of Event #4
03:26:01	Description of Event #5

Table 5.3 Example of Gantt chart investigation timeline.

		Time and duration				
ID	What	Start	Finish	03:00:00	03:30:00	04:00:00
1	Event #1	03:00:02	03:30:07			
2	Event #2	03:00:09	04:00:45			
3	Event #3	03:29:44	04:01:29			

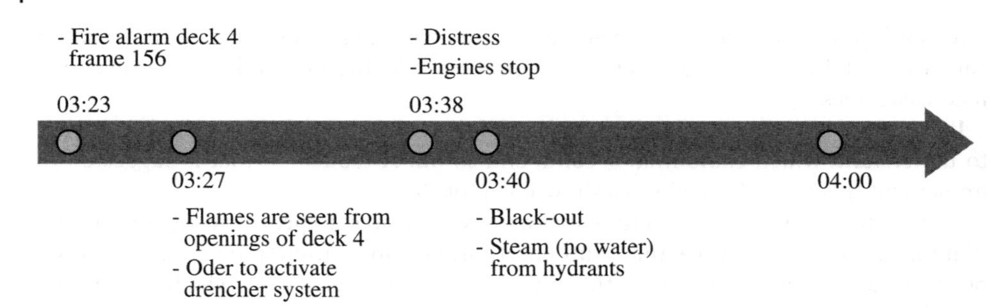

Figure 5.13 Example of timeline developed for the Norman Atlantic investigation (see Paragraph 7.2 for details).

the tenth of a second while a field operator's observation is far from being so precise and can be very approximate (e.g. "more or less at noon"). This leads to a timeline where data are used in combination with both precise and imprecise timing. Using this combination is a challenge for the investigator, since data need to be put in a chronological order at the end. However, a clear advantage of using this combination of data is that imprecise data can be detailed if coupled with more precise data. For instance, if the operator realised that during its intervention to close valve A, valve B was already automatically closed, then we can conclude that its intervention to close valve A was in a narrow window of time, more precise than the approximated value referred by the operator. Timelines combined with computer simulations are powerful tools to analyse the sequence of events and accurately recreate the dynamics of the incident.

When detailing a single event, i.e. a single building block of the timeline, it is suggested that the following four rules should be adhered to:

- Use complete sentences, avoiding fragmented information;
- use only one idea per building block (concatenate phrase should not be adopted);
- be as specific as possible (avoid qualitative terms and prefer quantitative assessment); and
- document the source for each event and condition, to assess the validity of the data.

For some complex incidents, it is suggested that parallel timelines are used showing the events sequences differentiated by location, actors, input or output variables, and so on. In any case, it is always recommended to put in relation the two or more timelines into a single, unitary chronological viewpoint of the incident. The timelines are important tools to assess the potential suspects in case of sabotage or malicious deliberative acts (which are not treated in this book, as has been clarified from the introduction). To have clear evidence about how a person acts, the timeline can be supported by a plot showing the movements: the combination of the temporal and spatial information may be extremely precious for those complex incidents where it is fundamental to know the exact position of a person within a certain time interval. Following what suggested by [11], timelines are usually constructed following this path:

- Identify the loss event. It needs to be defined specifically, according to what the investigation wants to focus on. If multiple loss events are selected, multiple timelines need to be created;

- identify the key actors (like people, equipment, parameters);
- develop building blocks for each actor, event, and condition. Then, add them to the timeline;
- generate questions and identify data sources to fill in the eventual gaps;
- gather data, according to what emerged from the previous step;
- add additional building blocks to the timeline, according to the results of the previous step;
- determine if the sequence of events is complete; and
- identify the causal factors.

It is self-evident that some steps of the timeline construction have points in common with the causal mechanisms analysis.

When developing a timeline, the investigator should always keep in mind what has already been stated about the time reversibility and irreversibility [10]. We briefly repeat here that according to the Newtonian standpoint, the trajectory of the events can be drawn towards both the future and the past. This is an assumption that is based on the idea that the only limit in the reconstruction of an incident is the effectiveness of the method used, since the knowledge is always fully available. But the most recent approaches, taking inspiration from the complex theory, claim that the precise set of conditions that characterise a complex system (like an accident is) cannot be exhaustively known. This happens because of the continuous changes and evolutions that affect the system and its relationships, following the adaptive nature of complexity. This second approach implies the loss of any effective predictive measures, since knowledge cannot be fully possessed. This is why an investigator uses both deductive and inductive methods, because they help him/her in moving in the time sequence of the events, looking for the causal link between them. Moreover, the simple selection of events to be arranged in a timeline is a transformation of the real story: it is not a description of what happened but a description of the most important events that happened. It is therefore crucial to be experienced in creating a timeline, since there is not a priori objective method to establish what is important and what can be neglected from the incident timeline sequence.

For investigation purposes, it is relevant to determine the conditions at the time of failure. This activity is in the middle between the evidence collection and the root cause analysis, and can be faced thanks to a timeline. The searched condition can be short-term (i.e. immediately before the failure) or long-term (i.e. an existing latent condition), depending on the peculiar incident. Knowing those conditions and having evidence correspondence is a key-part to validate failure hypothesis and the entire back-in-time reconstruction.

Timelines always comprise two sections: the events prior to the incident, and the incident itself. Sometimes, depending they may focus also on the events after the incident [12].

In conclusion, the timeline tool takes into account all the information required to start properly the investigation, also suggesting where it is necessary to focus the efforts. Its creation is one of the very first steps, to have a single manageable record of events and an introductory tool to the causal analysis and the root cause determination.

5.2.1 STEP Method

The Sequentially Timed Events Plotting (STEP) method is one of the most known formats belonging to those methodologies based on a time sequence. It can be used for both less and more severe incidents. The method, developed by Hendrich and Benner in 1987 [13], analyses the accident events using a systematic process view approach based on multi-linear events sequences [4]. It relies on the following assumptions:

- The accident (as well as its investigation) is not a single linear chain of events; instead, multiple activities take place simultaneously;
- the accident description is developed in a worksheet, using the "Building Block" format for data, i.e. a format to describe one action carried out by an actor (i.e. an event);
- the chain of events follows the rule of logic. Their flow follows the arrows; and
- productive processes and accident processes are similar (they both involve actors and actions), i.e. they are understood using similar procedures. An accident process starts with the transformation of the productive process into accident process and ends with the last event of the accident process itself.

The building blocks are organised in a worksheet, allowing a multi-linear arrangement and a full comprehension of the whole accident process. The STEP worksheet is a matrix. The actors are placed in rows, while the columns represent different time intervals, on a timeline. It is not important to have a linear time scale: what is important is to arrange the events in order, i.e. to correlate each other (which is before, which is after) in the timescale. An example of STEP-worksheet is in Figure 5.14.

An actor is a person or an item whose action affects the flow of events. There are two types of changes that an actor can generate: adaptive or initiating changes. Adaptive changes describe those actions that tend to re-establish the dynamic balance, disturbed by a certain factor. Initiating changes, instead, are those changes to which other actors must adapt. The action carried out by an actor (thus defining an event) is generally tangible, i.e. physical. However, if the actor is a person, an action can be also mental, not observable. Every action must be stated in the active voice, to clarify with no doubt who is its actor.

An example of a STEP-worksheet filled in with events (i.e. the building block) is shown in Figure 5.15, where a fatal car incident is described (the example is taken from [4]).

Figure 5.14 STEP-worksheet. Source: Adapted from [4]. Reproduced with permission.

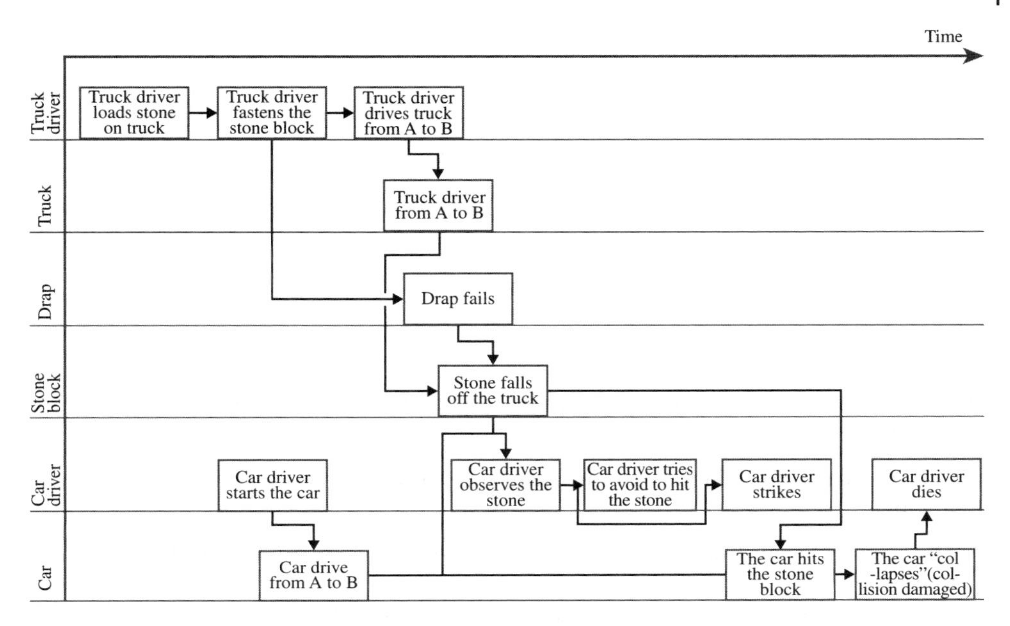

Figure 5.15 An example of STEP-diagram for a car accident. Source: Adapted from [4]. Reproduced with permission.

The example in Figure 5.15 shows the proper use of arrows to link events: an arrow connects the past event to the resultant event. Developing those links is one the most crucial parts in using the STEP method. Indeed, for each event, the investigator has to verify if all the identified preceding actions are sufficient to cause the result event or if other measures are necessary. It is a mental effort that requires practice to develop a certain confidence to answer the previous questions: it requires, for the investigator, to development of the capability to mentally visualise the actions and the actors, trying to reconstruct the "movie", and so the links.

When developing the STEP-diagram, gaps may arise in the reconstruction of the accident process. Indeed, it may happen that the investigator has not enough evidence to establish a connection between two or more events. In such a case, the BackSTEP technique is often used. Moving on the contrary respect to the traditional STEP method, the BackSTEP analysis starts from the right-side of the diagram and intends to reconstruct the possible links between a resultant event and its most probable preceding event(s). Following this strategy, more than one link is often found between the left and the right side of the STEP diagram. This suggests where the investigation should go in deep, to establish which of the proposed paths is/are the real one(s), overcoming this way the gaps previously found.

The STEP method also includes some rigorous testing to verify the structure of the resulting accident process. They are the row test, the column test and the necessary and sufficient test.

The row test (or horizontal test) is used to verify if additional building blocks are required for each actor. Instead, the column test (or vertical test) is used to check the sequence of events. Figure 5.16 shows the row and column tests.

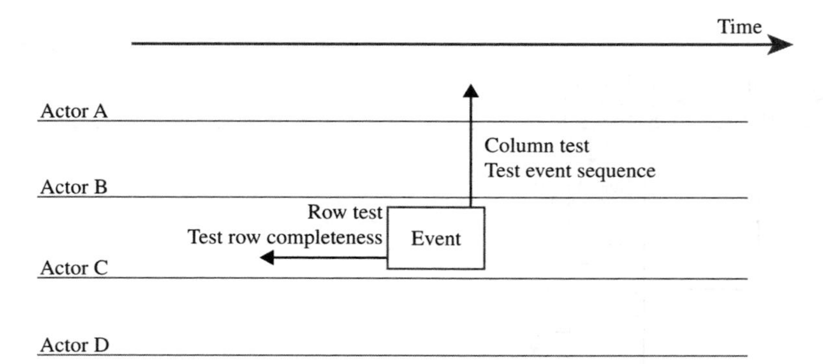

Figure 5.16 Row and column tests for STEP method. Source: Adapted from [4]. Reproduced with permission.

In particular, the column test prescribes to verify, for each building block, the following:

- The selected event must have occurred after all the events at the left of the selected one;
- the chosen event must have taken place before all the facts at the right of the selected one; and
- the chosen event must have occurred at the same time of all the events in the same column.

When creating the STEP diagram and placing the events in sequence, the investigator may ask if a certain earlier event is sufficient by itself to determine the later event or if other actions are also necessary. The necessary-and-sufficient test is applied to verify these uncertain situations.Obviously, if the earlier action is sufficient, then it is not required to collect other data and evidence: indeed this is the case of a necessary-only action, where further analysis is needed to provide additional information in order to reconstruct the whole process.

The STEP method also takes into account the development of recommendations about safety problems, being a useful tool to identify possible safety problems. The investigator analyzes a building block per time, and a single arrow per time: this approach allows him to identify the safety problems related to the incident process. The individualised warrant safety actions are then converted to recommendations for corrective action, which are marked properly in the STEP worksheet as shown in Figure 5.17. The development of safety recommendations is not inside the scope of this Paragraph: for further details, see Paragraph 6.2

When developing a STEP diagram, you could be pushed in finding "the cause", i.e. select a single event and label it as the only cause of the accident. This could result in poor attention towards other crucial causes that trigger the incidents. This is why we talked about "multi-linear" approach, to highlight the possibility of finding more than one single cause, according to the approach already diffused presented in this book. List multiple causes means to call attention to more than one problem, increasing the effective probability of avoiding the reoccurrence of the incident.

The interested reader can study in depth how to investigating accident with STEP, reading [13].

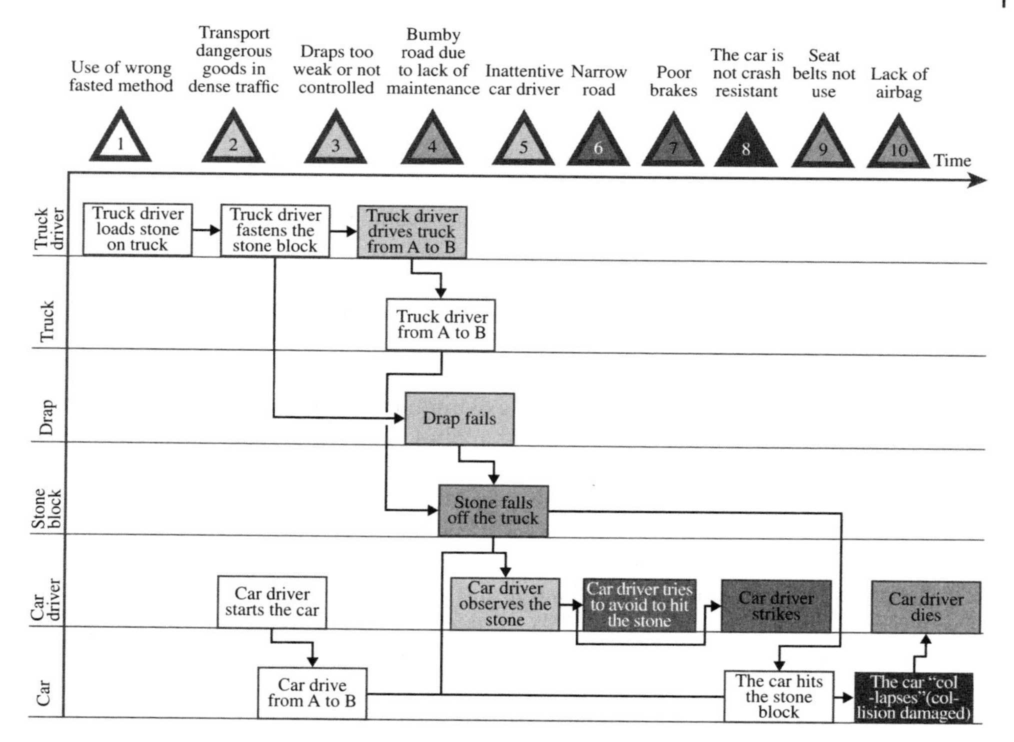

Figure 5.17 STEP worksheet with safety problems. Source: Adapted from [4]. Reproduced with permission.

5.3 Human Factor

Establishing the influence of the human factor on the causation of an accident is not an easy task. However, statistical analyses reveal that about the 50% of industrial accidents has the "human factor" as primary cause [14].

Finding such correlation means to understand if someone failed in their own role, or made the wrong choice. A proper analysis of the human factor must evaluate the level of fatigue and stress on the individual, assessing also the suitability of his/her training with respect to the assigned task and if the knowledge possessed was sufficient to make the right choice. The presence of distracting elements should be also taken into account, as well as the physical capability of the person in performing the analysed task, or the correct availability of the proper tools for the duty. Moreover, the question is generally posed about the possible deliberative choice of the individual in causing the failure (acting directly or not), because of eventual advantages for him/her self.

Understanding the interaction between humans and the process therefore becomes crucial. The analysis of the human factor spans over a variety of issues, including design, management systems and procedures. Having a database of incidents which have already happened and their investigations may help in providing useful elements to understand such a complex topic. The currently adopted approach in analysing human factors is more qualitative than quantitative, even if in the past some industries evaluated explicitly the human factors in their own PHAs.

According to [14], human factors are defined as "influences on human behaviour that may increase or decrease the likelihood of human error in a task". The Human Performance Evaluation Process (HPEP) is a method used to evaluate this influence [3]. It is used to study the personnel in their working environment, finding out the human factors that modify the probability of occurrence of a risky event. The HPEP aims to understand if someone failed to act, if the person was stressed and error-prone, if someone made the wrong choice, if the training was correctly carried out, if the person possessed the proper knowledge to act, if the physical suitability was evaluated, if he/she was distracted, or if it was a deliberative malicious act.

A further definition of human factors is given in [1]: "Human factors is the scientific discipline concerned with the understanding of interactions between humans and other elements of a system, and the profession that applies theory, principles, data, and methods to design in order to optimise human well-being and overall system performance". Including human factors in an incident investigation requires the management commitment. Human performance decrease when technologies, environments, and organizations are not designed to properly fit with the human capabilities. In the past, a limited attention was focused on the human factors and when issues were detected, the human was expected to adapt to the system. This strategy was almost ineffective, and today the approach is exactly its contrary: a system should ensure the worker's success, not his/her failure.

The environment affects the physical performance (e.g. poor lighting, extreme temperature, high noisy conditions), and, cascading, the mental performance and the decision making. Also, the technology affects the decision making (e.g. too many alarms may be perceived as the result of a not consistent control system), but also the agility, the safety, or the human perception of risk (e.g. a red light usually means something wrong, but a technology may use a green light to indicate an alarm situation). Finally, the organizational affects the teamwork, the knowledge (e.g. inadequate training) and the work practices (like schedules and procedures).

A well-designed safety management system takes into account the human factors when designing safeguards [15], which should be human error tolerant and ensure a diagnosis to correct catastrophic deviations. A great example of the human factor is alarm management: DCS allows infinite capabilities to add alarms, with the result of an alarm overload that is demanding to be managed by personnel. Therefore, alarm management gives the opportunity to prioritise alarms, associating the corresponding actions [16].

An interesting detailed study of what happens when machines and people are put together is discussed in [17]. The interested reader is encouraged to consult the reference.

The human factor is a wider concept that human error [18]: it is a multidisciplinary field of study to design equipment, process, device, and procedures to fit the human physical and cognitive abilities. It is much more than ergonomics. Three different human factors can be identified:

- Personal factors. They refer to inadequate capabilities, lack of knowledge or skills, stress, and fatigue;

- workplace factors. They include those weaknesses about supervision, engineering, maintenance, training, or procedures; and
- organizational factors. Here, human errors are seen as consequences (not as causes) of those inadequacies in the management system.

The attention on human factor is always high: indeed, the fact that human reliability is never 100% implies that all the administrative safeguards are far from being perfect [16].

It is therefore important for the investigator to know the human behaviour models. Some of them are now discussed, taking inspiration from [19].

In the first one, the mental models, the beliefs, and the values possessed by everyone will determine the thoughts which influence the behaviours, i.e. the way people act. Behaviors lead to a result, which may cause an incident. In order to change people's behaviours, it is therefore necessary to change their mental models, their beliefs, and their values. It is necessary to act on the invisible to have results on the visible (Figure 5.18).

According to another model, the mental process is activated by a stimulus, resulting in a response, i.e. the human behaviour. The consequence is the result. In accident investigation, the model is followed in reverse. Firstly, the analysis of the results (the incident) identifies what happened. Then, the response analysis clarifies how it happened. Finally, reconstructing the mental process, it is possible to evaluate the stimulus that activated the sequence, establishing why the incident happened (Figure 5.19).

In a further model, the successful performance is reached when internal and external factors affecting human abilities are met, as detailed in Figure 5.20. Similar to the previous models, if the model is used in reverse, it becomes a tool for accident investigation (Figure 5.21). Applying the basis of logic, AND gates become OR gates if they are crossed in the reverse direction. Thus, starting from the incident, it is possible to have a sort of predefined logic trees especially focused on the human factors.

Figure 5.18 Thought-behavior-result model. Source: Adapted from [19]. Reproduced with permission.

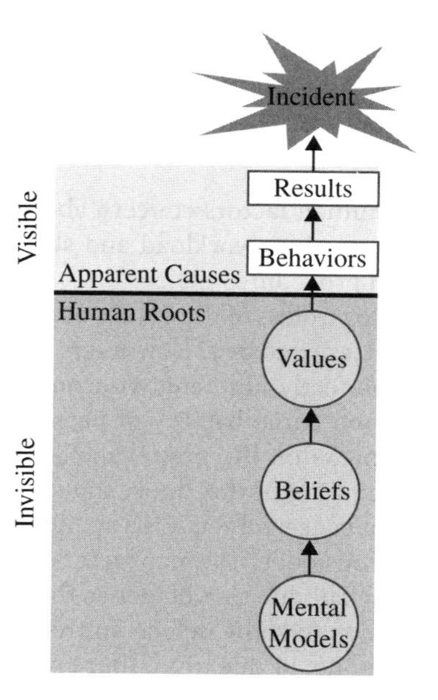

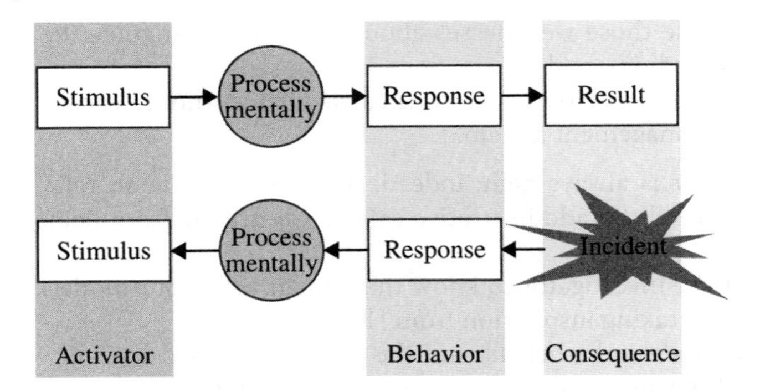

Figure 5.19 Stimulus-response model. Source: Adapted from [19]. Reproduced with permission.

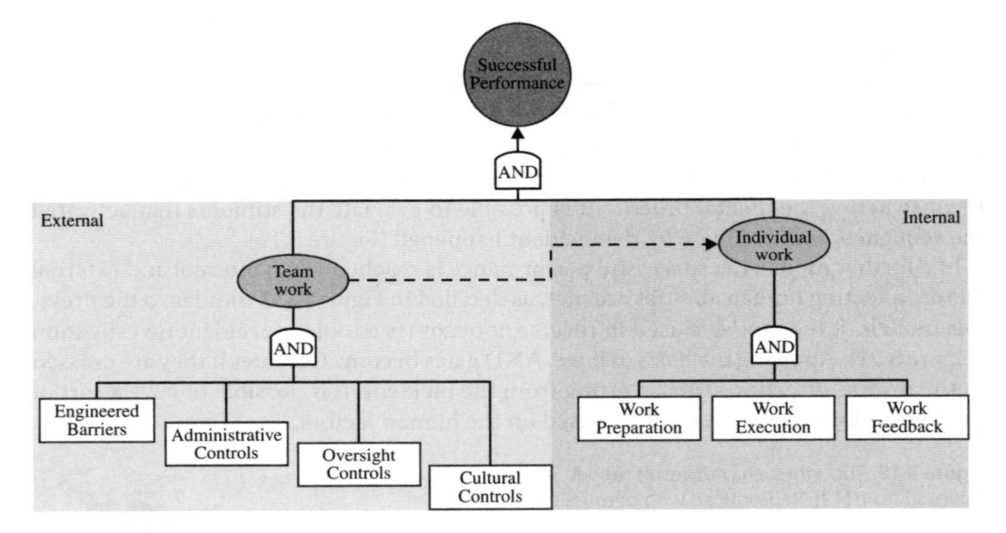

Figure 5.20 Two prongs model. Source: Adapted from [19]. Reproduced with permission.

Human factors concern about human information processing, system demands and automation, workload and staffing, interface, training, job and organizational design, and procedures [20].

Regardless of the details of the topic, here discussed only at its outer surface, it is clear that, as cited by [21], workers do not cause failure: workers trigger failure. This approach is perfectly adherent to the one adopted in this book. It is not a legal consideration about the potential liability of personnel acting incorrectly; rather it is a consideration that remarks on the proper incident investigation approach, as highlighted from the very beginning of this book. Indeed, the failure to follow an established procedure is not a root cause; rather, it is a symptom of a root cause. If this is recognised, then it is possible to consider the human factors in the incident investigation [1]. For example, an employee may fail due to a defect in the system to establish and share the standard procedures, or due to some defects in the document management system, or due to defects in the training, or due to a culture rewarding speed over quality, and so on.

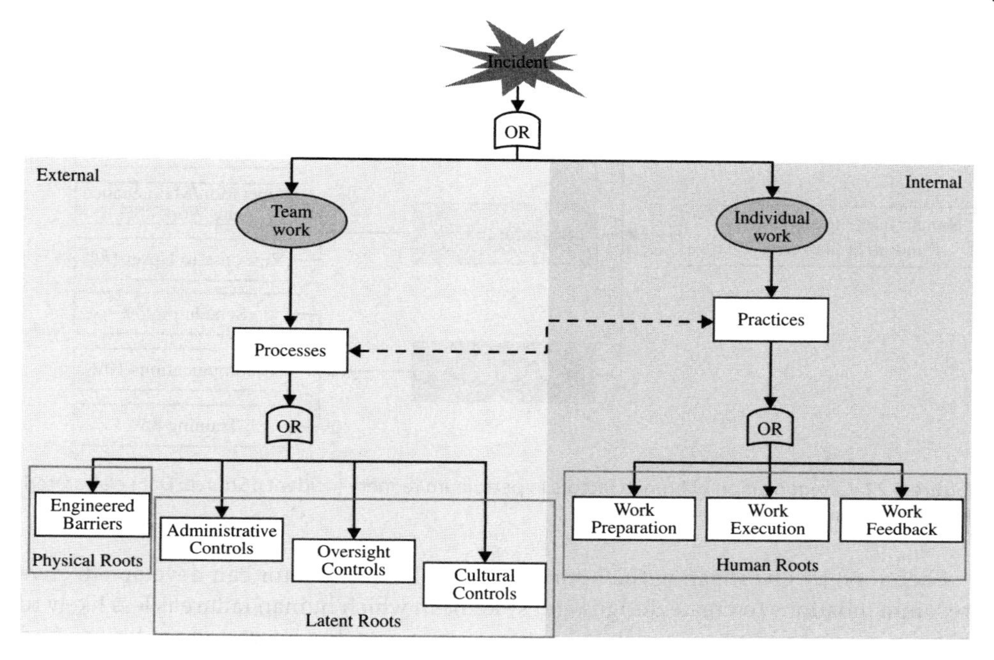

Figure 5.21 Two pronged model – accident analysis. Source: Adapted from [19]. Reproduced with permission.

At petroleum refineries, the human factors become a relevant and frequent cause of incidents. An interesting study to collect and analyse data coming from petroleum refineries incidents was done by Battelle Memorial Institute in 1999 by Chadwell et al., as mentioned in [14]. Data, collected from scientific journals, newspapers, and the internet, shows an interesting discovery. The incidents were divided into five categories, depending on the causes of the incident itself: equipment failure (no human error); random human error; human factors in facility design (environment, controls, equipment); human factors in procedural; human factors in management systems (training, communications, planning, and so on). The results of the study are impressive: 47% of identified causes involved human error. The majority of these (81%) has the contribution of human factors, while only the 19% of them is related with random human errors. The results, shown in Figure 5.22, fully demonstrate the importance of human factors in hazard identification.

In order to identify potential human factors, it is suggested that a checklist be used. Indeed, during a hazard identification session, like a HAZOP, hundreds of scenarios involving a human factor evaluation could emerge. Therefore, a checklist could be used to help the team in consistently identifying the hazards and evaluating the decrease or increase of the likelihood of the scenario related to the human factors. For instance, the hazard identification team can use the checklist during the HAZOP to discuss the scenario in depth; otherwise, the checklist can be utilised before the HAZOP study, helping the team in adopting a unitary view of the process respect to the human factors and favouring a better preparation of the team to the next hazard and risk assessments. A possible checklist is in Table 5.4.

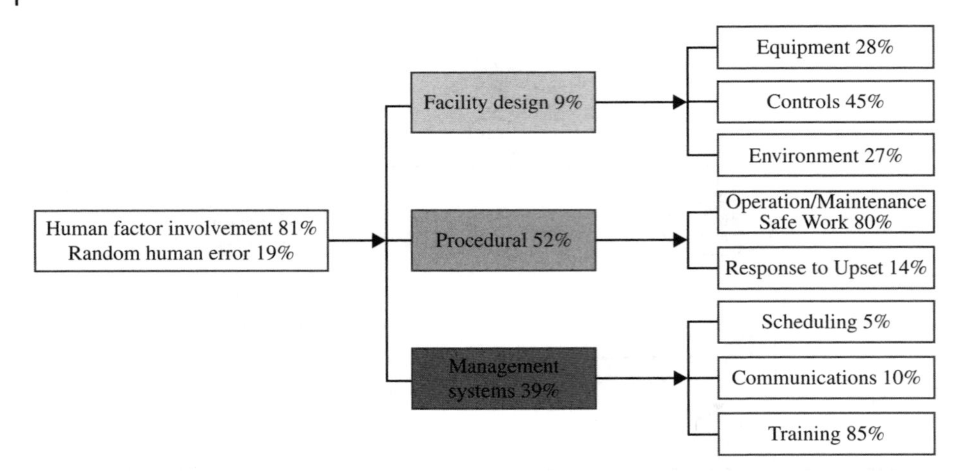

Figure 5.22 Categorization of human factors in petroleum refinery incidents. Source: Data elaborated from [14]. Reproduced with permission.

Therefore, by investigating the human factors, the investigator can develop effective recommendations to create designs and systems in which human failure is less likely to happen, being tolerant to human failure if it does occur. The interested reader can find additional information about human factors in [20], where the topic is discussed in deep.

5.3.1 Human Error

Following the definition given by [14], human error is a "departure from acceptable or desirable practice on the part of an individual that can result in unacceptable or undesirable results". The definition given clarifies that a human error is not only related to an operation carried out wrongly, but it also regards all the remaining parts of a company's structure (from management to front-line operators).

Two typologies of human errors can be identified:

- Intentional human errors; and
- unintentional human errors.

The former are deliberately committed, for example when a prescribed action is voluntarily not followed, even if not maliciously. These errors are generally performed by the maintenance personnel and produce a direct effect. The latter are errors where there is no intention to pursue a deliberative action in contrast with procedures or good practice, but they happen unintentionally. They can affect every level of the company (from management to operations and maintenance personnel), including also external decisions (i.e. taken outside the enterprise). Unintentional human errors may easily lead to an incident.

Going deeper in this classification, it is possible to group human errors into four categories, following the scheme in Figure 5.23:

- Involuntary or nonintentional action, like a switch tripped because the operator leaned against it;

Table 5.4 Example of human factors in process operations.

	Category	Human factor that reduces the likelihood of human error	Human factor that increases the likelihood of human error
Equipment	Label	Correctly labelled with uniform coding	Incorrectly or not labelled
	Access	Easy to be reached immediately	Hard to be reached
	Operability	Power-assisted operation	Hard to operate
Controls	Mode	Completely automatic	Several manual steps
	Involvement	Operator continuously involved	Operator involved at spots
	Feedback	Clear and immediate	Ambiguous or absent
Deviations	Alarm	Safety-critical	Many at the same time
	Coverage	At least 2 operators are always present	Operators not always present
	Time	No time pressure to act	Insufficient time to act
Transients	Procedures	Updated, accurate, complete	Obsolete, incomplete, inadequate
	Format	Graphic aids provided, details available	Hard to read, inconsistent
	Aids	Checklists	Tasks done by memory
Scheduling	Consistency	Permanent shift assignments	Inconsistent shift rotation
	Frequency	Routine task	Very infrequent task
	Intensity	Regular task, normal effort required	Extra effort required, more tasks in sequence
Communication	Field/Control	Communication with field	No communication with field
	Supervision	Frequent supervisory communication	No supervisory checks
	Emergency	Unambiguous and rapid alarm system	Confusing alarm system
Environment	Noise level	In office	In area with hearing protection required
	Climate	Indoors, climate controlled	Extreme weather conditions
	Visibility	No limitation	Foggy or other limitations

Source: Data taken from [14].

- spontaneous action, such as when a procedure has just changed and the operator, because of his/her background and sound experience, forgets to apply the new procedure;
- unintentional action (slips or lapse), like the activation of a different switch, because of distraction; and
- intentional action, such as when an operator voluntarily closes a different valve in respect to what ordered, thinking it is faster and equally effective.

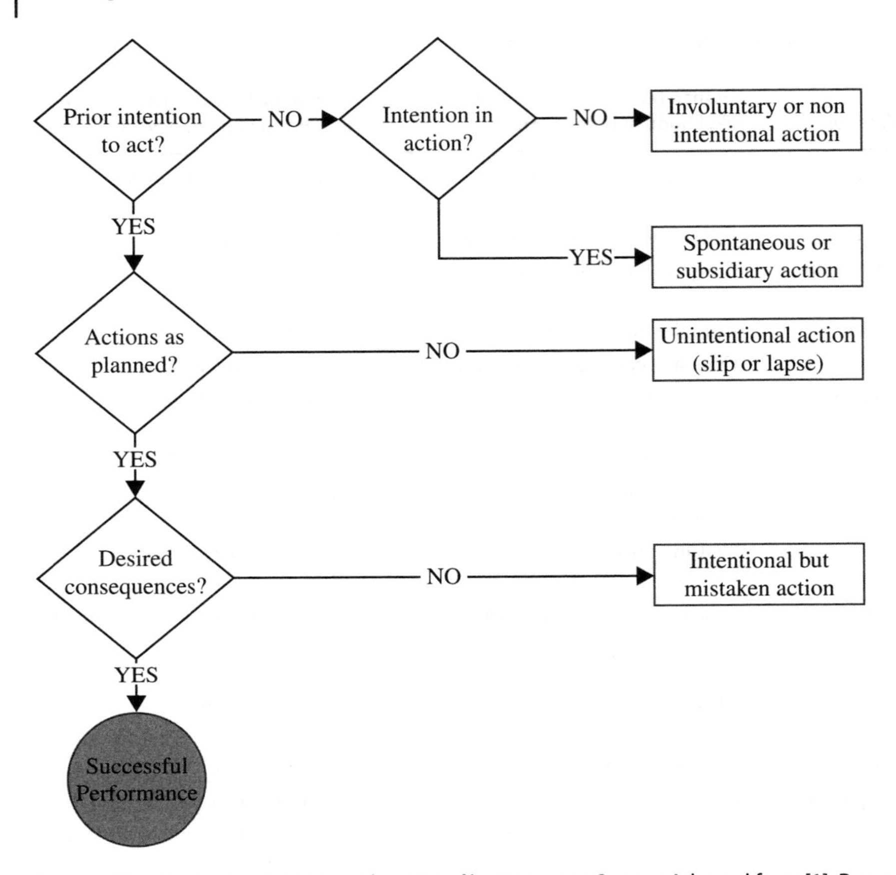

Figure 5.23 Method to determine the type of human error. Source: Adapted from [1]. Reproduced with permission.

A similar classification was proposed by Reason in 1990 [22]. It is summarised in Figure 5.24.

In the Reason's classification: the slips are those actions intentionally carried out, but incorrectly implemented; the lapses are those actions not carried out because of mere distraction; the mistakes are those actions correctly implemented but with an incorrect intention. Finally, violations are those actions deliberately carried out by someone who is aware that is acting differently from how he/she should act.

A similar human error classification has been also suggested by Rasmussen in 1983 [24], as discussed in [23] and [1], known as Skill-Rule-Knowledge (SRK) model (Figure 5.25). According to this classification, the human errors can be classified in:

- Skill-based behaviour. It includes those actions very little or no mental effort, being almost "automatic" actions, such as writing or cycling;
- rule-based behaviour. It regards those actions following an explicit procedure; and
- knowledge-based behaviour. It regards those actions involving a mental activity to solve a problem.

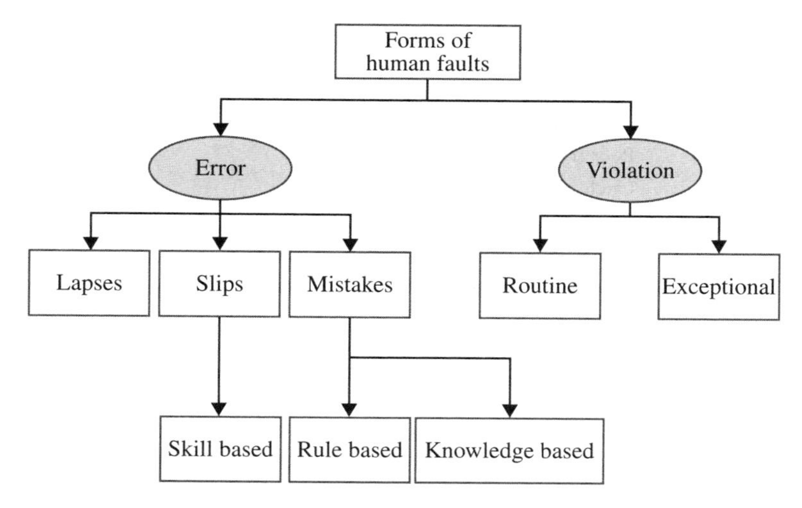

Figure 5.24 Reason's classification of human errors. Source: Adapted from [23]. Reproduced with permission.

Figure 5.25 Causes of human error.

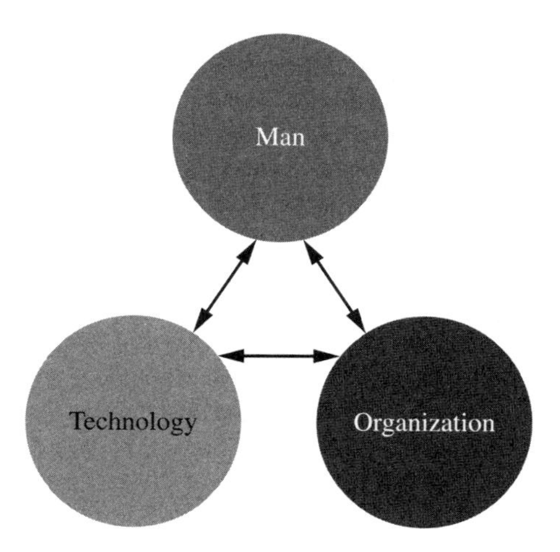

As shown in Figure 5.24, slips can be assimilated to skill-based errors, while mistakes can be related to rule and knowledge-based error.

It is also possible to classify the following additional human errors [1]:

- Errors of omissions. They refer to a failure to complete a task;
- errors of commissions. They refer to a failure in doing the right task;
- sequence errors. They refer to a failure in following a predefined sequence of actions; and
- timing errors. They refer to actions that are carried out with a wrong timing.

In the point of view of the change from a behavioural approach to an organizational approach, also discussed in [25], it appears useless to use human error as the actual causes of an incident, as already discussed and also underlined in [26]. Moreover,

Table 5.5 Human and management errors.

Human Errors	Management Errors
Inexperience	Poor training
Poor knowledge	Unclear instructions
Fatigue	Heavy workload
Complacency	Insufficient resources provided
"Make it work" attitude	Emphasis on deadlines

remembering the context of forensic engineering, it could be useful to distinguish among "really human errors", i.e. those caused by unsafe acts, and errors resulting from unsafe conditions, i.e. faulty equipment or poor working condition [17]. Clearly, it is a highly simplified approach, but it reveals useful when carrying out an incident investigation. In this distinction, unsafe acts are imputable to individuals while the unsafe conditions are attributable to the management. Possible human and administration errors are listed in Table 5.5. The simplified approach summarised in Table 5.5 is based on the assumption that the two categories are distinct; actually, they are not, since each affects the other. This link becomes crucial when an investigator intends to find the real root causes of an incident and it should be always kept in mind.

In conclusion, human errors can be seen as the interaction of man, technology, and organizational-based issues, as shown in Figure 5.25, which is self-explicative.

A powerful example, already discussed in this book, is the Deepwater Horizon disaster in [27]. The interested reader can consult [28] to know more about another important incident and its connection with human error: the BP Texas City Refinery disaster.

Detailed studies about human errors are in [29–32], which have been used as a reference in this Paragraph.

5.3.2 Analysis of Operative Instructions and Working Procedures

Talking about industrial accidents, it seems that some companies experience them on a regular basis while others not. Obviously it is not a matter of lucky, nor the workers of a company are more prone to trigger an accident respect than other company's workers. Simply, it is generally the number and the severity of mistakes that makes the difference. Making mistakes is absolutely normal, even if no one wants to pay for their consequences. Making mistakes is part of the learning process and avoiding them requires practice, that is to say. a period of time where doing something wrong is acceptable because there is enough freedom to experiment while enhancing skills. Indeed, many operative instructions and working procedures are far from being perfect the first time they are applied. It may require years to correct the mistakes that arise and to solve those problems that become evident only once the procedure is already valid. The adjustment is recursive: further enhancements are possible thanks to the detection and resolution of the mistakes previously found and solved. We have already discussed in this book how unrealistic is the "Goal Zero" policy of those companies that have zero tolerance for mistakes. Indeed, mistakes occur even by the most experienced, trained and motivated individual. The adoption of a system that has no tolerance for errors, pursuing punishment for who makes them, may cause a negative feedback and help to promote the occurrence of serious incidents. This may happen because, following that incorrect policy, mistakes are often hidden by the individuals who deny their occurrence. Moreover the company culture is oriented in blaming and scapegoating findings and this can encourage employees in sabotaging or doing malicious acts to get back to the management.

Obviously, a mistake is not a catastrophic event: by definition it is only a precursor of a probable undesirable future event. A mistake becomes an incident if it is not handled properly. If you imagine a task as a decision tree, making a mistake means taking the wrong branch: it depends on how the pathway continues if that error will turn into an incident or not. From this point of view, preventing an incident means to identify a possible mistake (a node of the decision tree where the incorrect branch can be taken) and correct it, following the right pathway and therefore avoiding the occurrence of the incident. This is what happens for quality controls in manufacturing industries, and similarly for safety control as is clarified next [17]. Indeed, old quality controls identified a mistake after is was manifested: at the end of the production line, some samples were taken and if deficiencies were detected (i.e. a certain parameter did not satisfy acceptance criteria), then the process was stopped and the product wasted. The stop continued until someone fixed the problem, and the quality control to the samples gave a positive result one more time. In the more modern industries, a new management approach leads to a different way to conduct a quality control, based on self-correcting methods. In this method, mistakes are not detected at the end of the production line (when it is too late to preventively intervene) but their correction is carried out at the point where the mistake is manifested. Feedback loops are therefore necessary at every significant step of the process: if the result of a decision process at a specific process step fulfils the requirements, then the process can continue to the next phase because no mistakes have been detected. Otherwise, if the requirements are not satisfied, then the process stops at that particular step, waiting for the corrective adjustment related to that peculiar step of the process. The concept of the feedback comparator is given in Figure 5.26.

What is described for product quality can similarly be applied to safety. Indeed, traditional management system compares the number of already occuring incidents with the average number for that year, which are published by industrial associations, and are identified as the performance goal level. Only then, does the management system look back at what was wrong, trying to implement those corrective actions to reduce the number of incidents. But it must be noted that this happens once the incidents have already occurred. It is like a quality control performed at the end of the production line, with a single feedback loop: it makes no sense, because the damage is already done. The problem is generally considered solved when the number of incidents is smaller than a set value that meets an acceptance criterium, generally established as an average for that particular process or industrial sector. On the contrary, the efforts of more modern safety management systems are to detect mistakes where and as they occur, correcting them immediately. This strategy guarantees that the mistake does not propagate across the whole system. In order to guarantee such high level of performance, the management system should be pacifically aware that:

- Mistakes are natural and humans make them, always;
- it is possible to predict errors and they can be managed to prevent incident occurring;

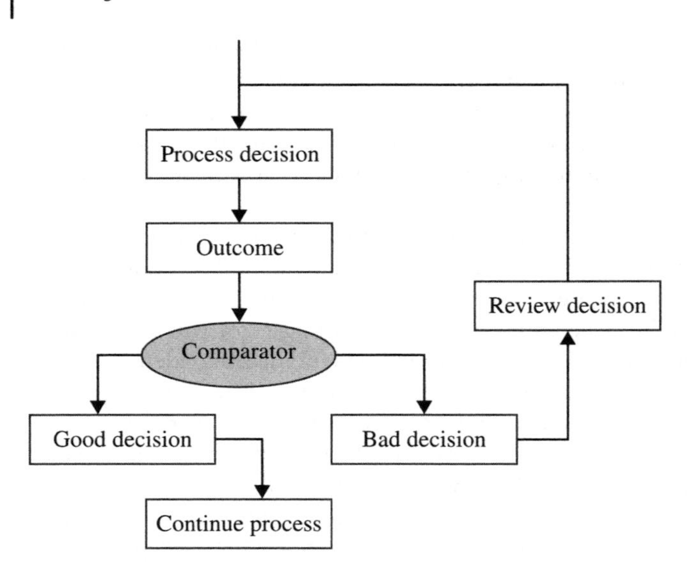

Figure 5.26 Self-correcting process step. Source: Adapted from [17]. Reproduced with permission.

- safety responsibility is shared among all the company's personnel, from the management to the employees; and
- the environment affects the human response: encouraging and positive reinforcing are desirable value for a company to ensure long-term results.

To establish if a human error is going to materialise (or, during an incident investigation, if it has actually happened), it could be useful to compare the Job Demand (JD) with the Job Ability (JA) of the person in charge of the specific task being analysed. Put simply, it can be assumed that an accident occurs if JD > JA, otherwise it does not. Demand and ability are intended as momentary, since it is sufficient for there to be a temporary change in their entity to trigger a sequence of events that leads to an incident. With the term "job demand" you intend the required skills, specific knowledge, procedures, tools, and resources, to have the ability to react to a condition or instruction that mutate during the job, the time constraints, the expectations. Reference has been made to momentary demand and capacity since they are susceptible to varience over time. Therefore, even if the job is carried out under the condition JD < JA, it may happen that a temporary decrease of the worker's ability brings about a real incident. This is particularly the case if the margin between JD and JA is small; if this margin is sufficiently wide, then an "off" day for the worker will result in zero consequence for the safety.

To repeat, it is the management attitude of the company that influences the level of job demand: for instance, providing the right tools, resources and ensuring a positive work environment (i.e. with necessary working conditions) are solutions to minimise the demands of the job. This helps in working under the JD < JA condition, that is minimizing the likelihood of incident occurrence.

When an incident occurs, it is crucial to investigate if and how much the human factors influenced the sequence of events that lead to the incident. A Man, Technology and Organisation (MTO) investigation can be carried out (Figure 5.25). This method is based on the assumption that human, technical and organisational factors must be equally studied in an incident [4]. The method was proposed by Rollenhagen in 1995 [33] and Bento in 1999 [34]. Briefly, this method brings together:

- An event and cause diagram, to reconstruct the dynamics of the events;
- a change analysis to describe how events deviates, increasing the likelihood of an incident; and
- a barrier analysis to identify those safeguards (mechanical and administrative) that failed or were missing.

Figure 5.27 shows an MTO worksheet. Firstly, the event sequence is developed horizontally, using a block diagram. For each event, the possible technical and human causes are found and placed vertically. Finally, the barriers that failed or that were missing (i.e. those that did not interrupt the incident sequence) are identified in the bottom part of the worksheet. Developing recommendations is within the scope of an MTO-analysis. They should be practicable and might be technical, human or organisational.

The MTO method takes also into account a checklist to identify the possible failure causes. A possible checklist, with a particular focus on human factors, was already presented in Table 5.4. Clearly, other checklists can be used.

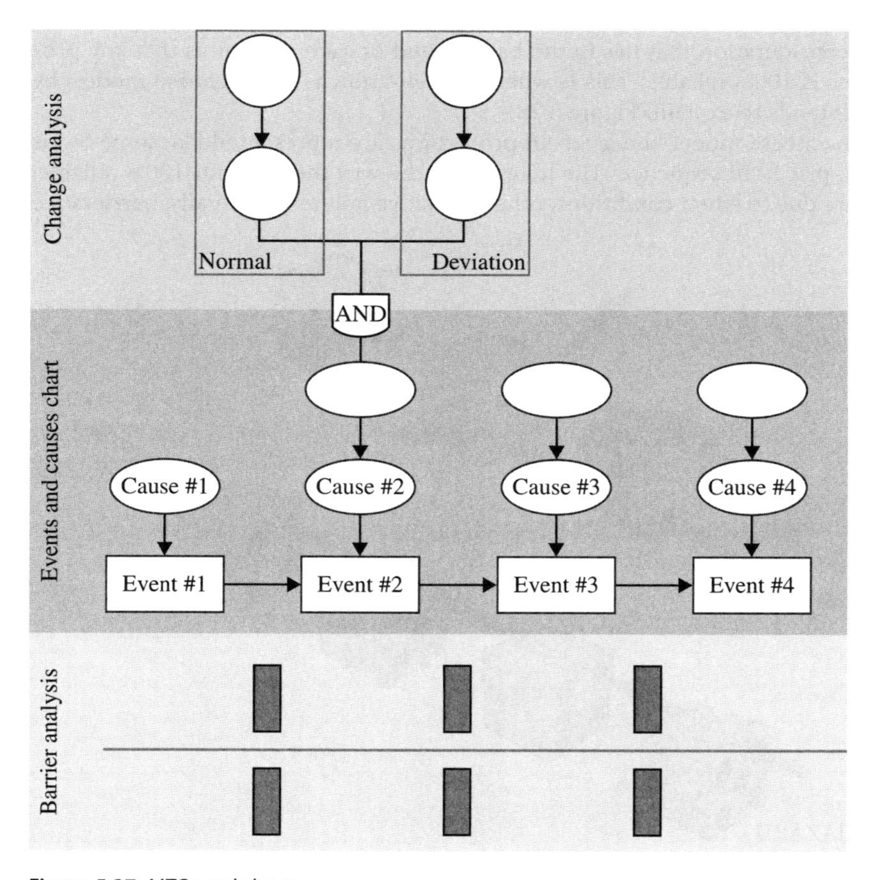

Figure 5.27 MTO worksheet.

5.4 Methods

Some methods are now described to conduct the analysis stage of the investigation workflow. Whichever is the method that the investigator intends to adopt, this phase of analysis has a general approach which should be commenced using a basic format that can be sum up in the following four steps:

- Develop, by brainstorming or a more structured approach, the possible incident sequences;
- eliminate as many incident sequences as possible, based on the available evidence;
- take a closer look at those that remain until the actual incident sequence is discovered (if possible); and
- determine the underlying root causes of the real incident sequence.

In order to develop the sequence of events of the incident, it is fundamentally necessary to determine:

- What was the cause or attack that changed the situation from "normal" to "abnormal";
- what was the actual (or potential, if a near miss) loss event; and
- what safeguards failed and what did not fail.

A general consideration that lies in the background of each method is that any protective barrier is 100% reliable. This is what the well-known "Swiss cheese model" by Reason [22] intends to explain (Figure 5.28)

In the Swiss cheese model, the layers of protections are represented like some Swiss cheese slices, placed in sequence. The holes show the way they are not 100% reliable: some holes are due to latent conditions, others to active failures. Generally, barriers are

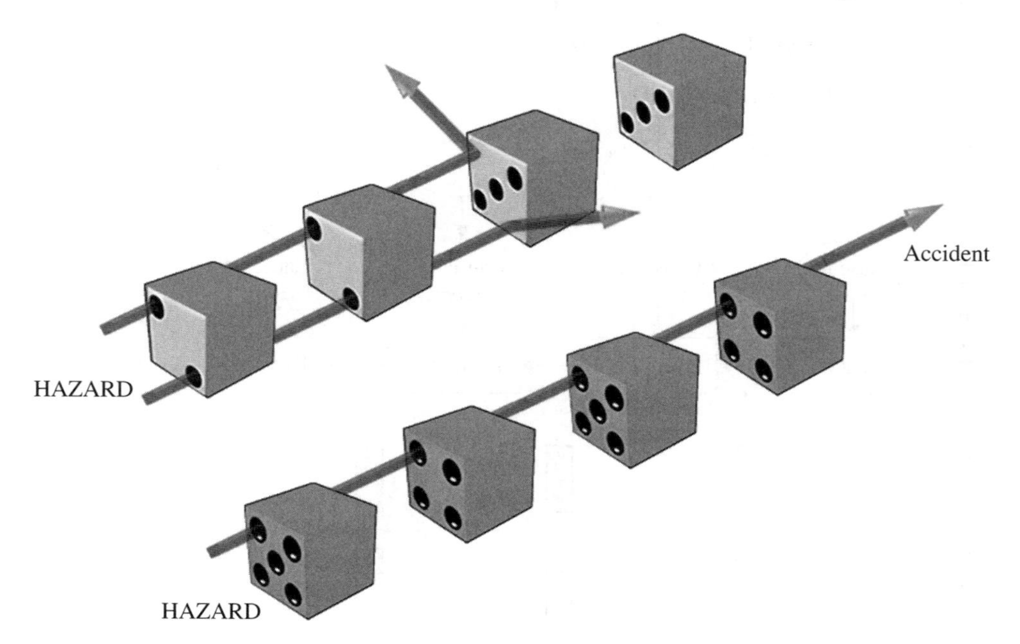

Figure 5.28 Swiss cheese model by Reason. Source: Adapted from [22]. Reproduced with permission.

put in place in order to have not an overlapping of the holes; this ensures that, even if each barrier is not 100% effective, the whole system is still safe. But under certain conditions, it may happen that the holes, i.e. the weaknesses of the barriers, overlap causing the actual transformation of hazards into an accident.

In the end, whichever is the adopted method, the investigator finds the most likely scenario that fits the facts, determines the underlying management system failures and develops layered recommendations.

Several methods for accident investigation are available, each with its strengths and weaknesses. A selection of these methods is presented following in this book. The discussed methods are widely used and significant literature is available to ensure a deep knowledge and give a sound reference to the reader. The following methods are discussed in this book:

- Expert judgment and brainstorming;
- Structured methods and approaches;
  - Pre-structured methods
    - Management Oversight Risk Tree (MORT)
  - Barrier-based Systematic Cause Analysis Technique (BSCAT™)
  - Tripod Beta
  - Barrier Failure Analysis (BFA)
  - Root Cause Analysis (RCA)
    - Introduction to RCA
    - TapRoot®
    - Apollo™
  - Quantitative Risk Assessment (QRA) derived tools
    - Fault Tree Analysis (FTA)
    - Event Tree Analysis (ETA)
    - Layer of Protection Analysis (LOPA)

5.4.1 Expert Judgment and Brainstorming

Once the evidence has been collected, the moment has been reached for the subsequent step of the investigation: generating hypotheses, with the aim to reconstruct the real sequence of events. Expert judgment and brainstorming are therefore essential stages that contribute to correctly addressing the development of the investigation during these early stages. For simple incidents, an experienced investigator might find the real scenario dynamics without further steps. Typically, for industrial accidents, this situation is unlikely to be enough for a complete understanding of the incident, but it is important to set valued starting points. During this phase, the evidence collected (and the related hypotheses) are verified through cross-checking with actual findings; at the same time, those facts (and evidence) not soundly based on the findings are eliminated and disregarded as contributing factors of the event. Therefore, the hypothesizing stage becomes useful both to select necessary and sufficient causalities and to establish the certainty of their involvement in the incident [9].

In order to generate correct hypotheses for an incident scenario, team work is highly recommended. The plurality of different points of view, because of the different backgrounds of the team members, helps in having an unbiased focus on the relevant aspects discussed by the team. The team, as already stated in Paragraph 4.2, is composed of a team leader (i.e. the investigator in charge) and additional members who are experts in particular disciplines related to the incident. The team tries to converge their findings into a shared and unique sequence of events: when it becomes difficult to reach such a convergence, because of different interpretations or ideas about the proposed hypothesis, then additional data collection is prescribed in order to have further elements for the discussion. At this stage, it is also possible to consider more than one hypothesis and then restrict the scope of the analysis during the incident investigation process.

Generating a hypothesis requires an open mind; objectivity is instead essential to provide a credible explanation of the accident to the outside world. This means that the reconstructed sequence of events will be based on facts, not on speculations. During the reasoning process, some hypotheses could not be accepted because of their little causal specificity or they imply remote factors that do not clearly identify a deviation from the normal state, or they are outside the control of the management (like natural events or malicious acts). A human and social sciences expert could be required to better understand those phenomena that could appear irrelevant for the trigger or development of the incident. In this topic, the level of experience is a fundamental factor to have a correct "diagnosis" of the incident (i.e. finding its root causes).

The easiest ways to generate hypotheses are brainstorming in teams and expert opinion/pattern recognition. These activities can also imply a more critical analysis including: matching facts and findings, building a timeline of events, reconstruct the event, generate possible scenarios. In order to establish what and how it happens, a full description of the event is required. Then, in the investigative phase, this description assumes importance to understand why the event occurred. To answer these questions, a number of "intermediate products" need to be created. They include: establishing the sequence of the event; identifying potential scenarios; identifying critical weaknesses; assigning priorities in the investigation issues.

Clearly, this stage is not error-free. The most recursive are errors in perceiving the facts (poor communication, altered perception because of the surrounding conditions, confused situation), analytical errors (due to poor skills, poor knowledge, or poor experience), decision-making errors (team affected by fixation, preferences that increase the complexity, tunnel vision), and action errors (incorrect or poor data gathering and analysis).

5.4.2 Structured Methods and Approaches

Two different approaches are available to carry out an incident investigation. A first method is the one looking for all the possible ways the undesired event happened using a timeline and a simplified logic tree approach. It is a structured approach, since it systematically helps in searching for all the underlying root causes [1]. Thanks to its structure (Figure 5.29), the method helps the investigator in reaching a sufficiently deep knowledge to not stop at the immediate causes. This method focuses on the logic tree approach, also using simplified methodologies.

The second method differs from the previous one because the creation of the logic tree is substituted with the casual factor identification and the subsequent usage of predefined logic trees or checklists (Figure 5.30). The reader should note that checklists, useful when related to human factor issues, are derived from logic trees. It is therefore

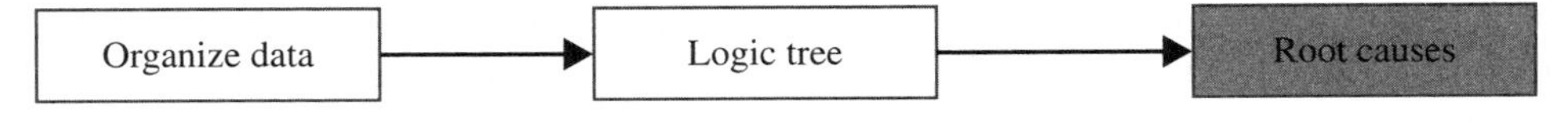

Figure 5.29 Workflow of structured methods.

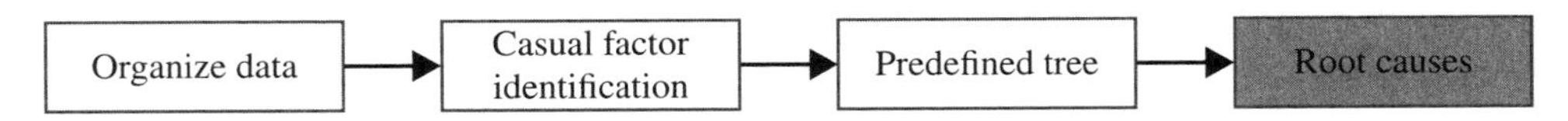

Figure 5.30 Workflow of pre-structured methods.

important to be familiar with logic trees and have a full knowledge of them. The predefined structure provides a systematic approach. Similar to the previous method, also this one recognises that a single incident may have multiple root causes and the adopted structure wants to identify those system changes that can reduce the probability of reoccurrence of similar events.

The adoption of structured and pre-structured methods improve the quality of the investigation, favoring the research for multiple causes and shifting the attention of the investigator from the immediate causes to the root causes. This objective becomes possible thanks to the common approaches the two methods have: to divide a single complex incident into multiple smaller occurrences to be then examined individually. The two approaches are also tools to test logic, determine if the identified causes are the root ones or not, helping the investigator in taking important decisions throughout the entire investigation process. Moreover, these approaches work with any analytical methodology. It is not the intention of this book to endorse one particular method respect than another. Instead, this book wants to be a guideline to the various alternative methodologies that are available to carry out an incident investigation. However, it is clear that structured and pre-structured approaches provide useful workflows not only to identify multiple root causes, but also to promote the corrective system solutions that avoid future reoccurrences. Therefore, in the context of the mentioned approaches, the guiding questions that an investigator should constantly ask him/her self, regardless the methods he/she intends to use, are:

- Why? Why? And Why?
- what are the root causes?
- was there a deficiency in the system that caused the condition to exist or to proceed?

During the last years, several techniques have been developed as tools for structured incident investigation analysis, based on the typical anatomy of the incidents and on the most recent theories about causality and human factors. All those techniques can be applied in different contexts and have been validated by the analytical experience in real cases. Even with their differences, those techniques have the following primary objectives [35]:

- Organization of information about the incident, before the facts collection;
- description of the incident causality and development of hypotheses to be further analyzed; and
- identification and formulation of corrective actions.

Therefore, they can provide useful support for the analysis and help in focusing on the significant causal aspects. Moreover, most of these techniques directly develop a useful structure to configure and highlight, with a rigorous approach, the relations between causes and effects. Finally, they allow developing easily the effective aids for communication and comprehension of the lessons learnt.

From the point of view of the involved logic, the techniques may be attributed to different fundamental approaches:

- Deductive;
- inductive; and
- morphologic.

The deductive approach relies on a logic path that, starting from the general, tends to reveal the particular. At the base of the use of this method there is the postulate according to which the system or the process has failed in some way. Once this is established, the components, operator, or organizational aspects that had a contribution to the failure are found. In other words, the deductive logics start from a point in the time sequence and looks back, with the aim to examine the previous steps (Figure 5.31).

The Fault Tree Analysis (FTA), discussed next, is a technique widely known and applied in safety analyses; it is a clear and typical example of deductive logic.

The deductive approach, on the contrary, pushes the logic process in following the path that, starting from the particular, tends to picture the general. In this sense, the analysis starts from the hypothesis that a determined fault or initial event has occurred and its effect on the functioning of the system is then evaluated. If compared with the deductive approach, which is strictly logic-sequential, the inductive one has instead the connotation of an exposition, where the entire structure of a system or a process is read in overlapping with the interesting event. Some techniques widely adopted in safety analyses, like the Failure Mode and Effect Analysis (FMEA) or the Hazard and Operability Study (HAZOP), or the Event Tree Analysis (ETA), are examples of techniques based on an inductive approach (Figure 5.32).

Often, starting from the general structure of the incident reconstruction determined using a deductive approach, it is necessary to use recourse to an inductive approach to delve into some causal details necessary to unequivocally demonstrate the cause-effect link. An example of a technique that represents this particular mode of hybrid application is the Cause-Consequence Diagram Method (CCDM), initially developed for safety analysis but not really widely known compared with the other methods previously listed.

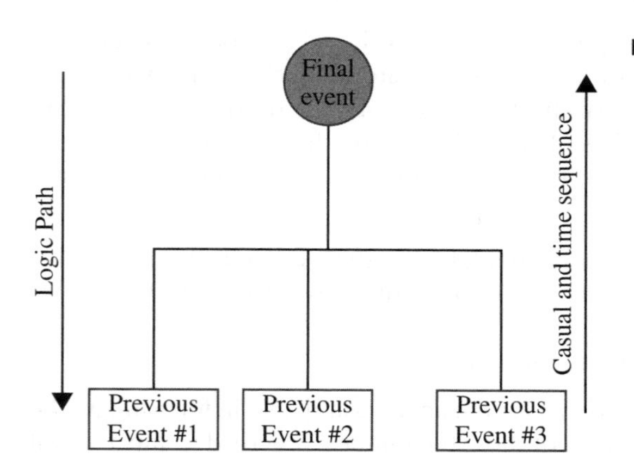

Figure 5.31 The deductive logic process.

Figure 5.32 The inductive logic process.

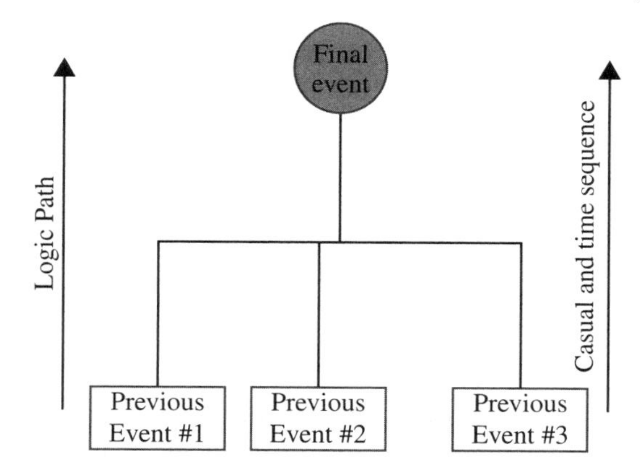

Another example is the BowTie methods, a hybrid method that, in extremely simplified words, links together a fault tree with an event tree (it is discussed next in detail). However, the hybrid approach can be useful in those accident investigations where the event sequence and the exact timing take on great importance.

It has been understood that a sound knowledge of the affected systems and processes from the analysts is necessary; using an inductive approach requires that such knowledge is particularly deep and internalised. Indeed, in the application of a deductive approach, a considerable in-depth analysis is gradually developed during the execution of the analysis itself and thanks to it; vice-versa, in an inductive approach the implementation of each logical step requires the preventive deep knowledge of the system or the process. Similarly, it is important that the analyst has a sound understanding of the adopted techniques, with their strengths and weaknesses.

In any case, one of the most critical aspects in the application of these techniques is a good coverage about the identification of the failure modes in the deductive approach and the malfunctioning situations (regarding the system or the process) in the inductive approach. The analyst has to consider how much detailed the analysis should be. An excessively extensive study might not bring any further advantages but it can make more the usage of the collected information more difficult, also for future demonstrations, conclusions or training.

Generally, the deductive techniques are more suitable for the identification and comprehension of the root causes, while the inductive techniques can be beneficial in properly addressing the application of the deductive ones, especially in very complex cases. The morphological approach for accident investigation pays close attention to the structure of the system (Figure 5.33), focusing directly on the hazardous elements already known, based on the nature and conformation of the system itself (critical operations and situations, hazardous surrounding conditions, operative parameters out of control, etc.), with the aim to highlight the most significant for the safety. Essentially, the attention of the analyst is addressed on all the known sources of hazards rather than in finding the various possible deviations or unusual events and then in analyzing their impact. Therefore, the effectiveness of the analysis depends strongly on the level of knowledge of the system and on the specific past experience: indeed, a wide recovery of the operative experience plays an important role. It can be said that the morphological approach is like a logic process, that is static from the sequential point of view, different from

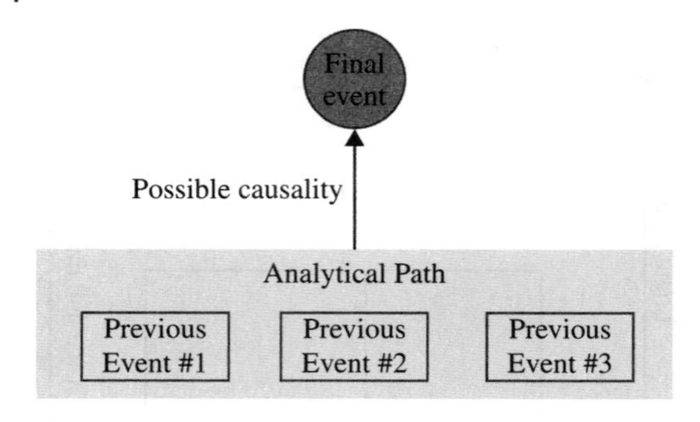

Figure 5.33 The morphological process.

the previous approaches tending to follow the timeline of the events in one of the two possible directions. The morphologic analysis can be supported by special techniques, which are usually adaptations of deductive or inductive approaches.

5.4.2.1 Pre-structured Methods

When the incident scenario has been reconstructed and the causal factors have been identified, it is time to look for the root causes. One way to perform the root cause analysis is to use pre-structured methods. They provide an organic and systematic approach for the discussion of the relevant elements about the incident. Generally, a pre-defined tree is a deductive technique, because it examines the past events (recursively asking "why?") that have been identified as necessary to produce the actual incident.

In a predefined tree, the potential root causes (listed exploiting the past experience) are categorised by matter (human error, equipment failure, etc.) and distributed according to a precise hierarchy that is represented by the structure of branches and sub-branches. Very often, logic symbols (such as AND/OR gates) are not shown in a predefined tree: however, each node of the tree between a single branch and its sub-branches generally represents an OR gate. Figure 5.34 shows an example of a proprietary predefined tree [36].

Differently from the procedure adopted to develop ex-novo a logic tree, an investigation team using a pre-structured method does not need to construct the tree. The analysts simply follow the logic path suggested by the pre-structured method, disregarding those branches that are not relevant to the incident. A considerable advantage in using a predefined method is that repeatability is ensured and it becomes possible to make comparisons between different investigators/investigations/companies, because a standard set of root causes is used for every incident, increasing consistency [1].

This aspect represents an excellent opportunity to find and report trends, using standard categories of root causes. From a company standpoint, this allows a better collection and analysis of data about incidents and near misses, in order to determine future trends that would not be visible from a single accident investigation. This is why some companies provide already structured methods for incident investigation, with a predefined categorization of the root causes according to their management system, thus helping to found eventual weaknesses easily.

Obviously, using a predefined method might reduce the lateral thinking capability of the investigators, who might not find his/her hypothesis in the provided list of root causes. This could happen even if a wide range of possibilities about the underlying causes is recently provided by the pre-structured methods. This apparent

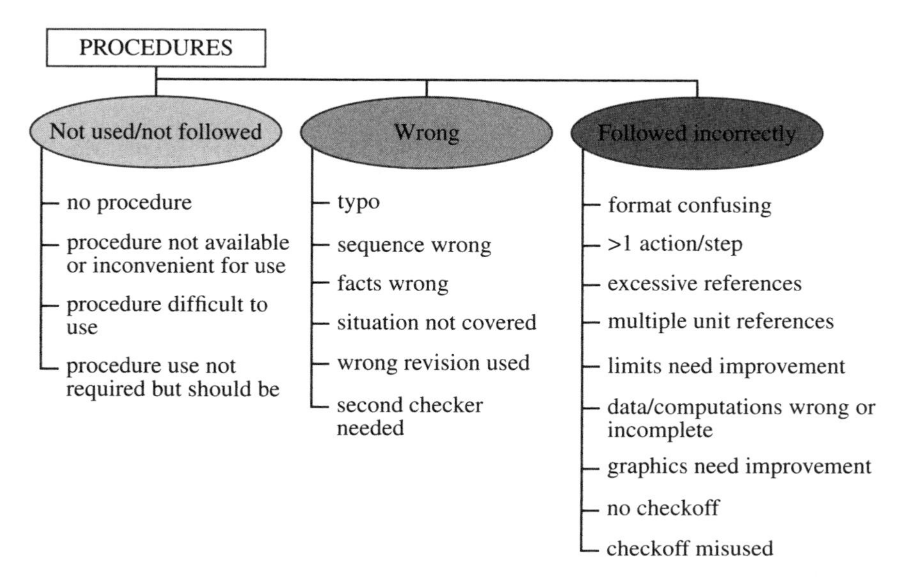

Figure 5.34 Example of root causes arranged hierarchically within a section of a predefined tree. Source: Adapted from [36]. Reproduced with permission.

weakness of the predefined methods is overcome by brainstorming activities (cited in Paragraph 5.4.1), an essential tool to develop a new root cause category not considered by the closed structure of some methods.

Management Oversight Risk Tree (MORT) The MORT technique (or Management Oversight and Risk Tree) was originally developed by the System Safety Development Center (SSDC) of the American Energy Research and Development Administration, with the main goal being to support incident analysis on workplaces and to provide information for the safety audits and solve some management issues. The MORT diagram is the key of the analytic process carried out; it is formally similar to a Fault Tree Analysis, but the basics rely on the concept of the flow of energy and barrier.

In MORT analysis, the following definitions are given:

- Event: an occurrence where a barrier to an unexpected flow of energy is inadequate or malfunctioning, without any damage or consequence;
- Incident: an unexpected flow of energy or an exposition to a particular environmental condition, causing damages or negative consequences.

Wherever there is the possibility that a person or a system get in touch with a flow of energy or an environmental condition that may cause damages, it is necessary to see to the isolation of the movement of energy or the environmental condition.

The development of a MORT diagram takes into account four different elements [35]:

- 1st element: it can be considered the equivalent of the top event in FTA and defines the general objective. It is the set of potential damaging energy flows and environmental conditions (sources of hazard);
- 2nd element: it is made of the people and the systems that are potentially vulnerable to the sources of hazard;

- 3rd element: it takes into account the possible malfunctioning or missing barriers and controls aimed at isolating the vulnerable objects from the sources of hazard; and
- 4th element: it is made of the precursor events.

The MORT analysis is based on a predefined structure of a tree (or generic tree), developed vertically from top to bottom. The structure, which is quite complex, must be considered as a standard checklist to use, as a reference, when performing the analysis. The generic tree contains about 100 problematic areas and 1500 possible causes and it is based on the historical experience and on studies carried out by experts in human factors. The detailed instructions on the procedures and the modalities to be followed when using this technique are available in specialised handbooks. Some automatic applications of the MORT techniques, based on software solutions, have been developed to remedy the excessive complexity of this method: examples of these simpler variants are the Mini-MORT, the SMORT (Safety Management Organizational Review Technique), or the PET (Project Evaluation Tree). Just like other structured techniques, MORT is not intended for direct application in the field, but it is seen as a tool for the analytic arrangement and examination of those facts and evidence collected in the field.

MORT is a systematic method to plan, organise and conduct an accident investigation. Essentially, a MORT tree is an organizational FTA that identifies causes and contributing factors by looking for the general causes and conditions which are systematically related to the incident. The undesired consequence, generally a failure or poor performance, is the starting point. The weaknesses and deficiencies that led to the undesired consequence are tracked, taking into account the safety programs in the organization that had to control them. The method allows identifying who or which program has to be considered the cause(s) for what happened. After the causes are identified, it is possible to assess and fix those parts that failed [3].

Typically, the MORT is based on five-stage accident sequence [16]:

- Background factors;
- initiating factors (underlying conditions);
- intermediate factors (e.g. environmental and hazard recognition);
- immediate factors (the trigger events or unsafe acts);
- resulting consequence results.

Elements in a MORT diagram are placed over three main branches [37]:

- The specific oversights and omissions related to the investigated incident;
- the assumed risks (known, but not controlled);
- the general characteristics of the management system.

The various elements are numbered and a set of questions is available for the analyst for each of the number reference.

Taking inspiration from [4] and [38], MORT is a graphical checklist that helps the investigators in finding the cause of the accident by driving their reasoning among the predefined structure, enabling investigators to focus on potential key causal factors. The top part of a MORT diagram is shown in Figure 5.35. Its usage requires a sound knowledge of the method and extensive training to go deep in the analysis of complex incidents. Moving down through the tree, level by level, investigators may found a deficient event, which is labelled as "Less Than Adequate" (LTA) in MORT terminology

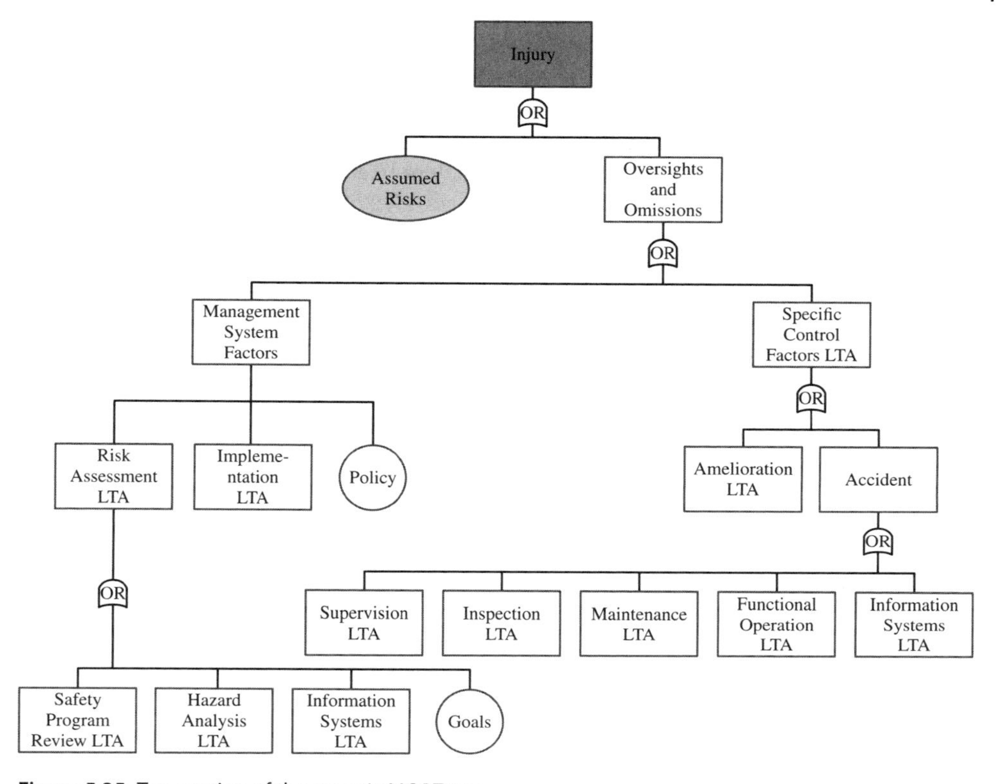

Figure 5.35 Top portion of the generic MORT tree.

and it is marked in red in the tree. Instead, an event that is "satisfactory" is marked green; unknowns are in blue. When the analysis is completed, it is possible to track the cause-and-effect path, from the undesired consequence to its root causes. The predefined structure of the tree allows a clear highlighting of the corrective actions that should be taken to avoid further reoccurrence of similar incidents. The chart is also used to assess the adequacy of control elements already in place.

The MORT chart here briefly presented is the key part of the whole MORT system safety program [19].

Here is an outline of how to apply the method:

- Obtain a copy of the MORT diagram;
- state the top of the tree, i.e. the undesired consequence of the incident;
- construct the tree using a deductive reasoning;
  - identify the elements that describe what happened (which barrier or control existed);
  - for each barrier or control, identify the managerial issues that explain why it happened;
  - color in red the LTA factors;
  - color in green the satisfactory factors;
  - color in blue those factors that have not enough elements to reach a conclusion;

- continue constructing the tree until root causes are reached;
- describe the identified problems;
- summarise the findings.

Being a structured approach, the analyst will see that parts of the tree cannot be applied to the peculiarities of the occurred incident. Conversely, some branches could be further developed to better evaluate some complex issues emerging from the diagram. However, it is clear that the MORT, like every predefined tool, aims at providing a means to assist, not to give additional workload. This implies that each element of the analytic tree should not be more complex than the subject being analyzed. An example of MORT chart for maintenance issue is shown in Figure 5.36.

5.4.2.2 Barrier-based Systematic Cause Analysis Technique (BSCAT™)

The materials used to write this paragraph are used by permission of CGE Risk Management Solutions (NL).

BSCAT™ is a relatively new methodology which takes its origin from the SCAT: the Systematic Cause Analysis Technique. Therefore, it is necessary to introduce the SCAT methodology before. SCAT is a well-known root cause analysis which takes into account the DNV-GL loss causation model. It allows to establish the hierarchy of the accident evolution, from the immediate cause to the root causes and was designed to help the investigator in using the DNV-GL loss causation model to real events. The SCAT chart was developed to analyze an event by means of standardised event descriptions.

The BSCAT™ method intends to apply the SCAT methodology to every single barrier separately, not only to the incident as a whole. Therefore BSCAT™ links together

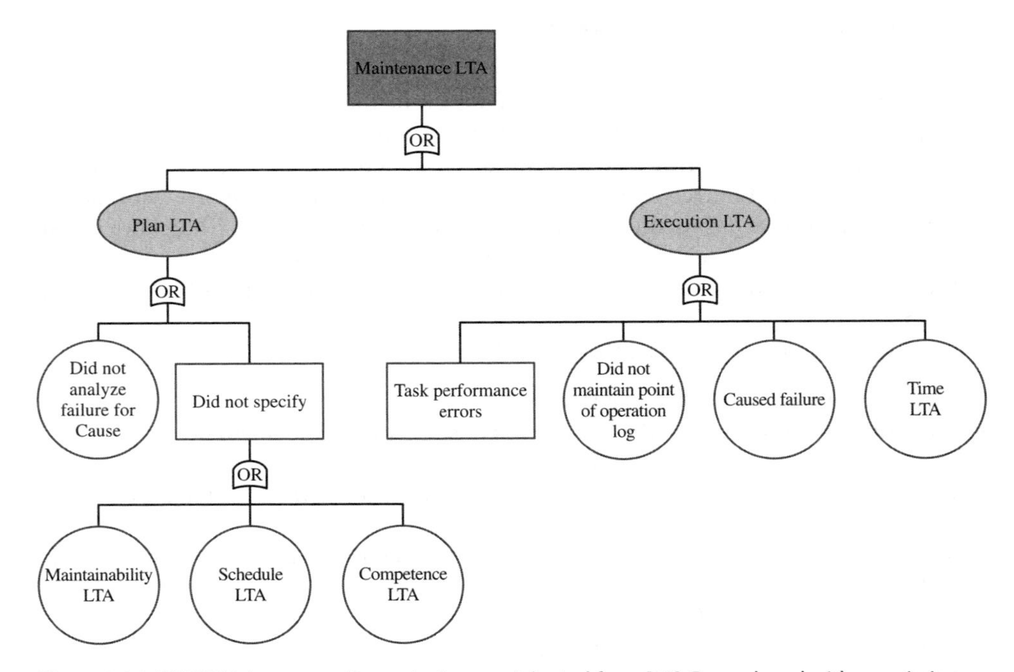

Figure 5.36 MORT Maintenance Example. Source: Adapted from [19]. Reproduced with permission.

the modern risk-based safety management approach with the systematic cause analysis for incident investigation. In other words, a deeper understanding of the incident is provided by an approach built on barrier-based risk management, while still identifying management system root causes. The letter B stands for "barrier-based", since each barrier is tested for why it failed. As a consequence, the BSCAT™ chart is a sort of upgrade of a SCAT one. The main difference between SCAT and BSCAT™ are shown in Figure 5.37.

The presented approach delivers a powerful and easy-to-understand causation diagram, developing internal knowledge, facilitating communication and ensuring that lessons are learned and linked to facility risk assessments.

An accident investigation through the BSCAT™ methodology consists of the following steps:

1. Evidence collection and preservation;
2. creation of a timeline/storyboard. It is a listing of the main important events and the relevant factor sorted in a temporal sequence. This is especially suggested for complex incidents, involving many people and systems (i.e. an evident complexity), helping to understand how latent issues (design aspects, unrevealed failures) affect the outcome;
3. identification of the key events and place them in an event flow diagram. The event types are shown in Figure 5.38;
4. identification of the barriers between the key events;
5. analysis of the barriers (what is gone wrong? How? Why?);
6. perform a BSCAT™ analysis on each barrier; and
7. generate the final report.

Steps from 2 to 5 are repeated for each hypothesis about the accident causation.

When assessing the state of each barrier, the icons in Figure 5.39 can be adopted to have a clearer view about them. The relation between barrier state and barrier lifecycle is shown in Figure 5.40. In order to decide the barrier state, the following logic should be followed:

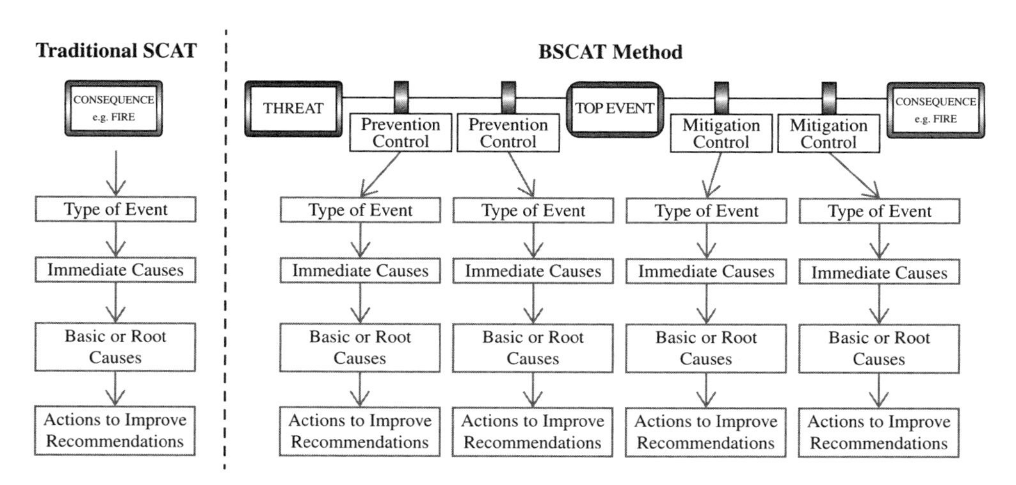

Figure 5.37 Difference between SCAT and BSCAT™ (Courtesy of CGE Risk Management Solutions (NL)).

Event | Threat | Consequence | Top event | Potential but not reached

Figure 5.38 Events types in a BSCAT™ diagram (Courtesy of CGE Risk Management Solutions (NL)).

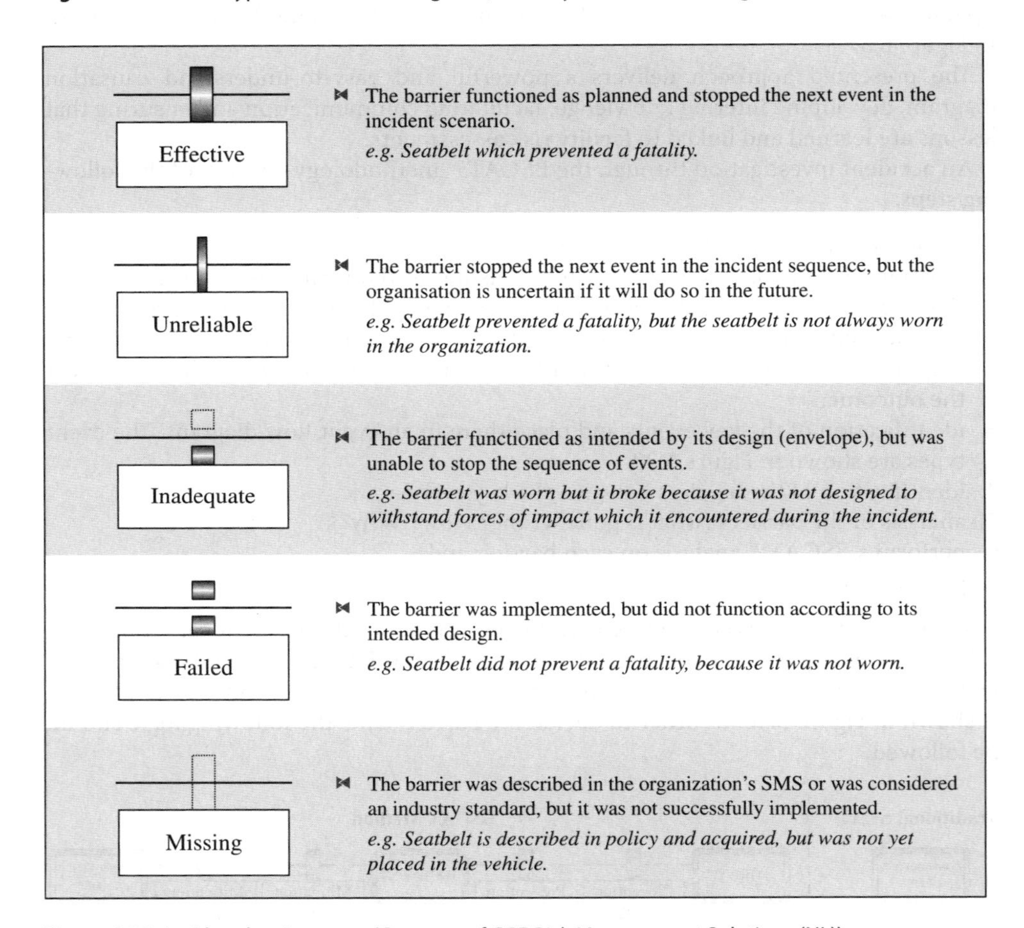

Figure 5.39 Incident barrier states (Courtesy of CGE Risk Management Solutions (NL)).

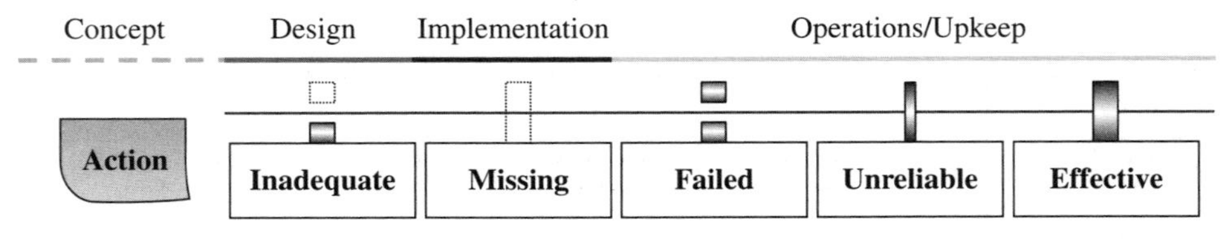

Figure 5.40 Relation between barrier state and barrier lifecycle (Courtesy of CGE Risk Management Solutions (NL)).

- Was the barrier described in the company's SMS or was it considered an industry standard? If no, take actions; if yes, ask the following;
- was the barrier implemented or could the barrier at one point perform according to its specification? If no, it is a "missing" barrier; if yes, ask the following;
- did the barrier function according to its intended design (envelope)? If no, it is a "failed" barrier; if yes, ask the following;
- did the barrier stop the next event in the incident sequence? If no, it is an "inadequate" barrier; if yes, ask the following;
- are you confident the barrier will stop the next event in the incident sequence in the future? If no, it is an "unreliable" barrier; if yes, it is an "effective barrier".

Two examples of BSCAT™ diagrams are in Figure 5.41 and Figure 5.42.

This technique is also used in combination with a bowtie, to give a wider view of the incident investigation [39]. The bowtie methodology is used for risk assessment, risk management and risk communication. The method is designed to provide a better overview of the situation in which certain risks are present, to help people understanding the relationship between the risks and organizational events. Risk in the bowtie methodology is represented by the relationship between hazards, top events, threats and consequences. Barriers are used to display what measures an organization has in place to control the risk. All these are combined in an easy-to-read diagram, as shown in Figure 5.43.

The word "hazard" suggests that it is unwanted, but in fact it is the opposite: it is precisely the thing you want or even need to make business. It is an entity with the potential to cause harm but without it there is no business. For example, in the oil industry, oil is a dangerous substance (and can cause a lot of injuries when treated without care) but it is the one the thing that keeps the oil industry in business! It needs to be managed because insofar as it is under control, it is of no harm.

Thus, as long as a hazard is controlled it is in its wanted state. For example: oil in a pipe on its way to shore. But certain events can cause the hazard to be released. In bowtie methodology such an event is called the top event. The top event is not a catastrophe yet, but the dangerous characteristics of the hazard are now in the open. This is the moment in which control over the hazard is lost. For example: oil is outside of the pipeline (loss of containment). Not a major disaster, but if not mitigated correctly it can result in multiple undesired events (consequences).

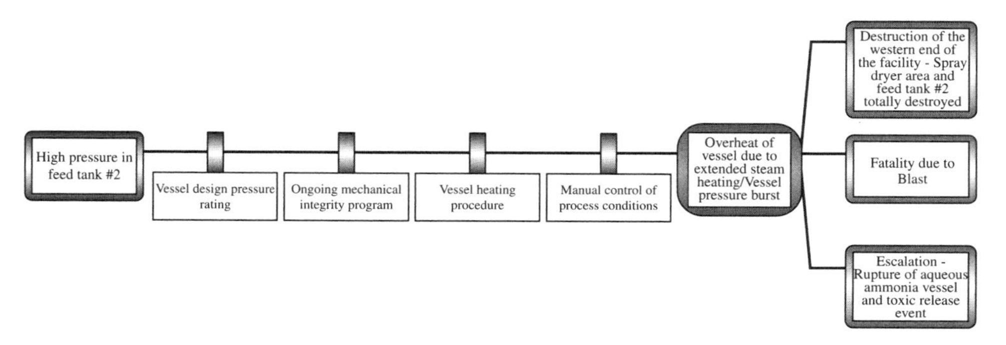

Figure 5.41 Example BSCAT™ diagram (Courtesy of CGE Risk Management Solutions (NL)).

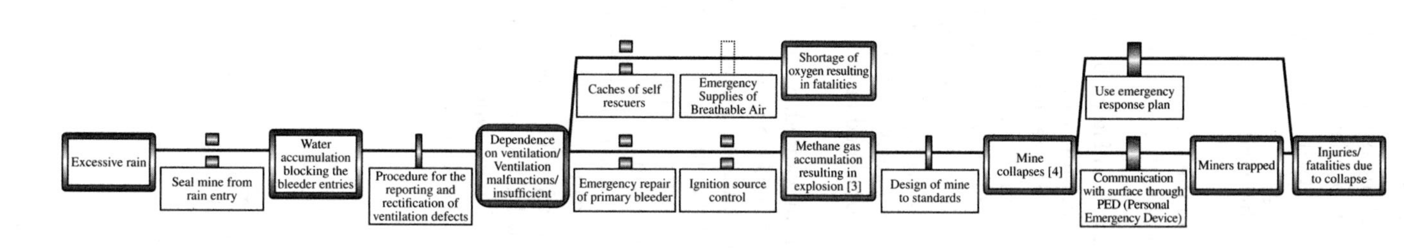

Figure 5.42 Example BSCAT™ diagram (Courtesy of CGE Risk Management Solutions (NL)).

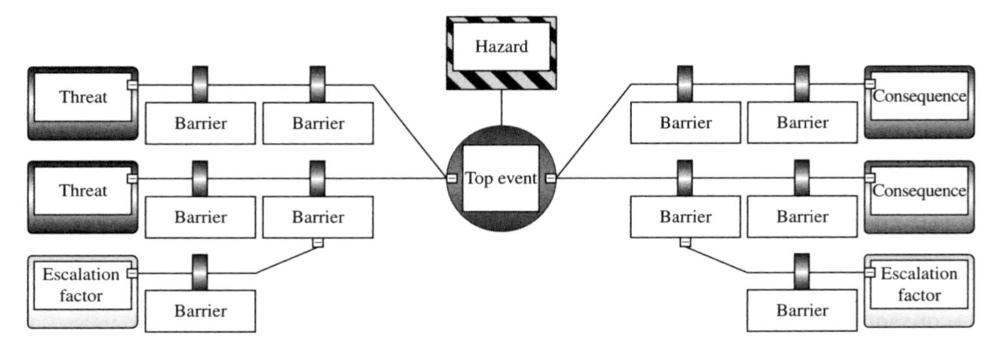

Figure 5.43 The bowtie diagram (Courtesy of CGE Risk Management Solutions (NL)).

Often several factors could cause the top event. In bowtie methodology these are called threats. These threats need to be sufficient or necessary: every threat itself should have the ability to cause the top event. For example: corrosion of the pipeline can lead to the loss of containment.

When a top event has occurred, it can result in certain consequences. A consequence is a potential event resulting from the release of the hazard which results directly in loss or damage. Consequences in bowtie methodology are unwanted events that an organization 'by all means' wants to avoid. For example: oil leaking into the environment.

Risk management is about controlling risks. This is done by placing barriers to prevent certain events from happening. A barrier (or control) can be any measure taken that acts against some undesirable force or intention, in order to maintain a desired state. In bowtie methodology there are proactive barriers (on the left side of the top event) that prevent the top event from happening. For example: regularly corrosion inspections of the pipelines. There are also reactive barriers (on the right side of the top event) that prevent the top event resulting into unwanted consequences. For example: leak detection equipment or concrete floor around oil tank platform. Note the terms barrier and control are the same construct and depending on industry and company, one or the other is used. In this book we will use the term barrier.

In an ideal situation a barrier will stop a threat from causing the top event. However, many barriers are not a 100% effective. Certain conditions can make a barrier fail. In bowtie methodology these are called escalation factors. An escalation factor is a condition that leads to increased risk by defeating or reducing the effectiveness of a barrier. For example: earthquake leading to cracks in the concrete floor around a pipeline.

Escalation factors are also known as defeating factors or barrier decay mechanisms – which term is used depends on industry and company. In this document we will use the term escalation factor.

After creating the basic bowtie diagram, there are several ways to work out the barriers in more detail. One good way is to identify and link the underlying management system activities to the barriers. This will tell what should be done to keep the barriers working, like maintenance activities on hardware barriers. Mapping the management system onto a bowtie also demonstrates in more detail how barriers are managed by a company. Furthermore, responsibilities could be attached to barriers, as well as a rating of their effectiveness and what type of barrier it is.

In conclusion, the following terms should now be familiar:

- The hazard, part of normal business but with the potential to cause harm, can be released by:
- a top event, no catastrophe yet but the first event in a chain of unwanted events;
- the top event can be caused by threats (sufficient or necessary causes);
- the top event has the potential to lead to unwanted consequences;
- proactive barriers are measures taken to prevent threats from resulting into the top event;
- reactive barriers are measures taken to avoid that the top event leads to unwanted consequences; and
- an escalation factor is a condition that defeats or reduces the effectiveness of a barrier.

Coming back to the combination of BSCAT™ and bowties, it is possible to reuse and link existing risk assessment information (bowties) and do full integration of incident investigation and risk analysis (Figure 5.44). If applicable bowtie diagrams are available for use during the investigation, you can bring events and barriers from the bowtie directly into your incident analysis diagram. This results in a better fit between incident and risk assessment analysis, which in turn allows the company to improve the risk assessment. Particularly for small incidents this is a significant advantage – it allows staff to analyze incidents in a barrier-based methodology with minimal training. Creating a barrier-based incident diagram requires training, but with this templated approach, all incidents which fit onto existing bowties, can be quickly analyzed using any incident analysis method, including BSCAT™.

After finishing the incident analysis, the incident data and recommendations can be linked back into the bowtie risk assessment and visualised on the barriers. Bringing all your incidents into a single case file allows you to aggregate barrier failures and lets users do trend analysis over multiple incidents and therefore it allows the company to visually see the weaker areas in its management system. This entire process allows gauging barrier effectiveness and availability based on real-world information extracted from the incident analyses.

5.4.2.3 Tripod Beta

The materials used to write this paragraph are used by permission of CGE Risk Management Solutions (NL).

Tripod Beta is an incident investigation methodology developed in the early 1990s. It was explicitly created, in line with the human behavior model, to help accident investigators model incidents to understand how the environment influenced the sequence of events and to discover the root organizational deficiencies that triggered the incident. Indeed, the idea behind Tripod is that organisational failures are the main factors in accident causation [4].

Using Tripod Beta, incident investigators model incidents in terms of:

- Objects (something acted upon, such as a flammable substance or a piece of equipment);
- agent of change (something – often an energy – that acts upon objects, such as a person or fire); and
- events (the result of an agent acting upon an object, such as an explosion).

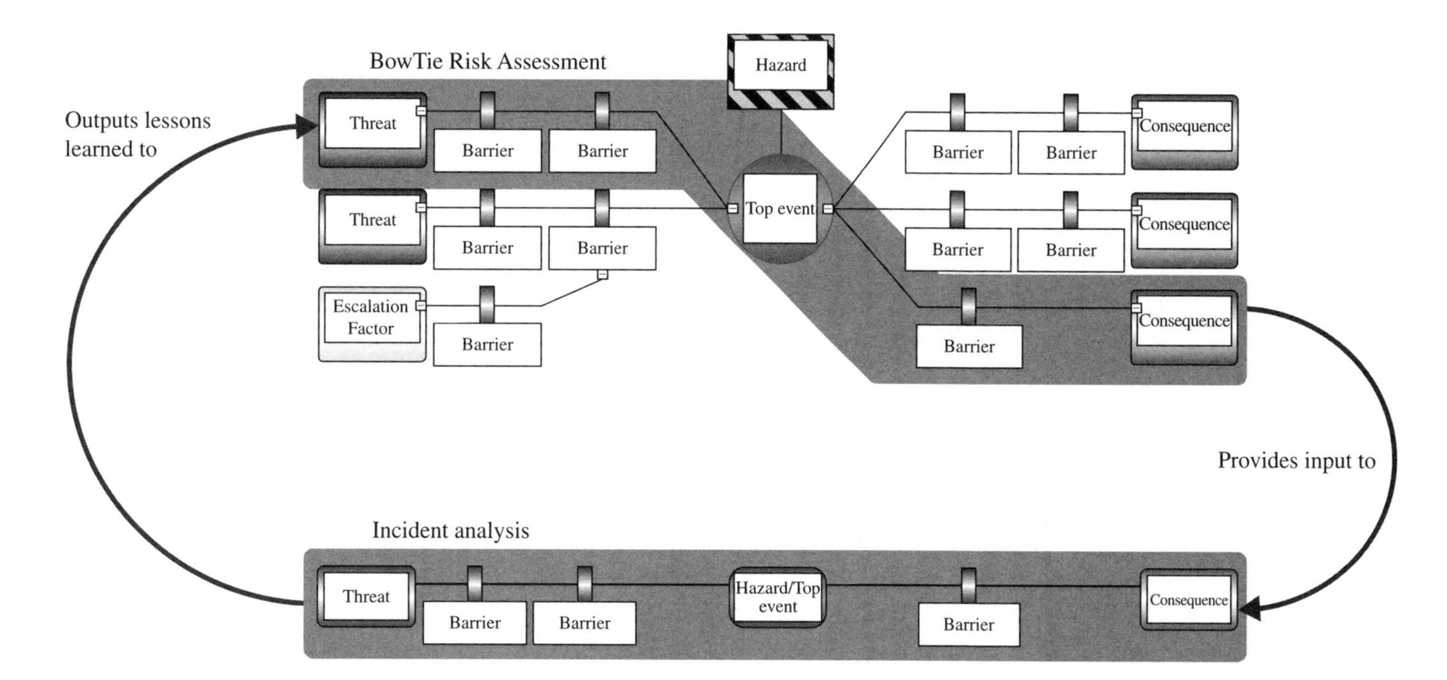

Figure 5.44 Bowtie risk assessment & incident analysis (Courtesy of CGE Risk Management Solutions (NL)).

Working back from the top event (the incident) allows a full understanding of what happened and how it happened. A set of shapes consisting of an agent, an object and an event is called a trio and is the basic building block of the Tripod beta method. Events themselves can also be objects or agents, allowing the investigator to chain these trios into a large diagram.

To understand why the incident happened, the next step is to determine what barriers were in place to prevent those objects and agents acting in the way they did and why they failed. Tripod Beta teaches looking at the immediate causes of the act that led to the incident, the psychological precursors to that, and ultimately the underlying organizational deficiencies that allowed those precursors to exist. An example of a Tripod Beta diagram is shown in Figure 5.45.

Performing a Tripod Beta analysis means following these steps:

1. Evidence collection and preservation;
2. creation of a timeline/storyboard. It is a listing of the main important events and the relevant factor sorted in a temporal sequence. This is especially suggested for complex incidents, involving many people and systems (i.e. an evident complexity), helping to understand how latent issues (design aspects, unrevealed failures) affect the outcome;
3. identification of the trios (agents, objects, and events) and link each other creating an event flow diagram. The combined node "event-agent" and "event-object" are the outcome of the linking process of an event to an agent or an event to an object respectively;
4. identification of the barriers between the event and agent or object;
5. analysis of the barriers (what is gone wrong? How? Why?);
6. perform a Tripod Beta causation assessment on each barrier; and
7. generate the final report.

The possible Tripod Beta appearances are in Figure 5.46.

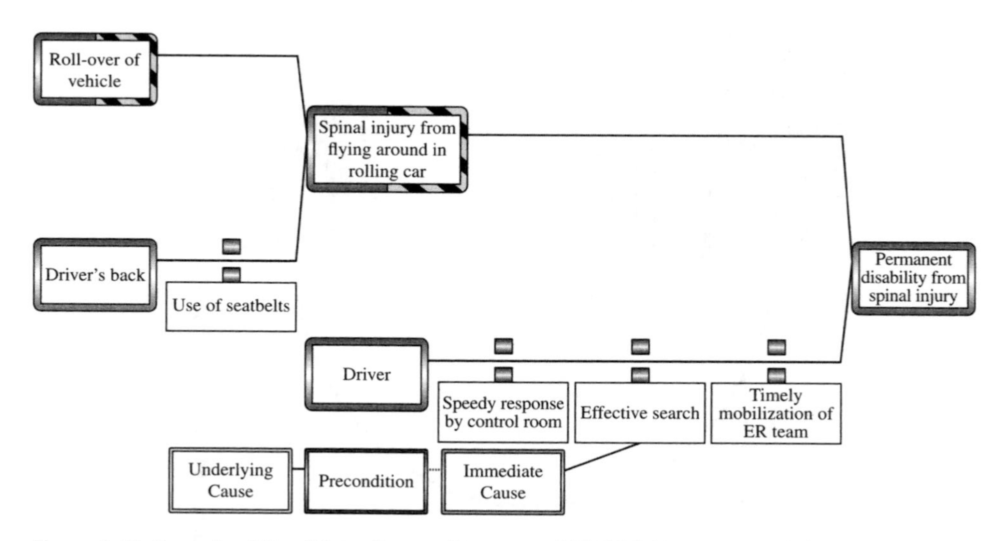

Figure 5.45 Example a Tripod Beta diagram (Courtesy of CGE Risk Management Solutions (NL)).

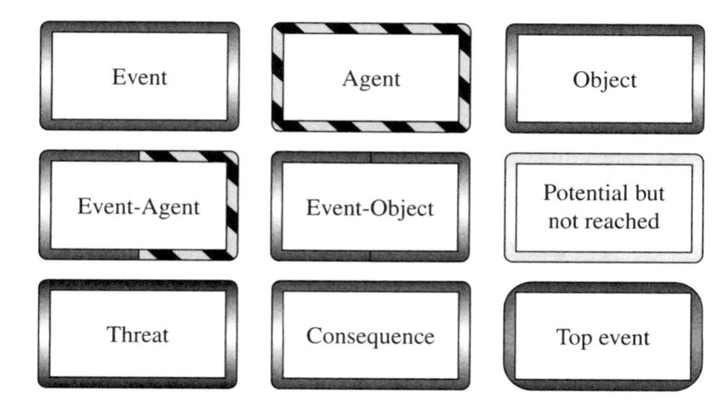

Figure 5.46 Possible Tripod Beta appearances (Courtesy of CGE Risk Management Solutions (NL)).

The state of each barrier can be drawn with the icons already showed in Figure 5.39 for BSCAT™.

The idea behind Tripod is that substandard situations do not just occur; indeed, they are generated by mechanisms coming from decisions taken at management level. These underlying mechanisms are the Basic Risk Factors (BRF). They cover a broad range of factors (human, organizational and technical) including psychological precursors like time pressure, poor motivation, and so on. Prevention of incidents is therefore performed by removing these latent factors. Examples of possible BRFs are in Table 5.6.

The Tripod Beta method merges two different models: the Hazard and Effects Management Process (HEMP) model and the original Tripod model. The accident mechanism according to the HEMP method is shown in Figure 5.47. This merging results in a computer-based instrument that exploits a menu-driven tool that guides the investigator in representing the accident dynamics.

Table 5.6 Definition of BRFs in Tripod.

Basic Risk Factor	Short description
Design	User-unfriendly tools or equipment
Tools and equipment	Poor quality, suitability or availability
Maintenance management	Inadequate performance of maintenance tasks
Housekeeping	Insufficient attention to keep the floor cleaned
Error enforcing conditions	Not appropriate physical performance
Procedures	Insufficient quality or availability of procedures
Training	Insufficient competences or experiences
Communications	Ineffective communications
Incompatible goals	Financial/production goals inconsistent with optimal working conditions
Organization	Inadequate or ineffective management
Defences	Insufficient protection of people, material and environment

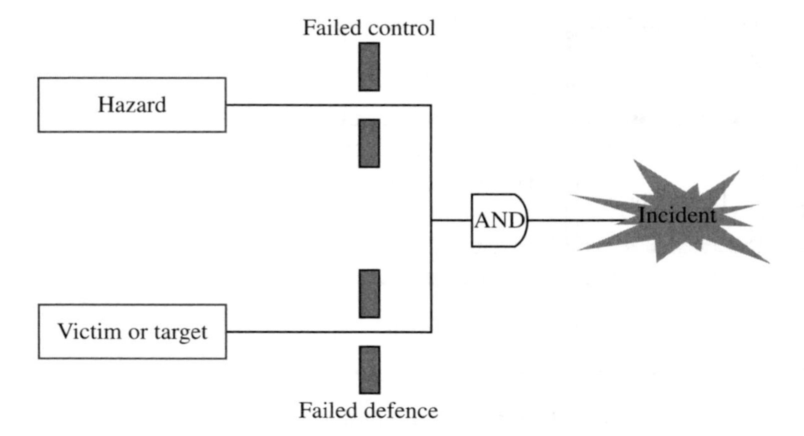

Figure 5.47 Accident mechanism according to HEMP method.

The Tripod Beta method follows the new way of investigating incidents; indeed, after the BRFs are identified, recommendations are developed in order to decrease or eliminate their impact. This means that the real source of the problem is faced, not simply the symptoms [4].

5.4.2.4 Barrier Failure Analysis (BFA)

The materials used to write this paragraph are used by permission of CGE Risk Management Solutions (NL).

Barrier Failure Analysis (BFA) is a pragmatic, un-opinionated, general-purpose incident analysis method. It has no affiliation with any particular organization. BFA is a way to structure an incident and to categorise parts of incident taxonomy. The structure has events, barriers and causation paths. Events are used to describe an unwanted causal sequence of events. This means each event causes the next event. There can also be parallel events that together cause the next event.

Barriers are used to highlight certain parts of our environment as being primarily designed to stop a chain of events. They are not necessarily independent, or sufficient. Since the unwanted events still happened, causation paths are added to explain why the barriers did not perform their function. The causation path goes three levels deep. The levels are simply called Primary, Secondary and Tertiary level. These labels can be changed, but the idea is that a barrier can be analyzed in three causal steps. It does not specify whether the analysis should end on an organizational level or not, although this is what would happen most.

Each level in the causation path can also be categorised. Because there is an infinite number of possible categorizations and a large number of different kinds of organizations, it is not possible to create a single definitive set of categories. Instead, the user can create custom categories. This is why any categorization should go through iterations, to add and remove categories as they become wanted or obsolete. Optionally, any organization has to make the categories specific to their context. This has a downside of not being able to make comparisons between different organizations, but that is not the primary goal of this methodology. It is better to use categories that are relevant than ones that are standardised.

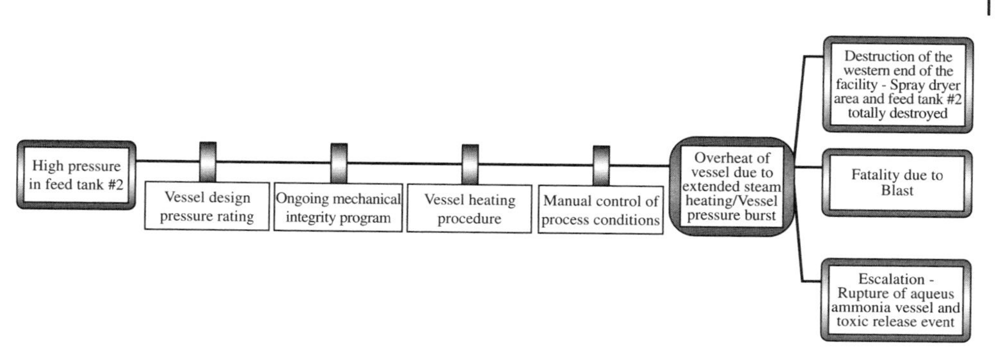

Figure 5.48 Example of a BFA diagram (Courtesy of CGE Risk Management Solutions (NL)).

Any organization should go through an initial period of testing and iterating categories. At some point a steady state should emerge that will capture most incidents. There will always be exceptions, but they should be exactly that, exceptions. Once exceptions happen more frequently, they stop being exceptions and should be integrated into the existing categorizations in a new iteration of the taxonomy.

The different appearance options for the events and the barrier states are the same as already shown previously for BSCAT™ (Figure 5.38 and Figure 5.39). Examples of a BFA diagram are shown in Figure 5.48 and Figure 5.49.

The core elements of a BFA diagram are shown in Figure 5.50.

Once developed, a BFA diagram looks like the structure shown in Figure 5.51.

The BFA is carried out in seven steps:

- Fact-finding (timeline)
- event chaining
- identifying barriers
- assessing barrier state
- causation analysis & categories
- recommendations and
- reporting/link to bowtie diagram

The first step is essential to get an overview by arranging the facts. It is suggested that the investigator should focus on gathering as much info as possible, beware of assumptions, focus on the facts, and use the timeline tool to organise the facts. A common pitfall is making recommendations at this stage, but it must be avoided.

The second step is the "event chaining" (Figure 5.52). Firstly, it is necessary to define what an event is: "a happening or a change of state, in which the incident sequence changes".

Two points are certain: the normal mode of operations and the moment of incident. The zoom level will depend on the scope: complex incidents typically require more events. It is important to not confuse defeated barriers with BFA events (Figure 5.53).

The third step is identification of the barrier (Figure 5.54). At this stage, the investigator asks: "which barriers should have prevented the next event?". A barrier should be defined in the normal/wanted state and should be able to prevent the event on its right. The barriers are put in the order of their effect.

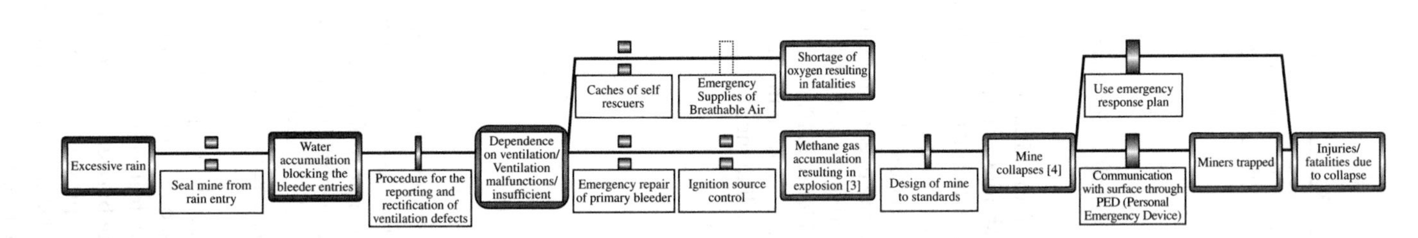

Figure 5.49 Example of a BFA diagram (Courtesy of CGE Risk Management Solutions (NL)).

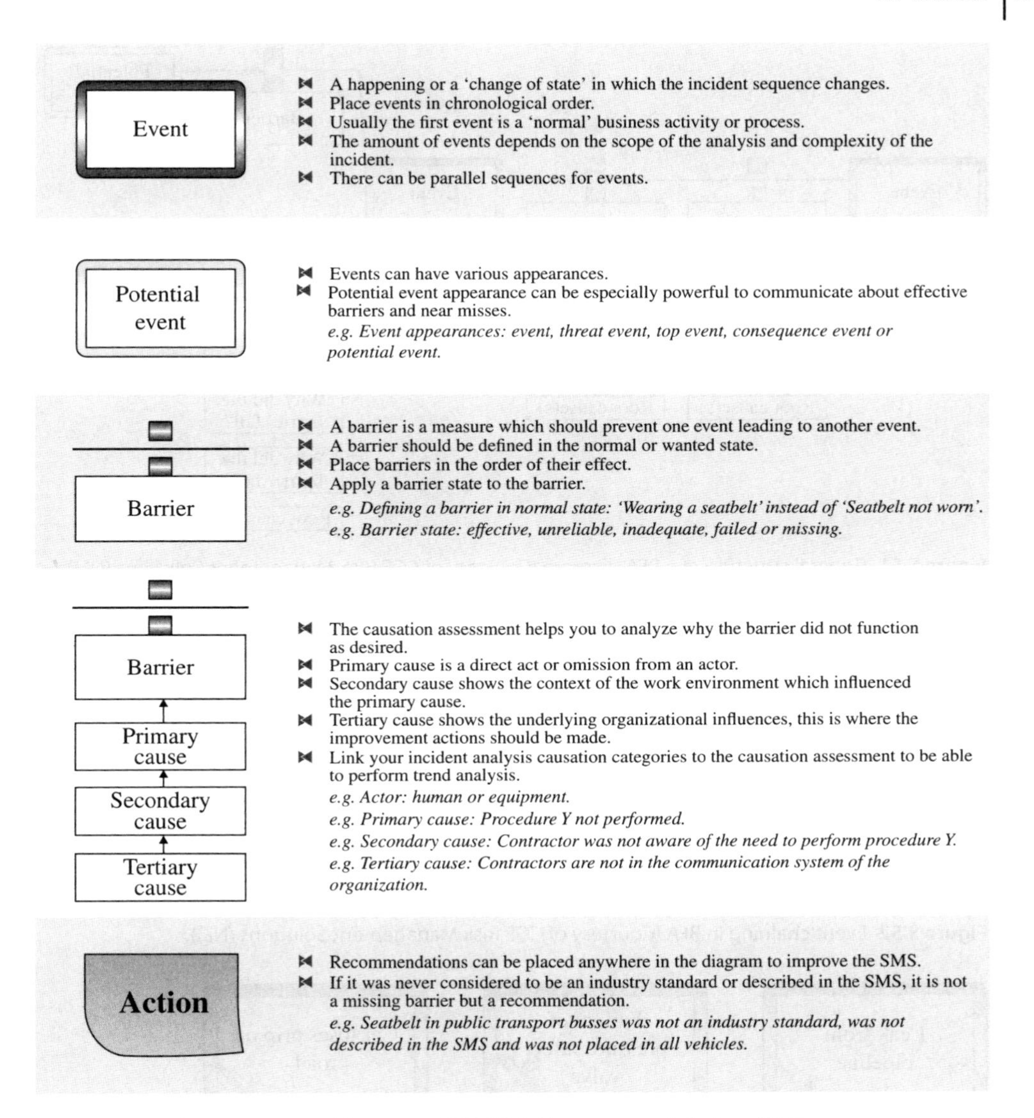

Figure 5.50 BFA core elements (Courtesy of CGE Risk Management Solutions (NL)).

Attention should be paid at this step, in order to avoid some common pitfalls, as shown in Figure 5.55.

The fourth step is the assessment of the barrier state, according to what already discussed for BSCAT™.

The fifth step is the BFA analysis (Figure 5.56). The aim of this step is to understand what caused the barrier to fail.

For each barrier, the analysis tries to find:

- Primary cause. What exactly happened? (Operational);
- secondary cause. Why did it happen? (Line management); and
- tertiary cause. How could the management prevent it? (Management).

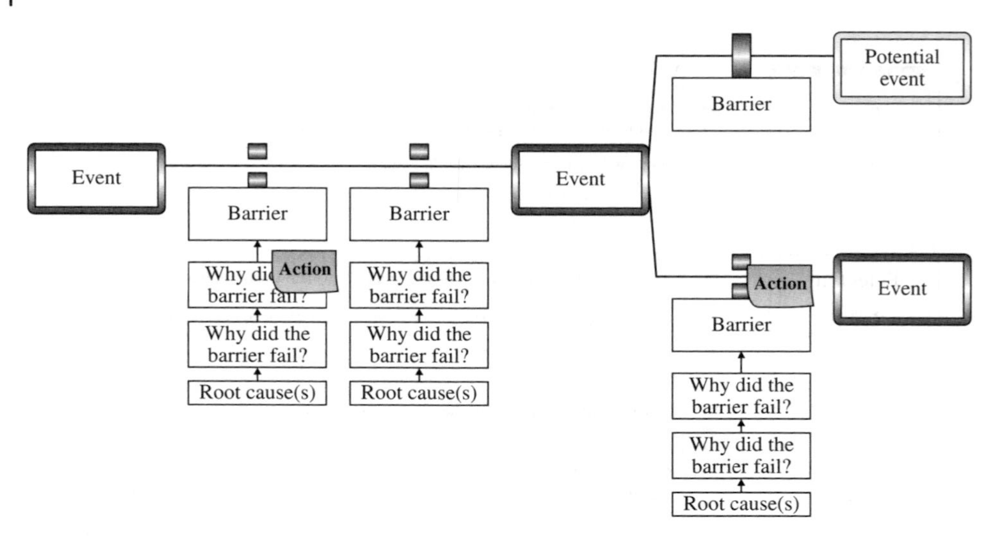

Figure 5.51 General structure of a BFA diagram (Courtesy of CGE Risk Management Solutions (NL)).

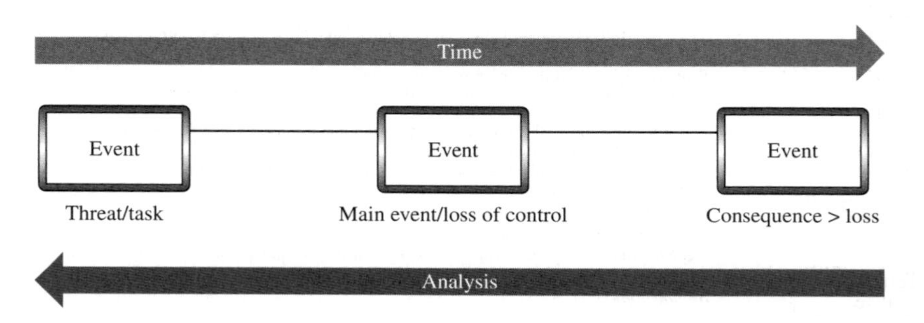

Figure 5.52 Event chaining in BFA (Courtesy of CGE Risk Management Solutions (NL)).

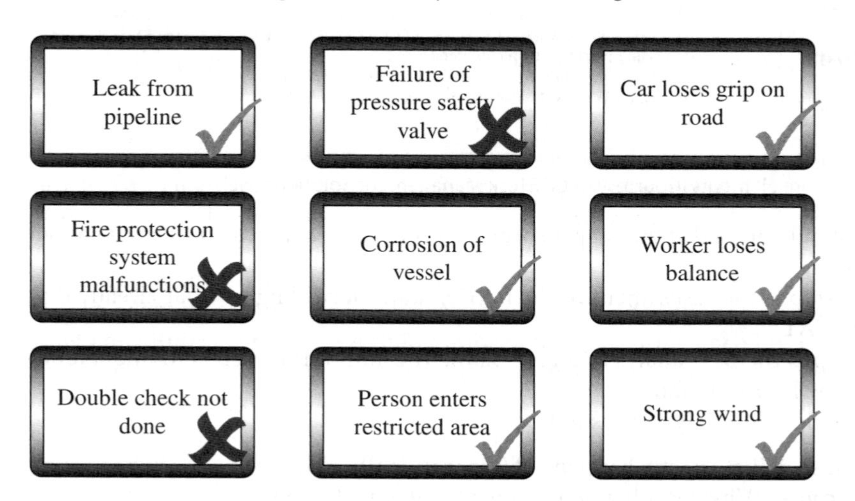

Figure 5.53 Defeated barriers are not BFA events (Courtesy of CGE Risk Management Solutions (NL)).

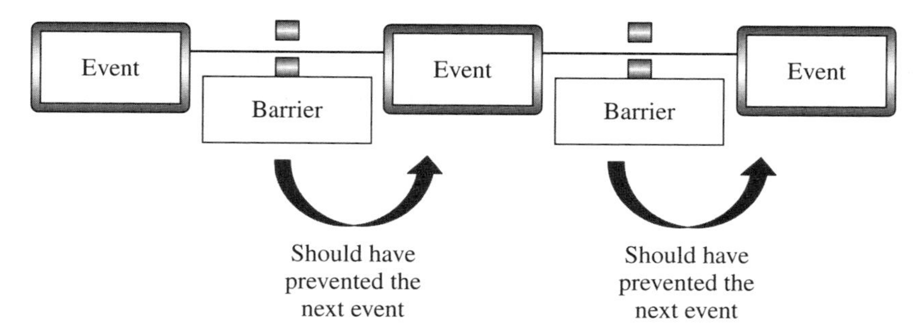

Figure 5.54 Barrier identification in BFA (Courtesy of CGE Risk Management Solutions (NL)).

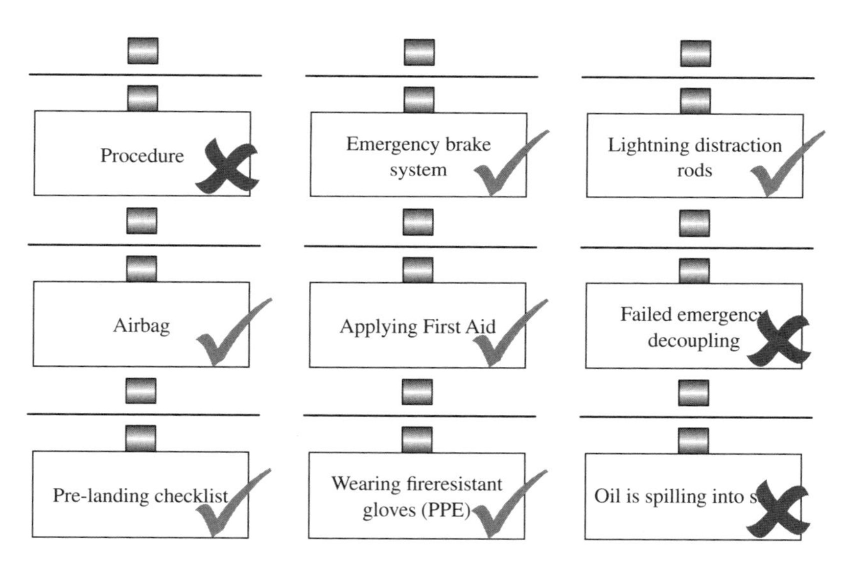

Figure 5.55 Correct and incorrect barrier identification in BFA (Courtesy of CGE Risk Management Solutions (NL)).

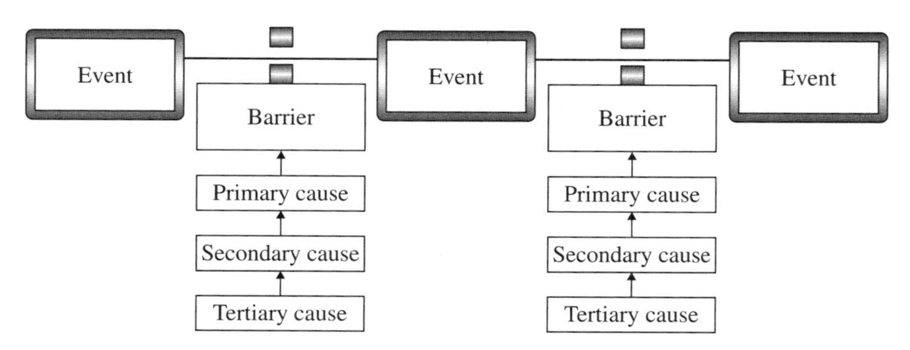

Figure 5.56 BFA analysis (Courtesy of CGE Risk Management Solutions (NL)).

The sixth step is the development of recommendations. How to prevent future accidents? Improving the safety management system. Actions should be formulated as tasks, assigning job titles to actions, and defining a target date. The topic is discussed in deep in the next Chapter. However, at this stage, it is useful to distinguish among:

- Short term solution: barrier level. Improve barrier quality before resuming operations again;
- medium-term solution: barrier level. Add barrier before resuming operations again; and
- long-term solution: organizational level. Correct management system/underlying cause.

The last step is the reporting and the eventual link with bowtie, as already discussed for BSCAT™.

5.4.2.5 Root Cause Analysis (RCA)

Root Cause Analysis (RCA) is among the most widely used incident investigation techniques. It is simple: it starts with the occurred incident and, continuously asking "why?", it goes down into the chain of events to find the root causes. The basic idea of drilling down to the underlying cause is adopted by several proprietary methods that share the common basic idea of RCA: keep asking "why" to find root causes.

When developing an RCA analysis, the investigator needs to know when to stop. Indeed, if the method is used without such a precaution, its result can be a useless jumble of elements, and the investigation team members could find themselves talking about theology or the two chief world systems by Galileo. To avoid this, categorisations and stopping criteria should be defined before starting with an RCA.

A powerful definition of cause analysis is given in [19], and it is here entirely cited:

"Simply put, cause analysis is the process by which you discover the invisible thoughts - mental! models, beliefs, values - that influence, and then produce, the visible behaviours/actions. The reasons need to be discovered so actions to prevent recurrence can be initiated to prevent future incidents. hue structured search is called root cause analysis. The underlying drivers or reasons are called root causes. Once your investigation allows you to fully integrate cause analysis techniques into a systematic methodology for analyzing and solving problems, you will be able to take the lead in decision-making and quality control that will be reflected in better and more consistent results for your organization."

In the next Paragraphs, three particular applications of the RCA are presented:

- Root Cause Map™;
- TapRoot®;
- Apollo™.

In the following introductory Paragraph, the basic ideas about the RCA are presented, so to have a common knowledge as the basis of the explanation of the proprietary method discussed in this book.

Introduction to RCA An extensive bibliography is available about Root Cause Analysis, being the most widely adopted approach to incident investigation. Taking inspiration from it, some basic concepts are presented in this Paragraph, highlighting the common hypotheses at the base of the technique, regardless the specific adopted methodology (which might be also proprietary).

Basically, as underlined by [4], the term "root cause analysis" is intended to refer to any techniques that identify root causes, i.e. those underlying weaknesses in the management system that eventually result in the incident and whose correction would prevent the occurrence of the same event and similar ones as well. The input data for the RCA are those coming from the previous investigation stages, when the investigator answered questions about what, when, where, who, and how the incident happened. With RCA, the investigation process adds new information: why the incident occurred. It is self-evident that accurate and comprehensive root causes derive from an exhaustive listing of causal factors.

The difference between traditional problem solving and RCA is that the second method is structured, thus developing effective conclusions and recommendations, focusing not only on individuals but also on all those factors affecting the performance of the task, such as the environment, the equipment, and other external factors. Instead, traditional problem solving does not possess that rigorous approach that identifies solutions connected to the causes of the incident [11].

The identification of root causes is undoubtedly the most challenging goal of an incident investigation. It requires the preliminary acquisition of all the causal factors. If the RCA is performed too early, poor conclusions will be reached, and ineffective recommendations will be developed. Actually, the identification of root causes is a double challenge for the investigator [9]: firstly, it is necessary to identify the remote factors; secondly, causalities, i.e. their influence on the incident occurrence, must be proven. This is why performing an RCA requires additional competencies for the investigators; indeed, recalling the influence of human factors, knowledge on social and human science should be possessed by the RCA performer. It must be observed how rare are these skills in a technician.

One way to pass from immediate to root causes is to look for safety barriers, which did not prevent the incident. This approach relies on models for risk assessment, with their own pros and cons. In particular, the limit is the weak capability to highlight the rationale behind the described actions. Therefore, comprehensive approaches should be used to also take into account the human and social aspects of the complex system.

Following the path suggested in [11], Root Cause Analysis consists of:

- Selection of a causal factor from timeline, or cause-and-effect tree, or any other causal factor identifier;
- brainstorming, to generate a list of potential management system weaknesses for each causal factor (investigation tools, such as the Root Cause Map™ in [11] can be used to stimulate thinking about potential root causes); and
- documenting the results.

In RCA, all the possibilities within the mission statement should be considered, avoiding the a priori exclusion of some of them. Since the analysis is focused on root causes, the management systems should be thoroughly investigated, questioning about those assumptions that are taken for granted during the proactive analysis just to save time. Indeed, every business is equipped with a management system, to ensure that the potential losses identified by the proactive analysis (like a PHA) would be low-frequency. But low-frequency is different from zero, and regardless of the efforts, incidents do occur. The output of reactive analysis, like an RCA, are therefore essential to drive the proactive analyses and the management system towards continuous improvement. Having an effective incident investigation procedure is not enough, if proactive analyses and management systems are not able to receive the outcome of an incident investigation. Therefore, RCA may question about:

- The management of change, both of technological solutions and procedures;
- the level of training of personnel;
- the accuracy of written procedures.

The method requires a certain level of judgement by the investigator: indeed, RCA cannot be assigned to neophytes. In order to perform an RCA, the investigator should [11]:

- Think creatively, to identify new failure modes;
- adopt a shared approach, to use knowledge of other people, inside and outside the company, experiencing the incident;
- think inquisitively, to be curious about how things and people work;
- be sceptical, to refuse poor explanations (i.e. those typically including the terms "everything, everybody, all, obvious", and so on);
- think logically, to test hypothesis with available data; and
- always remember the system, the macro, not only the micro, the details.

Apparent or Root Cause Analysis differ from the level they reach. The apparent cause analysis stops at immediate causes, exploring the causal factors; but only the root cause analysis goes deep into the underlying and root causes, as shown in Figure 5.57. Moreover, testing a root cause is not the same as testing an immediate cause, where tests, or simulation can be sufficient. A root cause could be validated by testing the analyses on the actors at its base.

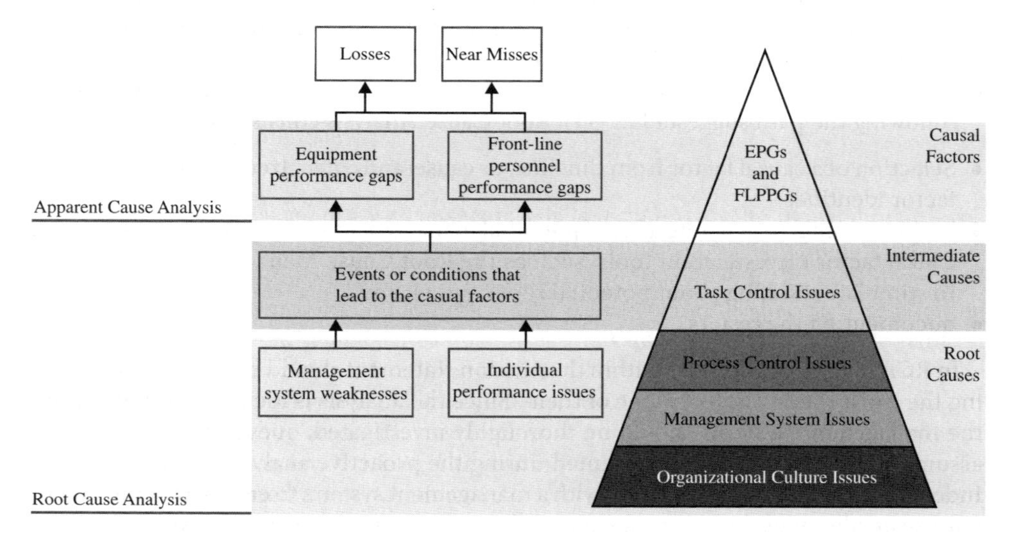

Figure 5.57 Levels of analysis.

Figure 5.57 also shows how the different levels actually affect different issues, on a scale from Equipment Performance Gaps (EPGs) and Front-Line Personnel Performance Gaps (FLPPGs) to organizational culture issues. Increasing the depth of the analysis, increasing the level of learning, and increasing the scope of corrective actions.

It is fundamental to distinguish among fundamental causes and root causes. If an incident occurred because someone slipped and fell, gravity could be seen as a fundamental cause. However, it is out of management control and therefore it is not a root cause. Indeed, a root cause deals with the weaknesses of the management system. From this point of view, the identification of a fundamental cause will increase the completeness of the investigation but it is irrelevant when the investigator develops the corrective actions to prevent further occurrences of the event. As cited in [11], root causes are "intended to be as deep as can reasonably be addressed with practical and measurable recommendations." Moreover, as already anticipated, there is rarely one single cause for an incident, and a single causal factor may have more than a single root cause associated with it.

Some common traps when performing RCA are the following:

- An equipment failure is seen as an event out of the control of management, because parts get old or simply work poorly. This is false, because equipment inspections, tests and maintenance are under the management system and prevent most failures. Moreover, defective parts should be detected by the quality management system;
- human performance failures are seen as out of the control of management, because "the procedure was right, the employee just made a mistake". This is false, because there are some correlations with the management system: correctness and sufficiency of training, accuracy of procedures, commitment to error, failures which have already happened but been overlooked, and so on; and
- external events, like natural phenomena, are obviously out of the control of management. However, this does not result in a poor management involvement. Indeed, the management system can minimise the risk associated with natural phenomena by reducing the magnitude of the probable consequences, for example through a proper structural design.

The main difference between RCA and other incident analysis methods is that RCA is not barrier based. Everything in RCA is an event, including those things that would be considered barriers or barrier failures in BSCAT™, Tripod or BFA. Therefore, whereas the barrier-based incident analysis methods like BSCAT™, Tripod and BFA can be mapped back onto the bowtie because their structure is similar, RCA cannot be linked back to a bowtie, because the bowtie structure depends heavily on identifying barriers, which RCA does not do.

When applying the Root Cause Analysis, to argument by analogy, examining incidents which have happened elsewhere and applying their recommendations to the incident being investigated, may have some limitations [12]. Firstly, attention should be paid to false extrapolation. Reasoning by analogy is useful, but it does not provide that thorough understanding of what really happened, because the assessment is no more fact-based but analogy-based. Secondly, stories from past experience are generally linear: first this happened, then this occurs, finally this is the result. The simplification of linearity may return a poor reasoning by analogy; indeed, events in an industrial incident are usually in relation to each other, creating a complex system. Finally, the storyteller possesses his/her own worldview: what is good, obvious, priority for someone cannot be for another. It is not about the inherent rightness or wrongness of statements: it is about their relative value, because of the person's point of view. Therefore, storytelling is encouraged, but it is important to be aware of its limitations.

For example, TIER diagrams [40] are one method to perform root cause analysis, discussed in [4] and mentioned in [38]. They help the investigator in finding not only the root causes, but also the corresponding management level that has the power to promote, implement, and follow-up the corrective actions. Another example of RCA is the PROACT® RCA, discussed in [41], where a proactive method to use the RCA is presented, not limiting its application only to accident investigation.

In conclusion, the reader should note that the RCA is a structured method to uncover the underlying factors, typically of undesired performance, even if it could also be used to investigate positive results. The interested reader may find additional information about RCA in [41].

Root Cause Map™ A powerful tool to investigate with RCA is the Root Cause Map™ provided by [11]. The Root Cause Map™ is a method developed by ABS Consulting, providing a step by step approach to RCA, using a predefined conceptual map, based on predefined categorizations.

Following the approach used by [11] with its Root Cause Map™, the starting point is the causal factors. For each causal factor, the investigator identifies its type, among the following categories:

- Software/equipment issues;
- front-line personnel issues; and
- external factors.

It might happen that a causal factor has a tolerable risk or its cause cannot be determined: in these cases, the root cause analysis stops and the next causal factor is investigated.

Going deeper in the analysis, once the causal factor types have been identified, it is possible to establish the problem category (Table 5.7).

Table 5.7 Causal factor types and problem categories.

	Causal Factor Type		
	Equipment/Software issue	**Front-line personnel issue**	**External factors**
Problems categories	Process/Manufacturing Equipment issue	Company personnel issue	Natural phenomena
	Software issue	Contract personnel issue	External events
	Material/Product issue	Third-party personnel issue	External sabotage and other criminal activity
	Utility/Support equipment issue	\	\
	Other equipment issues	\	\

Source: Adapted from [11].

With the exclusion of external factors' problem categories, a major root cause category can be assigned to the identified problem categories. Major root cause categories include:

- Design issue;
- equipment reliability program issue;
- documentation and records issue;
- material/parts and product issue;
- hazard/defect identification and analysis issue;
- procedure issue;
- human factor issue;
- training/personnel qualification issue;
- supervision issue;
- verbal and informal written communication issue; and
- personnel performance issue.

The identification of the major root cause category helps the investigator in finding the near root cause and, from it, the underlying cause. For example, for the major root cause category "design issue", [11] identifies three near root causes, with their underlying (intermediate) causes:

- Design input issue:
  - Design scope issue;
  - Design input data issue;
- Design output issue:
  - Design output incorrect;
  - Design output unclear or inconsistent;
- Design review/verification issue:
  - No review/verification;
  - Review/verification issue.

For each major root cause category, several near root causes can be identified, with a number of underlying causes. Once the intermediate cause has been identified, the root cause type can be identified among two options:

- Company Standards, Policies and Administrative Controls (SPAC) issue;
- Company Standards, Policies and Administrative Controls (SPAC) not used.

In particular, the first root cause type includes the lack of SPAC, or issue not addressed in it, or SPAC not strict enough, or confusing, or contradictory, or incorrect. Instead, the second root cause type includes being unaware of SPAC, SPAC enforcement issue or recently changed.

TapRooT® TapRooT® is a systematic process based on over 30 years of human factors and equipment reliability research. The TapRooT® 7-Step Investigation Process leads investigators beyond their current knowledge through the use of tools such as the Root Cause Tree® and Corrective Action Helper®. Following are the main principles of the process (Figure 5.58).

In Steps 1-2, the investigator plans the investigation and collects information using a SnapCharT®. A SnapCharT® is a simple method for drawing a sequence of events and

TapRoot® 7-Step Major Investigation Process		
Phases	Steps	Tools
Plan	1. Plan Your Investigation	**SnapCharT®** *Root Cause Tree®* *Equifactor®*
Investigate	2. Determine What Happened (Sequence of Events)	**SnapCharT®** *Equifactor®* *CHAP* *Change Analysis*
Analyze	3. Define Causal Factors	**SnapCharT®** **Safeguard Analysis**
	4. Analyze Each Causal Factor's Root Causes	**Root Cause Tree®**
	5. Analyze Each Root Cause's Generic Causes	**Corrective Action Helper®**
Fix	6. Develop Fixes	**Corrective Action Helper®** *Safeguard Analysis* **SMARTER**
Report	7. Present/Report for Approval	**SnapCharT®**

Figure 5.58 TapRooT® 7-Step Major Investigation Process. Source: Copyright © 2016 by System Improvements, Inc. Duplication prohibited. Used by permission.

is created with sticky notes or in the TapRooT® software. It is a method of documenting the facts and creating a visual for the next steps of the process: identifying the problems.

In Steps 3-5, the investigator examines the SnapCharT® and identifies mistakes, errors or equipment failures that directly led to or caused the incident or failed to mitigate the consequences of the original error (Causal Factors). A tool that is used to identify Causal Factors is Safeguard Analysis: hazards are forms of energy, targets are people, and safeguards are the barriers that keep the energy from making contact with the target. Causal Factors can be identified where the safeguards failed.

Once the investigators have identified all the Causal Factors, each one is analyzed separately through the Root Cause Tree® to determine the problems' root causes. The Root Cause Tree® is a predefined tree with Seven Basic Cause categories. Also built into the Root Cause Tree® is a 15-question Human Performance Troubleshooting Guide that helps investigators identify human errors and find their root causes. The investigator applies each Causal Factor to each branch of the tree starting at the top of the tree and working down the branches as far as the facts permit. The branches that are relevant are selected and those that are not relevant are discarded. After the root causes are identified, the investigator looks for Generic Causes (big picture problems related to culture, systemic, and organizational problems).

In Steps 6-7, fixes are developed using the Corrective Action Helper® as a guide, and Safeguard Analysis. Safeguard analysis may lead the investigator to remove the hazard or the target, install more reliable safeguards or fix the root of the failed safeguards. When the investigation is complete, TapRooT® tools are available in the software to present what was found to management and others.

The interested reader can find more information in [42].

The TapRooT® Basic Investigation Process (pictured in Figure 5.59) is a 5-step process providing tools to investigate low-to-medium risk incidents. (Major incidents are investigated using the TapRooT® 7-Step Major Investigation Process.)

In the following example, an investigator uses the TapRooT® Basic Investigation Process to investigate an incident with a fairly low consequence:

- Step One: Find out what happened and draw a SnapCharT®. The investigator begins by collecting information about the sequence of events that led up to the incident and draws a SnapCharT® (Figure 5.60).
- Step Two: Learn more or stop the investigation? Upon completion of information collection (after the SnapCharT® is complete), there is a decision point. The investigator may continue the investigation and determine Causal Factors (the mistakes, errors or equipment failures that lead to the incident or failed to mitigate the consequences of the original error). Alternatively, the investigator may decide there is nothing more to learn or decide the risk presented does not match the company's investigation criteria and stop the investigation.

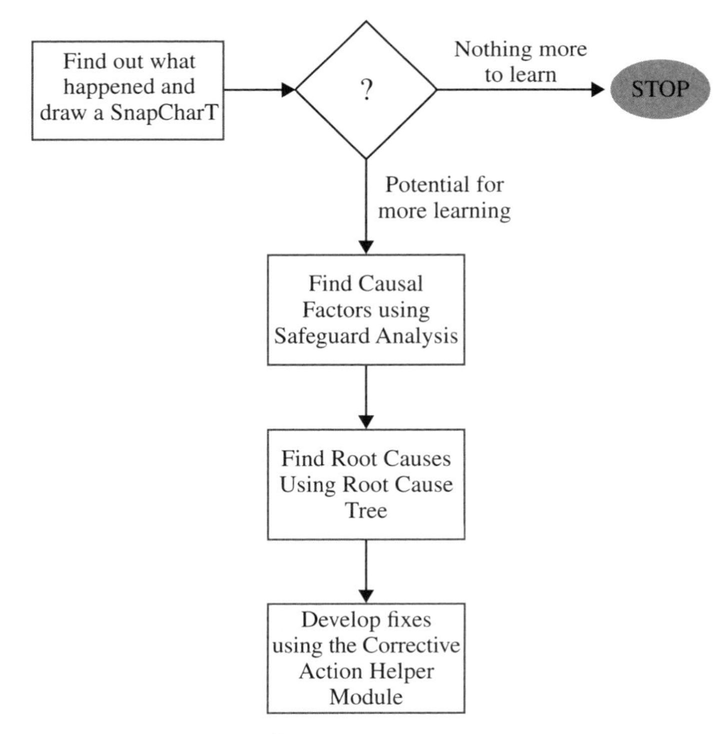

Figure 5.59 The TapRooT® Basic Investigation Process.

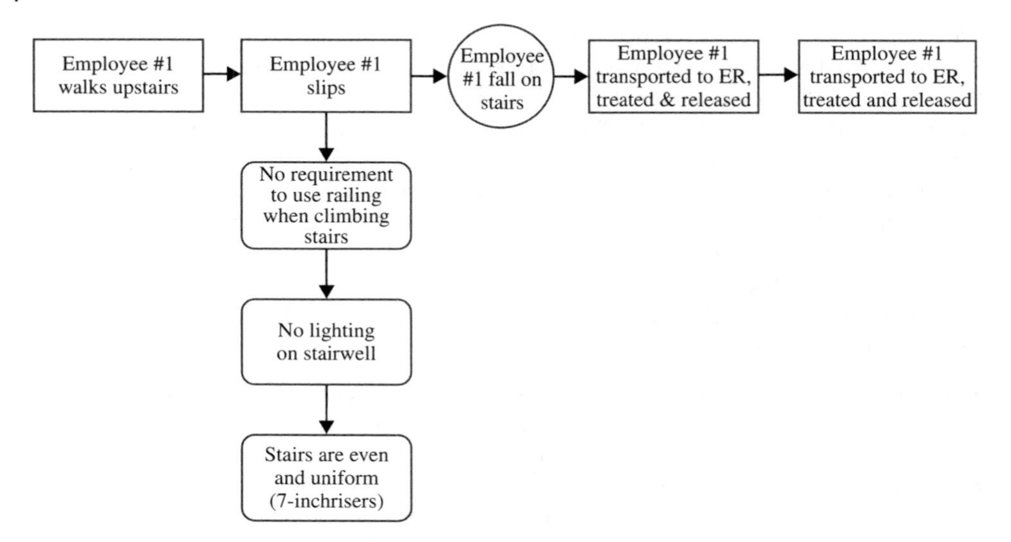

Figure 5.60 Example of SnapCharT®.

- Step Three: Find Causal Factors using Safeguard Analysis. The investigator then determines Causal Factors based on the information collected on the SnapCharT® using Safeguard Analysis. Safeguard Analysis is a 3-step analysis wherein the investigator can pinpoint the Causal Factors.
- Step Four: Find root causes using the Root Cause Tree® Diagram. Next, the investigator takes each Causal Factor through the Root Cause Tree® Diagram. Through a simple process of selection and elimination, "lights need improvement" was identified as one of the root causes in this example.
- Step Five: Develop Fixes Using the Corrective Action Helper® Module. The investigator then uses the Corrective Action Helper Module® to strengthen existing safeguards or develop completely independent additional safeguards.

The Corrective Action Helper Module provides ideas to fix generic (systemic) problems associated with the root cause too (Figure 5.61). Further, the Module provides references for investigators who need to dig deeper in determining innovative solutions.

Apollo RCA™ Methodology and RealityCharting® Software Disclaimer: Apollo RCA™ Methodology and RealityCharting® Software, intellectual property, trademark and registrations are owned by Apollonian Publications LLC a USA company [43].

While formally investigating the Three Mile Island Nuclear Generating Station in Pennsylvania, United States in 1979, Dean L. Gano founded Apollo Root Cause Analysis™ Methodology. Apollo RCA™ methodology supported by RealityCharting® software has been a proven methodology for 30+ years, and it has been taught to over 100,000 students worldwide in 11 different languages. While most other RCA methodology are people-centric methods that relies upon guessing, voting, cause coding or categorical thinking, Apollo RCA™ is a principle-based methodology that works every time on any event-based incident regardless of the user and or observer.

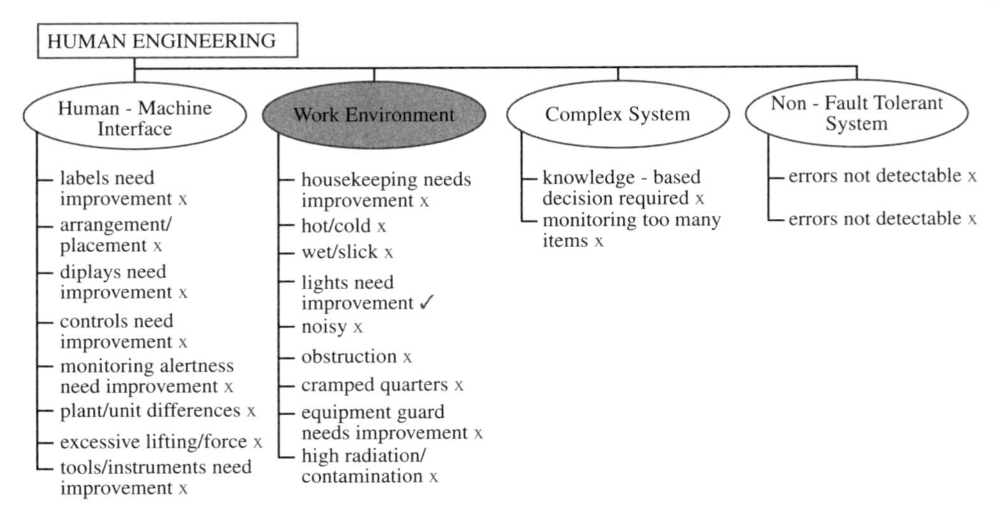

Figure 5.61 The Corrective Action Helper Module.

Apollo RCA™ Methodology process:

- Define the Problem: Develop a common understanding of the problem and its significance;
- create the Cause & Effect Chart (or Realitychart);
  - o determine the causal relationships;
  - o provide graphical representation of the causal relationships;
  - o support causes with evidence; and
  - o determine if causes are sufficient and necessary;
- identify Effective Solutions: Challenge the causes and determine the "best" solution(s);
- implement and track solutions: Implement and track solutions for effectiveness;

Determining and creating the cause and effect chart requires the understanding of the principle of causation. Apollo RCA™ Principle of Causations are:

- First: Causes & effect are the same thing.
- Second: Each effect has at least two causes in the form of actions and conditions.
- Three: Causes & effects are part of an infinite continuum of causes; and
- Four: An effect exists only if its causes exist in the same space and time frame.

As you learn more about Apollo RCA™ methodology principle of causation, you will begin to realise that when solving problems, it is not the root cause we seek, it is effective solutions that:

- Prevent recurrence;
- are within our control;
- meet our goals and objectives; and
- do not cause other problems.

Effective solutions are not necessarily at the end of a cause chain. These solutions can be anywhere in the causal chain, and they must be legitimately connected by evidence-based causes and causal relationships. Both Apollo RCA™ Methodology and RealityCharting® Software does this by creating a common reality that everyone can see and agree to, so the best solutions that will prevent and eliminate the problem are found and implemented.

Apollo RCA™ Methodology and RealityCharting® Software helps you conclude that things do not just happen – rather an understanding that complicated issues can be easily understood and shared by creating a common cause and effect chart or Reality-chart. The end result is the elimination of repeat events and a steady structured path to continuous improvement.

Following, it is an example of application of the methodology, used by permission from "The RealityCharting® Team". It is about the main line pump motor failure at an oil terminal.

Significance: This event led to multiple injuries and fatalities, due to fire and explosions, loss of primary containment with community and a catastrophic loss of corporate assets.

Methodology Applied: Apollo Root Cause Analysis™ Methodology and Reality Charting® Software was used to determine the causes that led to the incident with the objective of ensure this incident and similar incident will not reoccur in the future.

Analysis Summary: A catastrophic failure occurred when the motor shaft and associated equipment broke loose from the motor resulting in flying debris - as heavy as 40 lbs. - exceeding a distance greater than 400 feet. The pump motor failure occurred while the Project Manager was attempting to test the motor relay trip parameters - requiring the motor to run at full speed. The motor trip test was requested when it was discovered that a trip of the motor relays occurred the previous day during a motor rotation check. The trip went undetected because it occurred simultaneous to a planned stop of the motor and the trip indicators are concealed and not in proximity to the switchgear. The trip occurred when the motor current exceeded the trip settings - high current with no load. The Field Technician thought the motor starter was going to be "bumped" to check rotation. And a successful rotation check was conducted the previous day and no one anticipated the catastrophic failure.

Problem Analysis: The motor test was performed, however with the motor decoupled from the pump; the spool piece remained intact. The rotation check conducted the previous day was performed decoupled with the spool removed and the starter was "bumped" (quick on/off).

The unit is equipped with a flexible shim pack by design. The fact that the motor was decoupled from the pump, the unit is equipped with a flexible shim pack and the spool remained attached, then created a non-rigid configuration providing an imbalance/excessive vibration condition once the starter was engaged. The Field Technician did question why the test was being conducted with the spool attached. After some discussion with the other field technicians it was determined that it would be OK to check rotation with the spool attached. The decision was made based on the following; there was a change in work assignments the morning of the failure and they were not originally supposed to perform the task, they were not familiar with how the previous check was conducted because a different crew performed the decoupling of the pump and motor, the unit is equipped with a "soft start". The Project Manager nor the Inspector requested removal of the spool and it was the Field Technician's understanding that the motor was just going to be "bumped" to determine rotation. The Field Technicians - based on previous field experience - considered "bumped" to mean quick start-stop at low RPMs and never anticipated that the motor would reach full operating speed.

In short, the Project Manager had the full intention of running the motor at full speed to test the trip settings. However, the Field Technician thought the test was for rotation and the motor would not be brought up to full speed. A "bump" of the starter was communicated in the field prior to the test.

Once the unit was started under no-load conditions, within 5 seconds it reached a full operating speed of 3600 RPMs. Evidence shows full operating speed with the motor decoupled and spool attached caused the components to break loose from the unit. In addition to full speed, the failure has been caused by: material defect, internal flaw and a bearing failure. Physical evidence was gathered and analytical testing have been performed on the bearing components, shaft, bolts, seals, & oil that validated these failures.

As all communications were verbal; no written Standard Operating Procedure for commissioning of the unit or pre-start checklist list existed. The test was a non-linear sequence test resulting in a loss of continuity since there was a 2-hour work interruption between decoupling and starting of the motor. The representative providing instructions for the motor test and the crew decoupling the pump were different than from the previous day.

No work permit was issued to the contractors prior to the motor test or after the 2-hour work stoppage which is clearly a violation of HSSE policy. In addition, the PM, upon his own admission, did not follow HSSE procedure. In the haste to get the work started the work permit was "missed" according to the PM. Had the PM followed the HSSE procedures or utilised the Stop-Work doctrine and involved all parties in a pre-test discussion the decision to run the motor at full speed with the spool attached would not have occurred.

The Inspector was managing multiple activities - welding and excavating – on the morning of the incident. Operations personnel were aware the contractors were going to be on site. However, they were unaware the contractors were already on site and they were working without a permit.

There was a breakdown in communications resulting in several violations of policies concluding that the parties involved in the motor test were not clear on policy requirements, roles and responsibilities or were clear and chose not to follow them. The Inspector and Technician were unaware that a Contractor HSSE Guidebook - which outlines policy requirements, roles and responsibilities - existed. This highlights gaps in the contractor orientation process. The HSSE Guidebook should have been provided and discussed during their orientation process and prior to performing work.

An internal investigation was performed and was determined the site have systemic issues that reach beyond this location. The resulting Apollo RCA™ diagram is shown in Figures 5.62 and 5.63.

Reason© RCA REASON© Root Cause Analysis is an event inquiry and modeling process that is logic-based, objective, and repeatable. It is guided by rules that depict the event as a product of the causal factors of the organizational environment. As a result, REASON© pinpoints addressable internal causal issues for prevention, improvement, and control. The resulting model is validated for accuracy at each step and supports an objective quantification system that provides strong management decision-support guidance. Prevention options are measured, ranked, compared and documented for

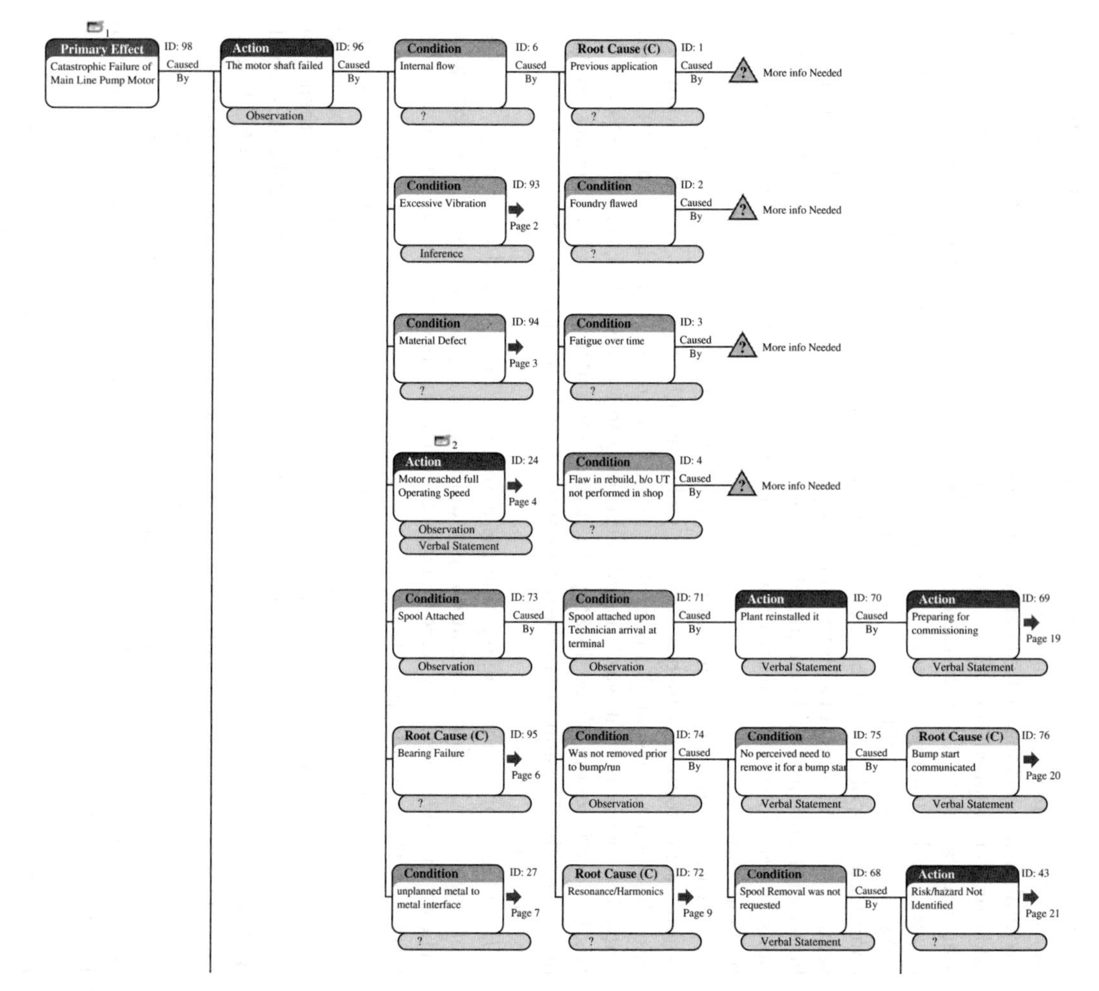

Figure 5.62 Apollo RCA™ diagram (it continues in Figure 5.63). Used by permission from "The RealityCharting® Team".

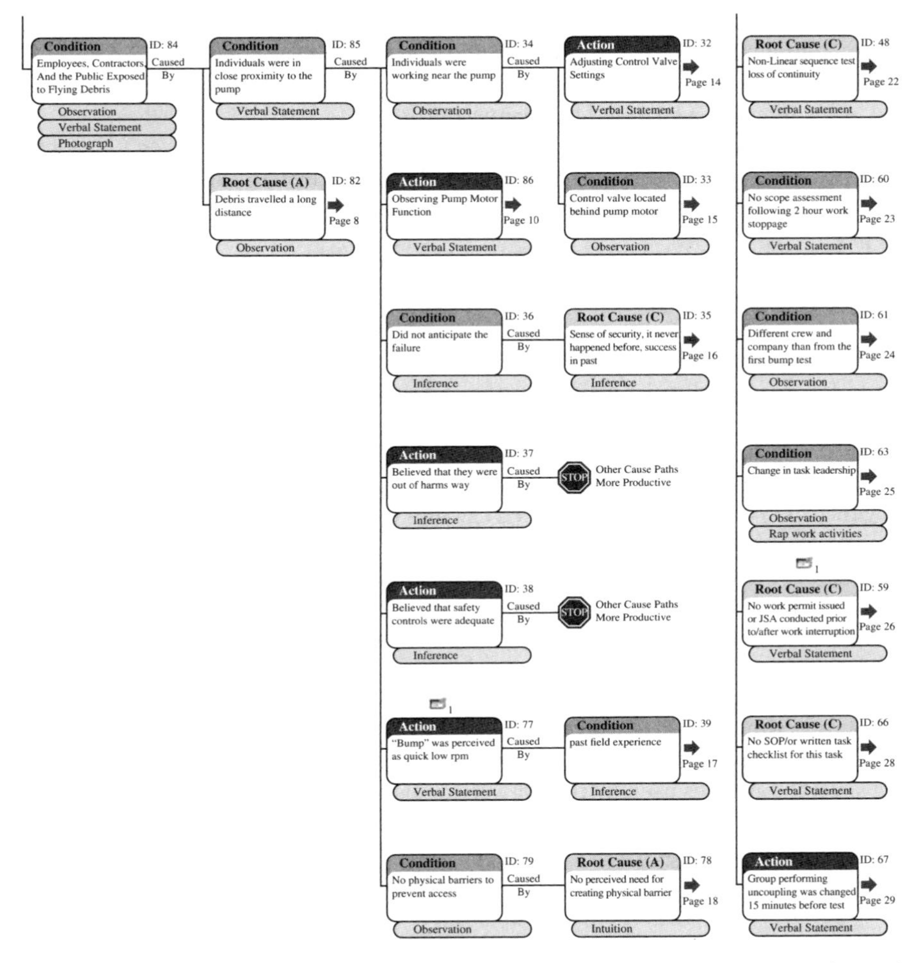

Figure 5.63 Apollo RCA™ diagram (it continues from Figure 5.62). Used by permission from "The RealityCharting® Team.

best and most cost-effective benefit. REASON© fills the gap in typical operations improvement approaches by providing objective, specific, actionable, and validated root cause solutions upon which to apply structured process management systems. The simple, iterative logic process provides criteria for accuracy and rejects subjective assumptions. The rigor of REASON's logic process escalates to the caliber of the issue being analyzed. As a result, it is suitable for use in both critical/complex events as well as issues that are simple and less costly. The process has been in a constant state of development and improvement for over 30 years. The REASON© system is the root cause analysis system utilised in the US/Canada investigation and analysis of the 2003 East Coast Electric Grid Blackout for the U.S. Department of Energy. The REASON© system was chosen for operations improvement and safety by the United Space Alliance at US Johnson and Kennedy Space Centers in support of the US Shuttle and International Space Station programs. It is routinely applied in operations across dozens of industries where it is essential to find the right answer the first time such as Los Alamos National Laboratory and US DOE nuclear waste facilities.

Here is an example of the application of the methodology ().

A co-worker who was seeking to provide assistance to a fellow employee (who was in distress) became injured when his foot became burned. The earliest causal issues for this event involve the company's installation of an outside lift system adjacent to the coke fuel pile.

The company installed the lift per the manufacturer's instructions provided.

Yet, when purchasing the lift, it was not a requirement of the company to detail the operating environment of the lift to the vendor. So, the vendor was unaware for the need to install a heavy-duty switch on the lift. The vendor also assumed that the lift was to be outfitted for 'regular duty.' As a result, the instructions that the company used for installation (from the vendor) called for a regular duty switch to be installed for the lift.

Additionally, because the company did not establish a policy concerning confirmation of such installation details, the company did not ask the manufacturer about switch specifics. Since instructions calling for an HD switch were not supplied by the lift manufacturer, and because the company did not ask the manufacturer about switch specifics, the company was unaware of the need to install HD switches. Then, since a switch was needed to make the lift operational and a regular duty switch was called for in the installation instruction, and since a regular duty switch was available to the installer, a regular duty switch was installed.

Consequently, as it was an extremely hot environment around the switch, and because the manufacturer did not design the regular duty switch to operate in such extreme heat, the switch failed within 6 months of operation leaving the lift unusable.

Among several purposes used by the lift was to raise a person to a height where he or she could perform a 'ring test' (a ring test is conducted by hitting the tank with a metal bar and listening for a hollow 'ring'.)

Given that the switch was broken, the platform could not be raised to the height required to reach the tank to conduct the ring test. The employee was able to stretch above his head and strike the chute near the tank in an attempt to do the ring test. But as there was loud steam noise by the chute, the employee could not hear the results from the test. So, this first attempt at conducting the ring test was inconclusive.

A positive ring test was needed and required to resume operations, but because a process of dealing with inconclusive tests had not been established, the employee attempted other methods of getting an accurate 'ring test' result.

As conclusive results were only attainable by hitting the tank directly, and as the rail was near the tank, and since the first attempt was inconclusive, and since there was not a policy for dealing with inconclusive ring tests, and because Bob needed the conclusive results of a ring test before resuming the operation, the employee decided to perform a ring test by climbing onto the rail to get near the tank.

Then, as the rail was over the chute, and because the employee lost his balance, he fell down the coke chute. Consequently, as a co-worker saw the employee fall down the chute, the co-worker believed he had fallen all the way to the coke pile.

In addition, because the co-worker's supervisor did not communicate the emergency procedures for such situations, the co-worker was not aware of the proper emergency procedures (to call for assistance and procure proper PPE before rendering assistance).

As the co-worker was aware that it was a long fall to the coke pile, he felt that the employee might very well be injured and need assistance. So, the co-worker decided to walk on the coke pile to get to the employee to render help.

The pile was routinely doused with water to control ignition and thus there were areas of the pile where the coke was floating in water (which becomes scalding hot by the coke).

Thus, as an unstable area of coke was in the path of the co-worker moving towards the employee, he stepped onto an unstable area of coke. So, the co-worker's foot slipped through the coke and into the water. Then, as boiling water was beneath the coke, and his foot was unprotected, his foot became burned.

Following, the REASON Interpretation.

Analysis of this investigation shows that it is valid to compare the identified Root Causes to each other, given a calculated Reliability of 96%. This event contains a typical mix of both conditions and actions.

In particular:

- The company has the opportunity to require purchasing to consider in detail operating environments when requisitioning engineered products. In terms of preventing this problem, this is the best option, removing 63.3% of this model;
- The company has the opportunity to establish a policy concerning confirmation of installation details. This is the second best prevention option. It eliminates 62.2% of this problem;
- Management has the opportunity to establish a safe procedure for dealing with inconclusive ring tests. Preventing this root cause is the third best option, and will deal with 47.0% of the causes that produced this problem;
- The supervisor has the opportunity to communicate the emergency procedures prior to employees working in this area. This action, the fourth best option, will remove 29.7% of this problem.

An example of Reason© RCA screenshot is shown in Figure 5.64.

5.4.2.6 QRA derived tools

Among the structured methods, there are some tools derived from the Quantitative Risk Assessment (QRA) methodologies. QRA is used when the Risk Assessment with a qualitative approach cannot be applied, because of the complexity of the system or the relevance of the identified hazards. Indeed, qualitative approaches do not require any numerical evaluations and are only based on simple qualitative tables to identify the likelihood and severity of a potential scenario, with the aim to label it with a risk. A basic example of semi-quantitative approach has been presented in Paragraph 2.6.

Instead, a full quantitative risk assessment is based on the numerical evaluation of both the likelihood and the severity of an incident scenario. The assessment of the severity is performed thanks to numerical simulation, using Computational Fluid Dynamics (CFD) software to model the system, its surrounding conditions and predict the consequence of a hazardous initial event on the system, in terms of damages for people, environment or business assets. The usage of these IT-technologies for an incident investigation is really challanging, since the simulated model must represent correctly the actual layout, operative conditions, process parameters, materials, and surrounding conditions. All these data are available thanks to the previous stage of evidence collection. For example, numerical simulations have been used to investigate the Norman

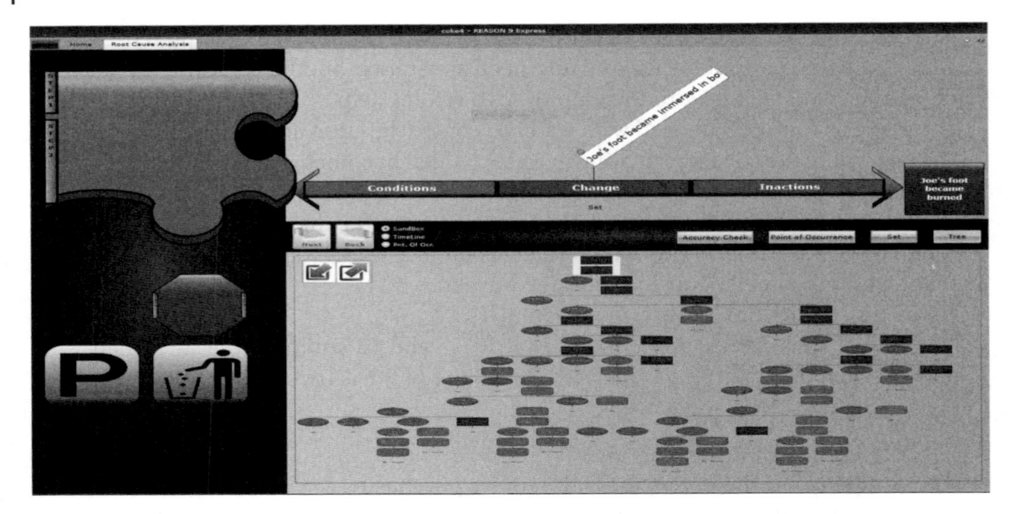

Figure 5.64 Example of Reason© RCA screenshot. Source: ©Copyright 2017, Skyline Software Solutions, reproduced with permission.

Atlantic Fire (Figure 5.65) or the Thyssen Kruup Fire (both the two incidents are discussed in deep in the next Chapter).

Talking about the estimation of the likelihood of a scenario with quantitative approaches, the following QRA derived tools will be discussed in the next Paragraphs:

- Fault Tree Analysis (FTA);
- Event Tree Analysis (ETA);
- Layer Of Protection Analysis (LOPA).

These event frequency techniques are based on rigorous logic and analysis, differently from other methods where the collective experience of the workers of a plant is the main base on which the hazard identification is carried out [14]. The ETA and the FTA may appear similar from a mathematical point of view, but they are intimately structured differently.

Fault Tree Analysis (FTA) This method, created in the Bell Telephone Laboratories in the early 1960s, intends to reconstruct the exact sequence of primary and intermediate events leading to a top event failure. It is therefore useful to recognise those situations that may give rise to undesired consequences when combined with specifically identified events. The main structure of a Fault Tree is shown in Figure 5.66.

The FTA is an analytical deductive technique to analyse failures. It focuses on a particular undesired event and attempts to determine its causes. The undesired event is known as "top event" in a fault tree diagram: it is generally a complete, catastrophic failure of the system under investigation [14]. The top event is the final effect but it is also the starting point of a Fault Tree Analysis. This explains why the formulation of the top event must be punctual and exhaustive: this will ensure the goodness of the outcomes provided by the FTA. The FTA diagram is a graphical representation of both parallel and sequential chains of failures that take the predefined undesired event (i.e. the top event) to occur. Usually, each fault in an FTA is the combination of system failures

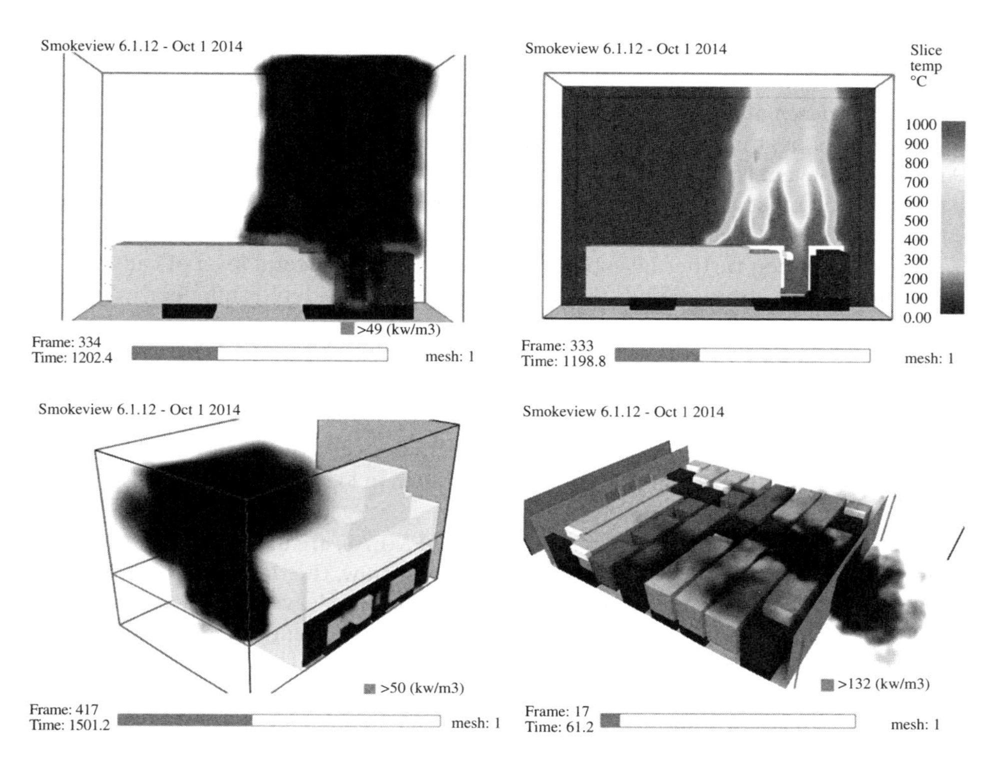

Figure 5.65 Numerical simulations in CFD to support the incident investigation of the Norman Atlantic Fire.

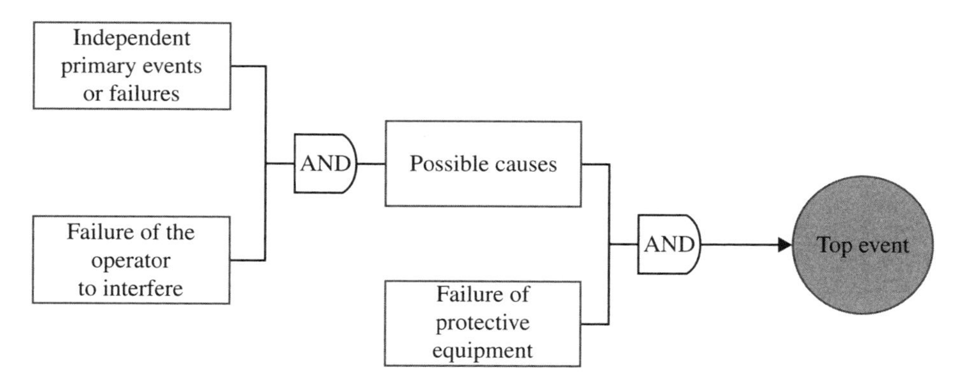

Figure 5.66 Basic structure of a Fault Tree.

(mechanical failures or human error) and the ineffective/missing/failed safeguards put on to stop the chain of failures but that revealed incapable of doing so, for a determined reason.

It is important to underline that a fault tree does not take into account all possible system failures or all possible causes potentially at the base of the event. Indeed, every fault tree is designed for a particular top event, that represents the starting point which will define the development of the rest of tree: only the contributing failure modes are indeed considered. Every single fault in an FTA is combined with the others through AND/OR logic gates. Other logic Boolean gates can be used, but generally they are not required. The usage of logic connectors is useful when a single event could be caused by one or more factors that must act at the same time.

After the top event is identified, the analysis of the faults proceeds level by level: firstly the possible and most general immediate causes are considered, always finding support in the collected evidence. Potential failures that are eliminated by the matching with the evidence are then further investigated, thus finding the second level of causes. The iterative approach continues until the found causes are considered sufficiently detailed to stop the investigation.

It could also happen that more than one path is found between the same faults at the origin and top event. When this happens, and the tree is fully drawn in a flowchart, the more realistic path between the final failure and a specific set of causes is called "minimum cut-set" and represents the shortest path between the two [3].

An example of a fault tree (in its upper part) is shown in Figure 5.67, taking inspiration from [4] and the Åsta railway incident.

Fault trees, which may be considered as reversed FMEAs, are used to guide the investigative resources in the most probable causes. Up to now, the description of the FTA seems to define a qualitative analysis. Even if it is possible to leave the fault tree without any number, significant advantages are taken if it is used in a quantitative approach. Data about the probabilities of failure are taken from historical databases (when available) or the guides provided by the manufacturers or independent publication. Obviously, there is a range of software that performs a computer-based fault tree analysis. The numerical data about the probabilities of human errors, component failures or environmental factors are combined using the mathematical rules for probability, depending on the logic gates on which multiple causes converge. For instance, an "AND" gate means that all the previous factors must be fulfilled to generate the subsequent event. From a probabilistic point of view, this is translated into a single probability of the subsequent event, obtained by multiplying all the probabilities of the single causes with each other (according to

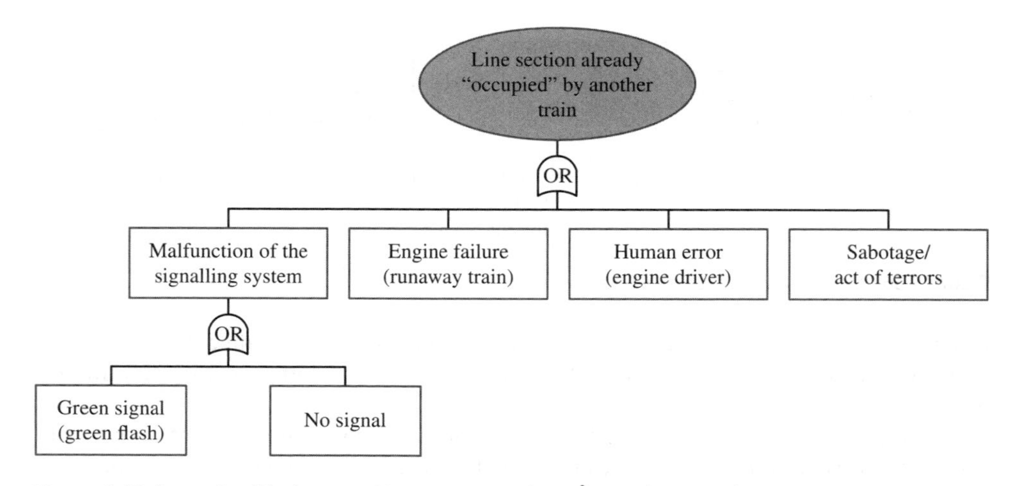

Figure 5.67 Example of fault tree, taking inspiration from Åsta railway incident. Source: Adapted from [4]. Reproduced with permission.

the combined probability rule). Instead, if the connection is through an "OR" gate, the likelihood of the resulting event is equal to the sum of the probabilities of the single causes. Continuing in in this way, the probability of the occurrence of the top event is found. In risk assessment, this information is combined with the severity of the event (for instance, obtained through numerical simulation, in a quantitative approach) to assess the final risk of the top event.

It clearly appears how FTA is an analytical tool for establishing relations; it does not provide any direct information about how to gather evidence [37]. The strengths of the fault tree, even when used in a qualitative approach, are its ability to break down an accident into root causes [38].

Here it is an outline of how to conduct a Fault Tree Analysis [19]:

- Develop the problem statement (i.e. the top event, the reason of the investigation);
- identify the first layer of inputs for the incident, considering basic components or procedures. Remember to consider all the possible inputs of failures for the considered equipment or procedures;
- define the relationship between the top event and the first layer inputs through a logic gate;
- evaluate each first layer input by identifying their second layer inputs;
- define the relationship between a single first layer input and its second layer inputs through a logic gate;
- continue with other layer inputs, until the required level of investigation is reached (typically, when the root causes are found, the iterative procedure stops);
- if required, gather additional information to complete, support or eliminate some branches or single inputs of the tree; and
- document and report the result of the FTA, also highlighting the minimum cut-set path and the probabilities related to it.

A further example is now shown, taking inspiration from [14]. Consider the flammable liquid storage system in Figure 5.68: it is kept under pressure by nitrogen and a pressure controller is used to maintain the pressure between certain limits otherwise an alarm is sent to the control room. The relief valve RV-1 opens to the atmosphere in case of emergency. Considering the tank rupture due to overpressure as the "top event": the corresponding fault tree is shown in Figure 5.69.

Event Tree Analysis (ETA) The Event Tree Analysis (ETA) determines the potential consequences in terms of undesired incident outcomes, starting from an initiating event (i.e. an equipment or process failure). The aim of the ETA is therefore complementary to an FTA goal. Indeed, an FTA explains how an undesired event can result from previous failures (allowing the root causes to also be found), while an ETA examines all the possible consequences of the undesired event.

The structure of a typical ETA diagram is shown in Figure 5.70.

This technique is among the most difficult to apply in practice. Indeed, meaningful results are only obtained if the undesired (or even desired) events, from which branches are created, are fully anticipated. It is therefore clear that the application of the method requires strong practical experience, in order to anticipate all the possible system events and to explore all the possible consequences of those events [14].

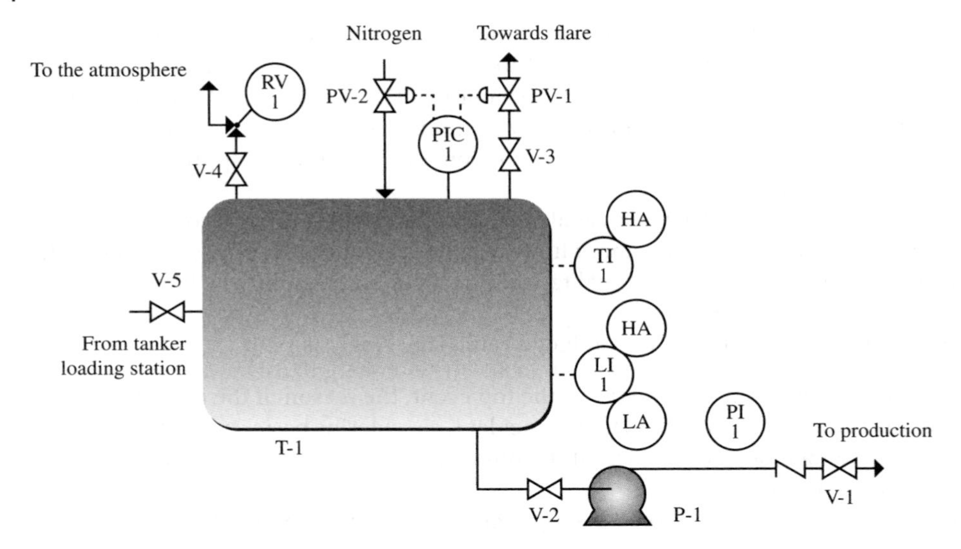

Figure 5.68 Flammable liquid storage system. Source: Adapted from [14]. Reproduced with permission.

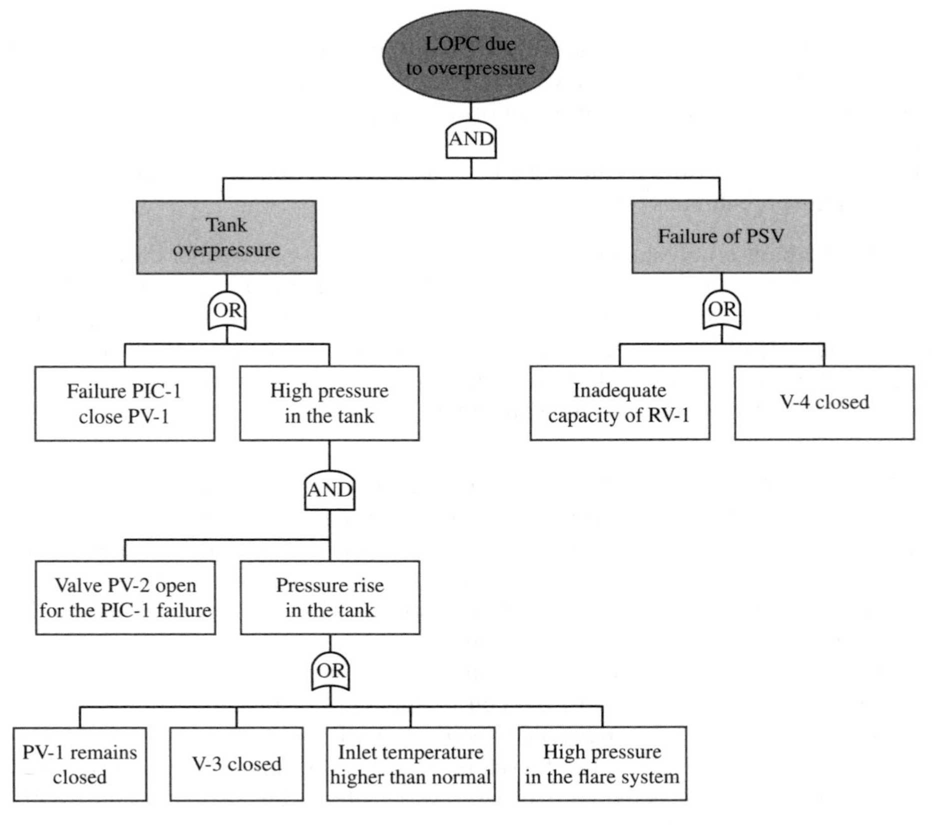

Figure 5.69 Example of FTA for a flammable liquid storage system. Source: Adapted from [14]. Reproduced with permission.

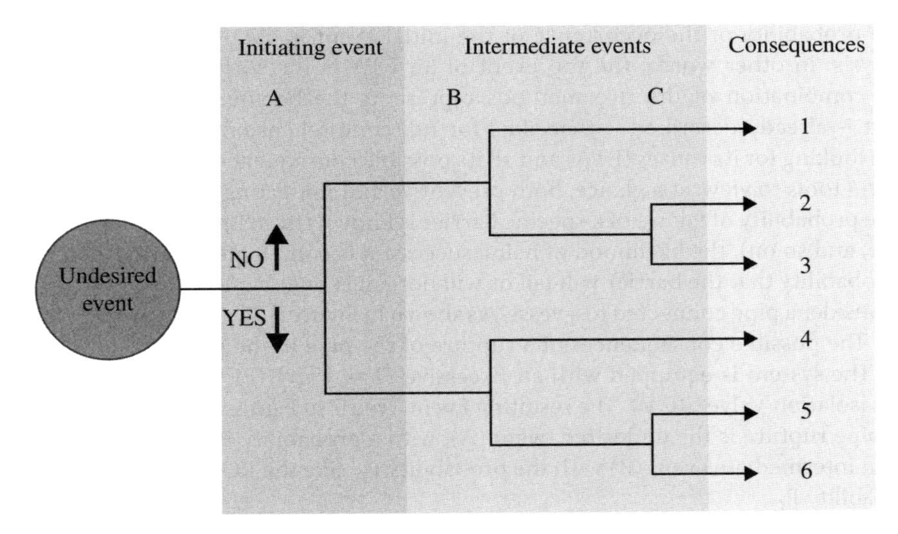

Figure 5.70 The structure of a typical ETA diagram.

The event sequence is defined by barriers that could be both successful or not. An example of ETA for the Åsta railway accident is shown in Figure 5.71 [4]. The tree underlines how likely was the occurred event. The ETA is an excellent method for risk assessment, being used to identify possible event scenarios. In an incident investigation, the actual incident path may be underlined among all the possible ones. For instance, the real incident path of the Åsta railway collision is highlighted in Figure 5.71 with a thicker line.

From a quantitative point of view, the frequency of the occurrence of each scenario is determined starting from the likelihood of the initial event and combining it with the probabilities of failure of the barriers put in position to create the nodes for diverging branches of the tree. As usual, combined probability rules are followed.

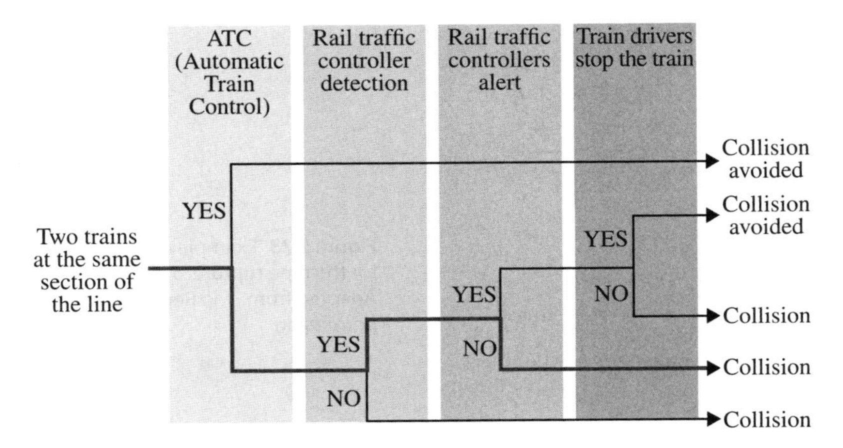

Figure 5.71 Event Tree Analysis for the Åsta railway accident. Source: Adapted from [4]. Reproduced with permission.

Often, the probability of the occurrence of the initial event is obtained from a Fault Tree Analysis: in other words, the top event of an FTA is the starting point for an ETA. The combination of the two methods represents the bowtie, the further risk assessment – already described – that aims for full comprehension of an undesired event both looking for its causes (FTA) and all its possible consequences (ETA). Bowties are powerful tools to view, at a glance, both preventive and mitigating measures [88].

Once the probability of failure of a specific barrier is known (from historical database, experience, and so on), the likelihood of being successful is complementary to the unit (i.e. the probability that the barrier will fail or will not fail is one).

Let us consider a pipe connected to a vessel, as shown in Figure 5.72, taking inspiration from [14]. The possible consequences of a rupture of the pipe in the point "P" need to be found. The system is equipped with an Excessive Flow Valve (EFV) and a Remote Controller isolation Valve (RCV). The resulting Event Tree is in Figure 5.73. In the Event Tree, the pipe rupture is the undesired event (A), with a probability P_A, while the EFV failure is an intermediate event (B) with the probability P_B, like the RCV failure (C) that has a probability P_C.

According to the resulting Event Tree, the probability of a continuous leak is given by $P_A \times P_B \times P_C$; the likelihood of a leakage until RCV is closed is $P_A \times P_B \times (1\text{-}P_C)$; finally, the likelihood of a minor leak is $P_A \times (1\text{-}P_B)$.

Layer Of Protection Analysis (LOPA) The LOPA is a standard method typically used as a risk assessment tool, that can also b also for incident investigation. Remembering what has already been discussed in Paragraph 2.3 and 3.7, different protection layers are in

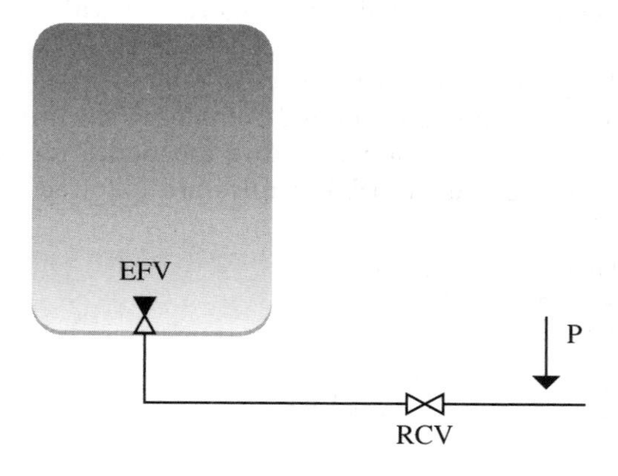

Figure 5.72 Pipe connected to a vessel. Source: Adapted from [14]. Reproduced with permission.

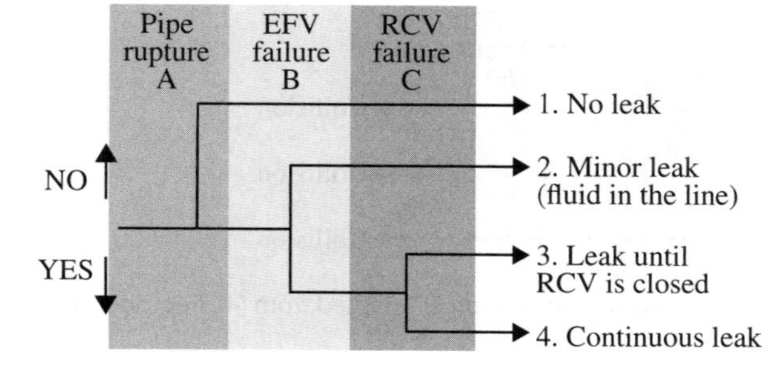

Figure 5.73 Example of Event Tree for the pipe rupture. Source: Adapted from [14]. Reproduced with permission.

place in chemical plants to reduce risks related to undesired events (Figure 5.74). Listed in order running from the inner layer to the outer layer, they are:

- Process design (inherently safety culture);
- BPCS;
- critical alarm and human intervention;
- SIF;
- physical protection (relief valves);
- post-release physical protection (like dykes, walls);
- plant emergency response; and
- community emergency response.

The aim of LOPA is to analyze the effectiveness of the proposed protection layers, comparing their combined effects with the risk tolerance criteria [44]. Indeed, LOPA uses orders of magnitude for both initial event frequencies, consequence severities, and the probability of failure of IPLs, approximating the risk scenario and determining if the existing protection layers are sufficient to mitigate risk below the tolerability limit.

Safeguards can be classified as active or passive and preventive (prerelease) or mitigating (postrelease). As already discussed in this book, all IPLs are safeguards, but not all safeguards are IPLs. A safeguard is any system, device, or action that can stop the chain of events following an initial event. In order to be an IPL, a safeguard must be: effective (having the capacity to take action in time), independent (avoiding the common causes of failure), and auditable (to demonstrate it meets the risk mitigation requirements). In particular, the EN/IEC 61511-3 establishes that:

- Each IPL must be independent from any other IPL;
- each IPL must be different from any other IPL;
- each IPL must be physically separated from any other IPL;
- each IPL must not share common causes failure with any other IPL;
- each IPL must be highly available (availability > 90%); and
- each IPL must be validated and auditable.

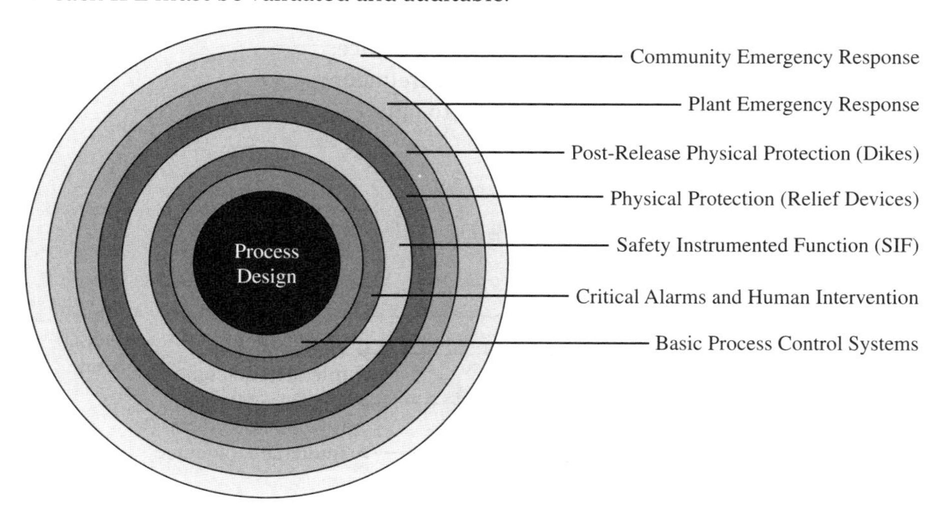

Figure 5.74 Layers of defence against a possible industrial accident.

This technique can be used throughout the process safety lifecycle [45]. It is generally used to examine those scenarios coming from other PHA tools, like HAZOP, and define the SIL targets to meet the risk acceptability criteria. However, it can be also used at the initial design stage, to evaluate alternative protections, or to identify the Safety Critical Equipment (SCE), that is to say the equipment used as a protection layer maintaining risk in the tolerable region. This often resulted in a decrease of the number of SCE [45], still maintaining safe conditions, in contrast with the old idea that adding equipment equals increasing safety. LOPA can be used also to identify the Critical Administrative Control (CAC), i.e. those operator actions or responses that are critical to maintaining risk inside the tolerable region. The method is also used to identify the ALARP risk scenarios.

It is clear that, in theory, a single layer of protection is itself sufficient to stop the incident sequence and prevent the risk scenario. However, no layer is 100% reliable, no one is perfectly effective. This is why a set of protection layers is generally identified to provide the required risk mitigation. If the risk is not tolerable, additional IPLs should be prescribed.

To perform a LOPA, the consequence categories, component failure data, and human error rates must be available. Obviously, in order to have consistent results, the risk tolerability and acceptability criteria must be defined before performing LOPA.

LOPA is not a fully quantitative method, but a simplified numerical approach to evaluating the effectiveness of the protection layers for a precise incident scenario. The fact LOPA uses numbers does not mean that it provides precise risk measurement: it only gives an approximation, that could be useful to make comparisons. However, its methodology can be seen in parallel with other QRA methods, like the ETA, as shown in Figure 5.75. The thickness of the arrow represents the frequency of the scenario, that is reduced by effective safeguards. It is clear how they share the same concepts at the base of the common reasoning.

Basically, LOPA consists of the following steps:

- Collect scenarios developed in prior studies, like HAZOP;
- select an incident scenario and evaluate its consequence;
- identify the related initial events (IEs) and their frequency;
- identify the related IPLs and their probability to failure;

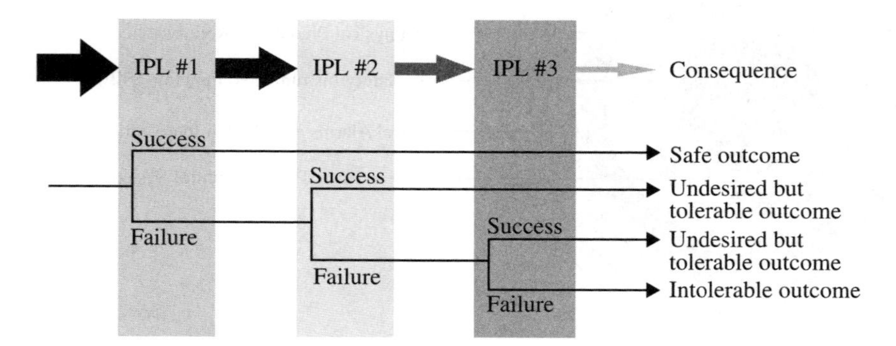

Figure 5.75 A comparison between ETA and LOPA's methodology.

- estimate the risk, combining IE's frequency, consequence severity, and IPLs data; and
- evaluate the risk and make risk decisions (evaluate if further risk reduction is required).

LOPA should not be confused with HAZOP: they are different techniques, with different goals. HAZOP is used to brainstorm the possible hazards and identify incident scenarios, whose risk can only be evaluated from a qualitative point of view. Instead, during LOPA the analysist uses a predefined scenario and estimates its risk in a quantitative way, even if approximated. From this point of view, LOPA is a complement of a HAZOP analysis.

The consequence analysis should take into account the nature of the scenario (LOPC, the release of toxic substances, fires, explosions) estimating injuries, fatalities, environmental damages or business losses. To do so, qualitative or quantitative approaches can be used. The former use a predefined categorization, identifying more severity classes depending on the quantity released, the amount of economic losses, or the qualitative evaluation of injuries and fatalities. The quantitative approach for consequence analysis, instead, usually requires complex models on computers, to evaluate the scenario, like the dispersion of a toxic cloud or the extension of a jet fire, using mathematical models. Great attention should be paid to hypotheses and surrounding conditions when using fully quantitative approaches for consequence analysis. When assessing the IE frequency, the analyst should take into account the different mode of intervention of an IPL: in continuous mode or in demand mode. Moreover, the failure probability must consider the time to risk, i.e. the adjustment to correct failure probabilities (expressed as occurrence per year) in fraction of years when the component is operating.

LOPA is one of the tools used to establish the SIL targets in the Functional Safety lifecycle. SIL targets are the quantitative measure of the required risk mitigation to meet the risk tolerability criteria. The SIL level is related to the PFD of the SIS performing a specified SIF. In particular, the relationship between PFD and SIL level has been already shown in Table 3.8. LOPA analysis allows the establishment of whether one or more SIFs are required and, if any, to allocate the SIL level to the SIF, on a not-generalistic approach, so as to achieve the required risk mitigation with smaller capital expenditure, if compared with risk mitigation based on generalistic approach (like assigning a SIL 3 reduction everywhere).

In conclusion, LOPA can also be used in incident investigation, being a power analysis and communication tool. It can be used to show how additional IPLs could prevent the occurrence of an incident, or to identify those scenarios sharing the same failed IPL and to show how to reduce the scenario frequency adding new IPLs.

References

1 CCPS (Center for Chemical Process Safety). (2003) Guidelines for investigating chemical process incidents. 2e. New York: American Institute of Chemical Engineers.
2 Usmani, F. (2014) Fishbone (Cause and Effect or Ishikawa) Diagram [Internet]. PM Study Circle - A PMP Exam Preparation Blog. 2014 [cited 31 October 2017].

Available from: https://pmstudycircle.com/2014/07/fishbone-cause-and-effect-or-ishikawa-diagram/
3 Noon, R. (2009) *Scientific method.* Boca Raton, FL: CRC Press.
4 Sklet, S. (2002) *Methods for accident investigation.* 1e. Trondheim: Norwegian University of Science and Technology.
5 Oakley, J. (2012) *Accident investigation techniques.* 1e. Des Plaines, Ill.: American Society of Safety Engineers.
6 Heinrich, H. (1931) *Industrial accident prevention: a scientific approach.* London: McGraw-Hill.
7 Bird, F., Germain, G., and Clark, D. (1985) Practical loss control leadership. DNV GL - Business Assurance.
8 Health and Safety Executive. (2004) *HSG245: Investigating accidents and incidents: a workbook for employers, unions, safety representatives and safety professionals.* 1e. London: Health and Safety Executive.
9 ESReDA Working Group on Accident Investigation. (2009) *Guidelines for Safety Investigations of Accidents.* 1e. European Safety and Reliability and Data Association.
10 Dekker, S., Cilliers, P., and Hofmeyr, J. (2011) The complexity of failure: Implications of complexity theory for safety investigations. *Safety Science,* 49(6):939–945.
11 ABS Consulting (Vanden Heuvel, L., Lorenzo, D., Jackson, L. et al.) (2008) *Root cause analysis handbook: a guide to efficient and effective incident investigation.* 3e. Brookfield, Conn.: Rothstein Associates Inc.
12 Sutton, I. (2010) *Process Risk and Reliability Management.* Burlington: William Andrew, Inc.
13 Hendrick, K. and Benner, L. (1987) *Investigating accidents with STEP.* 1e. New York: M. Dekker.
14 Assael, M. and Kakosimos, K. (2010) *Fires, explosions, and toxic gas dispersions. Effects calculation and risk analysis.* Boca Raton, FL: CRC Press/Taylor & Francis.
15 Mannan, S. (2014) *Lees' process safety essentials.* 1e. Kidlington, Oxford, U.K.: Butterworth-Heinemann.
16 Mannan, S. and Lees, F. (2012) *Lee's loss prevention in the process industries.* 4e. Boston: Butterworth-Heinemann.
17 Noon, R. (2001) *Forensic engineering investigation.* 1e. Boca Raton, FL: CRC Press; 2001.
18 Pasman, H. (2015) *Risk analysis and control for industrial processes.* 1e. Oxford: Elsevier Butterworth-Heinemann.
19 Forck, F. and Noakes-Fry, K. (2016) *Cause Analysis Manual.* 1e. Brookfield, US: Rothstein Publishing.
20 Strobhar, D. (2014) *Human Factors in Process Plant Operation.* 1e. New York: Momentum Press.
21 Conklin, T. (2012) *Pre-Accident Investigations.* 1e. Farnham: Ashgate Publishing Ltd.
22 Reason, J. (1990) The Contribution of Latent Human Failures to the Breakdown of Complex Systems. *Philosophical Transactions of the Royal Society B: Biological Sciences,* 327(1241):475–484.
23 Flaus, J. (2013) *Risk analysis: Socio-technical and Industrial Systems.* London: Wiley.
24 Rasmussen, J. (1983) Skills, rules, and knowledge; signals, signs, and symbols, and other distinctions in human performance models. *Systems, Man and Cybernetics, IEEE Transactions on.* SMC-13(3).

25 Dien, Y., Llory, M., and Montmayeul, R. (2004) Organisational accidents investigation methodology and lessons learned. *Journal of Hazardous Materials,* 111(1–3):147–153.

26 Kletz, T. (2002) *Accident investigation - Missed opportunities. Hazards XVI: Analysing the Past, Planning the Future.* Manchester: Institution of Chemical Engineers.

27 Hopkins, A. (2012) *Disastrous decisions: The Human and Organisational Causes of the Gulf of Mexico Blowout.* North Ryde, N.S.W.: CCH Australia.

28 Hopkins, A. (2008) *Failure to Learn: The BP Texas City Refinery Disaster.* CCH.

29 Taylor, J. (2016) Human error in process plant design and operations. 1e. Boca Raton: Taylor & Francis.

30 Woods, D., Dekker, S., Cook, R. et al. Behind human error. 2e. *Farnham: Ashgate.*

31 Reason, J. (2008) *The Human Contribution: Unsafe Acts, Accidents and Heroic Recoveries.* Boca Raton: CRC Press.

32 Dougherty, E. and Fragola, J. (1988) *Human reliability analysis: a systems engineering approach with nuclear power plant applications.* New York: John Wiley.

33 Rollenhagen, C. (1995) *MTO: en introduktion: sambandet människa, teknik och organisation.* Lecture presented at.

34 Bento, J. (1999) *Human-Technology-Organisation*; MTO-analysis of event reports. Lecture presented at.

35 Agenzia per la protezione dell'ambiente e per i servizi tecnici (APAT). (2005) Analisi post-incidentale nelle attività a rischio di incidente rilevante. 1e. Rome: APAT.

36 Paradies, M. and Unger, L. (2000) *TapRooT: The System for Root Cause Analysis, Problem Investigation & Proactive Improvement.* Knoxville, Tenn.: System Improvements.

37 Katsakiori, P., Sakellaropoulos, G., and Manatakis, E. (2009) Towards an evaluation of accident investigation methods in terms of their alignment with accident causation models. *Safety Science,* 47(7):1007–1015.

38 Sklet, S. (2004) Comparison of some selected methods for accident investigation. *Journal of Hazardous Materials,* 111(1–3):29–37.

39 Klein, J. (ABS Group) (2016) The ChE as Sherlock Holmes: Investigating Process Incidents. *CEP Magazine - Chemical Engineering Progress (An AIChe Publication).* 28–34.

40 US Department of Energy (1999) *Conducting Accident Investigations.* DOE, Washington DC.

41 Latino, R., Latino, K., and Latino, M. (2011) Root Cause Analysis. 4e. Hoboken: CRC Press, Taylor and Francis.

42 About TapRooT® [Internet]. Root Cause Analysis System, Training and Software by TapRooT®. 2017 [cited 13 November 2017]. Available from: http://www.taproot.com/products-services/about-taproot

43 Gano, D. (2011) RealityCharting®—Seven Steps to E ective Problem-Solving and Strategies for Personal Success. 2e. Richland: Apollonian Publications, LLC.

44 Franks, A. (2003) *Lines of defence/layers of protection analysis in the COMAH context.* Warrington: Ame VECTRA.

45 CCPS *(Center for Chemical Process Safety). Layer of Protection Analysis: Simplified Process Risk Assessment.* New York: Wiley; 2011.

Further Reading

Barry, T.(2002) *Risk-informed, performance-based industrial fire protection*. Knoxville: Tennessee Valley Publishing.

CCPS (Center for Chemical Process Safety) (2008) *Incidents that define process safety*. Hoboken: John Wiley and Sons.

CCPS (Center for Chemical Process Safety) (2013) *Guidelines for enabling conditions and conditional modifiers in layers of protection analysis*. Hoboken: Wiley.

CCPS (Center for Chemical Process Safety) (2015) *Guidelines for initiating events and independent protection layers in layer of protection analysis*. Hoboken: Wiley.

Dien, Y., Dechy, N., and Guillaume, E. (2012) Accident investigation: From searching direct causes to finding in-depth causes – Problem of analysis or/and of analyst? *Safety Science*, 2012;50(6):1398–1407.

Kirkcaldy, K. and Chauhan, D. (2012) Functional Safety in the Process Industry: a Handbook of Practical Guidance in the Application of IEC61511 and ANSI / ISA-84. 1e. Leipzig: Amazon Distribution GmbH.

Lentini, J. (2013) Fire Scene Inspection Methodology. *Encyclopedia of Forensic Sciences*. pp. 392–395.

Marszal, E. and Scharpf, E. (2002) *Safety integrity level selection: Systematic Methods Including Layer of Protection Analysis*. Research Triangle Park, NC: ISA, the Instrumentation, Systems, and Automation Society.

Scharpf, E., Thomas, H., and Stauffer, T. (2016) *Practical SIL target selection: Risk Analysis Per the IEC 61511 Safety Lifecycle*. Sellersville: exida.

6

Derive Lessons

6.1 Pre and Post Accident Management

The accent on the management, as diffusely stated in this book, reflects the recent perception of incidents where technical failures and human errors are seen only as immediate causes. Indeed, the historical background and the organisational context are also contributing factors to an incident, even if they are defined or generated long before the occurrence of an undesired top event which triggers the incident sequence [1]. This time interval, before the occurrence of the incident and where latent conditions represent its breeding ground, is sometimes referred to as "accident incubation period" [2]. This period of time should be shortened by the specific organisational context. In particular, it becomes crucial to point out the following aspect of management [3]:

- Process Knowledge Management;
- Contractor Management;
- Management of Change; and
- Emergency Management.

Let us start with Process Knowledge Management. Process knowledge is essential to possess an accurate understanding of all the possible risks: indeed from this knowledge, the entire risk-based process safety is then developed and the relative management set. Process knowledge uses a wide set of documents to form itself, like written technical documentation, calculations, engineering drawings (e.g. P&IDs), design/construction/installation standards, specification about the safety operational limits for the main process parameters, Material Safety Data Sheets (MSDSs), and so on. The process knowledge collects all these information and requires a set of activities to catalogue and make them available. But understanding the process also requires the possession of competency by users, as the key skill to properly understand the collected process knowledge information. It is clear that a proper management of this knowledge imposes effort from the early stages of the life cycle of the process and this continues when designing, evaluating risks, building, commissioning, operating. Process knowledge management corresponds to the Process Safety Information (PSI) of OSHA PSM and EPA RMP regulations; they require written information about the chemicals involved in the process, their hazards, the technologies and equipment used.

Contractor management is also an element of the OSHA PSM and EPA RMP sets of regulations. Talking about contractor management, its necessity appears to be clear, if we think about the number of contractors that are daily involved in the process industry

Principles of Forensic Engineering Applied to Industrial Accidents, First Edition.
Luca Fiorentini and Luca Marmo.

activities, especially during maintenance turnarounds. Using contract services and guaranteeing the safety of their operators is a hard challenge, since contractor personnel are usually unfamiliar with the hazards and risks of the operations they are called to carry out. This is the reason why a management system is required for contractors: there must not be any interference around the safety goals of the contract services and the company's personnel and facilities. Therefore, these contracted services need to be selected, acquired, used and monitored: in other words, a management system needs to be provided. The activities that are nowadays asked and provided by contractors extend from design and construction to personnel training. Exploiting third parties has its advantageous for the company, like using experts only when actually required or supplementing with these external resources the internal poor resources of the company during extremely demanding periods or increasing the number of workers without the costs of directly hiring new employees. The use of contractors is not only critical because of their exposure to unfamiliar process hazards, but also because they may introduce new hazards, due, for example, to chemicals that are different from the ones used in the process or x-ray sources. The activity of a contractor could also, unintentionally, damage or bypass the safety controls put in position by the company. In order to face these safety challenges, the company's contractor management system should:

- Verify the proper training of contractor personnel;
- provide the required information to the contractor, to guarantee a safe development of the contracted duties;
- check the contractor's safety records when selecting them; and
- decide upon roles, responsibilities and objectives for safety programs.

When introducing a change in the process, it is important to be sure that it does not bring in new hazards or even increase the pre-existing level of risk. The Management of Change (MOC) helps in this way. Firstly, a definition of "change" has to be provided: it is anything that is not a replacement-in-kind or, as defined by CCPS, anything that changes the process safety information. Therefore, the aim of the MOC is to evaluate the risk associated with a particular change and, eventually, mitigate that risk in order to become acceptable. The Management of Change takes into account not only the proposed changes to facility design, operations, or organization but it is also a managerial tool to notify all the affected people about the changes, to be sure that all the relevant documents are updated, such as technical drawings and written procedures, together with process safety knowledge. Being compliant with the review process provided by this management way prevents the possibility to have a process safety accident, by reducing this risk. MOC reviews are performed in operating plants and even more done during all the process life cycle by the company offices involved in planning and project design. The change can be suggested or requested from anyone in an organization: an individual, a project team, an R&D team. It is essential to establish what constitutes a change and therefore which interventions need to be managed by a MOC. The request for interventions is reviewed by qualified personnel who determine their impact on the risk level and, eventually, suggest how to handle that risk. The extent of the change affects the nature of the review process, including the number and the skills of the personnel called upon for to examine the request. Depending on the review, the change can be accepted, eventually with reserve, or rejected. A person external to the review team provides the final approval. OSHA PSM and EPA RMP regulations also concern the Management Of Change, including the following update of the Process Safety Information, procedures and information to personnel touched by the change. Together with MOC, companies should pay as much attention to the Management of Organizational Change (MOOC), also known as Organizational Change Management (OCM). It concerns changes in personnel, task allocation, organizational structure, policy and working condition (see, as further reading, the CCPS Guidelines for Managing Process Safety during Organizational Change).

Another relevant management issue concerns emergency. The scope of emergency management covers not only the concept of immediate response to "pull out the fire" (i.e. a response to undesired event like explosions, fire, or toxic releases), but it is also extended on the protection of people (both onsite and emergency responders), and the communication with internal and external stakeholders, including media. Indeed, emergency management includes: planning activities, resources allocation, continuous improvement of the emergency plan, training for employees and contractors about what to do and how to report, and providing effective communication to stakeholders when an incident occurs. The planning activities are usually carried out together with the Hazard Identification and Risk Assessment (HIRA) team, in order to identify those scenarios that require an emergency plan (See Figure 6.1).

Typically, the operations group is responsible for immediately response to the emergency (often by isolating hazardous materials or shutting down the process). An emergency plan should be developed through the following stages:

- Identification of the incident scenarios, considering the hazard already found by hira team. in the context of process industry, the classes of hazards related to the process are usually three: loss of primary containment (lopc), fires & explosions, and toxic vapour clouds;
- assessment of credible accident scenarios, to establish the types and the magnitude of the foreseen effects;

Figure 6.1 Emergency management is a crucial part of the overall safety management system. Source: (courtesy IPLOM S.p.A.).

- selection of the planning scenarios, depending on the extension of the affected area, the incident history in the industry, the types of potential effects;
- planning of the response actions. this step takes into account the recognition and reporting activities after the incident, the method to give an alarm, the necessity to equip the emergency responders and the people affected by the incident with personal protective equipment, the eventual installation of shelters, the evacuation routes, the assembly points, the command hierarchy to manage the people at the assembly point, and so on;
- planning offensive response, including firefighting preplans, definition of restricted-access areas, supporting communication, guidance to ppe choice, decontamination procedures, and so on;
- writing the emergency response plan, with the offensive response actions and informing about which resources (equipment, facilities, training, communication and coordination) are required;
- providing facilities and equipment. attention on the location of the equipment should be paid, in order to avoid useless increasing of the response time or increasing hazards in safely reaching the equipment;
- periodically testing of facilities and equipment;
- determination of the appropriacy of operator response (it has to be written in the plan);
- training of the emergency response team (ert) and keeping their knowledge always fresh and updated;
- planning communication, with employees, contractors, authorities, and other stakeholders;
- informing and training all the personnel; and
- reviewing the emergency response plan on a regular basis.

Even if it is true that an accident is an opportunity to learn, and improve management, it is also important to understand who wants to know. Many different actors want to know what happened, like the public or the police, but they do not usually need to learn a safety lesson, simply because their involvement in the industrial incident does not embrace the learning from experience process. Who truly do need to learn are those people who can make improvements and changes, depending on their role in the organization or in the community [4]. In particular, despite efforts to find root causes, it should be noted that "the potential for a repeat occurrence remains unchanged until recommendations are implemented" [5]. Implementation of recommendations is a necessary practice that could be asked also by regulations. The activities following the investigation process are: the initial resolution of the recommendations developed by the investigation team, the implementation of the accepted or modified recommendations, and the sharing of the lessons learnt from the investigation, according to the flowchart in Figure 6.2.

The lessons learned may have different origins: they can be from within the organization or from others. What is often disregarded are the cross-industry lessons. Indeed, there is the tendency to recognize only those incidents occurred in similar plants or, at least, in the same industrial sector. But the capability to reveal the root causes of an incident allows larger comparisons to be made, enjoying the lessons learned from

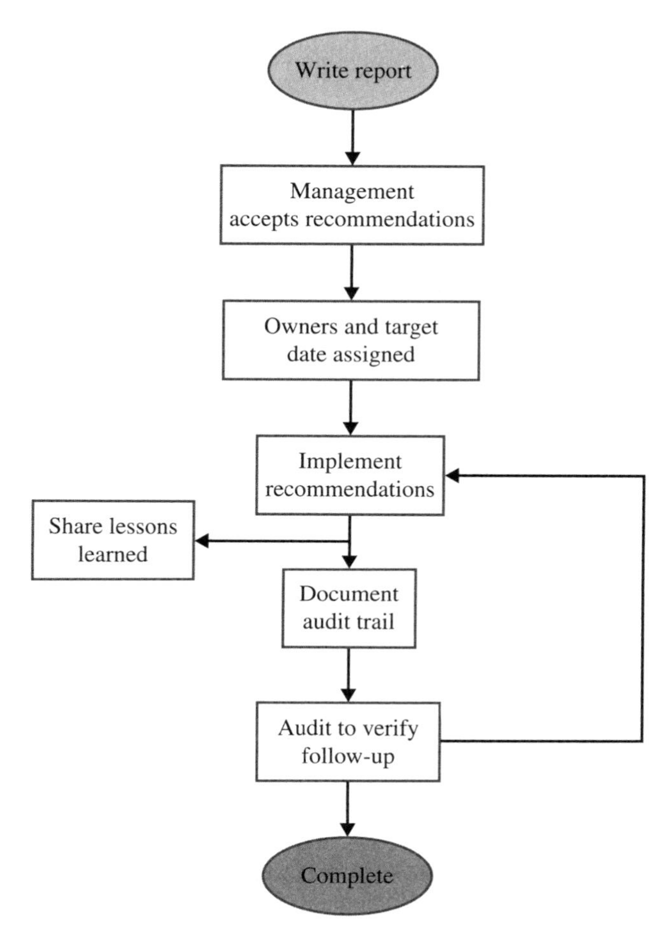

Figure 6.2 Flowchart for implementation and follow-up. Source: Adapted from [5]. Reproduced with permission.

incidents occurred in other business sectors. For example, being aware of this possibility, the Authors promoted safety lessons to a chemical company, starting from the Norman Atlantic fire: the universality of root causes guarantees a large audience for their lessons.

It is important to compare the facility's incident records with data from similar industries, in order to learn from the experience of others. Several databases are available to meet this objective, both national and international [6, 7]. For example, in Europe the Major Accident Reporting System (MARS) is a shared industrial accident notification scheme, that was established by the Seveso Directive in 1982. The database allows the examination the historical data related to the industrial accidents, in order to draw the lessons learnt and prevent the occurrence of future unwanted events or, at least, to mitigate their consequences. Therefore, once the investigation is concluded, according to the national law of reference, the investigator is encouraged to share its findings through this IT-solution.

Having in mind that changing the management means having recommendations, we briefly anticipated that the developed recommendations could be, essentially, of two types [8]:

- Change the design of the hardware or the process being involved in the incident; or
- change the behaviour of job performer.

But the most important products of an incident investigation are not the findings and recommendations developed from the study: they are the actions taken in response to them. A system must be in place to ensure all incident investigation action items are completed on time and as intended. The same system can be used also for hazard analysis and should include regular status reports to the management. Moreover, it is important to communicate the actions to the involved employees. For example, an action plan is generally used. It is a list of all the things that need to be implemented to improve safety, enhance the system, and avoid any reoccurrence of the event. Regardless of its format, it is an effective tool to explain the developed solutions to decision makers, management and peers.

The lessons learned and the findings obtained from an investigation are precious pieces of information that need to be shared to leverage the work of the investigation team. Typically, after an incident has occurred there are high expectations about the management of the company to proactively share information with all the interested parties. Actually, the number of incidents that reoccur constantly is a signal about the inadequacy of the sharing system of findings or of a poor implementation of what suggested by the investigation team as recommendations. The Concorde Air crash incident is an example of how ineffective the sharing system about previous near-miss events was [9]. It is undoubtedly the case that it is possible to improve the safety level in a company thanks to a learning culture oriented to the findings of incident investigations. Several sources of information are available, so it becomes necessary to select the relevant incidents to learn from. The case studies in Chapter 2 are a good starting point.

When assessing the management systems after an incident has occurred, it is important to distinguish among different levels of management [6]:

- Line supervision;
- facility management;
- executive management; and
- industry regulations and standards.

It is clear that the level of root cause analysis is in line with the management level. Following an event, the line supervisors generally perform a quick investigation to find the immediate causes of the incident. They have to question about the uniqueness of the incident or its repeatability in other contexts of the plant; the necessity to stop the production, as extreme safeguard to guarantee the safety; the need to implement an immediate temporary control to minimize the risk of a further occurrence, while the permanent solution is developed; the need to substitute the contractor who carried out a particular operation related with the incident or to change the maintenance program. At the line supervision level, the RCA is focused on the supervisor responsibilities, i.e. to guarantee the immediate restoration of the safety level and make short-term changes to avoid further reoccurrence of similar events. Aspects of the design, the management system or the company culture are outside their scope. At the upper layer there is the facility manager. He/she has direct control of the whole facility and a sufficient budget to implement what suggested by the RCA in a time horizon of 3-6 months. The facility manager is responsible for the implementation of the management systems, not for their creation/modification. This means that recommendations at facility management level are addressed for more efficient implementation of those systems. For instance, he/she may suggest a specific training or the improvement of the maintenance procedures; instead structural changes are not in his/her scope. The person who does have the power to modify the management systems is the executive manager. The time horizon (from months to years) and the capital expenditure discretion allow managers at this level to focus on cultural and human issues, going a step forward the management systems. The last management level actually concerns the industry regulations and standards. Indeed, the industrial community creates not only rules and regulations but also consensus standards written by external associations like API, ASME, and so on. Therefore, the results of an incident investigation are exploited by the community to improve those industry regulations and standards. At this management level, the time horizon is vast, typically many years. However, the related recommendations are very efficient and capable to create solid cultural changes, aimed at creating a unified process safety culture.

What is clear is that each level of the management needs a proper knowledge about the incident investigation policy, procedures and responsibilities. Having this governance structure means the need to have a convergence about how deep a root cause should be, the standard of quality accepted for recommendations and the different responsibilities of the personnel working under a certain manager [9]. To focus the attention on the management is generally correct, since the power to establish a policy and to allocate resources relies on this level and therefore the ultimate responsibility belongs to that position. From the incident investigation standpoint, some commonly shared management leadership attributes characterize the most successful facilities. These attributes are:

- A trained incident fact-finding team, capable of acting immediately after an incident;
- direct involvement of line supervisors in reviewing and approving the incident report;
- rigorous implementation of recommendations suggested by the incident report;
- sharing of findings and lessons learned;
- punctual reporting and understanding of near-misses; and
- prioritize the action items coming from the recommendations.

It is therefore important that managers at every level not only accept a recommendation but also follow their implementation to ensure it is as was intended by the investigative report. Auditing and follow-up verification are thus crucial for monitoring the implementation stage. The prioritization of the action items is often accomplished by the adoption of a risk matrix: it is the line management that usually establishes such acceptance criterion.

According to [10], much more can be done to improve the learning from experience process. Indeed, the fragmented literature on the topic mainly focuses the experience feedback process on the investigation method. But the experience feedback should also take into account what happens before and after the accident investigation. It is the paradigm of the continuous improvement for the incident investigation system, as also discussed in [5]. Starting from this idea, [10] promotes six quality criteria that should be included within the "lessons learnt", in addition to the peculiar contents of the investigation (i.e. in addition to the lessons learned from the uncovering of the root causes). They are:

- The initial reporting, i.e. how the feedback process can enhance the very first step of an investigation. indeed, a poor initial report may result in the choice to not investigate further in details;
- the selection methodology, i.e. the events selected to be investigated in deep should be those where as much information as possible can be extracted to develop the preventive recommendations;
- the investigation, i.e. the procedures and the methodologies adopted to carry out the task;
- the dissemination of results, i.e. the capability to effectively share the investigation findings with those who can use them to prevent a future occurrence;
- the preventive measures, i.e. those actions taken to avoid similar incidents; and
- the evaluation, i.e. the experience feedback process should be evaluated in order to be improved through another experience feedback process.

6.2 Develop Recommendations

When an incident occurs, the highest price has been already paid. It is therefore essential to try extracting the most valuable lessons from it. The investigation shows the areas of the risk assessments that need to be improved: the investigation team, once the root causes have been identified, develop those recommendations that can reduce the likelihood (or the magnitude) of a repeat incident. When the recommendations are submitted to the designated owners (typically the management), the responsibility for those actions is transferred from the investigation team to the organization's management, who must evaluate, accept, reject, modify and implement the proposed changes. Indeed, the team responsibilities stop at the development of the practical recommendations, submitting them to the management. It is then the task of the management to approve (or not) the recommendations, allocating the required personnel and economic resources for their implementation, and following up the derived action items, to be sure that the measures are implemented as expected. It is clear that until these changes are implemented, the risk profile will remain unchanged. Sometimes, immediate recommendations are developed even before the investigation is completed, in order to immediately face those hazardous factors that can be mitigated in a very short time [9].

Turning findings into recommendations is the analysis process of the learning experiences combined with their transformation into meaningful proposals. It is undoubtedly a real challenge. Recommendations are the most important product of the incident investigation: they are developed only after the analysis stage and the uncovering of the root causes. The corrective actions can be preventive or mitigating measures and have a different level from the socio-technical perspective. During this process a thorough understanding of the system is therefore necessary to develop meaningful recommendations, involving stakeholders and developing also a communicative strategy to share the lessons with them [4]. Recommendations should be formulated to address the following goals:

- Prevent the same and similar incidents from happening again;
- mitigate the consequences if a similar event should happen in the future;

- solve the knowledge deficiencies uncovered by the investigation;
- identify the weaknesses in the system and, especially, in its interfaces (each combination between the technical, human, and managerial sides of the system), which could be the weakest parts;
- strenght these weaknesses; and
- propose an early-warning system.

It is generally recommended to establish a specific time limit for responding to a recommendation. A recommendation should not be a prescriptive mandatory action item; instead, it is a good proposal idea based on incident investigation findings whose details – typically the technical ones – can be adjusted during the implementation phase. On the contrary, recommendations from safety or judicial authorities are mandatory.

Generally, there are two strategies for drafting recommendations:

- Restoring the initial safety level, which deviated from the normative level, dealing with the system "resilience";
- facing the deficiencies in the system and enabling it to change in the operating environment.

These two approaches can be seen in similarity with two different strategies for structural design. Basically, there are two ways to ensure that a structure will not collapse: making more robust structures, with solid materials and thick geometry, or using more flexible and light structures, leaving them free to deform under load conditions, without collapsing. The same approach is valid for drafting recommendations: the options are working towards robustness or allowing the system to adapt.

The application of recommendations follows this guideline:

- Owners who have the responsibilities of the activities affected by the recommendations must take them into account and take the appropriate corrective actions;
- before responding (accepting or rejecting) the recommendations, the primary responsible party (prp) needs to consider all the relevant information to manage the involved risks;
- responses to recommendations should be recorded. if the recommendation is rejected, a justification should be provided; if it is accepted, the related action plan should be attached;
- actions should be tracked from their proposal up to their completion;
- lessons learned should be preserved in the corporate memory, using a database for findings and actions (remember: organizations have not memory, people has);
- lessons learned should be shared across the industry sector;
- lessons must not be possessed by individuals but from the system, otherwise they will be lost; and
- recommendations should be used proactively, to enhance the hazards analyses and risk assessments. this can be done by a wide usage of database, which should not be used passively, but actively to develop continuous refinements.

Making recommendations is not an easy task. Help is given by [11] providing some aids about the overriding principles:

- Make safety (and security) investments on cost and performance basis;
- improve management systems;

- enhance the management and staff support;
- develop layered recommendations, especially to eliminate underlying causes (and hazards as well).

Enhanced application and sharing of lessons learned imply a full understanding of incidents and near-misses, responding so to prevent the same and similar unwanted events. The required critical knowledge communication needs a culture in which employees are driven to learn the best from those events [12]. Those companies with an excellent process safety performance do not just share the lessons, they take actions to document and respond to these learning opportunities. In order to reduce incidents, continuous learning should be pursued. To do so:

- Identify the lessons and be aware of the value of sharing them with others;
- use an efficient system to share these lessons; and
- embed the lessons in company's procedures/standards, checking if the existing equipment/process/procedures require modification.

This approach has its value in the rapid sharing of lessons, driving the improvement of safety company's standards and practices, supporting both safe and reliable operations. "Do something" is the motto driving the result of learning from lessons. Reports of major incidents are undoubtedly a source for critical lessons: every company should start their "learning from incident" process from someone else's incident, by consulting the available online databases and focusing on those incidents related to the same business sectors. Once the critical lessons and the actions taken have been identified, they should be communicated to the leaders.

In order to develop effective recommendations, lessons must be based on the root causes. If lessons are derived from immediate causes or contributing factors, the developed actions will be ineffective, being incapable of preventing future recurrences. From this perspective, recommendations (that are the last step of the forensic engineering workflow discussed in Chapter 4) can be developed on four different levels [6, 13]:

- Short term. they are the immediate corrective actions, usually related to the immediate causes. for example, if an incident occurred because, as immediate cause, a block valve was left in the closed position when it should be open, the short-term recommendation is to open and tag that valve, also in all the other plants of the company;
- intermediate. they refer to those actions requiring around three months to be implemented, which are addressed by the facility management, not requiring a substantial change of policy. for example, in the case of the incident previously used as example, the facility management may decide to conduct an hazard analysis to look for similar problems, ensures that contractors are aware of the newest procedures, provides them with formal training, removes the unnecessary offending valves not trusting on the tag-out system to leave them open;
- long range. recommendations at this level are related to the uncovered root causes. they regard the system. in the example above, assuming the root cause analysis revealed some weaknesses in the communication among owner, contractors and sub-contractors, the developed recommendation could be to evaluate and update the whole contractor management system; and
- industry-wide. major incidents can bring to the development of recommendations affecting the entire industrial sector. for example, the lessons learned from the major

incidents discussed in chapter 2 eventually resulted in industry-wide recommendations (e.g. to avoid the storage of hazardous intermediate products if not needed).

The best lessons have no value if there is not a formal process to share them throughout the company. Sending emails is a good starting point, but a robust system should use standard templates and formal sharing procedures. It is also important to share only the most valuable lessons, avoiding overwhelming people with too many not-critical incidents that provide only limited lessons.

The front-line personnel, who act to prevent incidents, belong to the operation unit. It is therefore important to regularly discuss the incident investigation status, findings, and action items with the operations personnel. Moreover, the incident outcomes should be also used during hazard analysis studies. Indeed, most proactive analyses are based on the understanding of what could be wrong. In this sense, the incident lessons are a valuable source of information to identify hidden scenarios that unit personnel might not have properly considered.

The evaluation of recommendations is a crucial step to examine if the proposed actions can effectively reduce the risk profile [9]. Indeed, it may happen that some developed recommendations actually create a new risky scenario or increment the pre-existing risk level, or are actually irrelevant to reduce the risk level (in the Authors experience, the last option is not uncommon when dealing with functional safety IPL, since who develops the action items sometimes might be unaware of the SIS requirements, as they are specified in the IEC-61511). Therefore, proactive risk assessment tools should be used to evaluate the potential risks in implementing a certain recommendation (for example, using nitrogen is a solution to make an inert atmosphere, but it increases the risks for asphyxia).

When developing recommendations, priority is given to those actions that could prevent the event: only then actions that mitigate consequences should be sought. The principles of inherent safety and LOPA can be applied when considering recommendations; in particular, remembering the onion-like structure of IPLs (Figure 5.70), remedies should firstly focus on the inner protection layers rather than on the outer ones. In other words, according to the different "priority" of IPLs, those recommendations targeting improvements to the inherent safety of the process design are preferred to those that simply prescribe additional barriers. Moreover, those recommendations leading to inherently safer designs limit the reliability of human performance, equipment, or maintenance program. Some strategies to implement inherent safer designs are:

- To reduce inventories of hazardous materials;
- to substitute chemicals with less hazardous materials;
- to intensify, i.e. reduce the reactor (and inventory) size; and
- to change, using a totally different process or method to achieve the same goal.

When drafting recommendations, information related to benefits by the implementation or potential consequences by rejecting should be included: this will help the management team in making their evaluation and final decision about the implementation of the recommendation. To do so, as mentioned in the previous paragraph, cost-benefit analysis can be implemented to evaluate the proposed recommendations, comparing the expected risk reduction benefits with the expected cost of implementation. Among the several financial parameters guiding the decision-makers towards the evaluation of recommendation, there is the Return On the Investment (ROI). It is the ratio between the total accident costs (including both direct and indirect costs) and the investment required to complete corrective actions and safety system improvements. For example, if a particular training session costs $10.000, and the total accident cost is $100.000, then the ROI is 1000%.

One method to prioritize implementation of recommendations is the evaluation of the cost/benefit ratios [13]. The benefits of implementing a safety recommendation are given by:

$$\boldsymbol{B = C_{PL} - C_{IR} - C_{RL}} \tag{6.1}$$

Where:

- $\boldsymbol{B}$ are expected benefits
- $\boldsymbol{C_{PL}}$ are the current expected costs of potential losses
- $\boldsymbol{C_{IR}}$ are the expected costs of losses that could occur while implementing the recommendation
- $\boldsymbol{C_{RL}}$ are the expected costs of potential losses after implementing the recommendation (residual losses)

In a detailed study, the time for implementation should be also taken into account, since "time is money". Instead, the costs for implementing a recommendation are given by:

$$\boldsymbol{C = C_{II} + C_{OI} + C_{SP}} \tag{6.2}$$

Where:

- $\boldsymbol{C}$ are expected costs
- $\boldsymbol{C_{II}}$ are the initial implementation costs (equipment, design, installation)
- $\boldsymbol{C_{OI}}$ are the annual costs for ongoing implementation (utilities, training, maintenance)
- $\boldsymbol{C_{SP}}$ are any special cost in the future (rebuilding, retraining)

Generally, those recommendations with a higher $\boldsymbol{B/C}$ ratio should be implemented first. When the cost-benefit analysis may cost more than the suggested recommendation, then it is implemented without performing a cost-benefit analysis.

In addition, the effectiveness of the implemented recommendations can be evaluated through the analysis of the process safety lagging and leading parameters, already presented in Chapter 2.

Generally, who will implement a recommendation is not the same person who wrote it: therefore, it is essential that the action items are written clearly, without any interpretative doubts. The recommendation text must not be wordy, to avoid ambiguities and misunderstandings. Indeed, the investigator should keep in mind that a clearly written recommendation has little opportunity to be misunderstood, opening different interpretations. A well-written recommendation is capable of transferring clearly the full responsibility and the ownership to the receiving department. The owner is fully responsible for the recommendations follow-up. As a rule of thumb, recommendations should include the reasons why they are developed, in order to be as exhaustive as possible. A possible format is: "In order to avoid X, Y should be done". Hard recommendations are written in specific and clear terms, while soft recommendations, usually starting with "evaluate/consider", allow greater flexibility for their implementation. It may happen that the investigation team has not sufficient detailed information to fully express an evaluation. In this case, recommendations asking for further details can be developed, like "Confirm that the aqueous mixture X is soluble in Y. If confirmed, do action n.1, otherwise do action n.2". For major incidents, the developed recommendations should be reviewed by the legal office of the company, in order to minimize the litigation exposures. Indeed, legal representatives are able to quickly identify inflammatory, judgmental, subjective, and damaging words in case of future litigation. From this perspective, the recommendations flowchart in Figure 6.2 can be detailed as shown in Figure 6.3.

Note that many companies distinguish among findings and recommendations, as the Authors do in this book: findings are statements of fact and do not include any suggested action, that is written as a recommendation. Generally, it is also distinguished among findings (which regards the known facts) and conclusions (which include judgements coming from the investigator's activities).

Proactive sharing of investigation results is encouraged, in order to amplify the potential benefits of the investigation. In some countries, like the US, this approach for sharing is mandatory, under specific conditions.

The developed recommendations are often risk-ranked, to drive the management in facing with priority the proposed solutions. In other words, their mitigating effect, that is to say their capability to reduce the scenario likelihood or magnitude, is evaluated. Recommendation status tracking is an essential key to an effective incident investigation

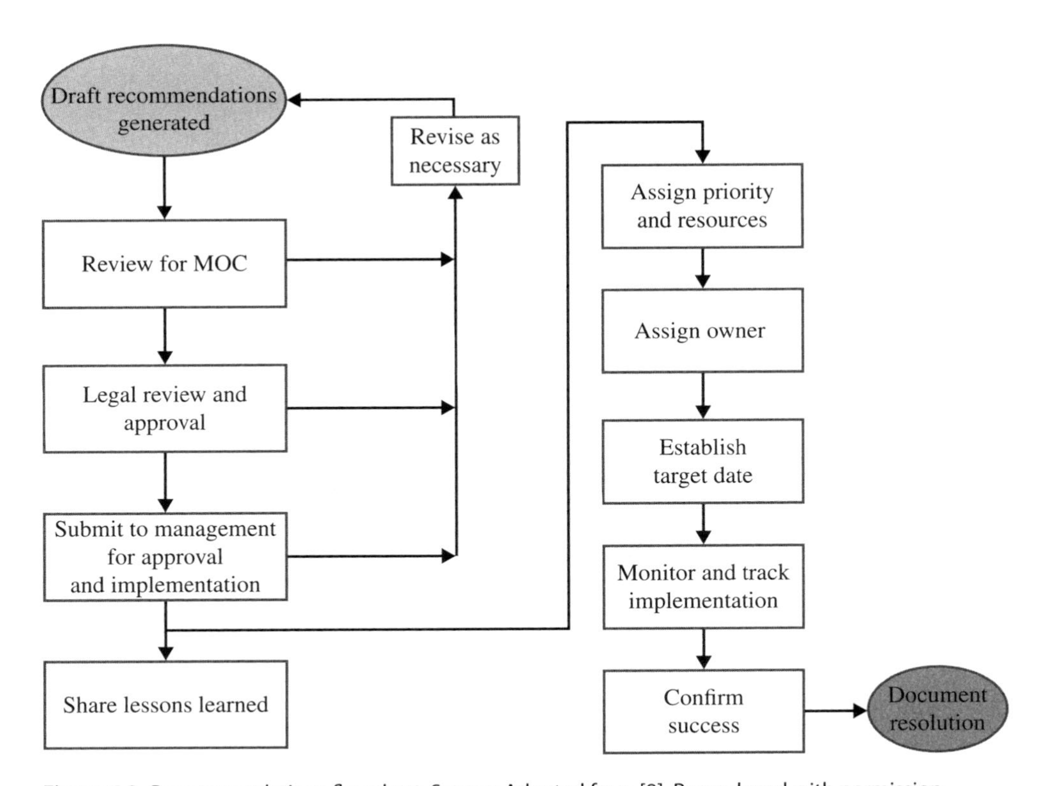

Figure 6.3 Recommendations flowchart. Source: Adapted from [9]. Reproduced with permission.

management system. Modified or rejected recommendations should be properly justified. A recommendation could be rejected because:

- A detailed analysis shows that it is not as beneficial as originally thought;
- additional information, not available originally to the investigation team, reveals that the problem is not so critical as it was expected to be;
- something changed, and the recommendation is no longer valid;
- an already implemented recommendation covers the objectives of another one, which is no longer necessary; and
- the suggested recommendation is yes beneficial, but not as much as required to mitigate the risk in the tolerable region.

Typically, a draft report for management approval is submitted by the investigation team, containing the proposed recommendations. Only after the management review, recommendations are assigned to the respective owners and target data are established. The owner of the action is not necessarily the same person who does the work: he/she monitors the progress of the risk control plan and can appoint a Primary Responsible Party (PRP) to do the work (i.e. to implement the action item). Periodic audits, aimed at the evaluation of the continuous improvement, paradigma of every management systems, are the occasions to verify the implementation of recommendations, to check if they were realized as intended. Those actions that are not closed within the target date are closely monitored, receiving special attention: indeed, the completion in time of action items is one of the leading indicators for a successful SMS. Management has an important role during this investigation stage. Indeed, it:

- Evaluates the feasibility and effectiveness of the proposed recommendations:
- schedules the implementation of the accepted ones;
- assigns the prp to implement the accepted recommendations;
- evaluates moc items;
- provides training to the affected personnel;
- documents the resolution of recommendations and track their completion; and
- extends the lesson learned from the incident to other areas, plants, facilities, or processes.

If decision-makers are not provided with enough information to make a judgement, recommendations are likely to not act upon. Therefore, it is important trying to anticipate the possible answers the decision-maker will ask, providing a detailed action items plan. Consultation with system owners (i.e. involved parties) on draft recommendations before publication leads to more practicable recommendations and a better likelihood of a more positive response [4].

Actually, an incident investigation is a reactive safety process, since it starts only after an unwanted event has occurred. However, the proposal of effective recommendations, spanning from immediate corrective actions to review on the management systems, allows extracting the best valuable lessons from the incident investigation, transforming it into a proactive safety process. Engineering and administrative controls are quite simple to develop. The real challenge is to convince the management to make changes: indeed, management typically recognizes the importance of the taken corrective actions, but if it does not understand the benefits, a successful implementation becomes improbable.

In order to reduce risk, recommendations can reduce the probability of occurrence or the magnitude of the event. From this perspective, there are different types of recommendations [5]:

- Those acting in the reduction of the likelihood (e.g. increasing maintenance programs);
- those acting in the reduction of the personnel exposure (e.g. decreasing the duration of exposure);
- those changing or-gate into and-gate, resulting in a lower frequency occurrence; and
- those eliminating or reducing the consequences (e.g. minimizing inventories of hazardous materials).

A recommendation will imply one of the following six hazard control strategies, as grouped by [14] (in order of priority):

- Elimination. the basic idea is that removing the hazard means having no incident;
- substitution. the hazard is substituted with a less hazardous condition, process, method, material (e.g. a toxic substance is substituted with a non-toxic one);
- engineering controls:
  - desing, e.g. to reduce the likelihood of an initial event;
  - redesign, e.g. to reduce the exposure of electrical circuits or dangerous moving parts;
  - enclosure, e.g. place a safety guard around a dangerous moving part;
- warnings, like signs and labels;
- administrative controls. they are typically combined with improved work procedures and practices. generally their goal is to reduce the duration of exposure to a hazard; and
- personal protective equipment (ppe). they can be required by law. however, they should never be considered an alternative to a barrier: their control over hazards works in conjunction with other control strategies.

The last three strategies are less effective than the first three. This happens because the last three relies on humans, who are naturally risk-takers: therefore, their barriers are inherently unreliable.

However, these six strategies apply for immediate causes, to face symptoms of an underlying SMS defeat or a root cause. Therefore, SMS improvements should be recommended (quoting [14]: "the most successful accident investigator is actually a systems analyst"). Example of this type of recommendations are:

- Include safety in the mission statement;
- improve safety policy, clearly establishing responsibilities and accountabilities;
- include safety checks in work process checklists;
- improve safety training program, including hands-on practice;
- include safety evaluation, together with cost, in the purchasing policy; and
- include supervisors and employees in the safety inspection process.

It is important to provide quality information when developing the recommendations: Quality In – Quality Out (QIQO). Otherwise, if not enough useful information is provided, the GIGO principle may apply: Garbage In – Garbage Out (GIGO). To do so, [14] suggests six questions that help to develop quality recommendations:

- What exactly is the problem?;
- what is the history of the problem?;
- what are the solutions that would correct the problem?;
- who is the decision-maker?;
- why is the decision-maker doing safety? what is motivating him/her?; and
- what will be the costs/benefits ratio of corrective actions and system improvements?

A good strategy is to provide the decision-maker with alternatives, to increase the probability he/she will choose one. For example, one option could be formulated regardless the economic constraints, while a second one considers limited funds.

Recommendations should be directly correlated to the causal factors and the root causes: if they are so developed, they will prevent recurrence of the incident. What can prevent a further incident is an effective recommendation; however, there is not a unique definition on what is effective, since it may vary from company to company, depending also on the risk tolerance/acceptability criteria, which are not universal [13].

From this perspective, recommendations can be developed on four different levels, depending on their depth:

- Addressing the causal factor. they typically face the equipment performance gaps (epg) or the front line personnel performance gaps (flppg). they are generally short-term recommendations;
- addressing the intermediate causes of the specific problem. they are short-term/intermediate recommendations that are effective in preventing recurrence of causal factors, but they do not address the root causes;
- fixing similar problems. these recommendations explore the extent of condition; and
- correcting the process that creates the problems. the recommendations developed at this level address the root causes.

Recommendations should be formulated in order to have a measurable completation criteria: for istance, it can be difficult to track the status of the recommendation "Provide a solution to mitigate the risk"; instead, it is easy to determine if the recommendation "Implement an interlock with a certified SIL2 to stop compressor when the low level alarm is activated in the tank" is completed or not, even if hard recommendations may sound more like prescriptions rather then performance-based solutions and prescriptive statements should be avoided, to leave the owner of the recommendation to act within strict boundaries, and not forcing him/her to accept aseptically the proposed correction.

According to the old concept of incident investigation, the task was considered over when the causes were found. In the point of view of a more modern approach to safety, this behaviour is now obsolete; indeed, the experience has shown how useless are the investigator's efforts if the incident investigation is not driven towards the necessary modifications and improvements of the SMS. This approach is now pursued both by companies and control authorities [7]. Going beyond the technical contents, a recommendation is considered properly formulated if:

- It refers to a root cause of the system, whose elimination solves the occurred problem;
- it clearly identifies the action to implement;
- it is feasible, flexible, and practical;
- it eliminates or reduces the risk (its likelihood or the consequences);
- a target date is defined;

- it identifies the responsible party for its actuation;
- it is congruent with the sms for changes (moc); and
- it is in harmony with the safety objectives of the company.

An action plan is the desired outcome of an investigation. Objectives should be SMART: Specific, Measurable, Agreed, Realistic, and with Timescales [15]. When assigning the priority, the risk level should be the driving factor. So, what cannot be left until another day is faced first. Financial constraints are often present, but it is not acceptable to not put in place measures to control severe risks for economic constraints.

Effective recommendations are capable of eliminating the multiple system-related causes of the incident [5]. Indeed, the recommendation process consists of the following steps (Figure 6.4):

- Select one cause;
- develop and examine preventive actions;
- perform a completeness test, to check if all the identified causes have been addressed;
- establish criteria to restart operations, together with the site and corporate management, and legal authorities (if involved);

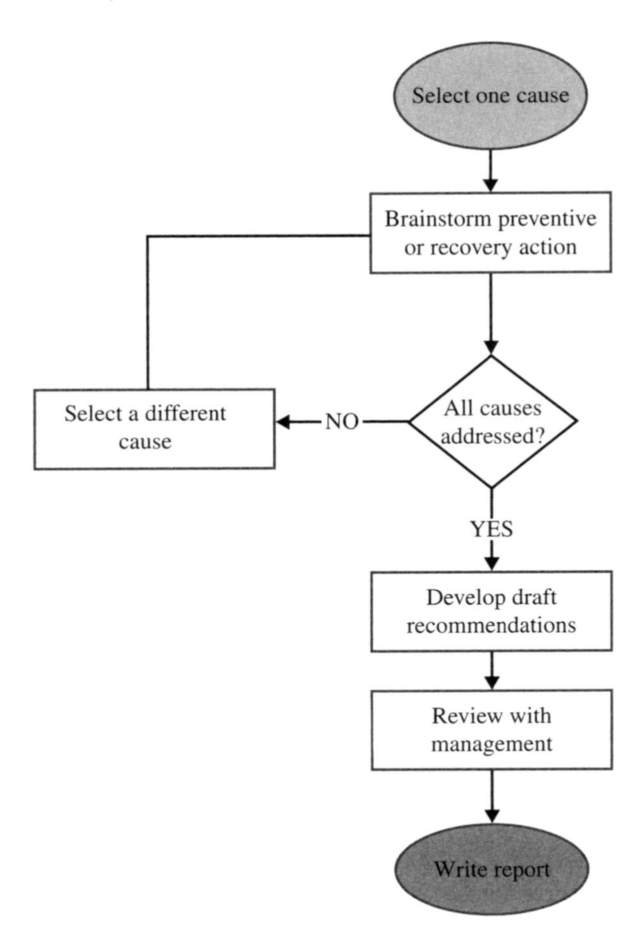

Figure 6.4 Workflow for recommendations and their monitoring. Source: Adapted from [5].

- present recommendations (they can be grouped by priority, system affected, cost, level of approval, and so on);
- review recommendations with management; and
- report and communicate recommendations formally.

To sum up, successful recommendations coming from the so called "learning process" [9]:

- Address root causes;
- are clearly stated;
- are practical, feasible, and achievable;
- add or enforce a layer of protection; and
- eliminate or decrease risks, acting on the likelihood, the consequence or both.

In conclusion, an incident investigation should very rarely result in disciplinary actions, since it is assumed that they are not part of the investigative scope. It should be noted that sometimes, the "no-action" recommendation could be suggested. This is often the case of risks that are evaluated through an ALARP study, still being tolerable for the company policy. Those recommendations requiring existing barriers to be reviewed to assess their effectiveness should also include the implemented action to carry out if the barriers reveal ineffective: indeed, the "review-type" recommendations are often incomplete from this point of view.

6.2.1 An Application of Risk Analysis to Choose the Best Corrective Measure

When developing corrective actions following an accident, it is stressed that the proposed recommendations should be effective. The following case study is proposed in summary form to define the process of selection of the improvements following an accident using risk analysis techniques. In particular, the choice between the possible alternatives of improvement is given by the solution that ensures a greater decrease of the probability of occurrence of the accidental event.

Following the event of in line detonation due to the arrival of oxygen on the blow-down manifold downstream of the refinery gasification reaction section, the possibility of introducing modifications is analysed relating to:

- The start-up procedures;
- the training of personnel who are entrusted with start-up of the system; and
- the inspection procedures of third-party companies involved in maintenance;

In order to reduce the probability of human error in the execution of operations related to start-up and maintenance of the Unit 300 (gasification and washing).

In particular, are analysed the causes (both due to "operational errors" and instrumental faults) which may involve the sending of oxygen to the Blow-Down System & Refinery Torches.

Other plant modifications are suggested which can be summarised as follows:

- The installation of a second valve (XV) on the fuel oil recirculation line in series with the oil recycling valve (V1) in the case where this second valve is not already present; and

- the installation of a fuel oil flow meter on the input line to the gasifiers with associated low flow alarm in the control room.

It should be noted that, with reference to the technological and safety adjustments envisaged by the licensee of the process, the plant configuration:

- Involves the presence of a second valve installed on the recirculation line (manual valve) for which was carried out an evaluation of the benefits expected as a result of the use of such a valve as part of the start-up sequence both in the case of use by the operator and considering automatic closing of the same in the start-up sequence; and
- it does not allow the installation (for plant layout reasons) of instruments intended to measure the fuel oil flow in input to the gasifiers for which was carried out an evaluation of the benefits expected as a result of the installation of instruments for measuring of the fuel oil flow rate on the recirculation line, fitted with oil high flow alarm in the control room.

Here is a brief description of the Unit 300 (gasification and washing) and of the start-up procedure.

The production of synthesis gas takes place in the gasification Unit 300, based on a licensed gasification technology.

The supply fuel oil that comes from the Unit 200 goes toward the T1 oil/steam mixer for mixing with high-pressure steam.

The mixture is sent to the process burners located at the head of the GAS1 gasifier.

The oxygen coming from the air separation unit through pipes is filtered and split into two currents:

- To Unit 300; and
- to Unit 510.

The reactants (steam, oxygen and the supply oil) are fed into the reactor chamber through the process burner. The oxygen is fed with a flow rate below the quantity needed for complete combustion of the supply. The moderation steam, premixed with the supply oil, mitigates the temperature in the reactor chamber, and reacts partially.

The gasification reaction is not catalytic but exothermic and the temperature of the gas at the outlet of the gasifier chamber varies from 1200 to 1450°C. The main products of the reaction are carbon monoxide, hydrogen, carbon dioxide, water vapour, methane and carbon black.

The gas mixture obtained from the gasification chamber passes into the cooling chamber through a tube immersed in the water. The mixture, coming into close contact with the cooling water, exits the GAS1 gasifier with a temperature of approximately 210°C.

The synthesis gas is sent to the scrubber connected to the gasifier, after mixing with the recirculation water coming from the scrubber itself. The first gas wash is performed in the scrubber.

The process burners, given the high temperature resulting from the reactions, are cooled by a coil fed with cold water.

The start-up operations of the gasifiers require heating of the refractory. For this purpose, each gasifier is provided with a pre-heating burner.

The start-up procedure of the gasifier following ordinary or extraordinary shutdown involves a sequence of automatic operations whose correct execution is verified by the operator.

The operations of pre-start (instrumentation tests on the valves and on the sequences) are used to verify that all the appropriate enabling conditions are met and, consequently, authorise the panel builder to launch from DCS the initialisation sequence.

The following sequence of operations includes:

1. Safety System Reset with a pressure to the gasifier below 4 barg;
2. Adjustment of the start-up flows of the moderator steam, oxygen and supply oil through the respective vent valves and of recirculation by the operator;
3. Replacement of the gas fuel burner with the fuel oil burner; and
4. Discharge with low-pressure nitrogen.

At the end of this part of the start-up procedure, following the verification that all the permissives are satisfied, the panel is authorised to launch from DCS the start-up sequence that includes:

5. Closure of the steam vent valve (V2) and subsequent timed opening of the main steam block valve (V3);
6. Opening of the fuel oil block valve (V4) and closing of the oil recirculation valve (V1) when the first is open 5%; and
7. Opening of the oxygen valve downstream (V5), subsequent closing of the oxygen vent valve (V6) and finally opening of the oxygen valve upstream (V7).

In order to estimate the frequency of the occurrence of the event that occurred on 13 March, 2010, a risk analysis was carried out, developed considering possible procedural and system changes mentioned in the introduction and to assess the envisaged benefits.

In particular, the analysis was performed taking into account:

- A) The procedures in place and the plant layout existing at the date of the event;
- B) the implementation of certain procedural and plant modifications, as detailed below:
  - B1) updating of the start-up procedure and related staff training;
  - B2) identification of the critical elements (and critical operations) of the system. Those elements (valves, instrumentation, etc.) Whose malfunction (failure, improper installation, incorrect maintenance) may result directly in a significant incidental event are considered as critical. Revision of the operational instruction i.o.001 relating to the maintenance management entrusted to third-party companies. A specific control plan for the critical safety components must be encoded by the third-party company that implemented the change. The verification of correct maintenance on the critical components must be performed in the presence of saras personnel and in any case the personnel of the third-party company that activates the quality control plan must be different from that which performed the maintenance.
  - B3) implementation of plant modifications which, in short, as part of the start-up sequence of the gasifier, ensure the automatic closing of no. 2 valves on the fuel oil recirculation line, the existing v1 and a second same type valve, in series. On the fuel oil recirculation line is currently installed a second manual valve (v8). This valve, after verification of feasibility, may be controlled automatically by the start-up sequence in order to adhere to the recommendations. Additional recommendations relate to the installation of a flow transmitter on the fuel oil input line

to the gasifiers alarmed for low oil flow: in consideration of the system layout, an equivalent modification is implemented which consists of the measurement by an alarmed transmitter in the control room for high flow of the fuel oil flow on the recycling line; and
- o B4) modification of the start-up procedure of the system in order to ensure closing by the operator, of the V8, in place of the automatic closing proposed.

The updating and adoption of the new start-up procedure and of the relative specific training of operating personnel identified in the safety analysis performed following the incidental event involve:

- The reduction of the probability associated with insufficient operative intervention in the event of malfunctions at the starting phase of the GAS1 gasifier;
- the reduction in frequency of occurrence of the accidental hypothesis of 1 order of magnitude; and
- A further increase of two orders of magnitude of the system's safety level can be achieved by modifying the maintenance management procedure carried out by third-party companies on components classified as critical in terms of safety, in the presence of internal staff at the refinery and in accordance with IO001, to be formalised as a procedure of the Safety Management System.

In particular, following the achievement of safety adjustments envisaged, the incidental hypothesis occurrence frequency relating to the "Sending of oxygen to blow down during start-up of the gasification reactor GAS1" would be of the order of magnitude of 10^{-5} oxy/year, falling within a probability class defined as "Unlikely" compared with a frequency of occurrence in the configuration subsequent to the first level of modifications that places the incidental cases in the probability class defined "Fairly improbable".

Finally, the adoption of plant recommendations, or the use, with automatic closure, of a second valve on the fuel oil recirculation line, a fuel oil flowmeter with alarm in the control room and the consequent further update of the start-up procedure of the unit, involves:

- Ergonomic optimisation of the operational measures thanks to the alarm signal;
- the reduction of the times of intervention by operators, in case of deviation of the operational critical parameters from the normal start-up conditions; and
- the hypothesis occurrence frequency reduction of sending oxygen to blow down during the start of the GAS1 gasification reactor.

The use, according to specific operating instruction, of the existing manual valve V8 at the starting phase by the operator in the field in place of the installation of a second block valve, also implementing the remaining plant recommendations, would be (with respect to the full adherence to the licensee's requirements) in a higher occurrence frequency by approximately 3 orders of magnitude. Within this context, the full adoption of what is defined by the process licensee is recommended and the involvement of these for the details regarding implementation of each of the recommendations identified.

The adoption of new blocks and alarms would reduce the incidental hypothesis occurrence frequency identified by an additional three orders of magnitude returning to a probability class defined as "Extremely improbable".

In order to assess the probability of operator error during the system start-up sequence currently recommended by the refinery manuals, the SLIM (Success

Likelihood Index Method) methodology has been applied. This technique consists of correlating the probability of operator error with certain factors (called PIF, Performance-Influencing Factor) that affect human behaviour.

In this case the significant factors are summarised below:

- Non-habitual and/or complex operation;
- noise or sources of distractions in general;
- control panel design;
- supports to work and procedures;
- training;
- experience of the operators; and
- group work.

The operations carried out by the operators to complete the starting procedure can be summarised as follows:

1. follow on the DCS panel the sequence of steps prior to start-up of the gasifier;
2. once all the appropriate permissives (instrumentation tests on the valves and on the sequences) have been satisfied, start up the system;
3. follow the sequence of valves opening and closing in order to verify correct performance;
4. check for a rapid increase in pressure and temperature in the gasifier and in the scrubber;
5. verify an increase in the torch flame (assistance to the panel builder for this operation);
6. check the level of the gasifier by acting on the quench ring flow rate, on the flow rate of blow down and on increasing of the pressure in the system.

For performing of the risk analysis, the SLIM methodology was used for evaluation of the probability of error in the performance of operations 3, 4 and 5 of the sequence.

Using operations 2 and 6 of the above sequence, the parameters A and B were estimated that characterise the equation of the SLIM method reported below:

$$\boldsymbol{Log(HEP) = A \cdot SLI + B} \tag{6.3}$$

where HEP is the probability of human error and SLI is the index of the likelihood of success achieved by the combination of the values assumed by the PIF.

On the basis of the PIF parameters defined for operations 2 and 6, the following equation was derived which links the operating error probability and the PIF, through the SLI:

$$\boldsymbol{Log(HEP) = -31.5 \cdot SLI + 14.4} \tag{6.4}$$

The equation obtained was used to estimate the probability of error in operations 4 and 5 of the sequence in the current configuration.

On the basis of the documentation available, operations 4 and 5 of interest are characterised by the following PIF in the current configuration (hereinafter referred to as "ANTE"), shown in Table 6.1.

The values referred to in the table above have been assigned in a relative scale between 1 and 9 in which the upper end is the optimum. For example, in the case of the first operation, value 9 represents the case of an habitual and non-complex operation.

Table 6.1 PIF (current configuration).

PIF	Value operation 4(ANTE)
a- Non-habitual and/or complex operation	2
b- Noise or sources of distractions in general	7
c- Control panel design	5
d- Supports to work and procedures;	5
e- Training	5
f- Experience of the operators;	5
g- Group work	5

There is a consequent probability of operator error in performing of the same equal to:

$$HEP_{ANTE} = P_{ANTE} = 1.46 \cdot 10^{-1} \tag{6.5}$$

In the first instance it is proposed to update the start-up procedure and the training of personnel on the system that includes:

- A written and detailed procedure that supports the interpretation of curves of temperature and pressure and the increase in torch flame to detect abnormal conditions and which describes the operations to be carried out to secure the system in case of the identification of critical issues; and
- training on the system start-up procedure in its entirety and on the procedure referred to in the preceding paragraph, in particular intended for all personnel involved in the operation before each start-up of the GAS1 gasifier.

Taking into account the modifications listed in the new configuration (A), operations 4 and 5 of interest are characterised by the following PIF, shown in Table 6.2.

The probability of human error which consists in failing to interrupt the start-up sequence in the event of anomalies was thus evaluated as:

$$P_A = 1.1 \cdot 10^{-2} \tag{6.6}$$

Table 6.2 PIF (A configuration).

PIF	Value operation 4(A)
a- Non-habitual and/or complex operation	2
b- Noise or sources of distractions in general	7
c- Control panel design	5
d- Supports to work and procedures;	**6**
e- Training	**6**
f- Experience of the operators;	5
g- Group work	5

In addition to the protections currently provided, during the start-up phases, the following were evaluated: the activation in the control room of a fuel oil high flow rate alarm in the recirculation line controlled by a new flow meter located downstream of the oil recirculation sectioning and the related update of the start-up procedure and of the training of personnel system that includes a written and detailed procedure that desxribes the operations to be carried out to secure the system following the high flow alarm signalling on the fuel oil recirculation line.

Adoption of the alarm mentioned above provides the panel builder with an improved support for evaluation of the correct performing of the start-up procedure.

This adjustment is equivalent to the licensee's recommendation to activate alarm signalling of low flow rate of load oil to the gasifiers, which cannot be achieved as the system layout does not allow positioning of measuring instruments on the supply line.

Taking into account the modifications listed in the new configuration, operations 4 and 5 of the procedure are characterised by the following PIF, shown in Table 6.3.

The probability of human error due to failing to interrupt the start-up sequence in the event of anomalies was thus evaluated as:

$$P_B = 3{\cdot}10^{-3} \tag{6.7}$$

The expected benefits are analysed, in terms of decrease in incidental hypothesis occurrence frequency, identified following the event in March 2010, by combining the likelihood of human error referred to above with the components failure accruals, referred to in the fault tree in Figure 6.5 to 6.7.

Figure 6.8, below, is a graphic summary of the occurrence frequencies obtained downstream of each modification previously listed.

The results are summarised in Table 6.4.

It therefore becomes clear how risk analysis is a useful tool in determining what recommendation to adopt in order to avoid the recurrence of incidental events.

6.3 Communication

Communication is an essential tool to powerfully share the lesson derived from an incident. The main obstacle here is that companies do not let others learn from their experience [16]. Generally they do not want to share their incident reports with anyone,

Table 6.3 PIF (POST configuration).

PIF	Value operation 4(POST)
a- Non-habitual and/or complex operation	2
b- Noise or sources of distractions in general	7
c- Control panel design	**6**
d- Supports to work and procedures;	6
e- Training	6
f- Experience of the operators;	5
g- Group work	5

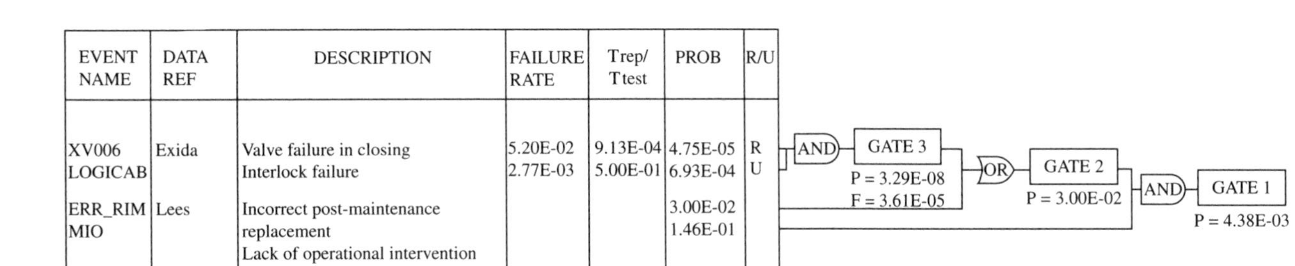

EVENT NAME	DATA REF	DESCRIPTION	FAILURE RATE	Trep/ Ttest	PROB	R/U
XV006	Exida	Valve failure in closing	5.20E-02	9.13E-04	4.75E-05	R
LOGICAB		Interlock failure	2.77E-03	5.00E-01	6.93E-04	U
ERR_RIM	Lees	Incorrect post-maintenance replacement			3.00E-02	
MIO		Lack of operational intervention			1.46E-01	

Figure 6.5 Fault Tree Analysis, current configuration (ANTE).

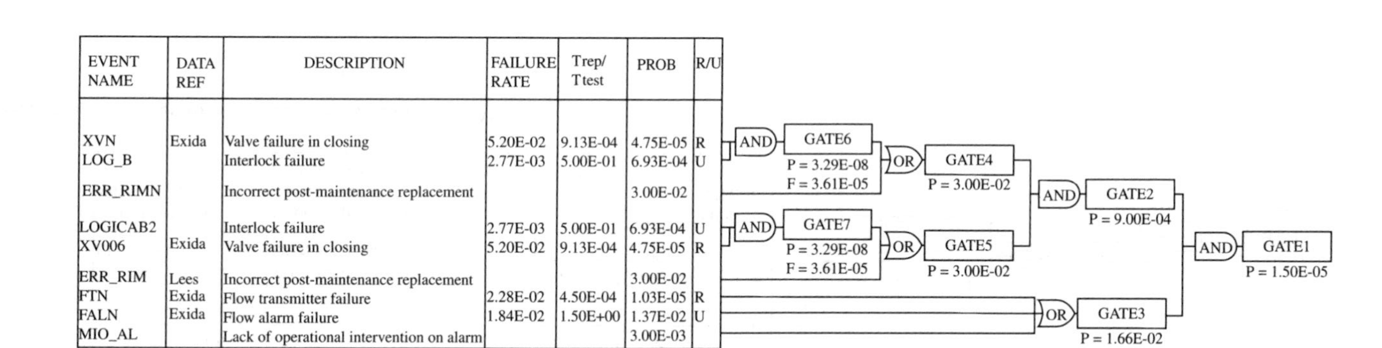

EVENT NAME	DATA REF	DESCRIPTION	FAILURE RATE	Trep/ Ttest	PROB	R/U
XVN	Exida	Valve failure in closing	5.20E-02	9.13E-04	4.75E-05	R
LOG_B		Interlock failure	2.77E-03	5.00E-01	6.93E-04	U
ERR_RIMN		Incorrect post-maintenance replacement			3.00E-02	
LOGICAB2		Interlock failure	2.77E-03	5.00E-01	6.93E-04	U
XV006	Exida	Valve failure in closing	5.20E-02	9.13E-04	4.75E-05	R
ERR_RIM	Lees	Incorrect post-maintenance replacement			3.00E-02	
FTN	Exida	Flow transmitter failure	2.28E-02	4.50E-04	1.03E-05	R
FALN	Exida	Flow alarm failure	1.84E-02	1.50E+00	1.37E-02	U
MIO_AL		Lack of operational intervention on alarm			3.00E-03	

Figure 6.6 Fault Tree Analysis, a better configuration (A-configuration).

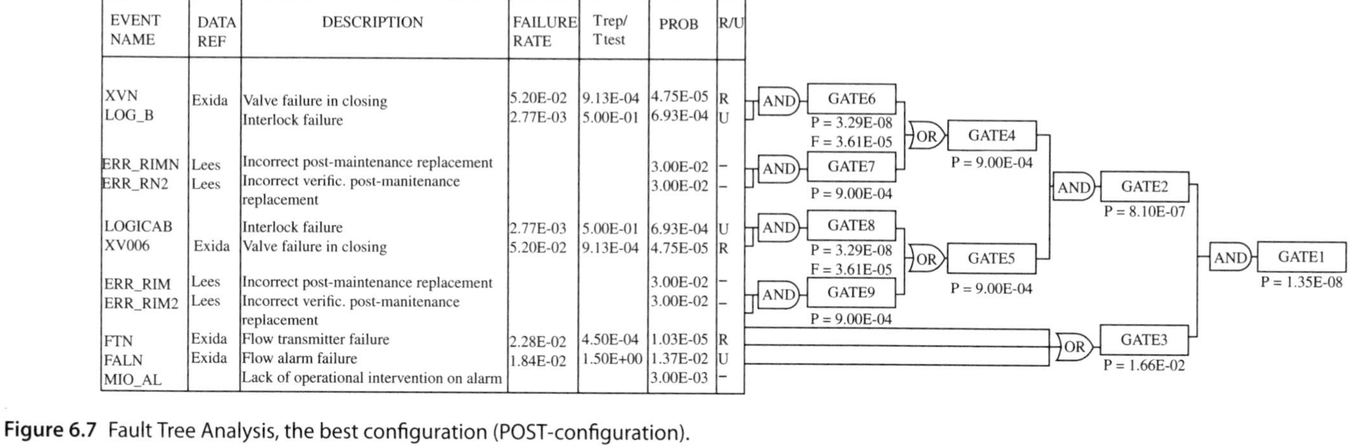

EVENT NAME	DATA REF	DESCRIPTION	FAILURE RATE	Trep/ Ttest	PROB	R/U
XVN	Exida	Valve failure in closing	5.20E-02	9.13E-04	4.75E-05	R
LOG_B		Interlock failure	2.77E-03	5.00E-01	6.93E-04	U
ERR_RIMN	Lees	Incorrect post-maintenance replacement			3.00E-02	–
ERR_RN2	Lees	Incorrect verific. post-manitenance replacement			3.00E-02	–
LOGICAB		Interlock failure	2.77E-03	5.00E-01	6.93E-04	U
XV006	Exida	Valve failure in closing	5.20E-02	9.13E-04	4.75E-05	R
ERR_RIM	Lees	Incorrect post-maintenance replacement			3.00E-02	–
ERR_RIM2	Lees	Incorrect verific. post-manitenance replacement			3.00E-02	–
FTN	Exida	Flow transmitter failure	2.28E-02	4.50E-04	1.03E-05	R
FALN	Exida	Flow alarm failure	1.84E-02	1.50E+00	1.37E-02	U
MIO_AL		Lack of operational intervention on alarm			3.00E-03	–

Figure 6.7 Fault Tree Analysis, the best configuration (POST-configuration).

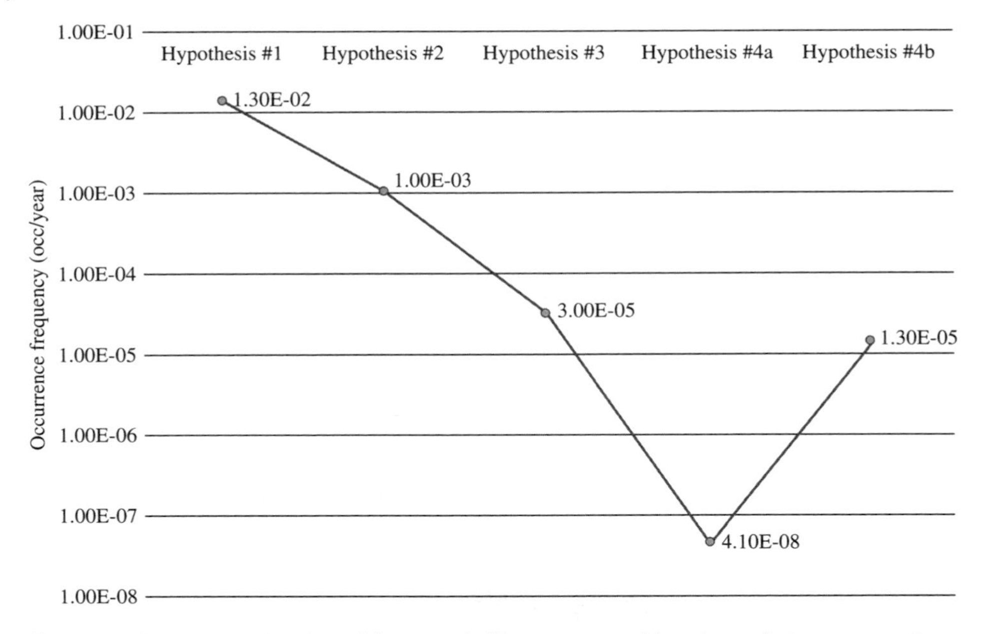

Figure 6.8 Frequency estimation of the scenario "Oxygen sent to blow down, during start up of reactor of GAS1".

even internally within the company, but this is not an effective policy to avoid the reccurrence of the incident. Instead, the circulation of essential messages, inside the company and outside, is crucial for many reasons. First of all it is a matter of morality: if we have information that might avoid another accident elsewhere, we have the moral duty to share that information. Secondly, sharing information could have a pragmatic advantage for the company: if we share, it is highly probable that the action is returned. From an economic point of view, knowing potential risks previously not considered will push our competitors in spending as much money as we do in safety. Another important point that should push the companies in sharing their information is that every accident affects the reputation of the whole industrial sector, which is the same for competitor companies.

Companies are not limited to documented lessons, but also respond to these opportunities. To do so, lessons are firstly identified and their value in sharing is recognized. Then, a system to efficiently share the lessons is used, without overwhelming the company. In the end, the lessons learned are embedded in the standards, procedures or practices of the company, checking if the existing processes or equipment need changes. The enhanced application of lessons learned relies deeply on the sharing and communication of critical knowledge [12]. This requires a culture driving employees to learn from many sources: not only incidents, but also near misses and jobs well done. Giving them the right priority is a challenge. Sharing lessons rapidly helps in driving the improvement in company standards and procedures, delivering process safety performance, and supporting both safe and reliable operations. Significant incidents or near misses are opportunity to encourage learnings: everyone inside the company should take personal action as a result of the lesson learned. Sharing is important; learning is the plus.

Table 6.4 Frequency of the considered incidental hypotheses

Hypotheses ref.	Description	Frequency of occurrence (oxy/year)	Probability class	Modifications made
1	Sending to blow down of oxygen during start-up of the GAS1 gasification reactor (*current configuration*)	$1{,}3 \cdot 10^{-2}$	Fairly likely	
2	Sending to blow down of oxygen during start-up of the GAS1 gasification reactor (following procedural modifications)	$1 \cdot 10^{-3}$	Fairly unlikely	Updating of start-up procedure and specific training of operators
3	Sending to blow down of oxygen during start-up of the GAS1 gasification reactor (following procedural modifications for third party company maintenance checks)	$3 \cdot 10^{-5}$	Unlikely	Maintenance of components classified as critical in terms of safety carried out by third-party companies in the presence of personnel at the refinery
4a	Sending to blow down of oxygen during start-up of the GAS1 gasification reactor (following plant recommendations)	$4{,}1 \cdot 10^{-8}$	Extremely unlikely	- Installation of a second valve on the fuel oil recirculation line - Installation of a fuel oil flow meter on the recirculation line with alarm in the control room - Updating of start-up procedure and specific training of operators
4b	Sending to blow down of oxygen during start-up of the GAS1 gasification reactor (without XV new installation)	$1{,}3 \cdot 10^{-5}$	Unlikely	Similar to modifications related to Hypothesis 4a according to the start-up procedure of the existing V8 manual valve in place of the installation of a second XV valve.

For example, the reports of major industrial incidents are a source for critical lessons that everyone in the process industry should know. Once identified the most appropriate major incident, perhaps looking for the one with similar processes or hazardous materials, its critical lessons are communicated, along with the subsequent actions, to the respective leaders. Obviously, the lessons should be based on root causes to be effective.

A formal process to communicate and share the lessons learned throughout the company is essential, otherwise the incident investigation's capability to enhance safety is wasted, even if the best investigation team has been recruited. Sending emails is a good starting point. Standard templates can be also used, to communicate lessons in a structured way. People should not be overwhelmed with too many lessons of minimal value: only critical lessons should be shared and communicated. The incident investigation status, findings and child action items can be discussed in regular meetings. They can be monitored using a software-based solution, to check if the expected results are achieved in the expected target date by the designated owner. The lessons learned from an incident investigation should be also used proactively, during the risk assessment studies. In Paragraph 5.4.2.2, it has been already discussed how the BowTie can be used as a tool to link these two sides of the coin.

6.4 Safety (and Risk) Management and Training

Risk management is the art of equilibrium between gains versus the risk of losses. This approach is taken in everyday life, when we are asked to evaluate an activity involving hazards, thus implying risks. For example, if we overtake a car in a congested road, it is supposed we have evaluated the benefit of overtaking and compared it with the risk of a possible collision. It is similar with the industrial incident and risk-based management system [17]. On the one hand, there are the costs of an incident, which are not only the direct economic loss for the company, but also the "cost" of people involved in the incident, the "cost" for eventual environmental damages, the "cost" for reputation. On the other hand, there is the cost of safety, which is the cost of investments and maintenance to keep the residual risk low.

In an extremely simplified approach, being out of the scope of this book, the cost optimization can be achieved as follows. As shown in Figure 6.9, the curve of the cost of the incident increases with the risk-level: higher the risk, higher the cost of a potential incident. Instead, the curve of the cost of safety decreases with the risk-level: higher risk scenarios, being a restricted number, often requires limited cost to be mitigated, while low-risk scenarios, which are the majority part of an industrial company, require a constant investment to improve safety, with higher total costs. The sum of the two curves defines the total costs curve, whose minimum represents the risk-based costs optimization criterion.

Safety training is among the most diffused prescription to prevent an accident. The number of action items developed after an incident occurred about the necessity to improve training actually reveals that training alone is not sufficient. It requires being effective. Some studies [18] demonstrate that the effectiveness of the training is affected by the workers' belief in its effectiveness. Moreover, they identify some weakness of the management systems in not providing a proper support on training, especially regarding PPE.

Knowing the management system perspective may be of interest before performing and after having performed a root cause analysis. Generally, when talking about "management", people are prone to think about few persons working in big offices at the last floor: these are only the manager and the term management system is not used to indicate only the management team [9]. Indeed, according to the approach adopted in this book, the management system is a term used to represent the totality of the administrative activities, from the management team to the front-line supervisor, who

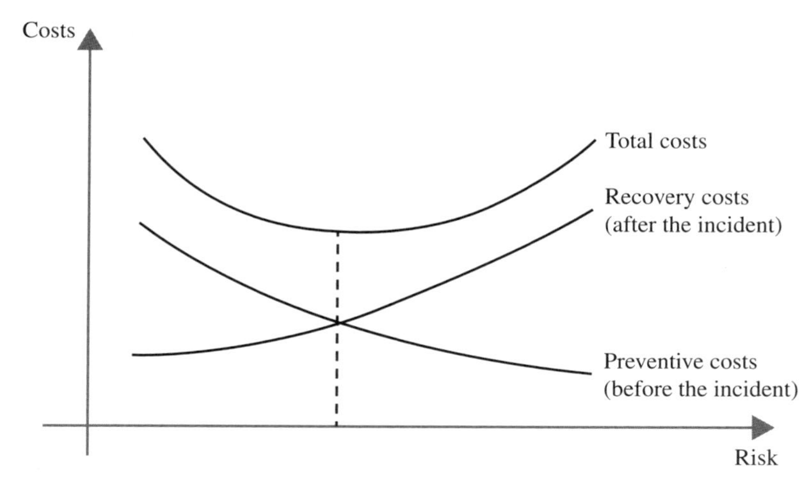

Figure 6.9 Risk-based cost optimization.

are required to deal with a particular task. A weakness in the management system, as here intended, is usually a root cause. The common components of safety and risk management are written procedures, training, performance indicators and objectives, and assigned responsibilities according to possessed competences and skills. The possession of an incident investigation system is itself a good example of the management system, where the final goals of the investigation and its expectations are clearly fixed and its effectiveness is assessed.

According to that expressed above, team preparation becomes crucial. All the potential investigation team members are formally trained, in order to be sure that a pool of investigators is always available inside the company. Depending on the specific topic, refresher training or certifications may be required. The eventuality to use breathing apparatus may also require pre-use medical evaluation and fitting test. Obviously, such training must be provided before the incident investigation; conducting them after may bring also a major legal consequence. A specific training should be provided to the investigation team leader, who needs to know the investigation methodology to be used and all the organizational aspects related to the duty (definitions, reports deadline and contents, general approaches to stakeholders, security chain for evidence, developing recommendations in line with the company policy). It is suggested to provide a refreshing training session to the investigation team members when the investigation begins, with a focus on the peculiar nature of the event. Arguments for this refresher training could be PPE, investigation methods, and emergency issues.

The occurrence of some major incidents, like the one discussed in Chapter 2 as Seveso, Flixborough, or Bhopal, was a leading factor to drive the education of engineer about safety and loss prevention, enlarging the academic curriculum with specific subjects. Probably, the greater efforts are dedicated to educating to the concepts of inherently safety design, which is the preferable solution, as already noted [9]. The concepts related to safety and loss prevention involve a number of basic principles: chemistry, thermodynamics, process safety, mechanics, hazard identification, risk assessment, functional safety, just to cite some of them. It is a complex topic, and this is way many efforts are addressed to safety (and risk) management and the related training. According to IChemE, in order to teach safety and loss prevention, the following topics, should be covered (this syllabus was provided in 1983):

- Legislation;
- management of safety;
- systematic identification and quantification of hazards, including hazard and operability studies;
- pressure relief and venting;
- emission and dispersion;
- fire and flammability characteristics;
- explosions;
- toxicity and toxic releases;
- safety in plant operation, maintenance and modification; and
- personal safety.

In addition, there are other two main topics that should be taught at least: environment protection and the human factors, being relevant in the modern approach to safety and loss prevention.

In conclusion the multidisciplinary approach required to carry out an incident investigation is the same required to train the engineers of tomorrow, who should have a clear link in their mind among the traditional subjects of chemical engineering courses and the safety and loss prevention topics.

6.5 Organization Systems and Safety Culture

The term "management" embraces those administrative activities related to a specific objective. Firstly we consider that the objective of the management system is the prevention of incidents, and the task is the conduction of the incident investigation. As noted in [9], the configuration of the incident investigation management system preliminary requires a corporate policy commitment to prevent incidents by investigating them, identifying root causes, developing recommendations, and implementing improvements. A written protocol is usually prepared and personnel of the organization is trained to complete the assigned action items. Three different investigation skill levels are generally present:

- A minimum baseline knowledge about reporting and general policy commitment, provided to all employees, contractors and visitors. It is generally provided with less than one hour of training and requires periodic refresh to ensure that a minimum awareness level is maintained;
- a second level, regarding line managers and those reviewing and approving draft investigation reports, who need to understand the acceptability performance target for incident investigation;
- a third level, regarding those who participate in or lead investigations, who must know how to do it.

Sometimes, a fourth level of competency is prescribed for experts and investigation team leaders, who should be competent also in some management tasks.

Typically, the magnitude of the incident's severity determines the necessity to report it or not. Increasing levels of severity are established, together with increased reporting and investigation requirements. Successful management systems take into account the feedback process, to evaluate and improve investigations on a regular basis. The accident investigation management system should be integrated with other SMS, like the PHA, training, auditing, MOC, and emergency response management. Taking inspiration from [9], a possible table of contents of an incident investigation management system manual is the following:

- Management leadership
  - policy
  - legal requirements
- definitions and categories of incidents
- reporting and notification requirements
- specific responsibilities
- preparing for investigations
  - criteria to select investigators and investigation teams
  - investigation training and refresh training
  - methodologies and tools
- conducting the investigation
  - internal and external communication issues
  - evidence recognition, collection, management, custody
  - evidence analysis
- root cause determination
- findings and conclusions
  - developing effective recommendations
  - review and approval of suggested recommendations
  - evaluating recommendations for moc
  - implementing action items
    - immediate measures
    - permanent changes
    - assigning priority
    - status tracking, target date and owner identification
    - verification for implementation
    - documentation of resolution
  - reports
  - implement lessons learned
  - sharing lessons learned
- continous improvement of the management system
  - monitoring effectiveness
  - periodic audits
  - update and changes to the management system
  - monitoring changes in regulatory requirements and corporate standards
  - incident trend analysis

The role of the management in the causation of accidents is underlined in [19]. Management is generally responsible for the overall conditions of the workplace, because it allocates resources for the tasks, the proper tools, the safety equipment, meeting the current safety standards and practices, it sets the standards to ensure adherence to safety standards, positively (or negatively) reinforcing practices by supervision, it selects personnel and qualifies it for a particular work task, it provides the proper training for the specialized work tasks, it implements procedures and standards to correct unsafe actions, and, in conclusion, it promotes general safety. The role of management has a great influence over the identification and reduction of accident precursors (i.e. those conditions favouring mistakes), the identification and reduction of initial events (i.e. those actions that can cause an incident), the development and promotion of both preventive and mitigating barriers.

Incident investigation is a reactive approach to enhance the safety performance of a system, since events are investigated after their occurrence. This approach is in opposition to the proactive approach, where efforts are dedicated to anticipating events before they occur, like for risk assessment. The two strategies are complementary to each other, and not replacements for one another. The substantial difference is that in a reactive approach the feedback process (that is to say the learning from experience) is used to prevent occurrence by taking appropriate corrective measures. However, the learning opportunities are also offered by the other performance indicators than the incidents: they are the safety perception, the rescue and emergency operations, the social safety issues. This means that the learning process identifies not only the system deficiencies (from the reactive approach) but also the knowledge deficiencies (from the proactive approach). This complementary nature allows the incident investigation to be seen as part of a system combining both reactive and proactive feedback loops, as shown in Figure 6.10.

A good strategy to enhance the benefits of incident investigations is to learn from multiple investigations [4]. Reviewing the output of several investigations can add value to identify those system weaknesses that a single investigation was not capable of uncovering. The analysis of multiple investigation reports helps to identify the recurring problems, assisting in giving actions the right priority. A learning opportunity is also offered by the incident investigation process itself, regardless of the content of the analysis. In particular, the feedback process may highlight some improvements for the evidence collection stage, or their preservation, or the interviewing of witnesses, together with a revision of existing causation models.

Another critical factor is related to the organizational memory. Indeed, humans have memory, organization not. The experience shows how the incident investigation is not sufficient to ensure that lessons are learned and retained in the collective memory. Specific measures must be taken to ensure the preservation of the lesson learned over time, from the viewpoint of a company. To use the words of Kletz [16], organizations have no memory. The actions to take to make sure that the lessons are not forgotten are the following:

- Recommendations should be followed up;
- lessons should be publicized;
- workshop should be held to keep incident fresh in the memory;
- incidents should be described in safety and loss prevention newsletter;
- a reference to the incident should be used in any documents containing the lesson learned from the incident;
- a database of case histories should be provided to personnel;

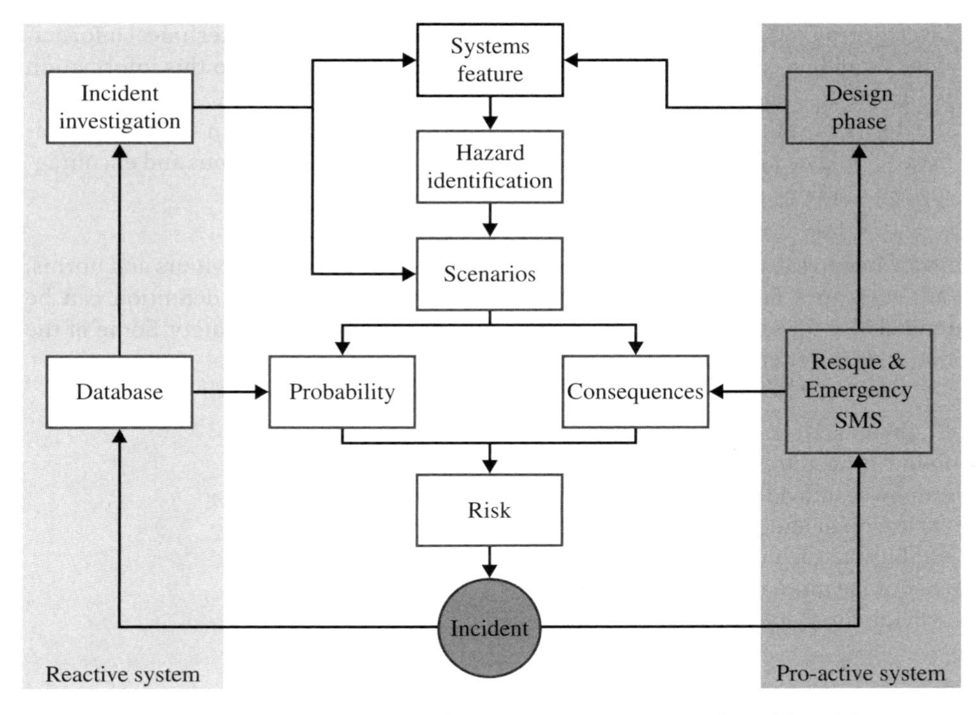

Figure 6.10 Proactive and reactive system safety enhancement. Source: Adapted from [4]. Reproduced with permission.

- management should not turn a blind eye to unacceptable working practices; and
- before removing equipment or abrogating a procedure, their original purpose should be found out.

Knowing the most recurrent themes in accident investigation may be advantageous: for example, among the most recurrent management defeat there are amateurism, insularity, failure to train personnel or to correct poor working practices. It is highly probable that an investigator will deal with them.

The learning culture is encouraged by different interests and standpoints of the actors, with their different perceptions and learning processes [4]. There is a common need to achieve the convergence in learning from an event (reactive approach) and creating new knowledge that can be used to improve safety (proactive approach). In order to pursue this learning culture:

- Lessons learned from practical experience should be integrated into the procedures involving safety (and risk) management;
- everyone at all levels should take part in the learning process, disseminating the gained knowledge and spreading the practice through the proper communication formats;
- communication processes should be established over a shared consensus among the stakeholders, ensuring that actors feel engaged in the learning process;

- investigation outcomes, both subjective experiences and objective technical information, should be collected systematically, ensuring that the access to this information is allowed to stakeholders, using a proper procedure; and
- the desire of stakeholders to learn the transferable lessons found in other investigations should be promoted, facilitating processes for cross-connections and encouraging the multidisciplinary exchange.

According to [3], the safety culture is a "common set of values, behaviours and norms, at all levels in a facility [...] that affect process safety". A similar definition can be extended also for safety culture in general, not only for the process safety. Some of the common features creating a safety culture are:

- Maintain a sense of vulnerability;
- follow procedures rigorously;
- empower individuals allowing them to reach their safety objectives;
- ensure open and effective communications;
- establish a learning environment;
- encourage mutual trust; and
- provide timely response to safety issues.

Probably, the maintenance of a sense of vulnerability is the among the most crucial factors. Indeed, it often happens that the good trend of the lagging indicators (that is to say, a low number of incidents per worked hours) may lead to one eye being closed over leading indicators, encouraging hazardous behaviours which ignore some actual risks. It concerns the capability of individuals, starting from the top management board members, to remain aware of risks regardless of the positive trend recorded: to put it simply, it is a matter of care being taken.

The distinction between "safety culture" and "safety climate" needs to be provided. It is difficult to give a structured definition of safety culture, even if it is studied widely; indeed, there is not a strong consensus on its definition [18] and on its boundaries. Safety climate concerns the attitudes an organization has about safety practices and regulations, and how they are perceived within the organization itself. The safety culture, instead, is the organizational configuration that a company follows to create the safety climate. Therefore, safety culture is crucial since it helps to determine how the individuals within a company approach the safety issues. In order to have a quantitative feedback about the individuals' way to deal with the safety issues, experimental methods involving sociological and psychological approaches are also used. The implementation of a safety culture rarely implies a change in the basic personality of an individual; indeed, its goal is to affect what can be changed effectively, like procedures and confidence. Nowadays, researchers are addressed to find a quantitative definition of safety culture, also considering the actual number of incidents (or at least injuries) recorded in a company [20], thus providing a quantitative tool to assess the reached level of safety culture.

The incident investigation is also the occasion to improve its link with the risk analysis process, as already mentioned previously: it's a great opportunity to create a more robust safety culture.

6.6 Behavior-based Safety (BBS)

By "Behavior-Based Safety", it is intended a group of methods to manage safety. They are based on the improvement of those behaviours being significant for workers' safety, with the objective to reduce the injuries consistently [21]. The BBS methodologies come from the "behaviour analysis", a branch of psychology explaining the human behaviour using the Skinner paradigm, that is to say correlating it with antecedent stimulus and consequent stimulus. The consequent stimulus models the probability of a future behaviour, when the antecedent stimulus is present.

Starting from the 1970s, different scientists developed methods applied to the occupational safety. The already mentioned Heinrich found how the 90% of injuries were attributable to human behaviour.

It must be considered that the human behavior and hazardous conditions (equipment, working places) are among a chain of factors leading to an injury, starting from the safety and health culture to the SMS, which are actually caused by the root causes. Since the first Heinrich's studies, the role of the management in ensuring safe conditions and behaviours has been depicted; however, the behavior is still central in this chain of factors.

During the years, the BBS methodology has been enriched with advanced management methods, cognitive psychology, and neurophysiology. This integration has been considered too forced [21], and the BBS has been hardly criticized because: behavior is determined by more complex factors than the ones of Skinner's paradigm; the behaviors in the work field cannot be modelled with BBS because they are extremely complex; it is necessary to work on the culture before modifying the behaviors (otherwise a behavior change will not be stable); looking at the behavior (immediate cause) is an act of myopia, since root causes are neglected; and looking at the behavior means to blame the worker for his/her injury.

Being aware of these critics, there are also some considerations acting in favour of the BBS. In particular, behavior is still the immediate cause of injury: not managing it properly is a fundamental weakness in the safety management. Moreover, the behavior is measurable. This is a crucial point. Some behaviors are hardly observable, but the majority of them are. This provides a clear representation of the safety and risk "boundaries" in a company. With data, it is possible to intervene proactively to correct risky situations before they generate incidents and injuries. At the same time, it is possible to evaluate the effectiveness of corrective actions, observing how a specific trend evolves over time. Another point of advantage is that behaviour is modifiable and can be improved. It is true that neglecting the cultural factors is an error, but the modifications at the culture have a big defeat: they are necessarily long-term. Instead, a manager wants that his/her employees are safe today! The faster way is to provide feedback that is coherent with this objective, listening to the suggestions to change the way people behaves, in order to improve safety.

The consideration that behaviors are responsible (i.e. the immediate cause) of the major part of injuries, deeply affected the legislation on health and safety, which pushed on training and information practices [22]. However, the effective safe behaviours of workers do not depend only on the training activities; indeed, they are also influenced by the surrounding environment (that is to say, the workplace, the stimulus, the interaction with colleagues and bosses, and so on).

In conclusion, the BBS is capable to:

- Identify specific behaviors that are not safe;
- explore the (root) causes of these behaviors;
- develop the recommendations to remove the identified causes; and
- provide leading parameters to predict future trends.

To be successful, these activities are carried out with a strong involvement of the workers, including all the employees, from the CEO to the front line operators. BBS is not based on an assumption: indeed, a successful BBS program is based on scientific knowledge.

According to [23], the BBS is a heritage of the Ford-Taylorism, where the independent variables are the productive process and the work organization: the worker must passively adapt to them. From this standpoint, some legislation, like the Italian D.Lgs 81/08, could consider illegal the BBS because of the recognition of the principles of ergonomics in the work environment. Indeed, its direct consequence is the adjustment of the workplace to the worker; it is exactly the contrary of Ford-Taylorism and BBS, which is its direct application.

6.7 Understanding Near-misses and Treat Them

Generally, major process incidents are preceded by warning signals, revealing themselves months, days, or hours before the incident. Sometimes, these symptoms may cause a near miss, also called "neat hit" or "close call". A near miss, as defined in Chapter 2, is an occurrence having the potentiality to result in an incident if the circumstances had been slightly different [5].

The investigation of a near miss follows exactly the same path already described for incidents, in order to determine the causes why it occurred. In particular, the weaknesses in the management system are identified, developing the corrective measures to fix them. The great advantage of near miss is that they provide free learning opportunities, differently from incidents where an actual loss is experienced.

The typical relationship among accidents, near misses and nonincidents (shown in Figure 6.11 and Table 6.5) actually depends on the definition of near-miss provided by a single company and on the type of loss [5]: for example, quality-related incidents, resulting from less severe excursions than occupational safety or process safety occurrences, have less near misses and nonincidents. Moreover, as the process gets simpler (that is to

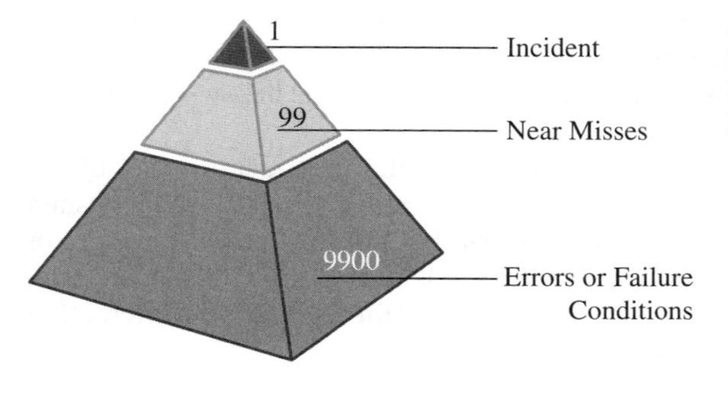

Figure 6.11 Relationship among incidents, near misses and nonincidents. Source: Adapted from [5]. Reproduced with permission.

Table 6.5 Comparative table for teaching differences between incidents and nonincidents.

Incidents	Nonincidents
Safety relief device open	Safety relief device found to be outside tolerances during routine inspection
Pressure reaches relief valve set pressure, but it does not open	Pressure excursion occurs, remaining within the process safety limits
High-high pressure trip/shutdown	High-pressure alarm
Toxic gas detector tripped/alarm	Toxic gas detector found ineffective during inspection/testing
Suspended crane load slips	Crane wire rope with defeats during pre lift checks

Source: Adapted from [5].

say operating conditions are closer to ambient and less layer of protection are needed), there are fewer near misses and errors per accident, thus fewer "symptoms" are available. Regardless this consideration, many chemical companies declare they have only one or two near misses reported for every incident. The reason is that there are some obstacles to near miss reporting.

When the potential consequences of a near miss overcome a certain threshold, it should be investigated. To do so, the company needs to specifically define what a near miss is and how to understand and treat it. In particular, in order to consider whether to investigate a near miss or not, the potential consequences need to be evaluated, also considering if they would have been more severe, due to slightly different circumstances/external condition (like weather), delayed detection, or a less experienced person performing the task [13]. The acceptability or tolerability of the potential risks should be assessed: if they belong to that region, it has no sense to carry out an investigation, since it will result in no changes. Similarly, if the assessment reveals that adequate barriers are already in place to protect the workers from these incidents, then an investigation would result in no changes, so it might not be necessary to investigate.

However, it is difficult to evaluate these criteria before an investigation is performed, since the judgment is based only on limited information. Therefore, the first investigation effort is focused on answering those questions and only then,if necessary, will the investigation continue.

The reason why near misses should be investigated is that they share the same causal factors and root causes of actual incidents. Therefore, investigating a near miss can help in preventing other near misses and incidents. To do so, near misses need to be reported. Unfortunately, some barriers are present for getting near misses reported. Indeed, a near miss is often only known to the personnel involved; therefore, they have the possibility to decide whether to report it or not. The typical barriers that discourage from reporting a near miss are the following:

- Fear of disciplinary action. to overcome this barrier the company must take and communicate its "no-punishment" approach to investigation, favouring the cooperation from the employees;
- fear of embarrassment. it is difficult to control if peers will embarrass the person involved in a near miss; unfortunately the organization has no control over it;

- fear of legal liability. this obstacle is overtaken involving the organization's legal staff in the investigation process, to limit the organization's legal exposure. the legal department should have to encourage the reporting activities, since the benefit coming from preventing incidents affect the legal exposure positively;
- disincentives for reporting near misses, like extra working activities (reporting, filling out forms, participating to meetings and interviews), with the potential consequence to increase the working hours;
- multiple investigation programs. if the organization has different procedures for reporting process safety, reliability, occupational, environmental, and business incidents, then the person in charge to report these incidents may be discouraged. it is preferable to have a single process for all the different incident typologies;
- lack of management follow-through. if the person reporting a near miss does not receive feedback from the management about taken actions, reporting can be seen as a waste of time;
- no incentive to report near misses. sometimes, receiving a reward for reporting near misses could encourage this practice. rewards can be money, hats, gadgets, and so on;
- apparent low return on effort to report. the scale between efforts and benefits could be seen to much unbalanced on the efforts if feedback is not provided. feedback is necessary to show what is done thanks to the near miss reporting; and
- lack of understanding of a near miss vs a nonincident. personnel can be confused about what should be reported and should not. therefore, it is important to provide clear definitions and procedures, ensuring that the knowledge has been correctly transferred.

The actions necessary to overcome these barriers may take from few weeks to years. In particular, those actions affecting the organization's culture may require long-term implementation, as required to change the perceived fear of punishment. The best solution is an effective incident investigation program: showing how a properly performed investigation is capable of improving the workplace and the working conditions is the best way to encourage personnel in reporting near misses.

The question of the purpose of near-miss reporting is also discussed in [9]. As also stated by van der Schaaf in [24], also cited in [9], near miss reporting should be carried out for three different purposes: modelling (focusing attention of new types of near misses for a qualitative insight); monitoring (focusing attention on already known near misses, for a quantitative insight); and maintaining alertness.

A near miss management system should be an integral part of an SMS, to establish clear practices and procedures about:

- Detection (recognition and reporting) of a near miss;
- its selection, depending on the purposes;
- description of the relevant human, hardware, and organizational factors;
- classification, according to the model adopted by the organization;
- computation, that is to say performing statistical analysis to check if recurrent factors are present;
- interpretation, i.e. developing corrective measures from statistics results; and
- monitoring, measuring the effectiveness of the implemented recommendations.

Different methods to investigate near misses are available in the literature. They are adherent to the methods already discussed for actual incident investigation, including logic trees, RCA, and barrier-based tools. This is why it is not necessary to provide specific tools for near misses, being the same already presented in Chapter 5.

References

1 Dien, Y., Llory, M., and Montmayeul, R. (2004). Organisational accidents investigation methodology and lessons learned. *Journal of Hazardous Materials* 111 (1-3): 147–153.
2 Turner B. (1978) *Man-Made Disasters*. 1e. London: Wykeham Publications.
3 CCPS (Center for Chemical Process Safety). (2016) *Introduction to process safety for undergraduates and engineers*. 1e. Hoboken: John Wiley & Sons.
4 ESReDA Working Group on Accident Investigation. (2009) *Guidelines for Safety Investigations of Accidents*. 1e. European Safety and Reliability and Data Association.
5 CCPS (Center for Chemical Process Safety). (2003) *Guidelines for investigating chemical process incidents*. 2e. New York: American Institute of Chemical Engineers.
6 Sutton, I. (2010). *Process Risk and Reliability Management*. Burlington: William Andrew, Inc.
7 Agenzia per la protezione dell'ambiente e per i servizi tecnici (APAT). (2005) Analisi post-incidentale nelle attività a rischio di incidente rilevante. 1e. Roma: APAT.
8 Forck, F. and Noakes-Fry, K. (2016) Cause Analysis Manual. 1e. Brookfield, US: Rothstein Publishing.
9 Mannan, S. and Lees, F. (2012) Lees' loss prevention in the process industries. 4e. Boston: Butterworth-Heinemann.
10 Lindberg, A., Hansson, S., and Rollenhagen, C. (2010). Learning from accidents – What more do we need to know? *Safety Science* 48 (6): 714–721.
11 Crowl, D. and Louvar, J. (2011) *Chemical process safety: Fundamentals with Applications*. 3e. Upper Saddle River, NJ: Prentice Hall.
12 CEP (2017) Enhanced Application and Sharing of Lessons Learned. *CEP Magazine - Chemical Engineering Progress* (An AIChe Publication). 57.
13 ABS Consulting (Vanden Heuvel L, Lorenzo D, Jackson L, Hanson W, Rooney J, Walker D) (2008) *Root cause analysis handbook*: a guide to efficient and effective incident investigation. 3e. Brookfield, Conn.: Rothstein Associates Inc.
14 OSHAcademy (2010) *Effective Accident Investigation*. 1e. Portland: Geigle Communications.
15 Health and Safety Executive. (2004) HSG245: Investigating accidents and incidents: a workbook for employers, unions, safety representatives and safety professionals. 1e. Health and Safety Executive.
16 Kletz, T. (2002) Accident investigation - Missed opportunities. Hazards XVI: Analysing the Past, Planning the Future. Manchester: Institution of Chemical Engineers, pp. 3-8.
17 Pasman, H. (2015) *Risk analysis and control for industrial processes*. 1e. Oxford: Elsevier Butterworth-Heinemann.
18 Mannan, S. (2001) *Lees process safety essentials*. 1e. Kidlington, Oxford, U.K.: Butterworth-Heinemann.
19 Noon, R. (2001) *Forensic engineering investigation*. 1e. Boca Raton, FL: CRC Press.

20 Payne, S., Bergman, M., Beus, J. et al. (2009) *Leading and Lagging: The Safety Climate-Injury Relationship at One Year*. Mary Kay O'Connor Process Safety Center International Symposium.

21 Introduzione alla behavior-based safety [Internet]. behaviorbasedsafety.org. 2014 [cited 14 November 2017]. Available from: http://www.behaviorbasedsafety.org/basi-behavior-based-safety/

22 BBS - Behavior-Based Safety [Internet]. Ali Srl. [cited 14 November 2017]. Available from: http://www.alisrl.it/bbs/

23 Frigeri G. Il feticcio della BBS (Behavior Based Safety): Perché non funziona - Euronorma [Internet]. Euronorma. 2013 [cited 14 November 2017]. Available from: http://www.euronorma.it/blog/il-feticcio-della-bbs-behavior-based-safety-perche-non-funziona/

24 Van der Schaaf, T., Lucas, D., and Hale, A. (1991). *Near Miss Reporting as a Safety Tool*. Burlington: Elsevier Science.

Further Reading

CCPS (Center for Chemical Process Safety) (2013). *Guidelines for Managing Process Safety Risks During Organizational Change*. Hoboken, NJ: John Wiley & Sons.

7

Case Studies

Some case studies are now presented, to give some examples of how to apply the concepts presented in this book. They are real cases extracted from the wide collection of investigations conducted by the authors. More specifically, they are:

1. A jet fire at a steel plant, to provide a real example about the logic trees and the qra tools;
2. a fire on board a vessel, to both show how to face complexity emerging in an incident investigation and to have an example about the difference between immediate and root causes;
3. a release of toxic substances from a process plant, to show an application of the causal factor diagram;
4. a refinery's pipe-way fire, to present the importance of the mars database and to show a compliance study with nfpa 550;
5. an external flash fire of pulverised sawdust during an emergency emptying of a 300 cubic meter silo after a smouldering combustion ignited inside it;
6. an explosion of a rotisserie van, to highlight the importance of digital evidence, like video frames;
7. a fragment projection with potential process pipes in a congested area damage;
8. a refinery process unit fire, to show how the importance of the imperative "to do something" after an incident, in order to prevent its reoccurrence;
9. a crack in an oil pipeline, to show the importance of a multidisciplinary approach when using fem analysis; and
10. a storage building on fire, to briefly discuss about the involvement in a fire of photovoltaic panels on a roof.

7.1 Jet Fire at a Steel Plant

7.1.1 Introduction

The general information about the case study is shown in Table 7.1.1 and is given in [1] and [2].

Stainless Steel coils are produced throughout the world in a multitude of industrial sites. The process is conceptually simple, whereas the mechanics of the machines are rather sophisticated. The process phases, which are well known, are melting, casting, hot rolling, cold rolling, pickling and annealing. The accident occurred at a pickling and

Principles of Forensic Engineering Applied to Industrial Accidents, First Edition.
Luca Fiorentini and Luca Marmo.

Table 7.1.1 General information about the case study.

Who	Steel plant
What	Jet fire
When	2007
Where	Turin, Italy
Consequences	7 victims
Mission statement	Determine the fire causes and dynamics
Credits	Luca Marmo (Politecnico di Torino) Norberto Piccinini (Politecnico di Torino) Luca Fiorentini (Tecsa s.r.l.)

annealing line, two operations that are usually conducted in the same plant. Pickling and Annealing (P&A) lines are conceptually very simple: steel coils have to be unrolled, then a thermal treatment is conducted in a furnace and chemical and electrochemical pickling are performed in a series of basins. After these treatments, the coil is re-rolled. The main technical challenges of the process derive from the need to run both the thermal and the electrochemical processes continuously, even when the coils have finished. In order to comply with continuous process constraints, the subsequent coils have to be welded, and this introduces a discontinuous process. As a consequence, some complications arise in the architecture of the lines. These lines should be provided with devices able to temporarily store the length of coil that must be supplied to the furnace and to the pickling section while the unrolling is suspended during welding. Further complications arise from the weight of the coils, reaching several Tons, depending upon the length and width, from the need to guarantee the correct traction of the coil, and from the need to move it over several hundred meters of process line while providing adequate position control.

The coil is handled via hydraulic systems which use mineral oil. This oil is not usually flammable, but it is of course combustible. Hydraulic circuits are fed with high-pressure oil. In this case, the pressure ranged from between 70 and 140 bar. Under these conditions, highly flammable spray/mists can originate from small leaks. As a consequence, a diffused fire risk should be recognised in P&A lines. Other sources of fire hazards are the flammable/combustible materials that accompany the coils which come from the lamination process. A paper ribbon is placed between the steel coils at the end of the lamination to prevent surface damage. The paper absorbs the lamination oil residue present on the coils and sometimes sticks onto it due to high temperatures and oil ageing. In this way, paper can spread along the inlet section of the P&A line thus adding combustible material and enhancing the local fire risk.

7.1.2 How it Happened (Incident Dynamics)

On December 6 2007 five workers were on the night shift (22.00 PM – 6.00 AM) on the annealing and pickling line. Another three workers were on the line either to substitute or train other workers. At 00.35, the line was restarted after an 84 min stop to remove some paper lost from a previously treated coil. As there was no automatic control system on the inlet section for the axial coil position, the coil, after some time, started to rub against the line structure, which was made of ironwork. The location of this scraping was identified just above flattener #2, while scraping occurred to coil #1. The rubbing lasted for several minutes, and, as a consequence, produced sparks and local overheating. A local fire, which involved paper and the hydraulic oil released from previous spills started from these circumstances. A small pool fire started in the flattener area, which is depicted in Figure 7.1.1 and 7.1.2, and subsequently spread to roughly 5 m^2, involving the flattener and its hydraulic circuits. At roughly 00.45, the workers realised there was a fire, and took some measures to fight it. First they stopped the entry section of the

Figure 7.1.1 Area involved in the accident. Right, unwinding section of the line, left, the front wall impinged by flames. Source: [1].

Figure 7.1.2 The flattener and the area involved in the accident. Details of the area struck by the jet fire, view from the front wall. Source: [1].

line, reduced the line velocity, seized some fire extinguishers and went close to the fire to attack it from at least two directions. After some seconds, they decided to also use a fire hydrant, so one of them walked to the fire hydrant and a second one handled the fire hose. At that moment, one of the several pipes of the hydraulic circuits involved in the fire (roughly 10 mm inner diameter) collapsed and released a jet of high-pressure oil from the pipe fitting. The pipe, which is depicted in Figure 7.1.3, was fed at a pressure of 70 bar from the main pump station, which was still running. As a consequence, a spray of hydraulic oil was released into the already existing fire.

The ignition of the spray was immediate, due to the contemporary presence of the pool fire near the release point, and this resulted in a severe jet fire that struck the eight workers directly or with its radiant heat effects. Figure 7.1.4 shows a map of the site with the presumed positions of the workers (no reliable witnesses could be found concerning this topic) and the extension of the area in which the jet fire took place. The length of the jet fire has not been determined precisely, since it hit the front wall that was located at a distance of more than 10 m from the broken hosepipe and some other structures in the nearby. The footprint of the jet fire is clearly visible on the wall in Figure 7.1.5. The spread angle of the jet fire was roughly 30, since some scattering occurred against the various equipment. The fire also spread backwards with respect to the hose direction, and in such a manner, it involved the large area indicated in Figure 7.1.4. This spread of the fire was determined by the interaction of the fluid released at very high velocity with a number of fixed structures located in the vicinity of the release point.

A total of 13 pipes then collapsed in a few minutes. Many of these pipes were under pressure and continuously fed by the pump station, hence provoking a huge spread of oil and of the flames. The pressure in the hydraulic circuit dropped, due to the huge oil leak, thus the intensity of the jet fire diminished very quickly. At the same time, the fire reached its maximum size (see Figure 7.1.4), as it was being fed by the released hydraulic oil that burned in a pool. From an analysis of the control system records, which are summarised in Table 7.1.2, the time scale of the events resulted to be those indicated

Figure 7.1.3 Details of the hydraulic pipe that provoked the flash fire. Source: [1].

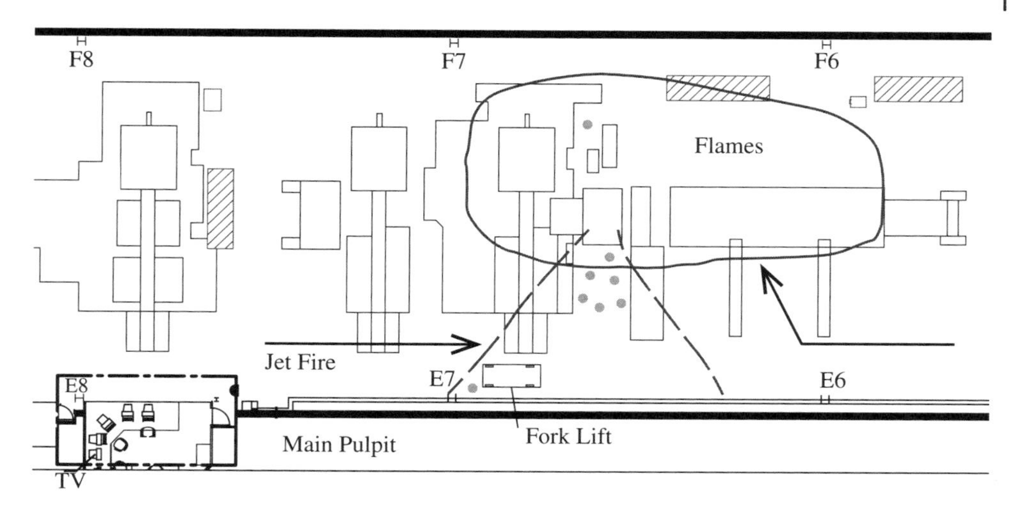

Figure 7.1.4 Map of the area struck by the jet fire and by the consequent fire. The dots represent the presumed position of the workers at the moment the jet originated. Source: [2].

Figure 7.1.5 Footprint of the jet fire on the front wall. Source: [2].

in Figure 7.1.6. The first pipe collapsed in a time interval of between 00.45′49″ and 00.48′24″; the pumps were stopped by the automatic control system at 0.53′10″, due to the low-level switch system on the basis of the oil level in the main reservoir. In this time interval, at least 400 liters of oil escaped.

The jet fire struck the eight workers who were fighting the initial fire. The worker who was close to the fire hydrant (see Figure 7.1.4) was sheltered by a forklift and only

Table 7.1.2 Record of the supervisor systems (adapted from Italian).

Time	Operator/Automatic	Meaning
0.31.05	O	Set coil thickness
0.31.10–0.31.20	O	Sending data to Mandrel 2
0.34.46–0.35.16	O	Start chemical section
0.35.43	O	Start Inlet section
0.35.46	A	Start confirmed by field sensors
0.35.48	A	Low flow – rinsing section
0.36.06	O	Low flow acknowledged by operator
0.45.45	O	Line speed set to 18 m/min (Group 5 events)
0.45.49	O	Start pump final rinsing unit
0.48.24	A	Low oil level alarm from hydraulic station (Group 6 events)
0.48.39	A	Lubrication fault – mandrel 1
0.48.39	A	Lubrication fault – mandrel 2
0.48.39	A	Flaw fault – mandrel 2
0.48.39	A	Low pressure – mandrel 1
0.48.39	A	Low pressure – mandrel 2
0.48.39	A	Loss of control – mandrel 1
0.48.39	A	Loss of control – mandrel 2
0.48.44	A	Loss of control – mandrel 2
0.49.53	A	Fault cable inlet section
0.53.00	A	Low oil level
0.53.10	A	Emergency stop

Source: [1].

suffered minor burns. Six workers were struck by the jet fire and suffered 3rd-degree burns covering from 60% to 90% of their bodies. They died over the following months. One, who went to the back of the plant to fight the fire, was trapped and died immediately. The fire spread to the machines and lasted for approximately 2 h, before the fire brigade from the National Fire Corps could extinguish it.

7.1.3 Why it Happened

A specific analysis has been conducted in order to understand the consequences of the accident and the level of risk for the operators. The fire scenario was modeled by means of a specific CFD calculation tool (Fire Dynamics Simulator, generally known as 'FDS', which was developed by the Building and Fire Research Laboratory e BFRL e of the U.S. National Institute for Standards and Technology) on the basis of the evidence and information collected during the investigation. The numerical simulation of the consequences of an accident is a useful and recognised methodology to estimate the consequences of accidental releases of hazardous chemicals in industrial premises in terms of thermal radiation, temperature rise, presence and extension of flames, smoke

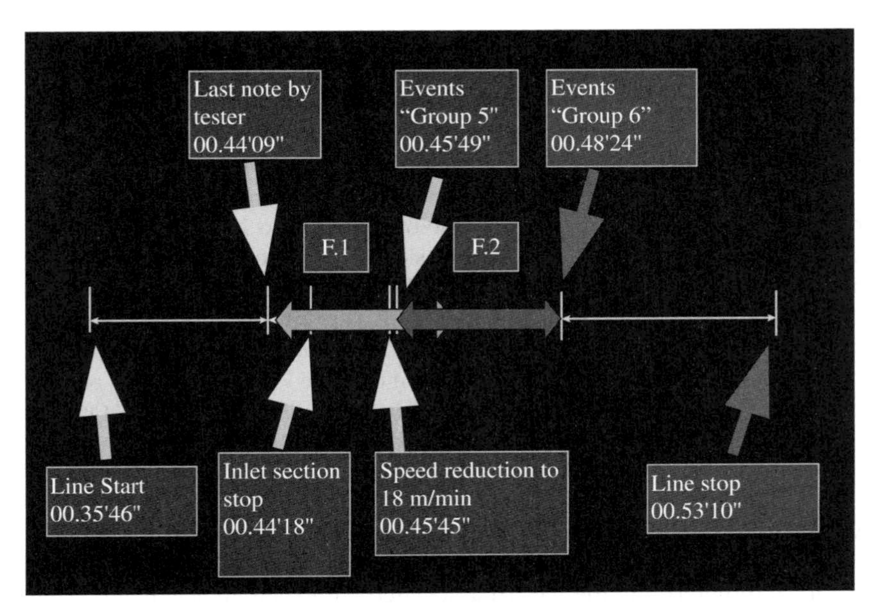

Figure 7.1.6 Timescale of the accident. F.1 is the time interval in which the ignition occurred. F.2 is the time interval in which it is probable that the workers noticed the fire. The group 5 and group 6 events are defined as in Table 7.1.2. Source: [1]. Reproduced with permission.

production, the dispersion of combustion products, and the movement of those species in the compartment/s under examination in order to verify what happened with a certain degree of certainty and to verify the modification of the consequences connected to the modification of the parameters that govern the accidental release. Simulation results can help technical consultants in the reconstruction of the accidental event. The well-known NFPA n. 921 standard (2008) recognises that fire behaviour numeric codes play a fundamental role for in-depth analysis in the forensic framework: both simplified routines and zone and field models are explicitly quoted. The 'FDS' chosen by the authors is, currently, one of the most specialised and frequently used codes to assess the consequences of a fire inside a compartment, even in industrial premises, and also for forensics purposes. An extensive amount of technical literature has been published by the authors of the code [3, 4, 5] and the same technical reference guide of version 5.0 of the code [5] presents a specific section (n. 2.3) on the reconstruction of real fires. A number of forensic activities are listed in this section (mainly concerning the reconstruction of the consequences and dynamics of real fires). One of the incidents dealt with the fire which occurred in the World Trade Center on 11 September 2011 ("The collapse of the Twin Towers"). In this technical consultancy, the NIST, on behalf of the FEMA (Federal Emergency Management Agency) investigated the danger of the release of flammable liquids in the form of sprays and this danger was also assessed by conducting a number of real tests. Those tests were in fact similar to the activity that was later conducted by the U.S. Navy to test the consequences of the accidental release of hydraulic flammable oil at high pressure at a real scale and which was described in detail in [6] and in a specific report [7] that qualifies the four main objectives of the tests: to investigate the consequences of fires from hydraulic flammable fluids in submarines; to investigate the potential for hydraulic fluid explosions; to estimate the event timeline; to acquire experimental data in order to allow a proper fire modeling to be used in engineering practice. In that report as well as in a subsequent paper the danger of hydraulic oil, even for releases of limited quantities of fuel, is described clearly, along with the description of the facilities used to simulate the release in the experiments. The real, full-scale experiments conducted by the U.S. Navy are comparable with the data used by the authors for the simulation of the Thyssen-Krupp accident, e.g. a pressure range from 69 bar to 103.4 bar, a released fluid with a combustion heat of 42.7 MJ/kg and a similar viscosity. With this data, the authors found it very useful to validate the use of FDS against the results of the U.S. Navy experiments and completed the dissertation with a specific example that showed full agreement of the simulation with the results of a series of experiments conducted by the U.S. Navy, characterised by a release pressure close to 70 bar. The heat release rate and thermal conditions in the compartments can reasonably be compared. On this basis, FDS has been employed to reconstruct the accidental release and fire that actually occurred at the Thyssen-Krupp plant in Turin. The details of the conducted simulations are not the scope of this Paragraph; a short description of the adopted workflow is given in the following sections in order to provide the readers with a clear picture of the procedure that has been employed by the authors as Technical Consultants of the Public Prosecutor's Office to determine the hazard associated with such an accidental event for the workers that died during the activities adopted to govern the emergency. The simulation activities helped the Authors to describe the consequences of the accident in order to define the real risk for the operators and, subsequently, to verify whether the calculated risk level corresponded to the level formally declared by the Owner (in the risk assessments required by law) to the Authorities Having Jurisdiction (AHJ) and to verify whether different scenarios could have exposed the operators to similar risks (e.g. with limited releases, with retarded ignition.). In particular, the authors quantified the consequences according to the legal requirements pertaining to industrial risks and identified the fire risk with respect to national law (see the threshold limits given by the national Decree dated 9 May 2001 in Table 7.1.3). Several analyses were conducted for both "jet fire" and "flash fire" cases, as defined from the extensive literature available, with the aim of comparing the results with the threshold values established by Italian law (Table 7.1.3). The simulation of the real case (i.e. a "jet fire") is described hereafter.

The simulation involved a preliminary reconstruction of the analysis domain. Several surveys were conducted to obtain a precise description of the tridimensional layout of the portion of the compartment that had to be investigated (dimensions: 12 m 10.8 m 11.2 m). The domain is presented in Figure 7.1.7 and in tridimensional form in

Table 7.1.3 Threshold values according to Italian regulations.

Accident	High fatalities	Beginning fatalities	Irreversible injuries	Reversible injuries	Domino effect
Fire (stationary thermal radiation)	12.5 kW/m^2	7 kW/m^2	5 kW/m^2	3 kW/m^2	12.5 kW/m^2
Flash-fire	LFL	½ LFL			

Source: [1].

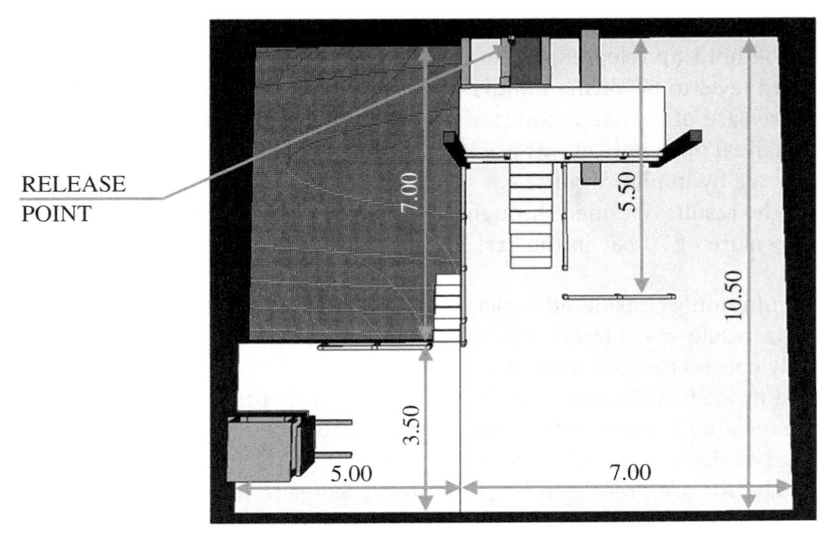

Figure 7.1.7 The domain used in the FDS fire simulations. Source: [1]. Reproduced with permission.

Figure 7.1.8. The release point and the main dimensions are indicated in the plot-plan in Figure 7.1.7, while the elevation in Figure 7.1.8 shows the model and the fork-lift that was located in the area where the flames spread. The release point was located at a height of 0.5 m and identified with a circular orifice (diameter equal to 1 cm) directed toward the front wall. The analysis domain was divided into a cubic cell mesh with 1 cm maximum length side.

The simulation of the "jet fire" considered an initial pressure of the hydraulic circuit of 70 bar, although a number of different simulations were run in order to verify the sensitivity of the consequences in the area where the workers were believed to be at the release time, with variations in the pressure range (up to 140 bar which is the design

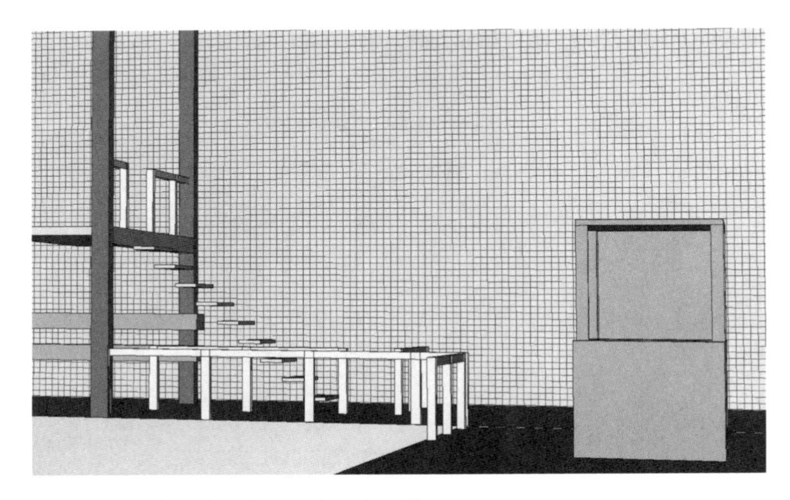

Figure 7.1.8 Simulated area, elevation [1].

pressure of the involved circuit) and a number of other parameters (e.g. direction of the release, total oil hold-up released, physical properties of the oil, etc.). This activity allowed the hazard level to be verified under different conditions and in particular to verify whether a release of a small amount of oil could have exposed the operators to danger (since a manual push button was present to limit the release via the isolation of the actuators of some hydraulic circuits).

An example of the results obtained through the use of FDS is given (the case considering a release pressure of 70 bar in the first instants from the release) in Figure 7.1.9 to 7.1.14.

P&A lines are commonly considered as plants at high fire risk. Nevertheless, the area at major fire risk is usually considered to be the annealing furnace where huge amounts of fuel gas (mainly natural gas) are used. Another huge fire that occurred in a P&A line in a plant located in Krefeld, Germany, in 2006, has shown that the fire risk can arise from the use of annealing basins and their covers made of plastic material.

Instead, the fire risk due to hydraulic circuits, in particular in the inlet zone of the line, where hydraulic circuits are present in huge numbers, seems to have been underestimated to a great extent in the present case, but also in general in the steel industry. The

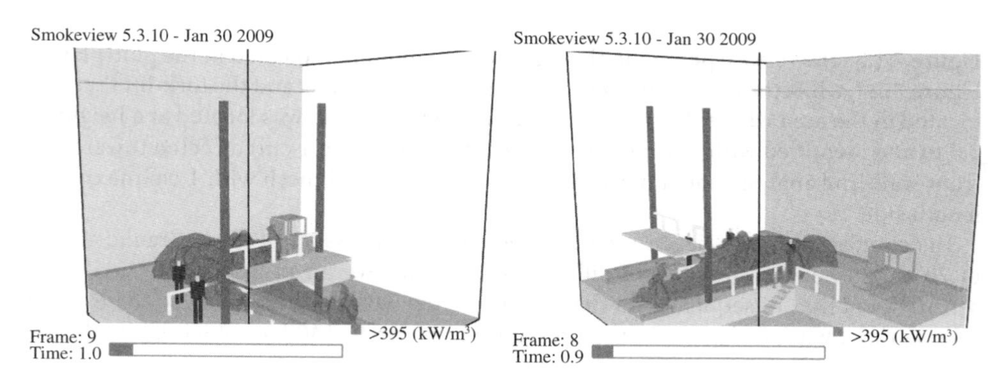

Figure 7.1.9 Jet fire simulation results: flames at 1 s from pipe collapse. Source: [1].

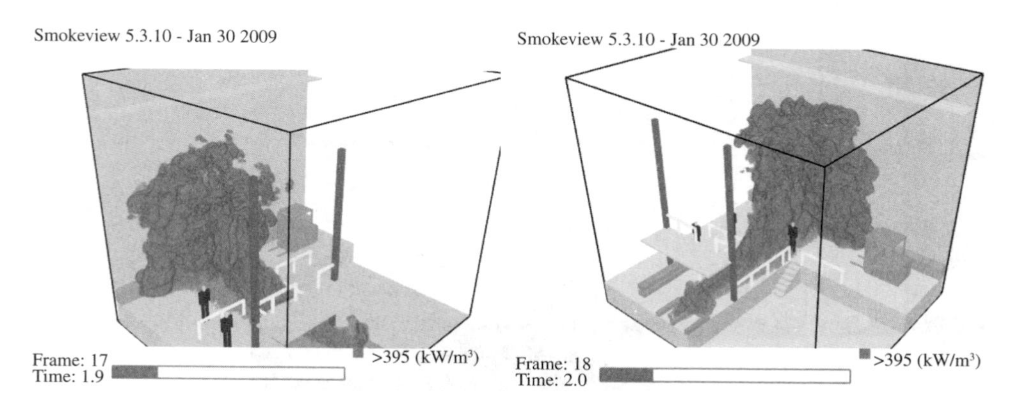

Figure 7.1.10 Jet fire simulation results: flames at 2 s from pipe collapse. Source: [1].

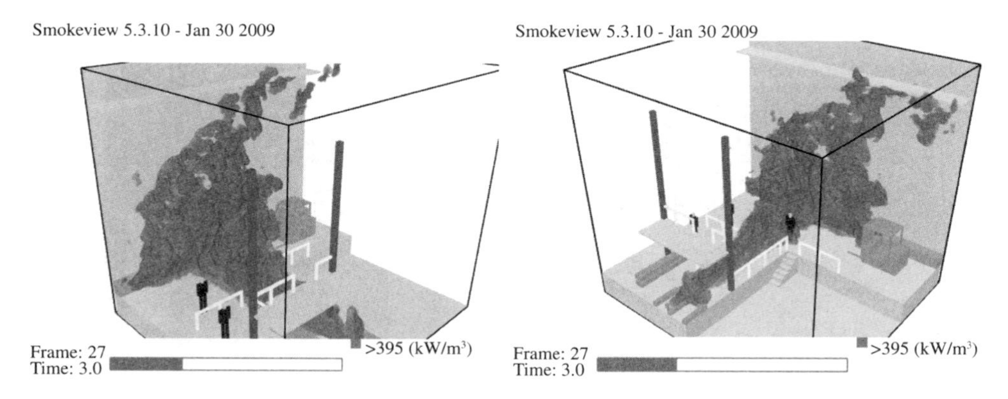

Figure 7.1.11 Jet fire simulation results: flames at 3 s from pipe collapse. Source: [1].

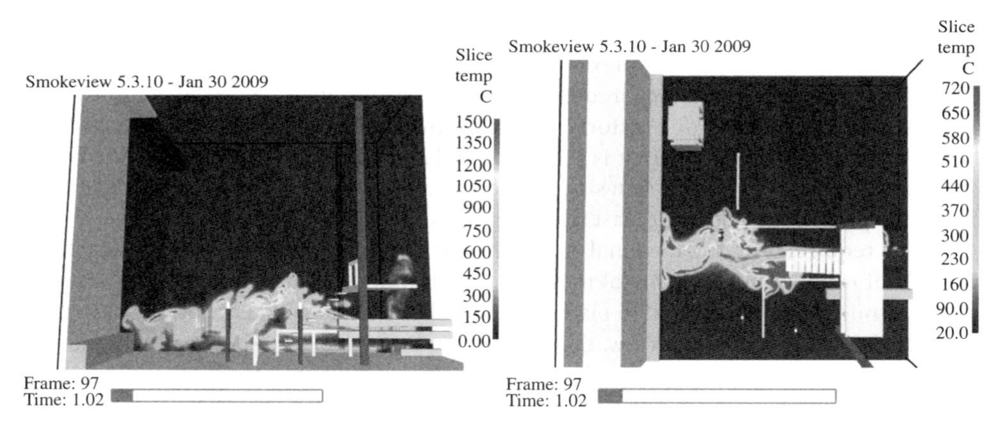

Figure 7.1.12 Jet fire simulation results: temperature at 1 s from pipe collapse. Source: [1].

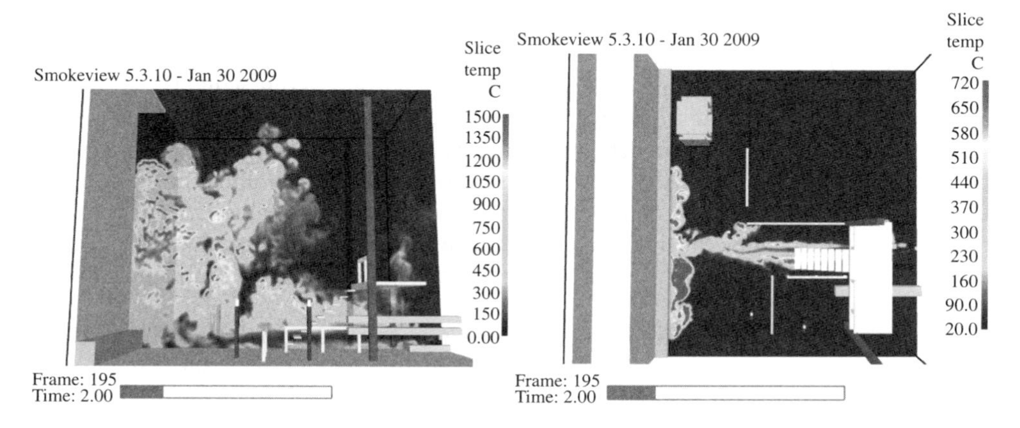

Figure 7.1.13 Jet fire simulation results: temperature at 2 s from pipe collapse. Source: [1].

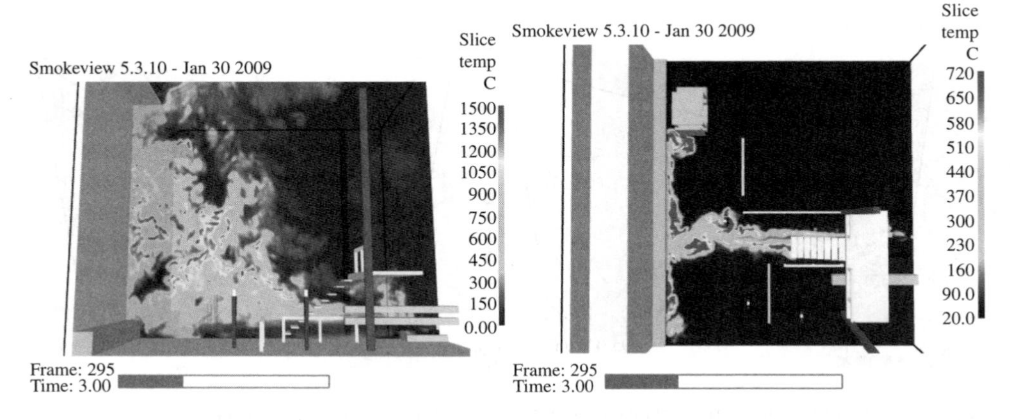

Figure 7.1.14 Jet fire simulation results: temperature at 3 s from pipe collapse. Source: [1].

inlet section of the line is a complex part of the plant, as it is composed of many devices that are activated by hydraulic circuits. Each of these circuits is generally composed of a couple of pipes, a hydraulic piston and a valve that is activated electrically. A simplified scheme of a hydraulic circuit is presented in Figure 7.1.15. The lengths of the pipes are mostly made of steel, but each single pipe is connected to a moving component (the hydraulic piston) and hence at least the last part of the pipe must be flexible. To accomplish this requirement, the terminal is usually made of a flexible, composed pipe, which is made of rubber and has a metal mesh that guarantees the mechanical performance. These terminals are connected to pistons and steel pipes via special fittings. Clearly this is, from the fire risk point of view, the weakest part of the plant. First the hydraulic oil is combustible. Second, hydraulic circuits quite frequently suffer from leaks. There are generally two sources: the piston seals are subject to wear, which can provoke local spills, while flexible pipes are subject to fatigue which can cause cracks that can lead to sudden leaks under the form of sprays or liquid jets. Due to the frequency of these leaks, and to the huge number of hydraulic circuits, which can reach as many as several tenths in the entry section, the environment can easily become "dirty" and prone to fire if a rigorous cleaning policy is not enforced.

Other sources of combustible material can also be present in the plant. The coils can, as in the present case, come from a cold rolling line. After cold rolling, coils are re-rolled with a paper strip between the metal coils to prevent surface damage. The paper should be recovered in the entry section of the P&A line to prevent its loss along the line. Sometimes the paper sticks to the metal and its recovery is almost impossible. In this situation, the paper is spread along the line or it enters the oven. The paper is usually impregnated by the oil used in the rolling unit. As a consequence, huge amounts of combustible materials can spread along the P&A line.

Ignition causes are also quite frequent. These are mainly due to mechanical or electrical faults or rubbing of the coil against some structural component. Scratching is far more probable in those plants, or in parts of them, where automatic coil position control devices are not present. The arc welding that is made to join each coil to the others is also a possible cause of ignition. However, this process is easy to control because welding

(a)

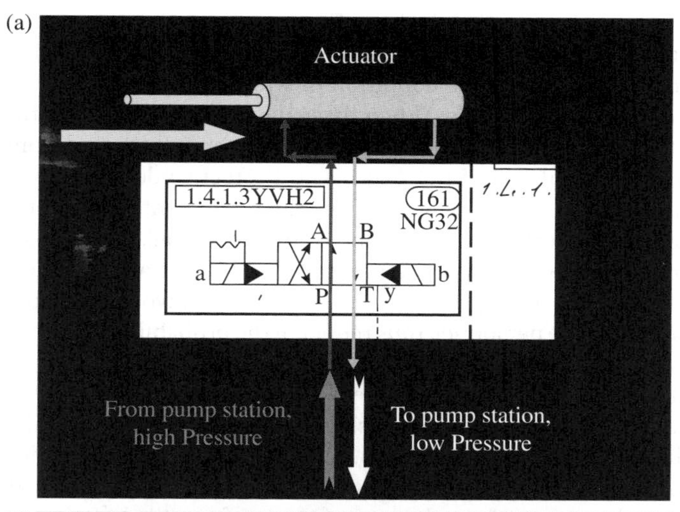

(b)

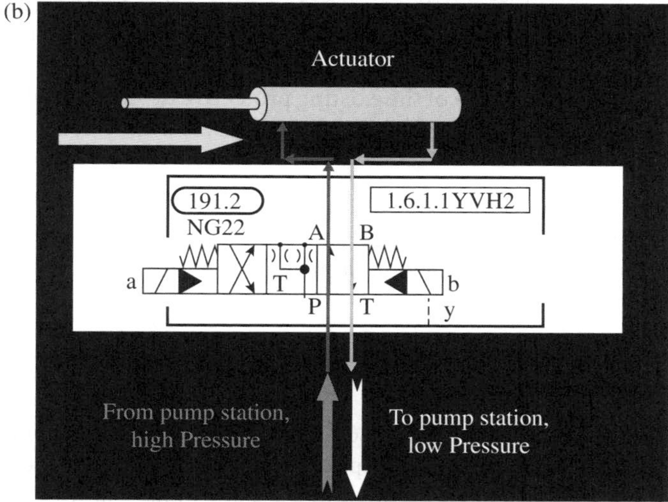

Figure 7.1.15 Scheme of the hydraulic circuits with two-position (a) and three-position (b) solenoid valves. Source: [1].

is usually done by an operator and an eventual fire can easily be detected. Mechanical faults can occur in elements such as ball bearings, which are present in large quantities in such plants.

7.1.4 Findings

An accident can be the consequence of a series of undesired events, with consequences on people, objects and/or the environment. The first element of the series is the primary event. There are usually many intermediate events between the primary event and the accident, which are determined by the reaction of the system and of the personnel. The dynamics of an accident generally start from a process failure, which is followed by the failure of automatic or manual protective devices. The common representation of this process, with logical trees, involves the use of logical gates (generally AND, OR).

Intermediate events are, in many cases, the condition in which the chain of events interacts with the action of protective devices. When these devices are successful, the chain of events is blocked, therefore the intermediate events correspond to conditions that contribute to decrease the likelihood of the Top Event. Protective devices can of course act either automatically or manually, which implies the implementation of procedures whose success depends on the level of training of the personnel. As a consequence, the expected frequency of a Top Event can be reduced by first and foremost adding protection devices, and then acting on the failure rate of the involved components or improving the training of the personnel, thus reducing the probability of human failure.

Hereafter the representation of the dynamics of the ThyssenKrupp fire is proposed using a fault tree assessment (Figure 7.1.16), in which also the INHIBIT gate is also used to represent the failure of the protective devices. This is substantially a variation of the AND door which is used in the case of protective means. The INHIBIT gate can distinguish between the entering events since the event entering from the bottom can propagate to the event at the top outlet if the side event, which is represented by the unavailability of a protective device, has already occurred [8, 9]. The dynamics of the accident is represented considering the failure of the existing protective devices and also of those that the plant was not equipped with, which are indicated in the grey boxes.

These are:

1. automatic shutdown of the hydraulic circuits;
2. automatic fire extinguishing plant;
3. automatic coil control position in the inlet section of the line; and
4. fire detection systems.

7.1.5 Lessons Learned and Recommendations

The TK accident that occurred in Turin, in December 2007 has offered some very important lessons about the fire risk for P&A lines. The dispute about the risk level of these plants was ongoing, with some technical experts (the minority) affirming that prevention and protection tools, such as automatic extinguishing plants, were necessary, while the majority of technicians declared that, in most cases they were not necessary. It seemed that, at that time, there was a general conviction that only some parts of the line needed specific fire protection equipment. The pump station (which is often located in a separate compartment together with the oil reservoir), was of course considered to be a high-risk zone. The oven zone was also considered to be at a high risk, considering the elevated temperature reached there and of the presence of large amounts of natural gas. In some cases, the pickling section was considered to be at a high fire risk, when the pickling pools and/or covers were made of plastic, as in the case of the fire in Krefeld.

However, the risk associated with high-pressure hydraulic circuits and with the potential release of a huge amount of oil because of pipe failure had been largely underestimated.

The present case is a clear example of how simple it would have been to adopt measures to reduce risks which would have prevented the accident from occurring, also on

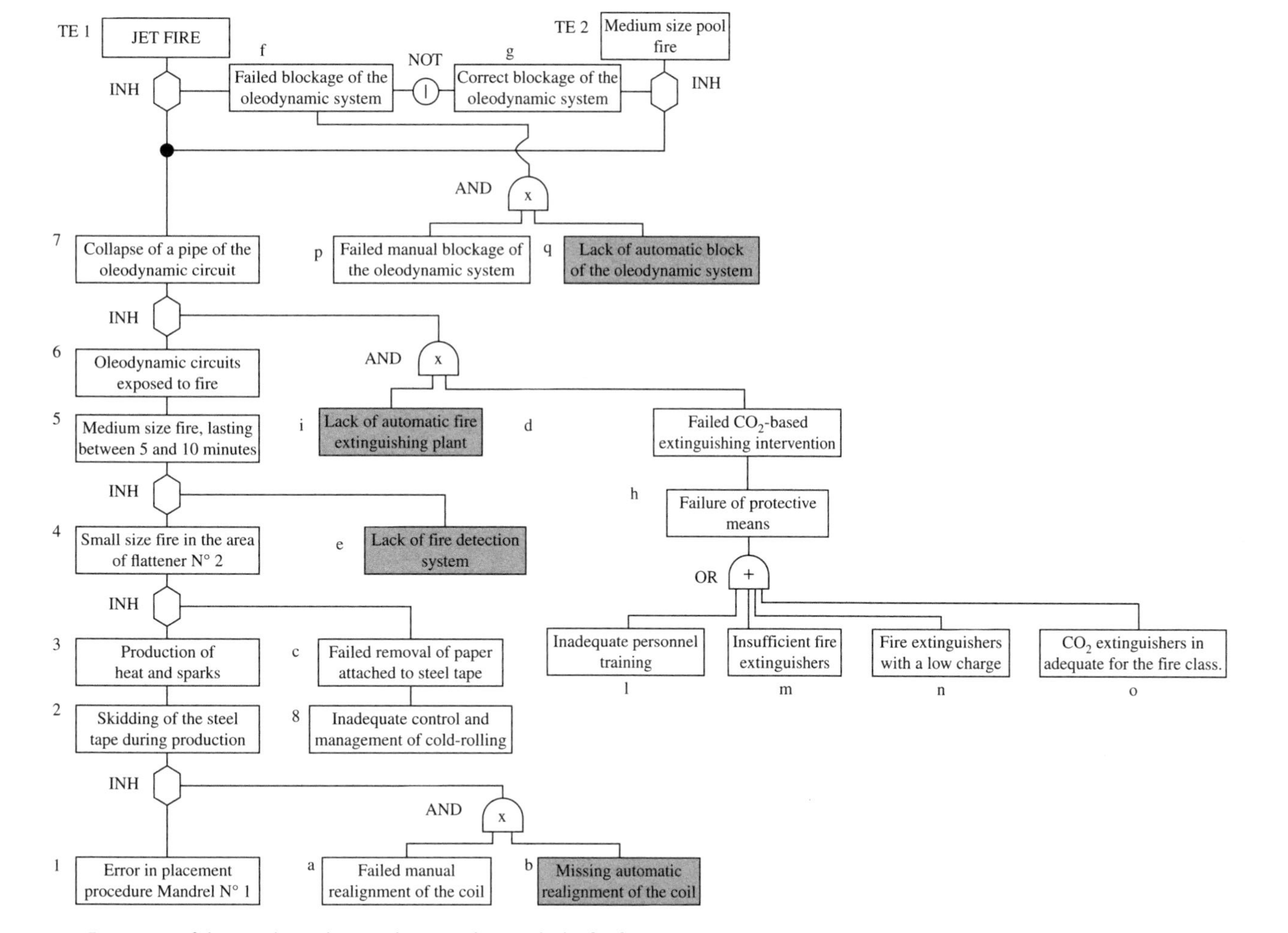

Figure 7.1.16 Event tree of the accident. The grey boxes indicate a lack of safety devices. Source: [1].

the basis of what has already been stated in the extensive technical literature available (e.g. the NFPA considerations in the well-known "Fire Protection Handbook" 1997).

In order to better understand the lessons that can be learned from this case, the dynamics of the accident was represented in the Fault Tree shown in Figure 7.1.16. It is easy to individuate the technical improvements which could have prevented the accident from the figure. Consequently, the following points can be put forward:

1. Coil position control. In the present case, the first automatic control position was located some tens of meters downstream from the coil unrolling station. This position corresponded to deflector roll N_1. Only manual position control devices were present in the inlet area. If an automatic position control system had been located some meters downstream from the payoff mandrel, the rubbing and consequent ignition would have been much less probable.
2. Housekeeping. This, in general, was insufficient at the moment of the accident. The spread of the fire in the early phase is a direct consequence of the presence of paper and oil on the site. As far as good housekeeping is concerned, it is generally recognised as critical to diminish the fire risk; in many fire accidents the housekeeping has been inadequate as a result. From a more general point of view, the cleaning of such plants is not a simple task, because of the complex location of the machines. Periodic cleaning by external enterprises, as in the present case, would not have been sufficient since the spread of paper cannot be forecast as it is something occasional whose occurrence depends above all on the control of the rolling phase.
3. Fire load and fire detectors. In the present case the zone involved in the accident was not equipped with fire detectors. The reason for this is that the fire load, according to a "traditional" calculation approach, was considered negligible by the technical expert who made the analysis. This accident in fact demonstrates all the limits of such an assumption. Although the fire load was very small, the potential for a huge release of combustible and hence for a huge fire was due to the conformation of the hydraulic circuits. If a fire detector system had been present, the workers would likely have discovered the fire in its early stages and would therefore have had a much better chance of controlling its dynamics.
4. Automatic fire extinguishing systems. This point is very similar to the previous one. No automatic extinguishing system, which would have limited the exposure of the workers to the fire, was in place. It is evident that the fire risk depends to a great extent on the chemical-physical conditions of the substances, as well as on the plant attitude to provoke a sudden release, which in turn depends on the structure and on the control strategy of the plants. In the P&A industry, there is extensive use of hydraulic circuits in areas in which ignition can occur at a non-negligible rate. This accident has shown that the perception of the fire risk of such units had been underestimated. Traditional hydraulic units, which are not equipped with adequate emergency stop devices (see the next point) are subject to the risk of jet fires due to sudden releases. The presence of personnel in these areas should be limited as much as possible especially when a fire breaks out.
5. Emergency stops. The emergency stopping of complex units, such as P&A lines, is a challenging task. The emergency shut-down of hydraulic units usually leads to a loss of control of the actuators, therefore uncontrolled movements generally occur. These, which can potentially involve very heavy loads, can easily cause damages and

injuries. Despite these aspects, this accident has clearly demonstrated that hydraulic units should be shut down when fires break out in order to avoid huge releases and the spread of fires. The lines are generally provided with different emergency stop levels, as in the examined case. The first level corresponds to a shutdown which is operated by the actuators, and consists of the solenoid valves being placed in the central position. This kind of intervention obviously only concerns those circuits that are equipped with three position valves. This strategy cannot be considered adequate to mitigate the risk of fire since some circuits, such as the expanding mandrel, which are equipped with two position valves, are not shut down. Hence, the risk of releases and of jet fires is still considerable. Moreover, in the case of a manual shutdown strategy, such as in the present case, the personnel could be involved if an accident occurs. It seems necessary to prevent the personnel from entering the area of the accident as much as possible, and for those who are present to leave the area, until emergency procedures are activated.

6. Emergency procedures. The emergency procedures in the case of fire were: "In the case of fire, if the person is trained, he should immediately start to fight the fire using the available equipment. When the fire is judged to be of "evident gravity", the personnel should call the surveillance personnel and then wait for the internal emergency team to arrive. The emergency team should be activated by the surveillance personnel. The emergency team, once on site, should turn the power off to the area, look for missing personnel and fight the fire with the available tools." In the present case, the procedure failed to prevent the accident because the fire was not evaluated to be of "evident gravity" by the personnel on the site. This case demonstrates, in a very evident manner, how it can be a critical task to judge the gravity of a relatively small fire in a complex industrial context such the one described here. If such a procedure fails in the most critical phase, in which the decision on how serious a fire is must be made very quickly, the personnel is exposed to serious risk. Procedures that do not involve, or at least limit personnel judgments as much as possible, should be introduced when a huge risk is detected.

7.1.6 Forensic Engineering Highlights

The importance of the CFD analysis has been deeply discussed in Paragraph 7.1.3. On the basis of the analysis of the results, which were obtained through the use of the official FDS graphical post-processor (Smokeview© by NIST) and from the record of the temperatures via virtual thermocouple devices (19 devices positioned at the height of man), several considerations have been made [1, 2]:

1. The jet fire involved the entire area opposite the release point in a direct manner and immediately created a serious risk for the workers that were in the area of interest;
2. in fact, conditions that could have led to fatalities were recorded almost immediately;
3. all the reference thresholds (Table 7.1.3) for fire thermal radiation were reached in the area (e.g. incident radiant heat, with a value of 200 kW/m^2, on the wall in front of the release point);
4. a risk arose from both the direct and non-direct effects of the jet fire and it was also related to the flame extension, thermal radiation and temperature rise in the compartment area;

5. the previous effects were recorded from the first instants after the release;
6. the combustion of the hydraulic oil was characterised by a huge amount of smoke and soot which made the conditions in the area worse;
7. the jet was vertically and horizontally fragmented as a result of the impact with the wall (and as a result of the impact with the main plant structures in the area). This fragmentation allowed a flame wall to build up that divided the compartments into two parts (thus creating problems for the emergency procedures) and it determined more serious conditions in the area than a similar jet without the presence of fixed obstacles. The amplitude and the dispersion of the jet was in line with the status of the compartment after the real fire (see Figure 7.1.1 to 7.1.5);
8. the fork-lift could have protected a worker located behind it from the effects of the jet fire: this has been confirmed through an evaluation of the damage to the fork-lift itself, which presented different levels of damage on the two sides, as shown in Figure 7.1.17 (the damage is coherent with the shape and effects of the simulated flames);
9. the release of hypothetically smaller quantities of hydraulic oil could have exposed the workers who were possibly located at a distance of 15 m to a serious risk (with limited effects compared to the real jet-fire that occurred but which could have significant consequences on people); the same considerations were made for a hypothetical flash-fire considering that the reference thresholds were reached in the same area (a distance of 15 m for 0.5 LFL from the release point in the case of the release of 500 cc of hydraulic oil); and
10. the real conditions could have been judged worse than then the simulated ones because the simulations only considered one single accidental release; it is very

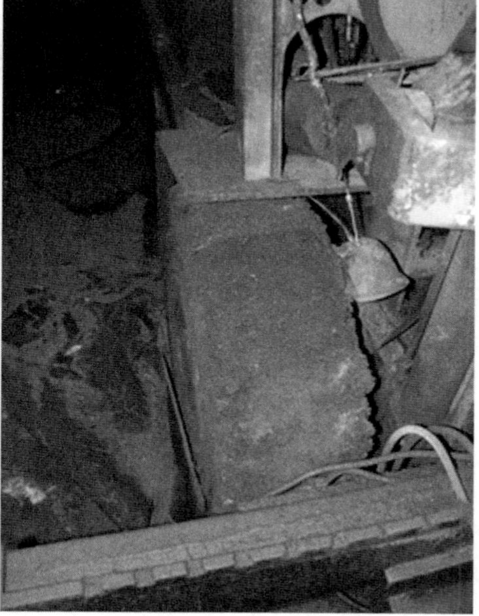

Figure 7.1.17 Damages on the forklift.

probable that the real evolution of the fire involved several subsequent collapses: the simulation of contemporary or slightly delayed releases from different sources would have led to a more significant impact, in terms of consequences for both accidental events (jet fire and flash-fire).

Among the forensic engineering highlights of this case study, there is also the logic path conducting from evidence to deductions. This effort is sum up in Table 7.1.4.

The sequence of the events has been reconstructed with a video, which was used to show the incident dynamics during the trial. The video environment has been built from the photos of the real incidental scenario, to highlight the site conditions. It reproduces the sounds during the work activities, and it uses 3D images to reconstruct the

Table 7.1.4 Summary of the investigation.

Activity	Evidence	Deductions
Site survey	Heat and fire damage. State of the coils, position. Scratching of coil against the carpentry (290 m). Paper spread along the line. Residue of carbonised paper in the area of the flattener.	Area reached by the jet fire. Area reached by the fire. Axial coil position not correct. Shift of coil N° 1 toward the side carpentry. Scratching lasted several minutes. Scratching occurred above the flattener. Combustible material in the area.
Documents on Risk analysis	Fire risk evaluation. The area was considered at medium risk according to Italian regulation.	No fire detection systems were provided.
Documents examination pertaining to technical description of plants	The complete inventory of the hydraulic circuits involved in the fire. Pipe state at the moment. Circuits working pressure were identified.	The pipe that collapse first one was identified, on the basis of position, direction, and because it was under pressure at normal conditions.
Witnesses	The size and position of the initial fire. The size and shape of the jet fire Fire growth rate (roughly).	Small fire on the flattener at the beginning. Fire grew in size after the first attempt to extinguish. Fire extinguishers unfit to control the fire. Sudden jet fire spreading "like a wave".
Electronic data	The timescale of the events.	Line start at 00.35'46". Speed reduction by workers at 00.45'45". The PLC lost the sensors located close to the flattener at 00.48'24". Line emergency stop (automatic) at 00.53'10" due to low oil level.

Source: Data from [2].

movement of the victims, on the base of the collected evidence (witnesses). Figure 7.1.18 collects some frames of the video.

7.1.7 References and Further Readings

References

1 Marmo, L., Piccinini, N., and Fiorentini, L. (2013) Missing safety measures led to the jet fire and seven deaths at a steel plant in Turin. Dynamics and lessons learned. *Journal of Loss Prevention in the Process Industries*, 26(1):215-224.
2 Marmo, L., Piccinini, N., and Fiorentini, L. (2012) The Thyssen Krupp accident in Torino: Investigation methods, accident dynamics and lesson learned. *Chemical Engineering Transactions*, 26, pp. 615–620.
3 McGrattan, K., Baum, H., and Rehm, R. (1998) Large eddy simulations of smoke movement. *Fire Safety Journal*, 30(2):161–178.
4 McGrattan, K., Hamins, A., and Stroup, D. (1998) Sprinkler, smoke & heat vent, draft curtain interaction. Large scale experiments and model development. Gaithersburg, MD: National Institute of Standards and Technology, NISTIR 6196-1.
5 McGrattan, K., Hostikka, S., Floyd, J. et al. in cooperation with VTT Technical Research Centre of Finland. (2008) NIST Special Publication 1018-5. Fire dynamics simulator (Version 5) technical reference guide. NIST - National Institute of Standards and Technology, US Department of Commerce.

Figure 7.1.18 Frames from the 3D video, reconstructing the incident dynamics.

6 Hoover, J., Bailey, J., Willauer, H., and Williams, F. (2008) Preliminary investigations into methods of mitigating hydraulic fluid mist explosions. *Fire Safety Journal,* 43(3):237–240.
7 Hoover, J., Bailey, J., Willauer, H., and Williams, F. (2005) *Evaluation of Submarine Hydraulic System Explosion and Fire Hazards. Ft.* Belvoir: Defense Technical Information Center.
8 Demichela, M., Piccinini, N., Ciarambino, I., and Contini, S. (2004) How to avoid the generation of logic loops in the construction of fault trees. *Reliability Engineering & System Safety,* 84(2):197–207.
9 Piccinini, N. and Ciarambino, I. (1997) Operability analysis devoted to the development of logic trees. *Reliability Engineering & System Safety,* 55(3):227–241.

Further readings

Cote A. (2008) *Fire protection handbook.* Quincy, Mass: National Fire Protection Association; 2008.
NFPA (National Fire Protection Association). (2017) NFPA 921: Guide for Fire and Explosion Investigations. NFPA (National Fire Protection Association); 2017.

7.2 Fire on Board a Ferryboat

7.2.1 Introduction

General information about the case study is shown in Table 7.2.1.

Regardless of the wide prescriptions of the maritime regulations, fire on board are still a significant cause of losses for human beings. This case study underlines how the structural weaknesses of the management system may affect the phase of incident management, which is amplified by the assumed condition of being at sea (longer time required for rescuers to arrive at the incident site). The outcomes of this fragility can be devastating, heavily affecting first the abandonment of the ship and then the rescue activities. This case study is explicitly focused on the fire investigation, ignoring all the other aspects related to the same incident, emanating from the complex world of maritime transportation.

Table 7.2.1 General information about the case study.

Who	Ro-ro pax ferryboat
What	Fire on board
When	2014
Where	Adriatic Sea
Consequences	9 victims, 14 lost
Mission statement	Determine the fire dynamics and the causal factors of the ineffectiveness of firefighting system
Credits	Rosario Sicari (ARCOS Engineering s.r.l.) Alessandro Cantelli Forti (RaSS NationalLaboratory)

7.2.2 How it Happened (Incident Dynamics)

The ferryboat Norman Atlantic was an Italian ship rented by a Greek company for ferry crossing between the two countries. The night of the incident, the route Patras – Igoumenitsa – Ancona was planned, and 55 crew members were on board. When leaving from Igoumenitsa to Ancona, at 11.28 p.m. on 27 December 2014, the cargo consisted of about 130 heavy vehicles, 417 passengers and 88 cars. The navigation was regular until 03.23 (UTC), when the fire alarm sprang into action on deck 4, near the frame #156. Because of a smoke sighting coming out from the lateral openings of the deck, a sailor was appointed to carry out an inspection on deck 4. He reported that the alarm was attributable to smoke coming from the auxiliary diesel engine belonging to a reefer truck, which was not connected to the electrical supply of the ship. After few minutes, at 03.27 a.m., the Master brought himself to the flying bridge deck on the starboard side and observed the flames coming out from the openings of deck 4. The 1st Engineer Officer activated the manual deluge system (known as "drencher"), following the Master's order. Meanwhile, the Chief Engineer Officer and his personnel abandoned the Engine Room because of the excessive smoke, while the two engines of the ship stopped definitively. The ship went in a black-out and the emergency generator, placed on deck 8, was incapable of providing energy to the emergency utilities, including the emergency pump. At the same time, the cooling team uselessly tried to cool the deck 5, but steam came from the fire hoses, instead of liquid water. The emergency management, especially during its first stages, was revealed as being chaotic. During the rescue operations, some passengers fell into the sea, while others threw the remaining life rafts into the sea, with no possibility of being properly able to use them. At the end of the Search And Rescue operations, 452 people were rescued, including 3 illegal immigrants; 9 victims and 14 lost in the sea were also counted. The ship was then towed by tug to Bari, where it was placed under the scrutiny of investigators.

The longitudinal section of the ship is shown in Figure 7.2.1.

7.2.3 Why it Happened

The reconstruction of the facts that led to the Norman Atlantic fire, including its dynamics and the research of the root causes, has been based mainly on the data from the Voyage Data Recorder (VDR), the testimonies from the interrogations, the documentation taken on board and from the ship-owner, the transcription of the audio communications, the census operations about the vehicles on garage decks and the collected evidence [1]. The investigation team, aimed at better conducting its tasks, was divided into 5 sub-teams to handle the following five topics, which findings were deemed to be critical from the outset: vehicles embarkation, evacuation and emergency management, fire dynamics, ship automation and onboard IT and electronic plant design. In this Paragraph, the topic addressed is "the fire dynamics", and the relevant investigation aspects related to it. The approach used was multidisciplinary, to face such a complex system (i.e. full of relations and interconnections, as the "ship" system is, because of the crew, a chain of command, procedures, alarms, plant-men interfaces, and so on). The method required constant comparisons and sharing of results among the different sub-teams. The investigation activities were aimed at finding the causes of the fire, the most probable source of ignition and its location onboard, including those elements that facilitated

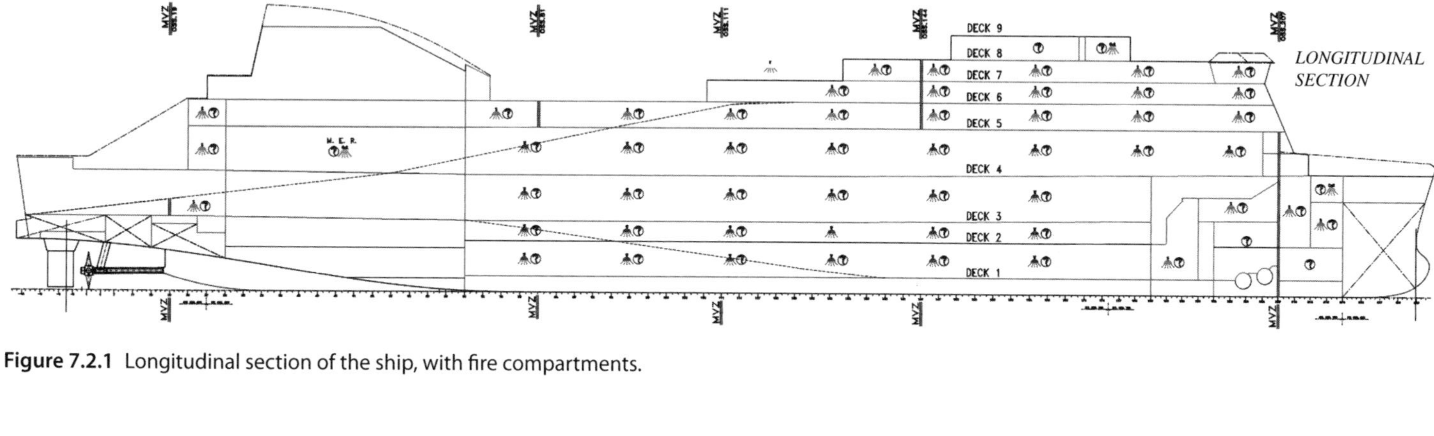

Figure 7.2.1 Longitudinal section of the ship, with fire compartments.

its propagation, to reconstruct the fire dynamics and the reasons why the safety systems were ineffective. The "conic spiral" has been the pursued methodology. The first useful information was extracted from the already-known data (available using the original documentation), to delineate the "stage zero" and to define a first distinction between what was necessary to be examined in depth through further investigations and what was not of interest. The investigation scope was narrowed down by repeating this basic step, focusing attention on the details which were progressively emerging. Regarding the fire, the investigation team was made up of B. Chiaia (Team Leader), L. Marmo (expert in chemistry and fire dynamics), L. Fiorentini (expert in advanced simulations of fires and explosions by means of advanced simulation as Computational Fluid Dynamics) and R. Sicari (expert in firefighting protection systems design and verification in the maritime industry). Moreover, the multidisciplinary approach often required the need to interface with A. Cantelli Forti (expert in digital memories for maritime apparatus).

The team performed a structured Root Cause Analysis (RCA) to investigate the Norman Atlantic fire. The recursive questioning of "why", starting from the "main event" (i.e. the Norman Atlantic Fire), led the team who were driving the investigation, including the collection and the analysis of the evidence, to find the immediate and the root causes of the incident. The logic tree has several ramifications (Figure 7.2.11), whose immediate causes embrace different types of human errors and whose root causes involve both design aspects and weaknesses of the fire safety management system. It is clear that the main event was the consequence of the failures of different sets of safeguards, which are now discussed.

The fire dampers for the garage ventilation were found opened, so favouring the fire propagation to the other decks apart from the 4th, where the fire started (Figure 7.2.2).

One contributory cause is the positioning of the local commands for closing the dampers: indeed the majority of them is placed on deck 4 (Figure 7.2.2), and only a limited number of them can be controlled remotely, in a safer position. The possibility of continuing to feed the drencher system (manual deluge) after the occurrence of the blackout is guaranteed by the emergency pump, which never started. This was because of the emergency generator that, even if its engine started, was not capable of supplying the energy to the final utilities, because of an electrical fault due to the propagation of the flames in other spaces of the ship that damaged the electrical cables.

Figure 7.2.2 Left: open fire damper of the garage ventilation. Right: local command at deck 4 for closing the fire dampers.

Moreover, it should be noted that even a correct supply of energy during the blackout at the emergency pump could not pump the water inside the garage deck because the intercept valve between the emergency pump and the drencher collector was found closed (Figure 7.2.3). The managerial causes (related to the internal procedures, habits, and so on), if any, of this singular context, will be probably clarified during the trial.

However, even if the intercept valve would be found open, the zones activated by the operator in the drencher room (i.e. the valve house, where the distribution of the drencher water is set) were wrong. Indeed the four zones activated were on deck 3 (Figure 7.2.4) while it is clear that the fire should be faced on deck 4 (as correctly ordered by the Master).

A possible contributory cause is the drencher plan (Figure 7.2.5), provided as documentation inside the drencher room to the operator that intends to activate the system. In this scheme, the decks of the ship are named with their English names (e.g. deck 4 was named "Weather Deck"), while the order given by the Master was to "activate the drencher at deck number 4". Therefore, there is not a full alignment between the order of the Master, that needs to be elaborated, and the documentation available in drencher room. Also, the plan contained some errors in locating the drencher room on the "weather deck" (deck 4) instead of the "main deck" (deck 3), where it actually is. The confusion increases if we think that a "weather deck" is, by definition, an open deck, while in all the technical drawings of the Norman Atlantic this terminology was referred to the deck 4, a closed deck with openings just below the open deck (deck 5).

However, even if the drencher were to be activated at deck 4, the operator opened 4 zones versus the maximum allowable of 2, according to the drencher manual, to ensure its extinguishing performances: this incorrect operation could be addressed as ineffective training. However, the timeline reconstruction of the event and the advanced

Figure 7.2.3 Closed intercept valve between the emergency pump and the drencher collector.

ZONA 4	n.42 ugelli B15SSP	Deck3 Garage P.te Principale	Fr.96 - 122	Locale Drencher
ZONA 5	n.42 ugelli B15SSP	Deck3 Garage P.te Principale	Fr.122 - 148	Locale Drencher
ZONA 6	n.44 ugelli B15SSP	Deck3 Garage P.te Principale	Fr.148 - 174	Locale Drencher
ZONA 7	n.36 ugelli B15SSP	Deck3 Garage P.te Principale	Fr.174 - 207	Locale Drencher

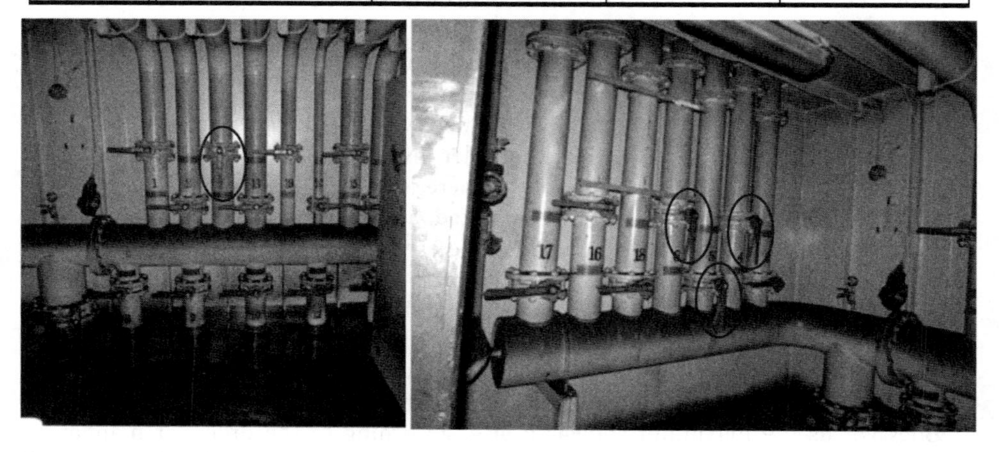

Figure 7.2.4 The valves opened in the valve house are those activating the drencher at deck 3 (instead of deck 4).

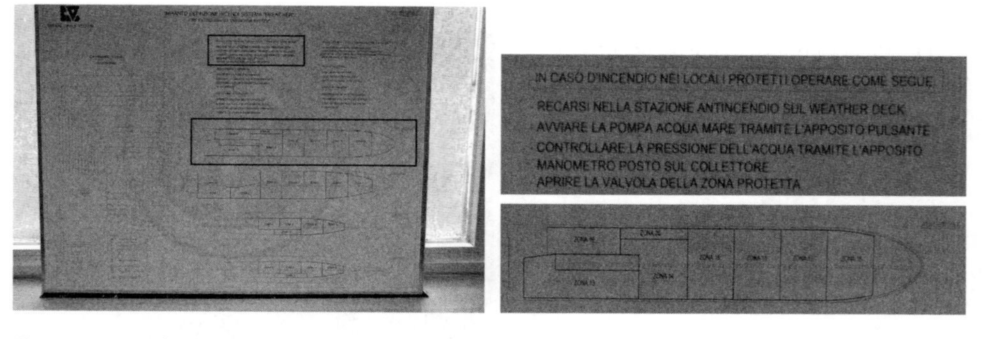

Figure 7.2.5 Left. The drencher plan located in the drencher room. Right. Details of the instruction on the plan.

simulations in Computational Fluid Dynamics revealed that even a correct activation of the drencher system would not have extinguished the fire, but only controlled it, because of its belated activation. The reasons for such late intervention are mainly attributable to a self-evident underestimation of the problem by the crew. Indeed, regardless of the alarms provided by the Fire Detection System, the order of the Master to activate the drencher was only given when the fire is already fully developed, and the flames were coming out from the openings of deck 4. The outcomes of the inspection conducted by the sailor at deck 4 are the main cause of this underestimation. Indeed the inspection was required because of some uncertainties over traces of smoke coming out from the openings of deck 4, which were confused with reflections of the sea. The inspection was also hurried because of the difficulties of the corpulent sailor in passing through the narrow spaces between the heavy vehicles.

At the end of the inspection, the sailor clarified that the alarms detected at the bridge were attributable to the smoke produced by the auxiliary diesel engine of a reefer truck. The crew members, being aware of this illegal practice, accepted it overestimating their capability to put such a hazardous situation under control. Being alerted by other alarms, the 1st deck officer asked the sailor to perform a further inspection, but this was deliberately never carried out.

The Root Cause Analysis revealed that the crew members agreed in having reefer trucks not connected to the electrical supply of the ship (Figure 7.2.6), because they embarked a higher number of reefers in respect to the number of available reefer sockets, violating the prescription of a correct embarkation for commercial purposes. Finally, the flames, the smoke and the alarms recorded are all consequences of the first hotbed in deck 4 that the sailor did not find during his inspection. The malfunction of an auxiliary diesel engine of one of the loaded vehicles can be considered the most likely cause of the ignition. This probability must be regarded as higher for those vehicles equipped with an auxiliary diesel generator at the service of the refrigerator system or at the service of the oxygen pumping in the water tanks for the transportation of live fish.

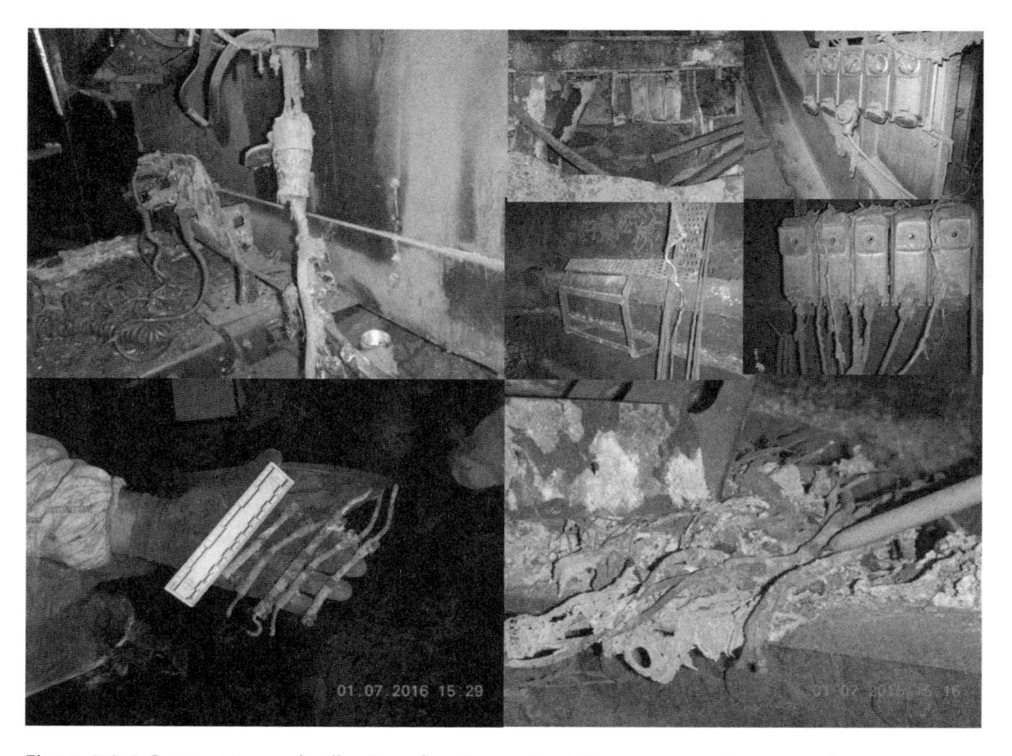

Figure 7.2.6 Recognition and collection of evidence about the power supply on board.

The usage of these diesel engines is forbidden inside the garage of the ship, because they should be used only when the vehicle is in motion, being cooled by air. The trucks, the oil in their tanks and the olive oil (including pomace olive oil, that in some cases may be flammable [3]) transported by some of them represented the combustible materials (Figure 7.2.7). The openings of deck 4 (Figure 7.2.8) continuously provided the oxygen, prompting serious questions about its design. The fire triangle was satisfied.

The complexity of the Root Cause Analysis shows the applicability of the Reason's Swiss Cheese Model: safeguards are not 100% reliable and when their ineffectiveness (i.e. their probability of failure on demand) align, then a hazard situation may evolve into an incident. It happened with the embarkation, the inspection, the drencher, the emergency pump, the emergency generator and the dampers, but, as the root cause analysis reveals, they are all attributable to an inherently weak fire safety management system.

The root cause analysis identifies some significant and intrinsic engineering weaknesses, also related to the high probability of human error, even if the drencher system, by the law, is compliant with the applicable regulations.

Figure 7.2.7 Localised bending of transversal beams and V-shaped traces of smoke on the bulkhead. The majority of the fire load is attributable to the olive oil tanks.

Figure 7.2.8 Lateral openings on deck 4.

The limited arc of time between the first alarm at deck 4 and the other decks is very short (e.g. 3 minutes between deck 4 and 5) but not incompatible with the literature [2]. Moreover, this rapidity also emerged from the numerical simulations, that have been carried out to validate the hypothesis advanced during the first stages of the investigation.

Four different simulations have been performed. The first one was used to calibrate the Heat Release Rate of a single isolated heavy vehicle, with a load comparable to that found on average during the census and unloading activities. The other three simulations, confined in a domain over the frame 156, have been addressed: study the expected outcome of the Fire Detection System, through its smoke and heat detectors; simulate the real fire on deck 4, taking into account the relative wind and all the vehicles inside the simulation domain; study the propagation of fire at deck 3. Some of the outcomes are shown in Figures 7.2.9 and 7.2.10.

The simulations, together with the physical, digital and documental collected evidence, allow the thermal stress propagation hypothesis, the timeline sequences of the main events, the capability of the drencher system and the time of activation of the fire alarms to be verified. The simulations reveal that only a prompt and correct activation of the drencher system over the area of the first hotbed would have allowed control of the fire and avoid its propagation. The missed prompt activation of the drencher

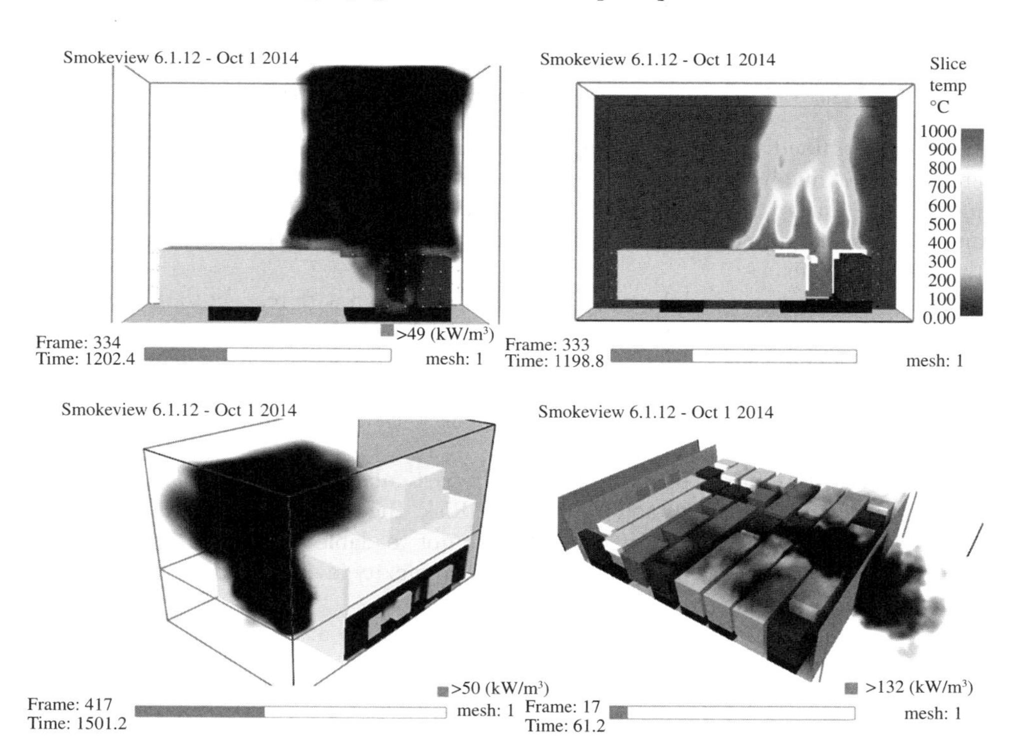

Figure 7.2.9 CFD simulations: single truck combustion and 3D pictures of the first instants of fire at deck no. 4, with smoke emission and flames from the openings on the starboard side of the ferryboat.

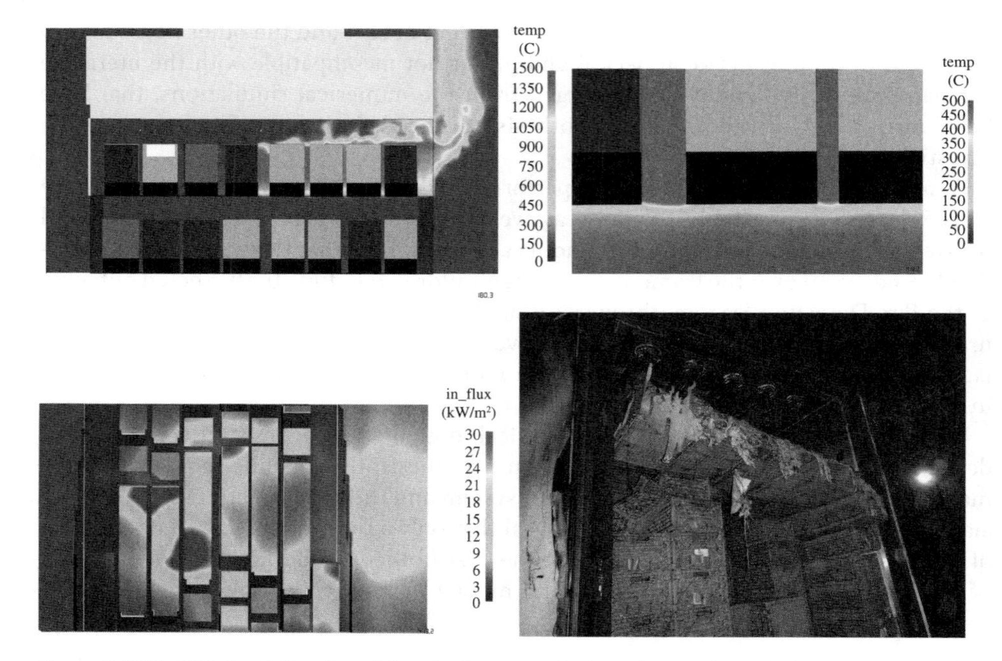

Figure 7.2.10 CFD simulation describing the heat transfer by radiation through the metal plate between decks no. 3 and no. 4. Conditions of the plastic boxes inside a truck on deck no. 3.

system determined a serious spoiling because the thermal regimes that tended to arise are capable of involving a significant part of the flammable material present in the area into the fire. It also caused critical damages to the structures, both for thermal radiation and for flame engulfment, with temperatures higher than the critical ones distinctive of the used materials.

The state of the areas and the timing of the investigation with respect to the event did not allow to find with any certainty the origin of the flames. The fire dynamics, extremely rapid, could be however compared to typical dynamics in heavy vehicle fleets.

7.2.4 Findings

The event can be depicted using a logic tree. In particular, the outcome of the RCA has been drawn using IncidentXP software by CGE Risk. The complexity of the investigation is reflected in the tree of Figure 7.2.11, that is not readable at this level of zoom but is presented to let the reader understand the complexity level of a root cause analysis performed on a real incident.

However, the adoption of the RCA allows investigating at different levels, depending on the mission statement established since the beginning of the activities. For example, a detailed RCA tree is shown in Figure 7.2.12, where the attention has been focused on the uncapability to fight the fire and the ineffectiveness of the drencher system.

Moreover, the timeline of the event has been reconstructed taking into account several source of evidence, like the Voyage Data Recorder, the fire detection system, the interviews, and so on. A part of the timeline, for readibility reason, is shown in Figure 7.2.13.

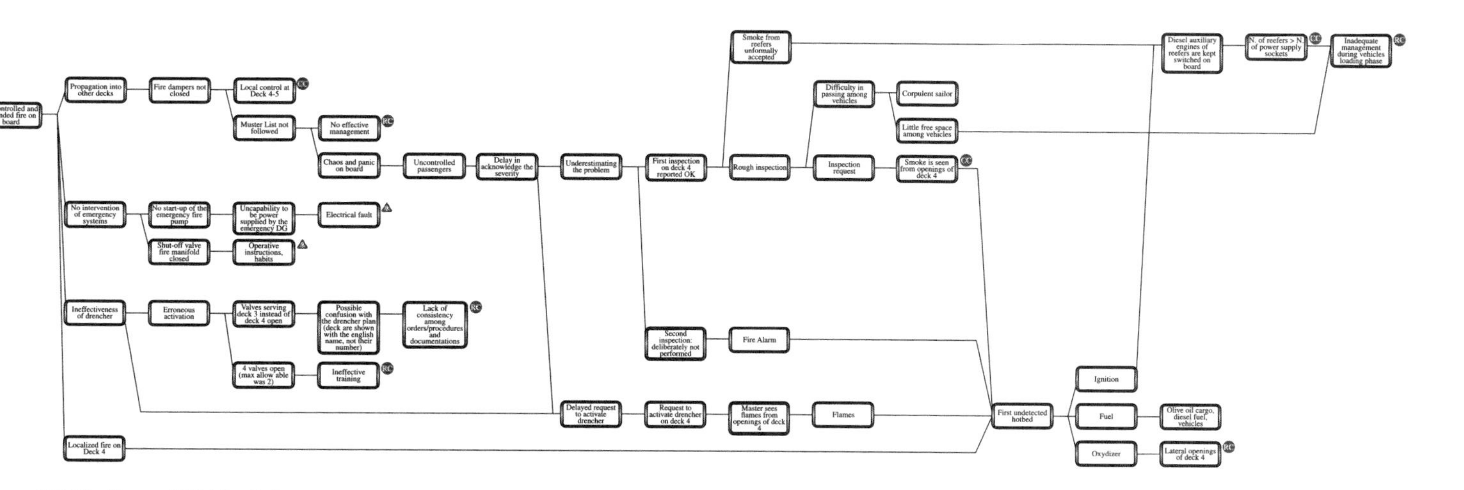

Figure 7.2.11 General RCA logic tree.

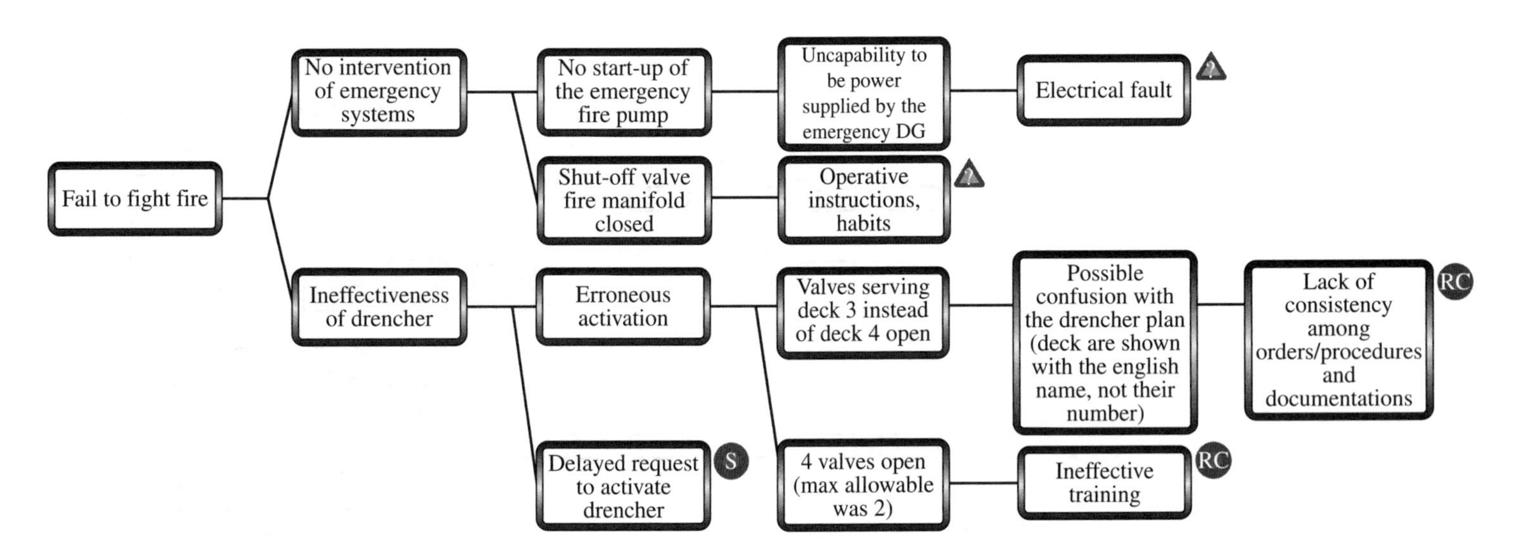

Figure 7.2.12 Detailed RCA logic tree.

	27/12/2014 15:39:48	27/12/2014 23:27:25	28/12/2014 03:23:05	28/12/2014 03:27:00	28/12/2014 03:38:00	28/12/2014 03:40 - 03:50	28/12/2014 04:39:01
Witness						Steam coming from hydrants, not water	
Voyage Data Recorder					Engines stop		
Fire Detection System			Fire Alarm at Deck 4, frame 156				
Master				Flames are seen from openings of Deck 4 Order to activate drencher			
Ferry boat	Departure from Patras	Departure from Igoumenitsa				Black-out	Arrival of the first rescuer

Figure 7.2.13 Part of the timeline of the incident.

7.2.5 Lessons Learned and Recommendations

Sometimes, the lessons learned from an incident run so fast to be implemented before the final legal judgement. This case study is an example. During a travel to Sicily, one of the Authors' collaborators sent us the photos shown in Figure 7.2.14. They clearly depict two positive reinforcements from the management that were never noted before. The photo on the left shows the prescription "shut down engine", that is written in big capital letters on the lateral wall of the main deck, where cars are parked. The photo on the right shows an information sign about the drencher system, where the distinction between "Section #5" and "Section #6" is clearly understandable: in this way, walking on that area, everyone in the crew is automatically reminded about the correspondence between drencher section number and the covered area.

This case study has been also used by the Authors as an example of incident investigation and management commitment to safety during a training session for a process industry. Indeed the lessons from the findings, as shown in Table 7.2.2, can also be connected and used by the process industry: this is the power of the universality of safety culture and root cause analysis.

7.2.6 Forensic Engineering Highlights

7.2.6.1 The Discharge Activity and the Evidence Collection

The evidence collection has been carried out during the discharging operation of deck 3 and 4. With this definition we intend for the controlled disembark operation of the vehicles inside garage decks 3 and 4 To be carried out These operations, complex for safety issues and interferences, allowed five objectives to be satisfied:

1. Know in detail the nature and the position of the vehicles on board, through crystallization of the numerous pieces of evidence collected during the operation by means of collection form specifically created for the census (Figure 7.2.15);
2. Explore the context, with the aim to know its peculiarity, with a particular focus on the aspects regarding the fire propagation;

Figure 7.2.14 Photos taken inside the ferryboat from Villa to Messina, 2016.

Table 7.2.2 Some lessons learned from the incident, written so that they can also be used in other business sectors, such as the process industry.

What	Finding	Lesson
Daily activity	1. Improper management of vehicle embarking	2. These activities must be always carried out respecting the procedures
Emergency procedures	3. Wrong activation of the firefighting system's valves	4. Emergency simulations must be carried out to: be trained; highlight enhancing possibilities
Alarm management	5. Underestimation	6. Always know the causes of the alarm 7. Do not bypass alarm system, without formal request and official approval
Emergency procedures	8. Incongruity between drencher plan and order given	9. Verify that orders and procedures are clear: they must not be interpreted/elaborated by the performer 10. Always keep always update procedures and plans
Emergency management	11. Passengers threw the remaining life rafts in the sea	12. Respect your role as established in the emergency procedure. If you have no task assigned, do nothing.
Working suitability	13. Corpulent sailor to inspect a poky place	14. Evaluate the physical suitability of the personnel to the assigned task
Design aspects	15. Semi-open deck	18. Always perform the risk analysis
	16. Local commands for fire dampers	19. Evaluate the interface man-machine, also in emergency context
	17. Limited memory of the Fire Detection System	20. Ensure adequate capacity to record data

3. Collect eventual physical or chemical evidence; and
4. Conclude that no further human remains have been found in addition to the ones already identified before the appointment of the investigation team.

Such a complex operation requires a preliminary planning phase, it being not recommended to leave the conduction of these operations to a not-standardised, unsafe procedure. Mainly, it is possible to distinguish three steps:

1. The census of the vehicle, its characteristics, and eventual physical evidence (this is the main phase to collect evidence);
2. The movement of the vehicle from on board to the quay, still under the direction of the investigation team; and
3. The disposal of the vehicle, which is removed from the quay (in this step there is no investigative interest).

During the discharge operations, it was necessary to collect those pieces of evidence, as shown in Figure 7.2.6, that were considered useful for the investigation, including: reefer cables, plugs and sockets, fuses, thermal engines used to insufflate air in tanks for alive fishes, and so on. Among the identified evidence that it was not possible to collect there were: the status of the valves of the firefighting systems, the conditions of the

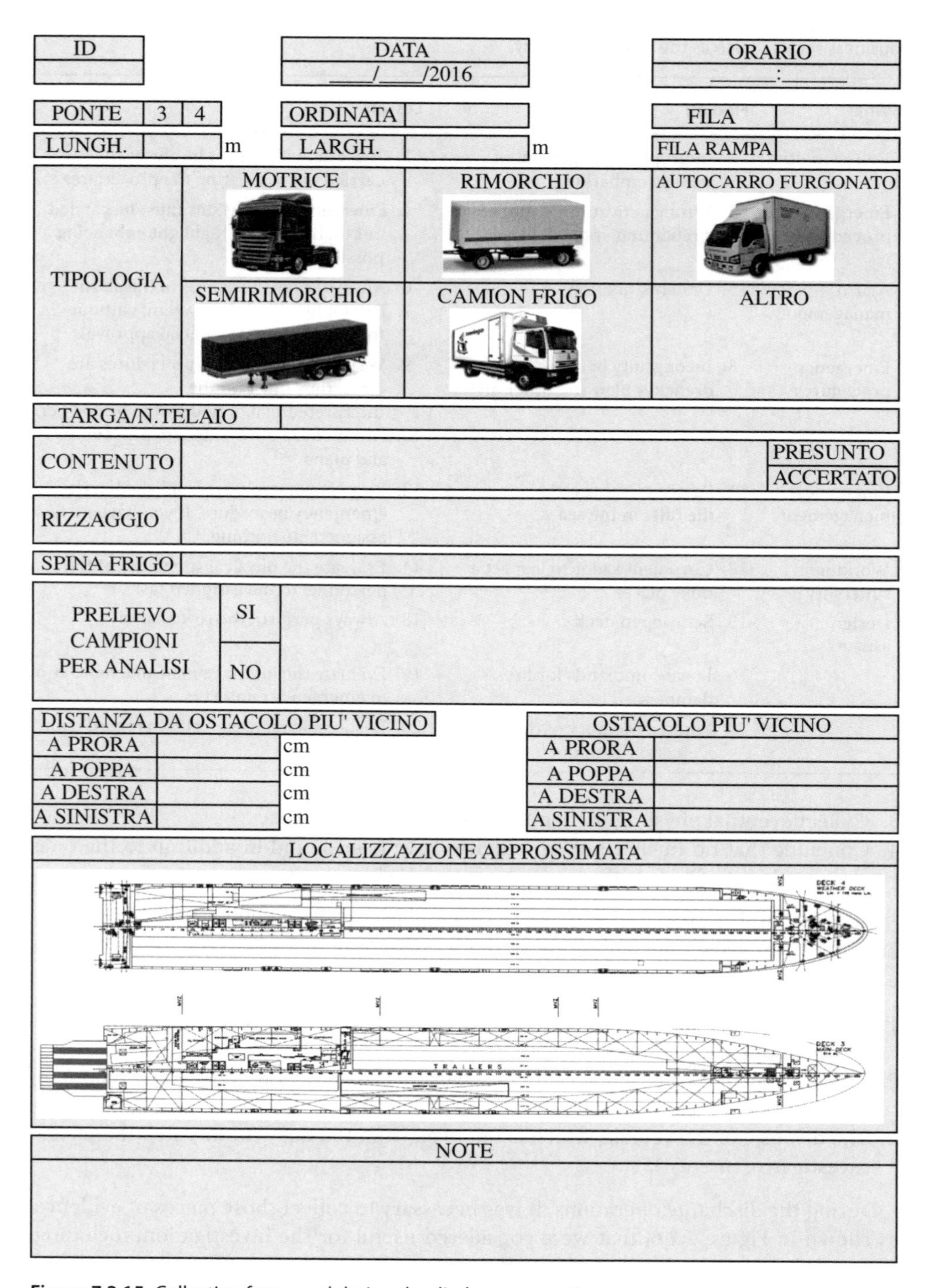

ID

DATA ___/___/2016

ORARIO ___:___

PONTE | 3 | 4

ORDINATA

FILA

LUNGH. ___ m

LARGH. ___ m

FILA RAMPA

TIPOLOGIA: MOTRICE | RIMORCHIO | AUTOCARRO FURGONATO | SEMIRIMORCHIO | CAMION FRIGO | ALTRO

TARGA/N.TELAIO

CONTENUTO — PRESUNTO / ACCERTATO

RIZZAGGIO

SPINA FRIGO

PRELIEVO CAMPIONI PER ANALISI — SI / NO

DISTANZA DA OSTACOLO PIU' VICINO

A PRORA		cm
A POPPA		cm
A DESTRA		cm
A SINISTRA		cm

OSTACOLO PIU' VICINO

A PRORA	
A POPPA	
A DESTRA	
A SINISTRA	

LOCALIZZAZIONE APPROSSIMATA

NOTE

Figure 7.2.15 Collection form used during the discharge operations.

fire doors, the status of the fire dampers, the structural damages survey of the bearing structures at deck 4 and 5. They provided, through reverse engineering processes, support to the hypothesis of ignition and fire propagation.

During the discharge operations, more than 4000 photos were taken. They were catalogued using tags, allowing to simplify the following step of finding. At the end of the discharge operation, 115 heavy vehicles were disembarked and registered. For every vehicle the following information has been recorded:

1. Typology;
2. brand;
3. model;
4. plate;
5. chassis number;
6. position on board;
7. cargo; and
8. connection to the power supply.

Because of the conditions of the garage decks, it was not always possible to have all the listed data. They have been managed through a database, in which the unique key was the progressive number assigned to each disembarked vehicle.

The evidence and the outcomes of their analysis were part of some attachments to the technical report:

1. The cargo plan at deck 3 and 4 (figure 7.2.16);
2. the vehicles identity records (figure 7.2.17);
3. the physical evidence identity records; and
4. the activity diary, containing the description of the operations carried out, day-by-day;

The collection of evidence by the investigation team, always performed in cross-examination, aimed at the acquisition of that physical evidence that required further laboratory analyses or to be preserved, as proof, from the disposal phase following the discharge operations. Evidence was collected on board and, once agreed on the necessity to pick them, they have been taken. Depending on their typology, glass jars, or plastic bags were used as containers. Evidence has been sealed and signs were written on the seal by the ones that were present at the collecting phase. Finally, the chain of custody decided that the evidence was given to the "Capitaneria di Porto", who kept them under surveillance.

7.2.6.2 Use of and Issues Regarding Digital Evidences

The Norman Atlantic ship was equipped with modern TLC systems and electronic aids for navigation. In this chapter will be mentioned only the systems that were more deeply investigated by the expert report. In particular, the VDR (Voyage Data Recorder) system model 100G2 produced by Rutter (now Netwave) company, the Fire Detection "AUTRONICA" system model BC-320" and the system for ship automation model K-Chief 500 from Kongsberg company. In addition, several digital memories were recovered on board: hard disks, both mechanical and solid state, internal and external. When possible, for twelve of these memories a bit-stream image was made in the presence of claimants. An appropriate methodology was followed for this kind of

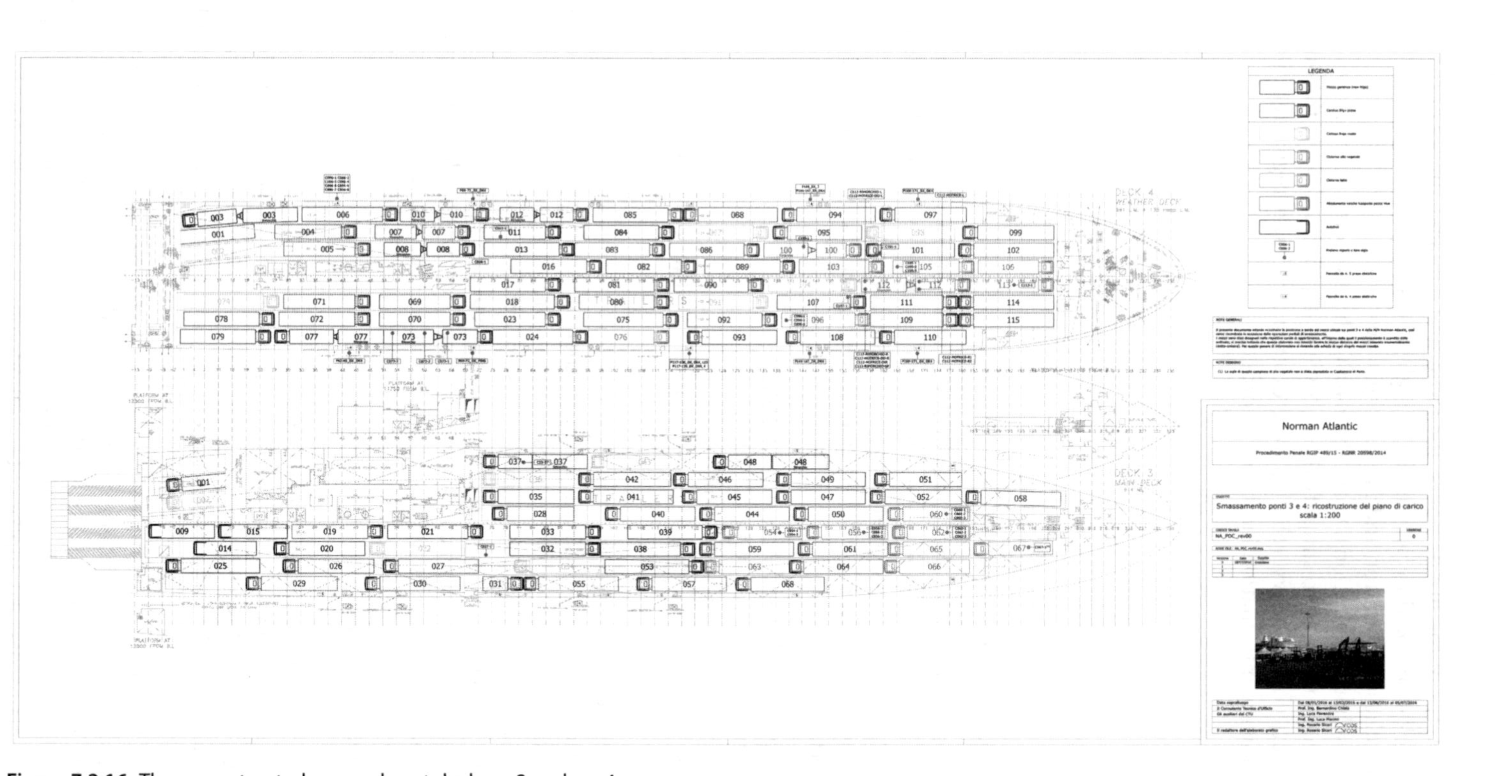

Figure 7.2.16 The reconstructed cargo plan at deck no. 3 and no. 4.

Tribunale di Bari
Procedimento Penale RGIP 489/15 - RGNR 20598/2014

Censimento dei mezzi a bordo dei ponti 3 e 4 della M/N Norman Atlantic eseguito durante le operazioni di smassamento

Data censimento	Orario censimento	ID
14/01/2016	10:26	019

LOCALIZZAZIONE

Ponte	Fila	Da ordinata	A ordinata
3	5	31	51

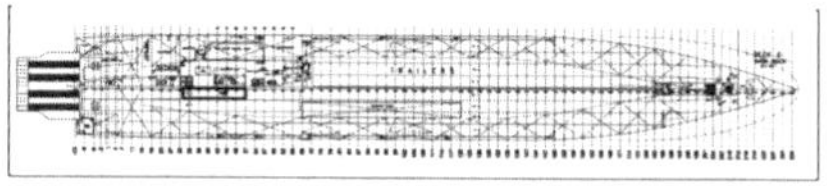

Orientamento motrice verso...
Prua

DATI IDENTIFICATIVI

Tipologia
Motrice + semirimorchio frigo

Marca motrice	Numero telaio motrice	Targa motrice
Scania	N.d.	N.d.
Marca semirimorchio/rimorchio	**Numero telaio semirimorcho/rimorchio**	**Targa semirimorchio/rimorchio**
N.d.	N.d.	N.d.

CONTENUTO
Pesche sciroppate

RIZZAGGIO RILEVATO
2 ganci a poppa

FOTO

ELEMENTI CIRCOSTANTI

	Elemento più vicino	Distanza [cm]
A prora	Altro mezzo	–
A poppa	Già rimosso	–
A dritta	Altro mezzo	0
A sinistra	Cattedrale	72

ALLACCIO ALLA SPINA FRIGO RILEVATO
N.d.

REPERTI PRELEVATI
No

NOTE
Plastica non bruciata a poppa, così come anche assi e tavole di legno

Data sbarco 14/01/2016

Compilata da Ing. Rosario Sicari

arcos

Pagina 37 di 230

Tribunale di Bari
Procedimento Penale RGIP 489/15 - RGNR 20598/2014

Censimento dei mezzi a bordo dei ponti 3 e 4 della M/N Norman Atlantic eseguito durante le operazioni di smassamento

ID 019

ALLEGATO FOTOGRAFICO

arcos

Pagina 38 di 230

Figure 7.2.17 An example of a vehicle identity record.

operations because digital data were subjected to high temperature and smoke during the accident. This methodology also includes the statistical analysis of disk health metrics ("S.M.A.R.T." parameters) which were aligned with the lifetime estimated to further confirm the origin of the parts.

7.2.6.3 Expected Performances of the Installed Digital Memories

Digital memories installed on board, some of which are governed by international standards and are therefore tested and subject to periodic tests, are intended to be a prime source of evidence for the reconstruction of the accident dynamics. Systems involved in generating and collecting these data should be designed to create information archives that model real-life events occurred during the accident, and shall be available before collecting and using other survey elements. The immediate availability of objective and ordered information, both in time and in their exact location, would have allowed the team of experts to work on a skeleton without limitations that characterise summative testimonials based just on memories or beliefs. In following sections, failures of systems installed on board, that have compromised almost entirely its effectiveness, will be highlighted.

7.2.6.4 The VDR (Voyage Data Recorder) System

The M/N Norman Atlantic equipment provided a VDR system, the 100G2 of the Rutter company (Figure 7.2.18). The "Performance Standards" to be respected are dictated by IMO (International Maritime Organization) Resolution A.861 (20), which integrates inside the IEC61996. The amendments to A.861 (20) to be considered are Resolution MSC.214 (81) while Resolution MSC.333 (90) is not applicable as it was installed before June 2014. The primary aim of this device, defined by the aforementioned provisions, is to record and maintain information regarding position, movement, physical state, command and control data, and all boat alarms for a range of time not less than the last 12 hours of operation. These data can be visualised by a software named "VDR Playback" (Figure 7.2.19). The system consists mainly of the following three components: "DPU", "FRM" and "DAU". The DPU (Data Processing Unit) represents the system where all the signals are stored (audio, video and NMEA messages), collected and processed; it also deals with the data distribution to all other storage devices (i.e.: Final Recording Medium, USB memory, and an external disk called Remote Storage Module or "RSM"). This module includes a battery powered unit that comes into operation in case of black-out. The Final Recording Medium (FRM) is a hardened storage unit that is resistant to impact, fire or sinking and is also known as a "black box". The Data Acquisition Unit (DAU) is a module for digitising and/or converting data streams (audio, video and NMEA) from the sensors installed on the ship and returning them back to the DPU module (Figure 7.2.20). The hard disk installed in the DPU and the Remote Storage Module (RSM) were visually highly damaged by the fire and an attempt to recover data was possible only for abroad specialised companies which were already identified in the preliminary investigation. During a first attempt from the Prosecutor's experts team, with the help of the installer company as well as the annual review station, the content of the FRM capsule was found to be devoid of all audio tracks containing voice recordings which had happened on the Norman Atlantinc deck. The fire detection panel looked equally damaged and did not have any redundancy of the very small internal memory except for a thermal paper print module. Later on, it was verified that this memory

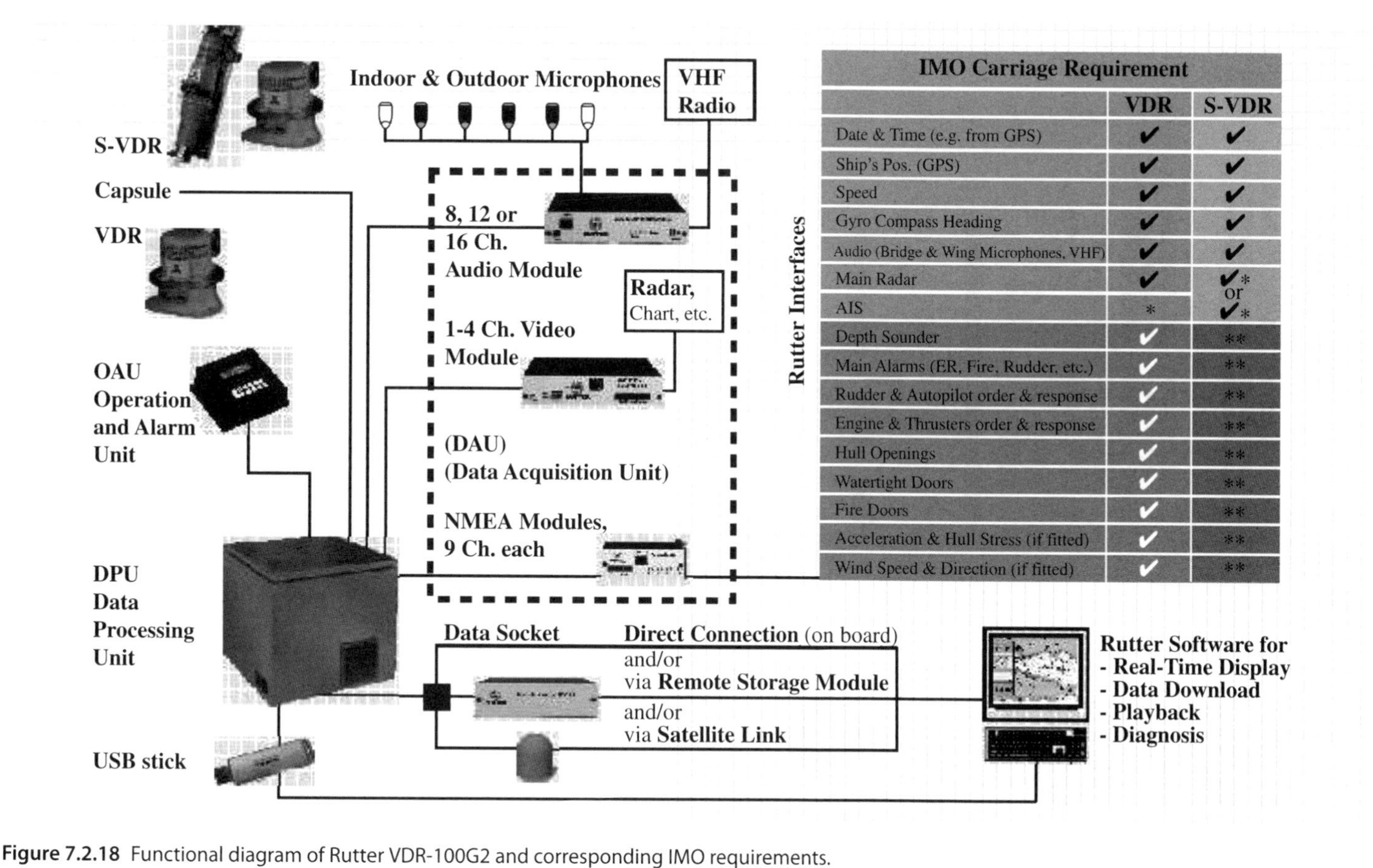

Figure 7.2.18 Functional diagram of Rutter VDR-100G2 and corresponding IMO requirements.

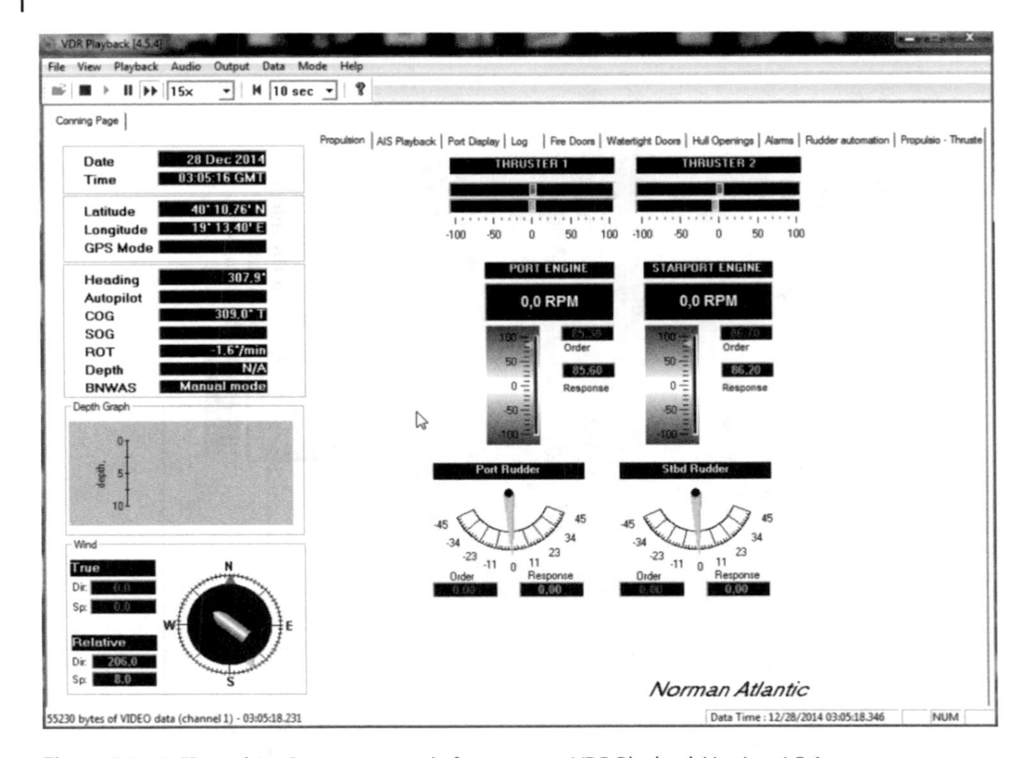

Figure 7.2.19 "Propulsion" screen example from system VDR Playback Version 4.5.4.

had not been damaged but it was still useless, as it was not large enough to register the minutes containing the beginning of the fire because it was of a circular type (so it is overwritten). The usability of data sources is therefore extremely critical and, by pursuing the necessary procedural mechanics that requires the presence of parts for these reading operations (potentially unrepeatable), all activities, that are not executable within the national territory, have been discarded. For what concerns the VDR system, a new data extraction procedure was performed, while the Fire Detection panel was analysed by reading its unique memory. Both these extractions could not be carried out with human tolerance guarantees covered by forensic copies because copies would have required so called "CHIP OFF" procedures (mechanical removal of non-volatile integrated memory incorporated in the media) judged statistically more risky with respect to a safe reading "read only".

7.2.6.5 Data Extraction from the "Black Box" (i.e.: FRM Module)

Upon repetition of the procedure described by the manufacturer, the extracted data were only half of the amount expected (2Gb) from the module features and the lack of all deck sounds was confirmed. Extracted configuration files led to the hypothesis of a partitioning error during installation that may have confused the reader software, which was explored using different versions. It was also figured out in an internal log of a DMM (Data Mangement Module) replacement procedure made on 30th August 2011, following which a reconditioned replacement module was probably installed. This module was provided with a different license than the original one, which also included the

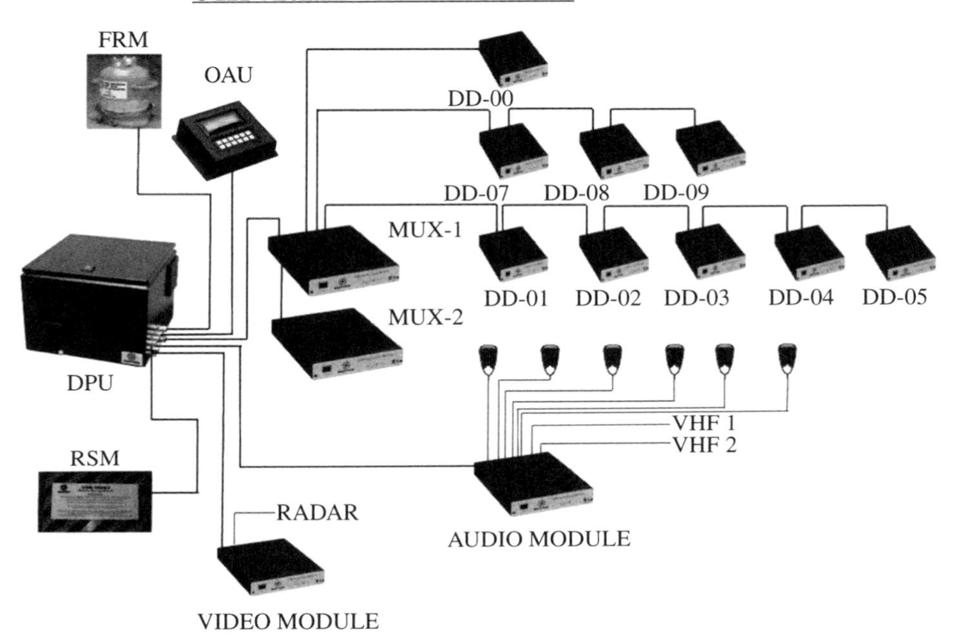

Figure 7.2.20 Connections schematic between DPU and the partially undocumented Data Discrete acquisition Units.

rescue on the burned RSM disk. Therefore, there is no one, but two additional memories on which a recovery of the data (DMM and RSM) could be possible. A FRM reading strategy, not expressly provided in the manual, was worked out. This procedure consists of connecting the FRM to a DMM, as if it was operating normally on board and after having effectively tampered with writing software to the module. This work hypothesis has come to reflect on how the capsule had positively passed the mandatory Annual Performance Tests. For the release of the Certificate of Compliance (IMO MSC.1/Circ. 1222 Ref. T4/8.01), it is expected that only a few minutes of data will be downloaded from the FRM module in order to be sent to the manufacturer. Instead, the entire dataset, or all the data in the form, must still be downloaded and validated locally by the "Service Agent". During some extraction attempts with direct link to the FRM, the procedure was intentionally interrupted by obtaining 27Kb of previously unexpressed data, to prove that the dataset could have been written correctly. The last obstacle to the success of the extraction was the reconstruction of a working DMM module: those made available to the installer were likely to fail, because the Intel processor TDP (Thermal Design Power) was not respected by the active ventilation on the heat sink.

7.2.6.6 Analysis and Use of Extracted Data

The audio data have been digitally processed by at least two specialised companies, since the voices seemed to be overwhelmed by noises, alarm sounds or frustums with just acceptable results. A timely transcript of the recorded period of interest to the left was delivered. A timely transcript of the recorded period of interest for the accident was delivered. Other extracted digital data were analysed with the aid of "replay" software, that represents all the parameters recorded by the ship. Despite the functionality of this software and the completeness of the information represented both being subject to annual certification (and the requirements of MSC.214 (81)), many data were not correctly reported including: engine speed, all wheel data, real wind (only the apparent one), speed over ground, information about the autopilot, echo sounder, watertight doors and fire extinguisher, all unusable and undocumented alarm data. Therefore, it was decided to use the "raw" NMEA data (format that follows the IEC 61162 standard) and to represent them through one or more tables. Using a table allows the selection of the columns of interest for a given period and the comparison between them or the drawing of graph representations. To make this conversion, it is necessary to develop dedicated software, a parser, which has produced several tables, the most complete of which includes the entire time span between 19:46:55 of 26 December 2014 and 14:58:50 of 28 December 2014 and 4,653,381 lines and over 200 columns (without duplicates). However, such "parser" reads as source data those that comply with the NMEA 0183 standard (simple 7-bit text files, with lines ending with special characters <LF> and <CR>), then it extracts ones which must first be converted from the proprietary format of the FRM through software provided by the manufacturer, the "Data Conversion Utility". The entire extraction and conversion operation was repeated several times and with different hardware and software systems, but the fundamental Data Conversion Utility seems to be irremediably malfunctioning, as the output data have: offset problems, checksum-free strings (control characters) that should have been discarded, entire lines missing, moved, casually duplicated, strings without header, presence of multiple LF characters before CRs, and several special "BELL" characters that should not be present. This anomaly makes it unreliable to read all "short" events, such as opening or closing a fire door or an instant silenced alarm. We find out that about 14% of the recorded data was missing after the conversion. Of the remaining 86%, converted and entered in the table correctly, it cannot be said that there was a simple interpretation: the documentation of the "non-standard" strings, that is, those implemented directly by the manufacturer of the specific apparatus and not provided by the NMEA standard, it was not immediately availabile to the ship owner. As far as documentation of "contact" data is concerned, that is, alarms that only cause the type of binary information "on" or "off", the documentation took more than a year before being provided to us. The installation company, which is also "Revision Station" should have been able to decode all the strings, including the owners, since it has to certify that the system is functioning properly every year and that all compulsory data required by IMO already mentioned Resolution A621 (20) are regularly recorded and stored in the VDR, because the MSC 1024 Circuit charges the controller to maintain the data decoding system (See Figures 7.2.21 to 7.2.23).

7.2.6.7 Documentation Analysis of the Fire Detection System

To conclude, the documentation provided was incorrect and did not reflect the links actually encountered by the data analysis: many contacts were mistakenly moved or exchanged during the installation and macroscopic changes to the fire detection system

INTERNATIONAL MARITIME ORGANIZATION
4 ALBERT EMBANKMENT
LONDON SE1 7SR

Telephone : 020 7735 7611
Fax: 020 7587 3210
Telex: 23588 IMOLDN G

E

Ref. T1/2.02

MSC/Cire.1024
29 May 2002

GUIDELINES ON VOYAGE DATA RECORDER (VDR) OWNERSHIP AND RECOVERY

Ownership of VDR information

1 The ship owner will, in all circumstances and at all times, own the VDR and its information. However, in the event of an accident the following guidelines would apply. The owner of the ship should make available and maintain all decoding instructions necessary to recovery the recorded information.

Figure 7.2.21 Extract from MSC/Circ. 1024.

```
1  ÖBELACKNULEOTNULBSNULACKNULENQNULNULNUL@ETX*EOTNULNULNULKNULNULNULNUL5NUL
2  NULSOHNULNULDC4NULNULNULVT%‡D24$HEHDT,309.2,T
3  NULSOHNULNUL<NULNULNULVT%‡D24$VWLOG,10.0
4  NULSOHNULNULPNULNULNULVT%‡D24$HNROT,007.5
5  NULSOHNULNULPNULNULNULVT%‡D11$GPDTM,W84,,00.0000,N,00.0000,E,,W84*41
6  NULSOHNULNULdNULNULNULVT%‡D15$DIS00,101111,111111,000000,001000*5E
7  NULSOHNULNULdNULNULNULVT%‡D25$SDDPT,0010.3,0000.0,20*7B
8  NULSOHNULNULxNULNULNULVT%‡D12$TIROT,006.1,A*3C
9  NULSOHNULNULŒNULNULNULVT%‡D25$PFEC,Alarm,0,0*6F
10 NULSOHNULNUL´NULNULNULVT%‡D11$GPGLL,5332.7077,N,00005.4859,E,060459.00,A,D*67
11 NULSOHNULNUL´NULNULNULVT%‡D25$PFEC,xdr,,200*60
12 NULSOHNULNULÜNULNULNULFF%‡D15$DIS01,111010,000000,000000,000000*5F
13 NULSOHNULNULCANSOHNULNULFF%‡D11$GPVTG,214.7,T,217.3,M,0.3,N,0.6,K,D*24
14 NULSOHNULNULTSOHNULNULFF%‡D11$GPZDA,060459.00,08,06,2006,00,00*62
15 NULSOHNULNUL¤SOHNULNULFF%‡D24$PADRT,030.0
16 NULSOHNULNUL¸SOHNULNULFF%‡D24$IICTS,309.2
17 NULSOHNULNULÌSOHNULNULFF%‡D12$HEHDT,309.2,T*27
18 NULSOHNULNULàSOHNULNULFF%‡D16$ANL00,2413,2421,801,0,2,-12,-12,-12*57
19 NULSOHNULNULàSOHNULNULFF%‡D24$PASVW.10.0.A
```

Figure 7.2.22 Example 1 of RAW data from FRM with bogus characters.

```
12/27/2014,14:45:57,25,PRRS,4,0.9,0.6,0,0.9,16448,0.3,0,0.9,0,0,0,1A^M
12/27/2014,14:45:59,25^G^G,PRRS,4,0.9,0.6,0,0.9,16448,0.3,0,0.9,0,0,0,1A^M
12/27/2014,14:46:00,25,PRRS,4,0.9,0.6,0,0.9,16448,0.3,0,0.9,0,0,0,1A^M
```

Figure 7.2.23 Example 2 of RAW data from FRM with bogus characters.

were communicated by the ship owner, one of the defendants, only upon requests for explanations regarding the wrong links. These changes have not been updated in any documents officially deposited in the Naval Registry. Nevertheless, reliable documentation would not have been enough to uniquely indicate the exact origin of the fire since the sensors were grouped into "loops" or rings with resilient wiring that, however, covered entire bridges or even more than one. The alarm and the pre-alarm specific for the activated sensor are recorded only in the small and insufficiently protected memory of the anti-fire system, while at VDR, according to regulations, only the indication of which loop is in an unspecified alarm state.

7.2.7 References and Further Readings

References

1 Chiaia, B., Marmo, L., Fiorentini, L. et al. (2017) Incendio della motonave Norman Atlantic: indagini multidisciplinari in incidente probatorio. IF CRASC 17. Milano: Dario Flaccovio Editore; pp. 129–140.
2 Ingason, H. and Lonnermark, A. (2005), 40:646–668.
3 Marmo, L., Piccinini, N., Russo, G. et al. (2013) Multiple Tank Explosions in an Edible-Oil Refinery Plant: A Case Study., 36(7):1131–1137.

Further Readings

ANAS S.p.A. Direzione Centrale Progettazione. (2006) Linee Guida per la progettazione della sicurezza nelle Gallerie Stradali.
Bigi, S.B. (2018) Gli strumenti ingegneristici a supporto dell'indagine forense in caso di incendio. Caso studio: la propagazione dell'incendio a bordo della M/N Norman Atlantic, dallo studio della dinamica occorsa alle "lessons learnt".

7.3 LOPC of Toxic Substance at a Chemical Plant

7.3.1 Introduction

The general information about the case study is shown in Table 7.3.1.

This incident described here really happened. In this paragraph, the name of the company that experienced this incident is substituted with "Company", in order to preserve the privacy.

7.3.2 How it Happened (Incident Dynamics)

At the Italian Company's site of "City", on January 20th 2017, at 4.00 PM, 24 tons of 2,4-DiChloroBenzoTriChloride (next written as 2,4-DCBTC, CAS 13014-18-1) spilt because of the rupture of a Teflon bellows. 2,4-DCBTC is an irritating and noxious substance for aquatic organisms. The bellows was between the manual valve at the bottom of tank D6003 (used to store the finite product) and the pump P6004, used to recirculate at the tank and to load on car tanks from D6003. Both the tank D6003 and the pump P6004 are inside the photo-chlorination unit, that is part of the ChlorineAromatic plant (next indicated as CLAR).

Table 7.3.1 General information about the case study.

Who	Chemical plant
What	LOPC of 2,4-DiChloroBenzoTriChloride
When	2017
Where	Italy
Consequences	24 tons toxic release. Not relevant environmental consequences
Mission statement	Determine the causal factors and the root causes of the incident
Credits	Giovanni Pinetti (Tecsa s.r.l.) Pasquale Fanelli (Tecsa s.r.l.)

The chlorinated liquid released inside the curbed area of CLAR plant generated slightly acid vapours (due to hydrochloric acid) that have spread from the point of spillage to the adjacent areas, before dispersing without causing health issues for Company's personnel. The environmental consequences were not relevant.

The site management submitted the final report of the incidental event, to the authority in charge. In the same report, the Company communicates to the authority that an RCA has been commissioned to examine what happened, finding the major causal factors and, basing on them, to identify the root causes and the relative possible enhancing actions to avoid the recurrence of the event.

7.3.3 Why it Happened

The immediate cause of the incident is the failure of the Teflon bellows, installed in 2002, with three waves and two enforcing metallic rings. Interviews with the operation and maintenance personnel have been conducted by the RCA analysts, in order to collect data and information about the incident, according to the methodology established since the beginning of this investigation. In particular, the interviews regarded: an assistant of CLAR, an operator of CLAR, a mechanical maintenance supervisor. The interviews have been conducted following a face-to-face approach, with no external people other than the interviewer and the interviewee, on the base of a set of predefined questions. Before and after the interviews, the RCA analysts met the director and the top management board to clarify the terms of reference of the investigation, its boundaries, its objectives, that are the identification of the causes of the damaging of the Teflon bellows, and the identification of the possible action items. The information collected through the interviews and the meetings are the base of the RCA.

The documentation used as inputs for RCA includes:

1. Communications from/to authority;
2. safety report;
3. incident report;
4. operation data (tabular, trend, and events from dcs);
5. design documents;
6. planimetry;
7. specifications;
8. analytical data;

9. msds;
10. bellows data sheet;
11. procedures;
12. emergency plan; and
13. photos.

The RCA conducted led to the causal factors diagram shown in Figure 7.3.1. According to the RCA methodology, for each identified causal factor, the RCA analysts went back to one or more root causes. The root causes were identified by the RCA analysts, focusing on supervision, management, procedures, design, engineering, construction, hardware and software, systemic factors, inspection, tests, communication issues, and other objective causes (subjective causes excluded from RCA regards the personnel's competency and adequacy, the company culture, and every cause aimed at looking for and identifying individual responsibilities).

The immediate cause of the incident is the LOPC of 2,4-DCBTC from the bellowss in Teflon, placed on the suction line of P6004 and very likely attributable to the thermal expansion of the product trapped and solidified after closing the steam tracing of the suction and discharge of pump P6004. Once the steam tracing was re-opened in the late morning, the solid product liquefied irregularly, because of different factors, such as the material of the pipe (nickel), material of the bellowss (Teflon), material of flanges, piping thickness, flange thickness, material of the gaskets, surface of steam tracings, not-insulated parts, not-steam traced parts, insulation thickness, and so on. Therefore, some fluid was trapped inside the casing of pump P6004, between suction and discharge. Indeed, even if pump P6004 was working (with the manual valves open both in suction and discharge) only when the solid inside the valve casing at the bottom of D6003 melted, the temperature at the bottom of D6003 started raising, because of the contact between the temperature sensor at the bottom of D6003 and the parts of liquid being hotter and hotter because of thermal stratification. The recorded thermal trend at the bottom of D6003 is also the evidence of the solidified product, because the pump (50 m^3/h), recirculating with no interruption for 33 minutes (ignoring the three attempts to recirculate in the previous three hours) would have to recirculate the entire volume of product (approx. 16 m^3), making temperature uniform.

The thermal expansion of the product, when passing from the solid state at the ambient temperature (the maximum temperature on the day of the incident was 8 °C) to the steam tracing temperature (140 °C) is equal to 9-10%.

Therefore, the trapped product and the heating with non-uniform melting on the suction and discharge line of P6004 generated a thermal expansion, with axial force on the bellowss and, once the maximum limit was reached for axial expansion of 28 mm, with radial force, which led to the rupture of both the two reinforcement rings and the rupture of the bellows itself.

A further contribution to the thermal expansion was given by the magnetic drive pump P6004 that, working with the discharge closed (because of the solidified product at the discharge section), dissipated the hydraulic power equal to 2 kWh/h in the product trapped inside the pump casing, for 33 minutes.

The particular friability of the solid product 2,4-DCBTC and the open impeller of P6004 have actually impeded the thermal switch to trip for the overload of the starting motor of the pump in the presence of solidified product.

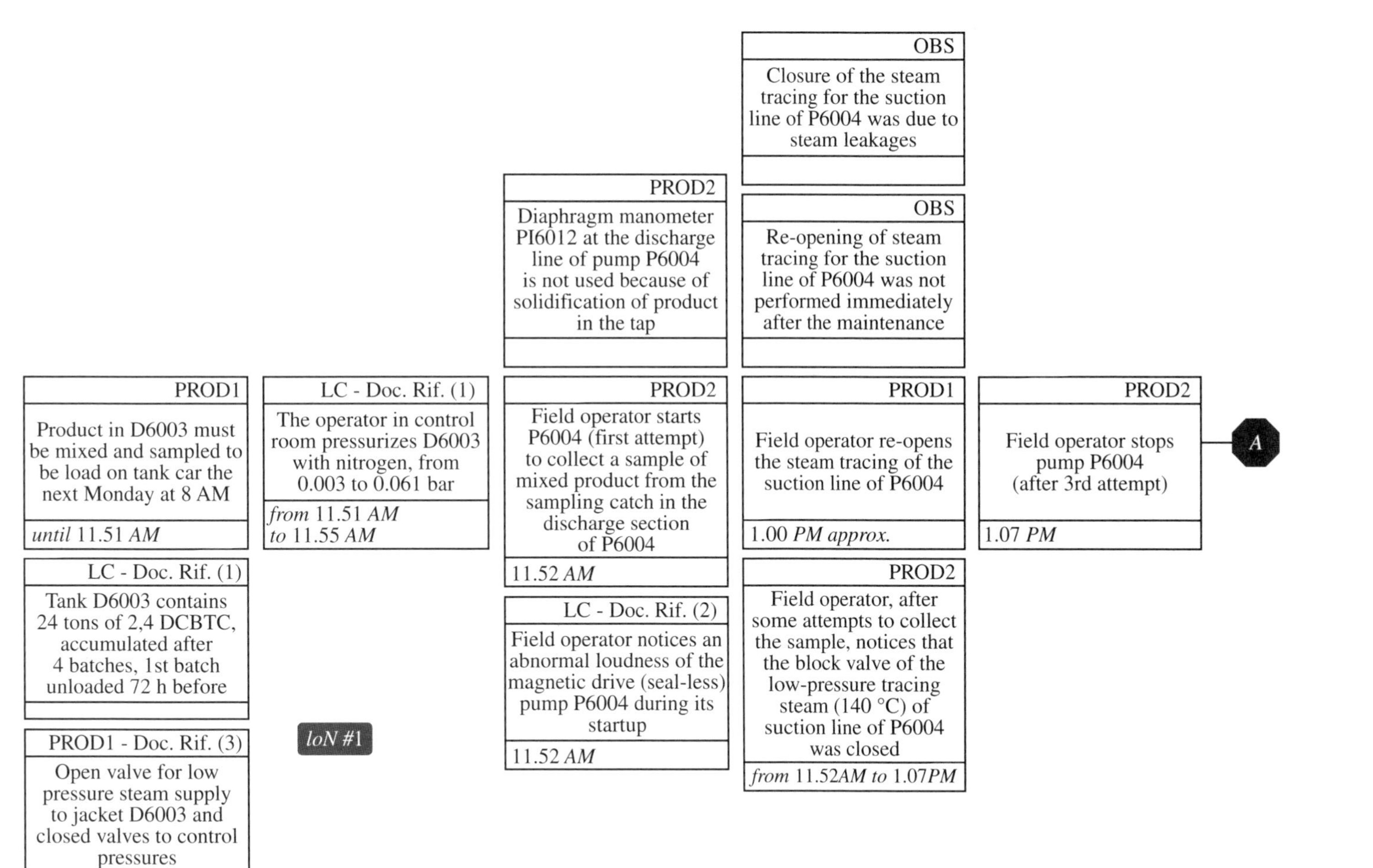

Figure 7.3.1 Causal factors diagram (part 1/4).

7.3.4 Findings

In the causal factors diagram in Figures from 7.3.1 to 7.3.4, events are depicted in boxes, including the incidental event itself, the initiator events, the incidental dynamics events, the eventual conditions (e.g. failures or lack of protections), the concomitant factors, the conditions depicted from the interviews, as well as notes, observations, and logic conclusions by RCA analysts. Each box of the causal factors diagram contains the source of both the data/information used and the recorded (or estimated) time, when possible, of the single event.

Here are the outcomes of the investigation, as shown in the causal factors diagram.

Causal factor CF#1
Rupture of the Teflon bellowss on the suction line of pump P6004 from tank D6003.

Intermediate and root causes

Rupture of the Teflon bellowss for wear and tear (not confirmed hypothesis)
The damaged bellows did not have wear and tear evidence, in addition to the ones determined by the overpressure for liquid expansion.

Component in critical position at the bottom of D6003 (Teflon bellowss) not under periodic inspections;

1. Teflon bellows at the bottom of D6003 was not included as a component in critical position in the list of critical equipment attached to the procedure PS119 and, consequentially, not included in the periodic maintenance plan;
   1. SMS did not identify the specific component as critical.

Rupture of the Teflon bellows for improper operation (not confirmed hypothesis)
Complete solidification of the product inside the valve at the bottom of D6003, the Teflon bellows, the suction pipe of P6004, the manual valve at the suction of P6004, the casing of the pump P6004, the manual valve at the discharge of P6004, the discharge pipe of P6004 and the recirculation line of P6004 for the sampling catch;

2. Block of the steam tracing of the suction and discharge line of P6004, for maintenance operations (elimination of steam leakages) carried out according to usual condition, without having drained the line itself. The voluntary closure of the tracings happens solely in the following cases:
   1. Maintenance (closure and re-opening of tracings managed by the permit to work);
   2. Closure for the end of the winter (only for products with a low melting point and so not applicable to the considered pipe).
      Moreover, steam can be involuntarily loss in tracings for:
   3. Out of service of steam (e.g. stop of thermal central);
   4. Out of service of the steam manifold supplying tracings (e.g. for maintenance on the manifold);
   5. Malfunctioning condensation discharger from the tracings.
3. Late re-opening of steam tracings on the suction and discharge line, after the maintenance operation. If the re-opening was performed before, product would be not solidified completely inside the valve at the bottom and the entire pipe, consequentially allowing the overpressure (due to the liquid dilatation towards the tank) to vent;

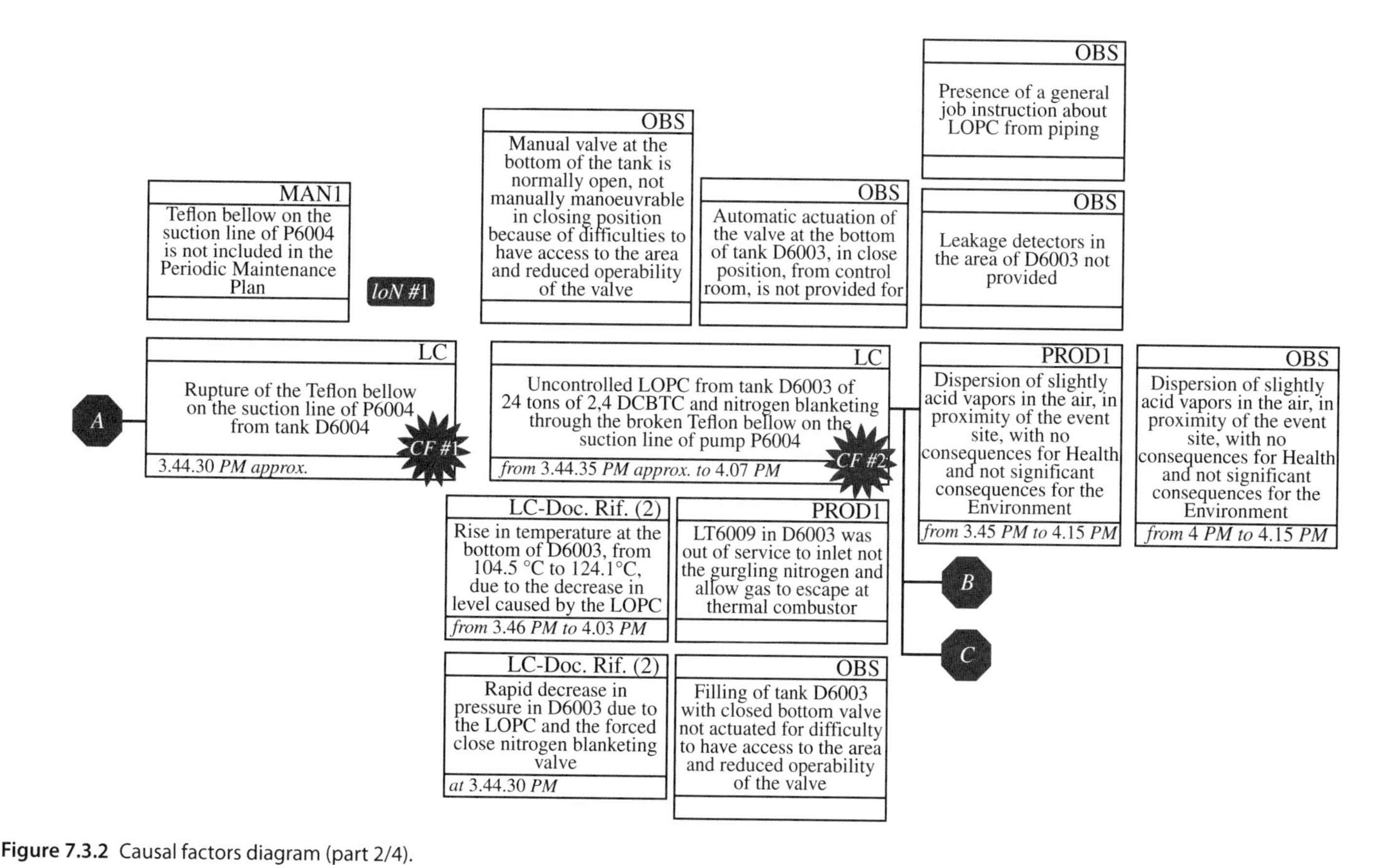

Figure 7.3.2 Causal factors diagram (part 2/4).

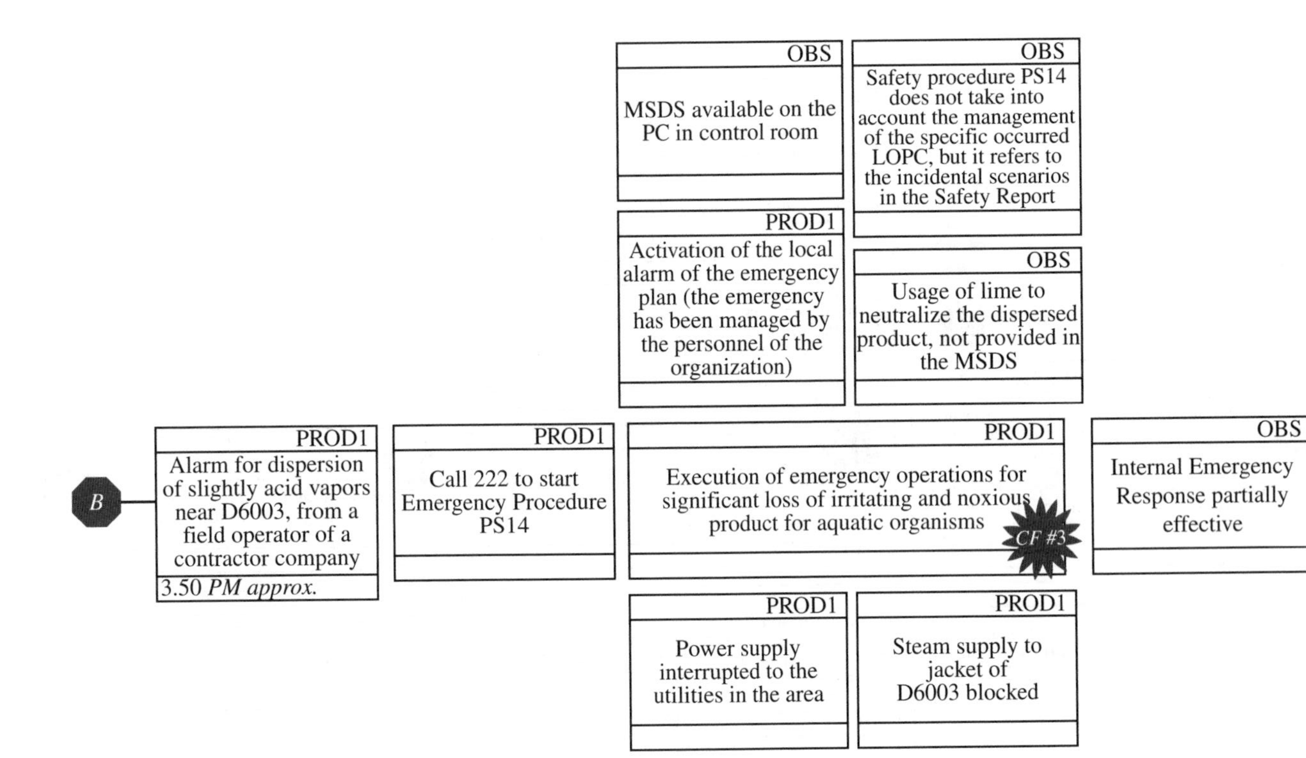

Figure 7.3.3 Causal factors diagram (part 3/4).

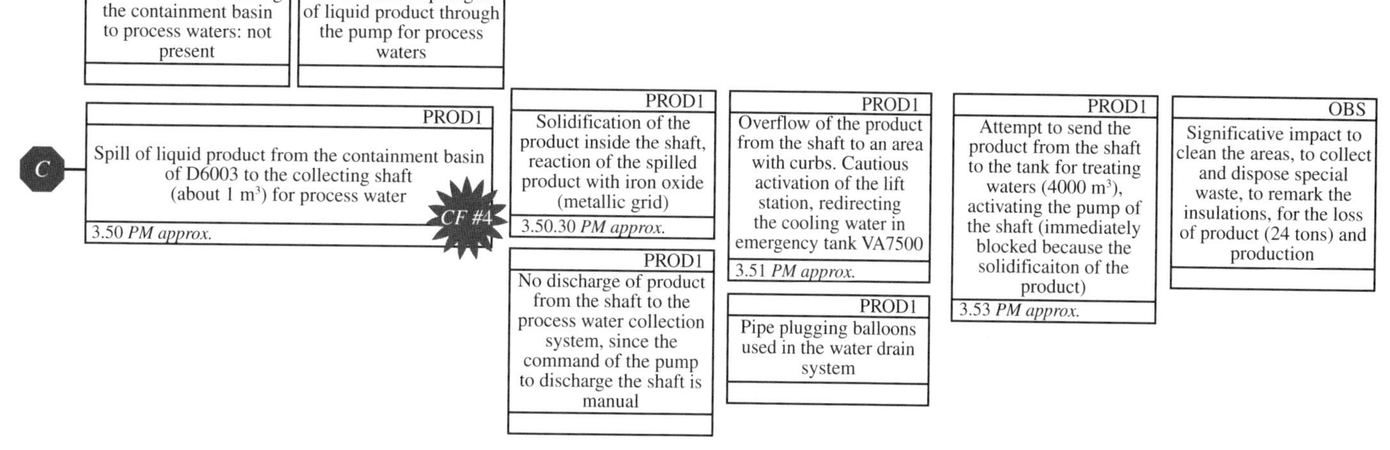

Figure 7.3.4 Causal factors diagram (part 4/4).

4. Permits to work for the elimination of steam leakages from the tracings are emitted in a generic way;
   1. SMS does not completely detail the restart phase after maintenance operations.

Rupture of the Teflon bellows for axial/radial/angular movement, exceeding the maximum design limits (rejected hypothesis)
The hypothesis was rejected by the mechanic maintenance manager because a significant misalignment was not found upstream and downstream the Teflon bellows after the incident (after 15 years in operation without issues of this type).

Rupture of the Teflon bellows for erroneous design of the piping and/or the tracing system
Piping design did not take properly into account the possible freezing of the product inside itself and, in particular, the unfreezing phases of the piping were not analyzed and completely considered during the design;

5. The piping does not have vents for partial freezing condition near the weakest point (the bellows);
6. The piping tracing is not sufficiently boosted in those points where the pipe exchanges less heat (bottom of the tank, valves);
   1. SMS does not reserve particular attention to the freezing and unfreezing of the products with a high melting point during the design stage.

Causal factor CF#2
Uncontrolled loss of containment from tank D6003 of 24 tons of 2,4-DCBTC and nitrogen blanketing through the broken Teflon bellows on the suction line of pump P6004.

Intermediate and root causes

Delayed detection of the LOPC in area D6003
Leakage detectors not provided in area D6003;

7. Not expected leakage detection in area D6003;
   1. SMS did not include the area of D6003 in the systems to detect a leakage.

Delayed detection of the level drop in D6003
Level transmitter LT-6009 in tank D6003 was put into service only periodically, to minimise the vents of gurgling nitrogen containing chlorine and vapours of HCl from tank D6003 to the thermal combustor. However, no alarm would be intervened for the level drop in D6003, since the emptying of tank D6003 is a usual operation during the car tank loading and, therefore, no alarm thresholds are set, except that for the empty tank;

8. The change of the management of the level in D6003 was not correctly detailed in the procedure about the MOC of vents in the photo-chlorination unit;
9. The MOC system was not correctly applied to the change of the management of vents in D6003;
   1. The SMS did not include this managerial intervention into the MOC system.

Impossibility to rapidly block the LOPC from the Teflon bellows on the suction line of pump P6004
Impossibility to close the manual valve at the bottom of D6003 to limit the LOPC from D6003;

10. It was not possible to perform what was prescribed by the internal procedure IL CLAR 91 "operative procedures about the significant incidental hypotheses", like the unexpected loss of liquid product due to significant loss from the bellows, for impossibility to close the valve ad the bottom of D6003 rapidly and from safe position;
11. The valve at the bottom of the tank is hardly maneuverable and precautions were not considered to manoeuvre it differently;
    1. The SMS did not identify the specific case, with the product at a high temperature and with difficulty to manoeuvre the block valve because of reduced ergonomic operability.

Causal factor CF#3
Emergency operations for significant loss of irritant and noxious product for the aquatic organisms were partially "procedured".

Intermediate and root causes

Emergency operations partially "procedured"
The MSDS of 2,4-DCBTC does not detail the interventions for significant losses of product;

12. Safety procedure PS14 "Internal Emergency Plan" does not highlight generic events of loss of containment, in addition to the scenarios in the Safety Report; indeed, it refers to the operative instructions and to what prescribed by IL 91 CLAR about the generic events;
    1. The SMS does not address the safety procedure PS14 towards generic issues not expressly defined as incidental scenarios in the Safety Report.

Causal factor CF#4
Spill of liquid product from the containment basin of D6003 to the collecting shaft (about 1 m^3) for process water.

Intermediate and root causes

Drainage of liquid product from the containment basin of D6003 to the collecting shaft (about 1 m^3) for process water
No drainage valve to block liquid product from the containment basin of D6003 to the collecting shaft (about 1 m^3) for process water;

13. Tank D6003 was considered as plant receiving tank and so the containment was intended the whole plant curb and not as storage tank with its containment basin;
14. The Safety Report and the procedure PA13 "Monitoring of storage tanks and relative containment basins" do not take into account tank D6003 among the storage tank and, consequentially, do not consider the relative controls also con the drainage valves;
    1. The SMS did not identify the specific case of tank equipped with containment basin.

7.3.5 Lessons Learned and Recommendations

For each root cause of a causal factor, the analysts identified the possible actions for safety improvement, defined as recommendations, aimed in avoiding reoccurrence of the event (or, at least, in reducing its likelihood) and/or in decreasing the severity of its consequences.

Recommendations about Causal Factor CF #1

Identify the Teflon bellows at the bottom of D6003 in the list of critical equipment, and add its control in the Periodic Maintenance Plan;

Insulate the bellows to preserve heat, for a quick run of the periodic inspection (see above);

Verify that the low-pressure steam tracing (140 °C) is not in direct contact with the Teflon bellows to minimise the risk of plastic rupture or rupture for deformation due to excessive thermal gradient;

Modify the pipe in suction so that the part after the bellows might have the possibility to vent an eventual overpressure towards the tank. Modify the piping tracing so that the melting of the product would occur in a more homogeneous way, inside the pipe itself;

Reduce the low-pressure steam temperature to the jacket of tank D6003, in order to avoid rapid and significant increases of temperature of the bellows (thermal shock) at the start of the product recirculation;

Evaluate the possibility to further detail the filling of the permits to work regarding a pipe with product having a high melting point.

Recommendations about Causal Factor CF #2

Use leakage detectors in the area of tank D6003;

Consider the remote actuation of the valve at the bottom of D6003, with opening and closing commands from DCS and emergency closing command in field (safe zone), and limit switches;

Consider a permissive logic to start pump P6004 when the automatic valve at the bottom of D6003 is open (limit switches opening);

Evaluate to change the management of vents of D6003 and the consequent management of the level transmitter LT-6009 in tank D6003.

Recommendations about Causal Factor CF #3

Include, in the work instruction IL 91 CLAR, the operative procedure about the significant loss of 2,4-DCBTC caused by the rupture of the Teflon bellows;

Include the type of interventions for significant LOPC in the MSDS of 2,4-DCBTC;

Include the reference to the generic events, as already listed in IL91 CLAR, in the operative instructions of the safety procedure PS14 "Internal Emergency Plan".

Recommendations about Causal Factor CF #4

Install the drain valve for the containment basin;

Verify the presence of tanks with containment basin which are excluded from the monitoring modules of the basins attached to procedure PA13;

Insert the management about the containment basin of D6003 in the procedure IL CLAR 97 "Management of the unit water".

7.3.6 Forensic Engineering Highlights

By adopting a structured accident investigation it is feasible to identify the underlying causes which led to accident itself. To get the necessary answers the right questions shall be arisen for a thorough investigation of the causes aimed to correct the systematic and non-systematic faults.

The structured approach includes the assessment of the incident or accident scenario by collection of the data from DCS (process variables trends from DCS historian, first-in/first-out, event logger, alarms log, closed loop tuning parameters, refreshing time, others), technical interview with field operators and shift supervisors, organizational and procedural investigation based on predefined Questionnaire (not submitted before the interview) and interviews with selected middle and top management representatives, sample data analyses, high profile specialist technical investigations (such as loss of integrity reports following a fatigue rupture, historical accidents literature on similar plants or same or equivalent hazardous substances, design review extended to basic and detailed engineering packages, involving process licensor for specific process issues. The accident investigation shall include the Process Control System, Emergency Shutdown System, Fire and Gas System and Unit Control Systems hardware and application program potential systematic errors. A thorough accident investigation shall include an extended collection of digital pictures taken at accident location, including any detail relevant to damaged components. Video may be useful as well. Vendors expertise shall be involved in case of accidents involving Vendor's equipment and packages.

In the reference real case the accident involved a significant loss of containment of highly toxic material, which was caused by an expansion joint rupture on bottom line of a storage tank. The technical investigation required by the Regional Authorities was based on a process design review assessing the weakest points of design. After the execution of this review technical interviews were planned and carried out with the plant personnel, the shift supervisor, and other middle managers. The interviews were based on predefined questionnaire based on the technical information available on the accident and the process design review. the interview involved the shift supervisor (under duty during the accident shift), the field personnel and the middle management responsible of Operation & Maintenance procedures and working procedures. The interviews did not involve more than one Company representative. The top management was excluded from attendance of interviews, but that's absolutely a key point, which authorised the personnel to provide any details of the accident suitable to identify the hidden causes of the accident itself. The final results were impressive, since all the causes were identified, finding out at the end of investigation that the actual cause of accident - the rupture of an expansion joint - was just the result of a long list of deviations, including the main following issues, found out by a top-down approach:

1. Expansion joint not subject to inspection and leak testing;
2. missed definition of the expansion joint as critical item, subject to wearing and stress due to p/t cycles operation resulting in widely exceeding the useful life time allowable limits;
3. construction not consistent with good engineering practices (just for instance: steam tracing applied directly to the expansion joint instead of installing a winterizing box, steam tracing executed without following a steam tracing construction specification, manual isolation valve of storage tank not accessible, missed isolation valve on containment basin drainage, missing a secondary containment of tanks area, missing a spillage collection pit in safe area, a water spraying system barrier for toxic cloud missing, a toxic gas detection system in the area missing;
4. p and id not consistent with the "as-is" installation leading to wrong technical decision making process:

5. process variables monitoring and trending missing;
6. qa/qc documentation unavailability;
7. operation & maintenance log books missing;
8. field operator communication with shift supervisor lack;
9. toxic handling procedure missing;
10. emergency response procedure lacking;
11. safety logics aimed to minimise the risk of loss of containment of toxic material missing;
12. emergency logics to isolate the loss of containment of toxic material missing;
13. emergency response linked to specific accident missing;
14. pa/ga missing;
15. operation & maintenance procedures missing;
16. previous operability issues not solved by middle management decision making process;
17. operation practice based on alarms activation and full manual operations; and
18. risk analysis not executed.

The above underlying causes of the accident can be approached one by one and step by step by assigning intervention priorities, but the main issue in the decision process mechanism is not the intervention priority but rather the investment cost impact, so the easiest way is to postpone the intervention so introducing the time elapsing as a worsening factor.

The final cost of accident, in the above case without personal injuries, is much higher than the interventions made necessary by the above listed issues even taking into account the industrial insurance cost savings, the legal issues, the issues with the Authorities, the issues with the Trade Unions, the issues with Community and last but first for importance the reputation of the Company.

7.4 Refinery's Pipeway Fire

7.4.1 Introduction

The general information about the case study is shown in Table 7.4.1.

Since the years of construction of the petrochemical pole of Melilli-Priolo-Augusta (around the 1950s), there was not a significant sensibility towards the urbanistic and territorial compatibility issues, that were regulated, for those aspects of industrial risk, until many years later in 1999 (D.Lgs 334/99 – European Seveso II Directive) and 2001 (D.M. 9/5/2001), with a ministerial decree aimed at the regulation, according to the Seveso legislation, of these aspects since the phase of creation of the local urbanistic rules. In this area, a rapid urban development was observed, without any control for the areas next to the establishments, while the main terrestrial communication infrastructures (rail and road network) divided the industrial establishments for long distances, creating high-vulnerability points in case of incident but, at the same time, being too much difficult to be modified. The incident here described happened on April 30, 2006, when a mixture of hydrocarbons caught fire because of a leakage of crude oil from one of about 100 pipelines crossing a subway.

Table 7.4.1 General information about the case study.

Who	Oil refinery
What	Pool fire and BLEVE
When	April, 30th 2006
Where	Syracuse (Italy)
Consequences	Highway closed for 53 days, 14 injured people, more than 25M euros of total direct incident costs
Mission statement	Highlight safety criteria and improvements
Credits	Salvatore Tafaro (Italian National Fire Brigade)

The Authors highlighted how it is important to compare the facility's incident records with data from similar industries, in order to learn also from the experience of others. Several databases are available to meet this objective, both national and international. For example, in Europe the Major Accident Reporting System (MARS) is a shared industrial accident notification scheme, that was established by the Seveso Directive in 1982. The database allows examining the historical data related to the industrial accidents, in order to draw the lessons learnt and prevent the occurrence of a future unwanted event or, at least, to mitigate their consequences.

Some of the contents of this example are taken from the MARS report [1]. MARS database can be consulted online for free at https://emars.jrc.ec.europa.eu.

7.4.2 How it Happened (Incident Dynamics)

On April 30, 2006, at 3:40 PM, in one of the three refineries of the industrial pole near Syracuse, near the subway of ex SS114 (highway Siracusa – Catania), a leakage of crude oil occurred from an insulated pipe with a significant diameter (DN 500), which was then ignited, setting it on fire and extending flames both the upstream and downstream of the subway itself, also involving a part of the pipe-way near some LPG tanks.

Approximately 3 hours after the leak detection the onsite emergency response plan was triggered, performing the following:

1. Mobilisation of the onsite fire brigade;
2. isolation the pipelines starting with the leaking one, as a precautionary measure; and
3. closure of road ex SS114 passing through the establishment.

The onsite fire brigade mobilised initially two emergency response vehicles, spraying foam on the area where the smoke was generated. At 5:42 PM the shift commander of the onsite fire brigade alerted the public fire department in Siracusa, once he understood that the fire was out of control. After a few minutes another fire front developed from the entry of the subway in road Nr 9/0 (uphill), this fire was also fought initially by the onsite emergency response services, with other two firefighting vehicles.

At 6:10 PM the public fire brigade arrived from Siracusa and took the command of the emergency response management. At 6:50 a first BLEVE explosion occurred in another pipeline line containing probably light hydrocarbons (Figure 7.4.1), and was followed by successive explosions of pipelines in the same trench, all due to the overheating of the products inside the pipeline consequent to the heat irradiation of the fire in the trench (Figure 7.4.2).

Figure 7.4.1 Damages of the piping uphill the road. Gash caused by BLEVE.

Figure 7.4.2 Some damaged pipes downwards the road. There is also the pipe of the fire system.

The situation inside the subway at the moment of the fire (as shown in Figure 7.4.3) highlights a disorderly arrangement of the piping, not taking into account the different typology of the substances inside the pipes (e.g. flammable substances near toxic ones, or near the fire system pipe). The arrangement left no possibility to operate in case of verification and/or maintenance; moreover, a proper active fire protection system was absent, excluding the general fire-fighting system made of hydrants.

The subway and its trench contained about 100 pipes crossed by oil products and other hazardous substances, as well as fluids of utilities (including steam, nitrogen, water, air, and so on). The pipes were not owned by the same company; indeed, in the pipe-way there were the pipes of four different companies.

Since the very beginning of the fire, the seriousness of the incident was clear and therefore the fire brigade from different districts of Sicily was asked to intervene, working for more than 80 hours, in addition to the local district fire brigade and the company fire brigade.

During the fire-fighting activities (Figure 7.4.4), the fire brigade personnel found difficult to win the fire for two main reasons: the lacking water resources, for the unavailability of the fire water system in the establishment that was damaged by the fire; and

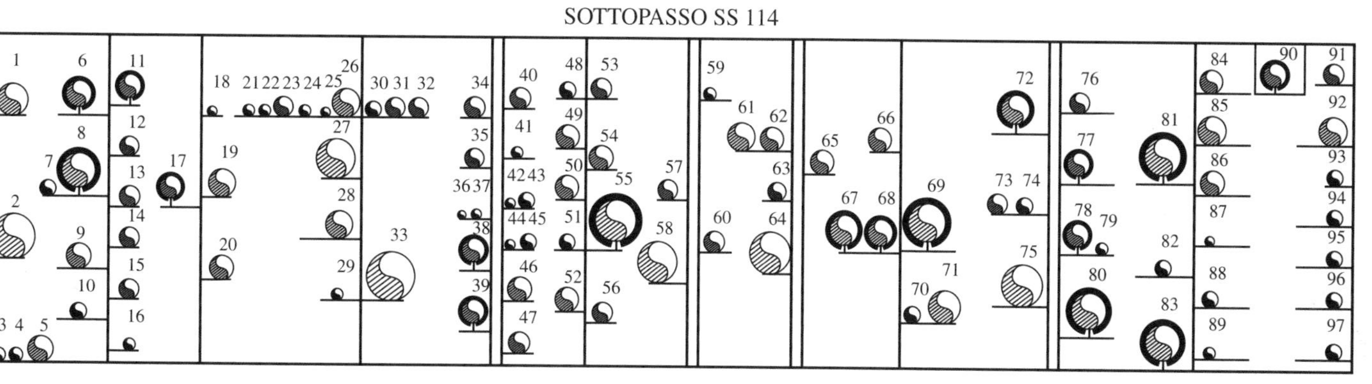

Figure 7.4.3 Transversal section of the subway before the incident. Taken from [2].

Figure 7.4.4 Photos of the extinguishment operation. Used by permission.

the fact that many pipes engulfed by the fire, especially the ones having major diameters, had a huge hold-up between two consecutive blocking valves, between 100 and 350 m^3.

Some of the main effects of the fire were:

1. Dense and extremely pollutant smoke; fortunately, because of the favourable weather conditions, there were no bad consequences for the surrounding residential areas;
2. the runoff of the product dispersed from the piping along the pipeway's trench for a significant length, such that the fire extended downhill the trench, with no possibility to contrast this effect with appropriate containment systems;
3. unexpected collapse of some pipes, especially those containing light hydrocarbons, causing not-serious injuries to some firefighters; and
4. escape of flammable gas/vapours in correspondence of the flanges, because of the lacking seal of the gaskets due to the reached overpressure.

At 6:55 p.m. the decision was taken to shut down all installations inside the establishment, as a precautionary measure. The Prefect ordered the closure of all gates of the establishment in accordance with the Offsite Emergency Plan, this caused the interruption of all vehicle circulation in the area around the accident place for several days. The circulation on the Siracusa-Catania railway was interrupted during 48 hours, and the ship circulation in the port of Augusta was interrupted during approx. 36 hours. The road ex-SS114 was closed (Figure 7.4.5) during 53 days, waiting for the results of the static structural tests and the substitution of the damaged asphalt covering.

Fire extinguishing operations lasted for 48 hours: the fire damaged seriously the pipelines uphill and downhill the subway, as well as aerial pipelines passing parallel to the trench or crossing it. Further limited damage occurred to the subway. During the accident 4 members of the onsite fire brigade, 2 plant operators and 8 members of the public Fire brigade were hospitalised for more than 10 days due to burns, contusions and/or intoxication. The offsite emergency has not caused health consequences to the population living in the neighbouring Communes. The major of Priolo ordered the closure of the public schools for one day and invited the population to stay indoors immediately after the accident as a precautionary measure. The on-site costs of the incident are both related to material losses (22M euros) and response, cleanup, and restoration activities (5.65M euros). The off-site costs are only related to material losses, for a total amount of 50000 euros. "Rebuilding" refers to the repair of the subway, the substitution of the pavement, the road signs and the guardrails of ex SS114 damaged by the fire on an approx. 100m long road section.

Figure 7.4.5 An helicopter view of the area. Used by permission.

The fire extinguishing operations managed to confine the fire in the area corresponding to the subway without extinguishing the fire completely, in order to completely eliminate the presence of hydrocarbons from the hardly accessible trench, where hydrocarbon vapours could have ignited again at any point. The incident attracted national media interest.

In order to restart the activity, it was necessary, for the AHJ, to find the technical corrective measures that the management of the society would have to adopt within a defined target date, in order to avoid the recurrence of the event, finding the direct and indirect causes, together with the factors having influenced negatively the dynamic of the incident (for instance, amplifying its effect).

7.4.3 Why it Happened

The hydrocarbon leak, which originated in the above-mentioned pipeline, formed and extended pool with a length of approx. 60 m with respect to the leak point, also due to the slight slope in that section. A total amount of 830 t of hazardous substances was involved in the accident. The amount of hazardous substances potentially involved is derived from the calculation of the quantities of products contained in the isolated segments of the pipe bundle, and corresponds to the hold-up of the pipe bundle, which corresponds to approx. 2200 m^3. The estimation in tons is complicated by the different densities of the products [1].

The causes of the accident are of technical nature: concerning the causes of the leak, it has been determined that the pipe was perforated due to corrosion processes which occurred externally on the pipe surface [1].

In particular, the incident report states that it is likely that the localisation of the fissure, with respect to the point where it formed, is linked to one or more of the following factors:

1. Localised damage in the original pipe coating;
2. material defect in the original pipe coating; and
3. critical operative conditions (of the pipe section in which the fissure occurred) linked to the placement of the pipe near the ground and its exposition to atmospheric events (sea air).

The company declares that the pipe was periodically inspected; the last inspection of the pipeline had been performed in February 2005.

It is not possible to affirm that the maintenance of the pipe in question was insufficient, but it is pointed out that the pipeline examined had been built more than 40 years ago, and was bought from another company in 2002, that did not provide the technical documentation on maintenance operations on the piping bundle prior to the sale. This circumstance has not allowed, according to the technical consultancy report, to verify the compliance with the technical norms on visual tests, inspections and maintenance of the piping system concerned.

Concerning the cause of the fire, the company made the supposition that the contact between the vapours formed from the spilled hydrocarbons with hot spots of high-pressure steam pipes, reaching up to 280° steam temperature, in correspondence of the subway, may have contributed to the expansion of a vapour cloud; the vapour cloud ignited on an ignition source downhill from the subway were the first fire was detected.

Concerning the successive BLEVEs, which caused major damage to the persons involved in the emergency response operations, these are related directly to the permanent heat irradiation of the fire (which lasted approx. 48 hours), which overheated other pipelines containing hydrocarbons also in gas form in the pipe string.

Concerning the direct involvement of the company, the same company states that the event may have had less serious consequences if the onsite fire brigade would have been alerted immediately at leak detection and not three hours later, as effectively happened. This assumption is confirmed by the technical consultancy report, which affirms that if the onsite fire brigade would have been alerted more rapidly the damages caused by the fire could have been contained.

The Human error indicated relates to the faulty application of the Emergency procedures foreseen in the onsite Emergency Plan. The fault during the Repair operation (dismantling of the pipe insulation layer) relates to the failure to adopt appropriate safety measures during the operation. The design fault relates to the inadequate design and rationalisation of the pipe bundle.

The analysis carried out by the national fire brigade, who followed a structured investigative approach, as claimed by international guidelines for process industries, like CCPS and NFPA 921, found the following shortcomings:

1. The lacking division by category of piping according to the hazardous substance. All the pipes crossed the subway, with no attention to eventual incompatibility among the different hazardous processed substances, especially in case of incident;
2. insufficient number of hydrants and monitors, especially in the pipe-way. the pipe-way, during the fire, was protected insufficiently against fire, for hydraulic

performances and number of hydrants and monitors, also because of the serious damages experienced by the main firefighting pipes, which were in direct contact with the pipes dragged into the event;
3. impracticability of the area in which the event occurred. the area directly interested by the fire does not allow the fire brigade to come closer and fight the fire from a favourable position, because of the high density of pipes in the pipe-way respect to the available surface;
4. difficulty to find promptly the pipes containing hazardous substances. the pipes were not tagged such that the type of substance and the operating conditions were known;
5. out of reach fire hotbeds, with the normal firefighting systems;
6. vast and uncontrollable fire areas, because of the lacking of curbs and any containment systems. both the pipe-way and the subway were not equipped with containment systems. this favoured the extension of the fire, drastically reducing the extinguishing power of the foam, under the runoff action of water;
7. impossibility to preserve the foam bed, running off downwards, inside the trench;
8. high hold-up, because of the rare block valves. the major parts of the pipes crossing the subway and the pipe-way were equipped with block valves placed at the far end of the pipe, very often longer than 1000 m. this implied a remarkable volume of substance inside the pipe, especially for those with diameter larger than dn 250;
9. overpressure under the effect of the heat in the parts of intercepted piping, and their subsequent rupture. the pipes, engulfed in the fire and containing volatile products (lpg, virgin nafta, and so on), once intercepted, experienced an increase in pressure and volume of the substance, but they were not equipped with trv (thermal relief valve), which would have vented the overpressure. this caused a series of mechanical collapses of the pipes, because of bleves and the subsequent fireball; and
10. the aerial crossing of the firefighting pipe was damaged by the fire.

7.4.4 Findings

The MARS accident report identifies the following causal factors [1]:

1. Organizational
   1. training/instruction;
   2. design of plant/equipment/system;
   3. maintenance/repair;
2. plant/equipment;
   1. corrosion/fatigue;
3. human;
   1. operator error.

For a better understanding of the global strategy provided to increase the fire safety level, an extremely intuitive qualitative representation is shown in Figure 7.4.6, based on the conceptual tree provided by the standard NFPA 550 [2], already discussed in Paragraph 3.7. Starting from an incidental scenario, this representation allows identifying those critical aspects particularly significant for a fire and the compensative areas that can be hypothesised consequentially in order to reduce the magnitude or the probability.

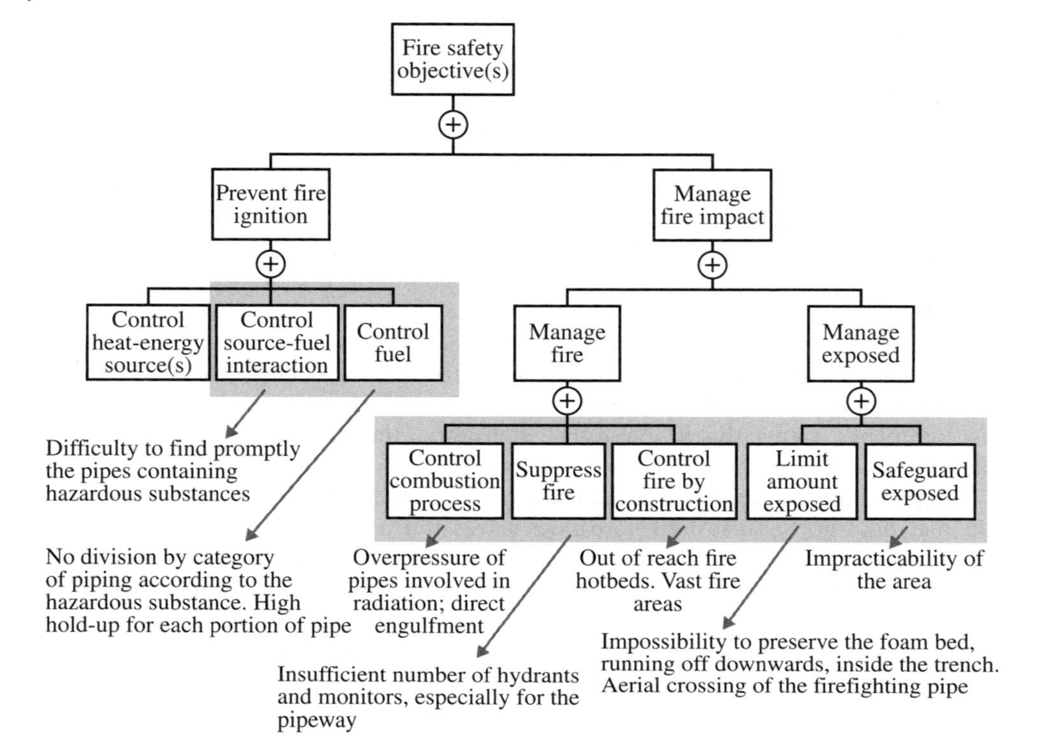

Figure 7.4.6 Graphical visualization of the found shortcomings. Source: Adapted from [2].

The comparison of the tree before (Figure 7.4.6) and after (Figure 7.4.7) the incident allows to immediately visualise the application of the method to the peculiar situation. In particular, Figure 7.4.6 shows the found shortcomings while Figure 7.4.7 shows the corrective measures developed after the incident analysis. From the graphical representation, it emerges that the identified strategy against fire promotes a global action of mitigation and that the different actions are effective for more than a single critical aspect. Indeed, it is highlighted how the incident has origin from a series of shortcomings (both in the design and management phase) that can be addressed to the major parts of the aspects contributing to a fire.

On the base of a preliminary identification of the fire strategy, a specific risk analysis (QRA) has been carried out in order to:

1. Verify the congruency of the outcomes of the analysis respect to what really happened;
2. measure the magnitude of the incident;
3. determine the risk reduction for each preventive or mitigating recommendation; and
4. verify that the proposed strategy would have met the objective (an accepted level of risk).

Among the immediate actions taken by the companies of the petrol-chemical pole, there is a census of the pipes along the entire industrial pole, allowing to identify the

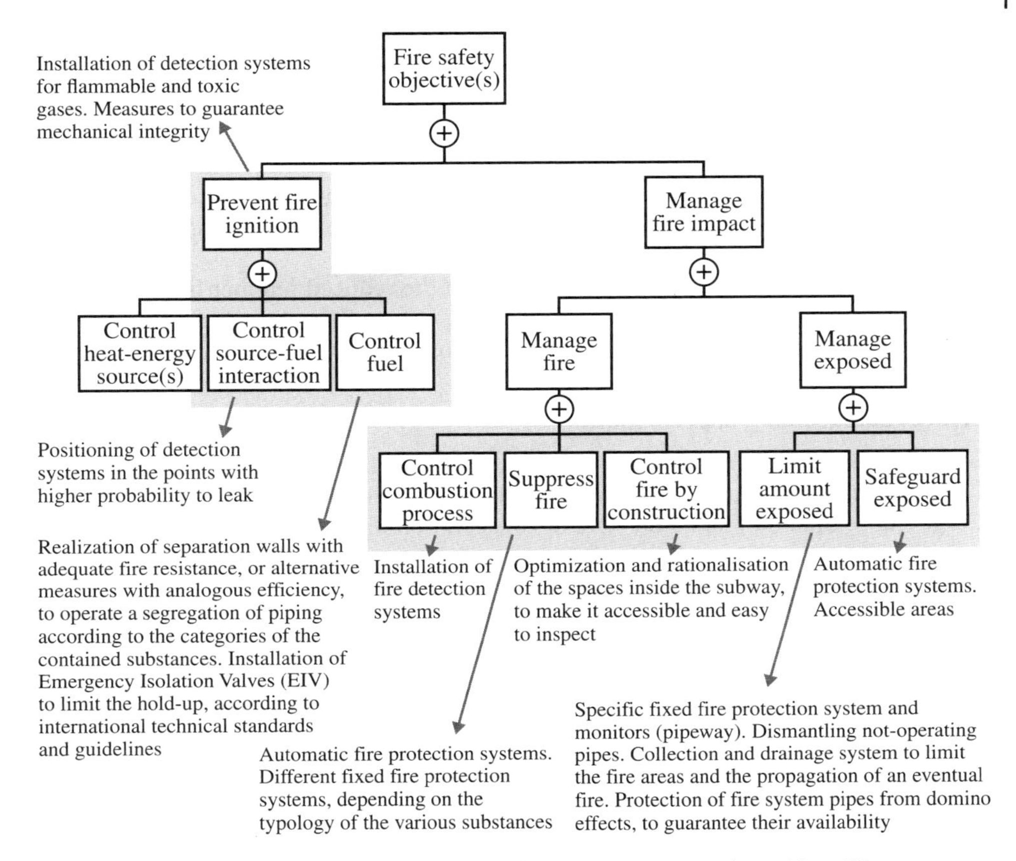

Figure 7.4.7 Graphical visualization of the defined fire strategy. Source: Adapted from [2].

critical nodes (crossing sections) and, consequentially, the relative safety measures. The performed activity allowed to increase the safety level, playing on well-defined, understandable and measurable criteria, avoiding useless cost related to a "tout-court" enhancing safety approach.

7.4.5 Lessons Learned and Recommendations

According to the MARS accident report, the theme of the lessons learned concerns both the causes (Plant/Equipment and Organisational) and the emergency response. This part is taken from [1].

Immediate lessons: critical equipment aspects encountered onsite the establishment immediately after the accident:

1. The pipe bundle disposition inside the trench and also inside the subway was such to cause extreme difficulties during the emergency response operations;
2. there was a large number of closely packed pipes, containing different kinds of liquids (and in particular hazardous ones), which were not immediately identifiable and traceable;

3. the presence and vicinity in the same bundle of pipes containing hazardous substances and pipes with high pressure/temperature steam;
4. mix up of the pipe layout of utility pipes (and in particular the extinguishing water pipe) with the product transfer pipes, this circumstance has caused the overpressure in the isolated pipe tract exposed to the heat irradiation from the fire and the consequent explosion of the pipes and interrupted the extinguishing water supply from the pipe inside the bundle and damaged also another suspended extinguishing water pipe crossing the pipe bundle;
5. high hold-up levels of the tube bundle, due to the reduced number isolating valves on the pipelines; and
6. inadequate gradient arrangement, absence of retaining curbs and of adequate water drainage systems.

Critical management aspects found:

1. Incomplete compliance with the onsite emergency plan: the onsite fire-brigade was alerted only three hours after the crude oil leak detection;
2. age (more than 25 years) and state of conservation of the pipe triggering the event, with respect to progressive corrosion processes encountered, which finally caused the leak;
3. the company stated that periodical inspections had been performed according to the technical "piping inspection code" api 570 on the event triggering pipe; and
4. these periodical inspections were part of the establishment's maintenance program, according to which the last inspection on the above-mentioned pipe had been performed in february 2005. following the accident the company asked a specialist company to assess the inspection methodology employed by the company and to propose possible improvements.

The company has realised the total reconstruction of the pipe bundle over a length of 300 m, according to a project integrating the requests formulated by the Technical Regional Committee of Sicily and additional considerations formulated by the company.

The reconstruction and modification project was divided into several phases:

1. Phase 1: reactivation of the product loading - unloading wharf;
2. Phase 2: startup of the pipelines and plant parts which were not damaged or functionally affected by the fire; and
3. Phase 3: restoration and modification of the damaged pipes.

The reconstruction project of the pipe bundle included in particular:

1. Rationalizing the piping layout, according to risk categories, inside of dedicated trench ducts;
2. division of the subway in segments with fire-resistant segmentation plates such to separate the pipes;
3. according to the substance categories transported;
4. passive fire protection of the steel structures supporting the pipe bundles inside the subway;
5. illumination, accessibility and inspectability of the subway;
6. installation of: adequate retaining curbs inside the trench such to limit the extension of fire inside the trench, paving and drainage systems connected with the establishment sewage system;

7. increasing the number of extinguishing monitors, assuring the protection and coverage of the entire trench;
8. installation of further emergency isolation valves (eiv) on all critical pipelines, the signals of the valves are transmitted to the control room and to the safety room of the area; and
9. installation of thermal reaction valves (trv) on all isolatable pipeline sections.

The company evaluates to extend the above mentioned technical improvements to the other subways of the refinery, depending on other factors from the results of the evaluation by the Technical Regional Committee.

Corrective actions in the management:

1. On June 21, 2006 a meeting of the safety and health committee was organised, with the goal to analyse the event and point out the lessons learnt and the measures to be taken.
2. increase the frequency of training courses for the personnel by internal training personnel;
3. foresee specific updating courses for the training personnel, and specific training courses for the department foremen and the management of the company;
4. improve the circulation of the contents of the safety report such to improve the safety consciousness of the personnel;
5. identification by the department foremen of sensitive aspects needing improvement in the safety management system (sms), e.g. incompleteness or excessive complexity. of the procedures foreseen by the sms.
6. identification of behavioural dynamics in the personnel indicating reduced attention to safety and setting up a safety prise; and
7. increase the frequency of training courses for the personnel in particular concerning the respect of emergency response procedures, following the cascade principle.

Measures to prevent recurrence:

The reconstruction project of the pipe bundle included in particular:

1. Rationalizing the piping layout, according to risk categories, inside of dedicated trench ducts;
2. illumination, accessibility and inspectability of the subway;
3. installation of flammable gas detectors; and
4. installation of thermal reaction valves (TRV) on all pipeline tracts which can be isolated.

Measures to mitigate consequences:

1. Passive fire protection of the steel structures supporting the pipe bundles inside the subway;
2. installation of adequate retaining curbs inside the trench such to limit the extension of fire;
3. installation of diversified fire extinguishing systems for each trench-duct section inside the trench, and drainage systems connected with the establishment sewage system;
4. increasing the number of extinguishing monitors, assuring the protection and coverage of the entire trench;

5. installation of further emergency isolation valves (eiv) on all critical pipelines, the signals of the valves are transmitted to the control room and to the safety room of the department; and
6. installation of thermal reaction valves (trv) on all isolatable pipeline sections.

7.4.6 Forensic Engineering Highlights

The 2005 safety Report foresaw an accident scenario similar to the one occurred, i.e. the rupture of a gasoline and/or LPG pipeline forming part of a pipe bundle inside a trench in-between a subway and the railway. For the rupture of a gasoline pipe the extension of the damaged area had been estimated in maximally 30m, corresponding to a heat irradiation of $3kw/m^2$ (reversible damage area), involving either road ex SS114 and the railway depending from the source location. Therefore the effects of the heat irradiation foreseen in the accident scenario of the Safety Report are compatible with the effects of the accident which occurred really. The management faults occurred during the event, i.e. the faulty application of the emergency response procedures foreseen in the onsite emergency response plan, points out the need to improve the training of the personnel, with a particular focus on emergency management.

One of the most interesting highlights is the application of the risk analysis as a tool to find the proper safety criteria. Indeed, a specific risk analysis was carried out before the complete reconstruction and adjustment of the pipe-way, uphill and downhill the subway. The realization of the new works was preceded by a safety study conducted in the form of safety report. The working group, together with the technicians and the consultants of the company, developed the criteria to adjust the subway and the pipe-way.

Developed criteria for the subway adjustment

To write the subway adjustment project, the following technical criteria have been considered:

1. Realization of separation walls, with adequate fire resistance or alternative measures with analogous efficiency, in order to realise a sectorialization of the pipes depending on the transported substances;
2. installation of detection systems (flammable gas, toxic gas, heat), depending on the substances inside the pertinent sectors, activated by reaching a pre-alarm threshold and an alarm threshold, subdued to the firefighting/cooling systems protecting the single sectors;
3. installation of extinguishment/cooling fixed firefighting systems;
4. optimization and rationalization of the spaces inside the subway, in order to make it accessible and easy to inspect, in addition to being well-lighted;
5. guarantee the mechanical integrity of the lines (design conditions must include the most severe operative condition and the TRVs); and
6. add the emergency isolation valves (eivs) in the pipes uniquely crossed by hazardous fluid, upwards and downwards the subway, according to the best practices internationally acknowledged.

Developed criteria for the pipe-way adjustment

1. All the pipes inside the pipe-way that are no longer used have been removed. Along the pipe-way, the in-service pipes have been placed respecting the segregation criteria inside the subway; and
2. a drainage system to collect running off liquids has been planned over the entire area of the pipe-way, immediately upwards and downwards the subway. The drainage system is used to isolate the subway from possible releases along the near trench and to limit the extension of the fire areas. The system to collect the fluids have been realised by sumps and an underground manifold.

Protection of the fire system pipes
The fire system pipes have been protected by the fire, where possible, Where the pipes cross the fire areas or are next to them, they have been buried or equipped with proper protection.

The layout
The risk analysis identified the safety criteria to adopt at the reconstruction stage. This step lasted almost one year and led to a deeply modified layout, as shown in Figure 7.4.8, where the actual rationalization of the piping and the enhanced safety measures can be observed.

7.4.7 References and Further Readings

References

1 EUROPA - eMARS Accident Details - European Commission [Internet]. Minerva.jrc.ec.europa.eu. [cited 15 November 2017]. Available from: https://minerva.jrc.ec.europa.eu/en/emars/accident/view/5a988875-96e2-2e88-5700-099e5b52f5bc

2 Fiorentini, L., Rossini, V., and Tafaro, S. (2012) L'incendio di una pipe-way di raffineria: l'indagine di un incidente industriale rilevante per il miglioramento della sicurezza. IF CRASC 12. Pisa: Dario Flaccovio Editore.

Further Readings

CCPS (Center for Chemical Process Safety). (2003) *Guidelines for investigating chemical process incidents*. 2nd ed. New York: American Institute of Chemical Engineers; 2003.

Dattilo, F., Puccia, V., Fiorentini, L., Rossini V, Tafaro S et al. (2008) L'applicazione alla "Fire Safety Engineering" di strumenti dell'analisi di rischio per aumentare l'efficienza dello studio e l'ottimizzazione del livello degli interventi sul progetto antincendio. Atti Convegno Valutazione Gestione Rischio. Pisa; 2008.

Fiorentini, L., and Marmo, L. (2011) *La valutazione dei rischi di incendio*. Roma: EPC Editore; 2011.

NFPA (National Fire Protection Association). (2017) NFPA 550: Guide to the Fire Safety Concepts Tree. NFPA (National Fire Protection Association); 2017.

NFPA (National Fire Protection Association). (2017) NFPA 921: Guide for Fire and Explosion Investigations. NFPA (National Fire Protection Association); 2017.

Tafaro, S. (2011) *Un caso studio per la valutazione del rischio industriale* – Incendio in una pipe-way di raffineria. Roma: Corpo nazionale dei Vigili del Fuoco – Direzione Centrale per la Formazione; 2011.

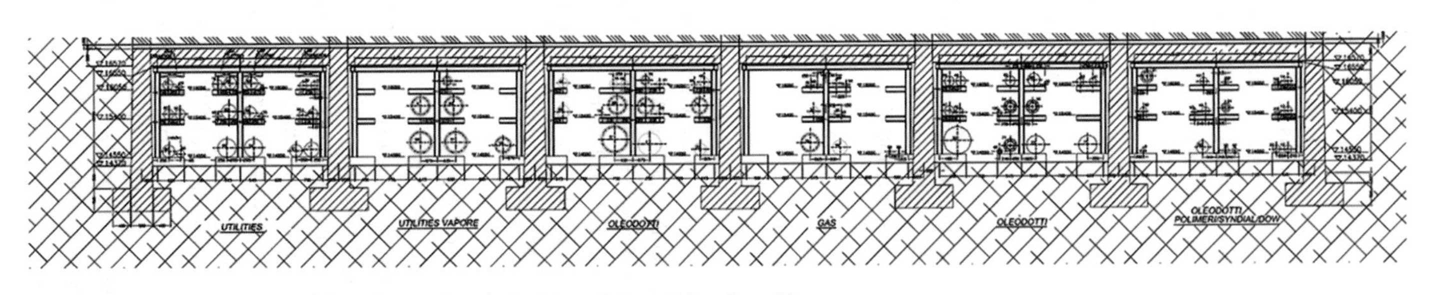

Figure 7.4.8 Transversal section of the subway after the incident. Source: Taken from [2].

7.5 Flash Fire at a Lime Furnace Fuel Storage Silo

7.5.1 Introduction

The general information about the case study are shown in Table 7.5.1.

Lime furnaces are historically widely diffused throughout the world, in a large variety of different technological level configurations. The lime production is an historical building technology achievement for masonry, nowadays a set of technology improvements and environmental requirements evolved an once artisan production in a full industrial process.

The chemistry of the process is essentially related to the heating at high temperature (900 °C) of calcareous rocks as illustrated in Table 7.5.2:

The process is performed in a current of air inside a special type of kiln to obtain a decarbonation reaction, with the well-known set of reactions:

$$\mathbf{CaCO_3 + Heat \leftrightarrow CaO + CO_2}\ (\mathbf{+760}\ \boldsymbol{kcal/kg}) \tag{7.5.1}$$

$$\mathbf{MgCO_3 + Heat \leftrightarrow MgO + CO_2}\ (\mathbf{+723}\ \boldsymbol{kcal/kg}) \tag{7.5.2}$$

$$\mathbf{CaCO_3\ MgCO_3 + Heat \leftrightarrow CaO\ MgO + CO_2}\ (\mathbf{+723}\ \boldsymbol{kcal/kg}) \tag{7.5.3}$$

They represent the reaction that, industrially, occurs in the limestone burning kiln when starting from the mineral coming from the quarry and supplying a prefixed heat quantity. The dissociation of carbonate ($CaCO_3$) in an oxide (CaO) and in carbon dioxide (CO_2) is obtained.

The reaction is endothermic and heat is supplied by the use of a fuel. They are necessary 760 kcal/kg to decarbonate the $CaCO_3$ and 723 kcal/kg for the $MgCO_3$.

This heavy energy supply was historically obtained in different ways, but all involving combustion processes at various technology levels. In our case study, probably to comply

Table 7.5.1 General information about the case study.

Who	Lime Furnace Fuel Storage
What	Flash Fire
When	2014
Where	Province of Padova, Italy
Consequences	1 victim, 1 badly burned
Mission statement	Determine the fire dynamics
Credits	Vincenzo Puccia (Italian National Fire Brigade)

Table 7.5.2 Chemical substances involved.

Calcium Carbonate	$CaCO_3$
Magnesium carbonate	$MgCO_3$
Dolomite	$CaCO_3\ MgCO_3$
Hydraulic Limestone	$CaCO_3\ MgCO_3SiO_2\ Fe2O_3\ Al_2O_3$

with environmental and budgetary constraints, the furnace was fueled with pulverised sawdust in a fluidised bed combustion unit, stored in some silo of capacity from 300 to 600 cubic meters, equipped with filters to reduce the dust emission to atmosphere.

The fire risk had been actually considered in the plant design, indeed the silos were equipped with various devices, as an external cooling system, an internal water sprinkler system with open heads placed close to the silos ceiling, and lastly a carbon dioxide reserve, to be injected on the tip, just over the ceiling and under the bag filter house. It's important evaluate the specification of each protection plant, as well as the scenario of design and the fire/explosion suppression aim.

Carbon dioxide may be suitable for extinguishment of smoldering silo as proper saturation to the desired level for a long time is guarantee. To this purpose, the silo should be almost sealed to prevent CO_2 to escape or oxygen to enter.

The internal sprinkler system, furthermore, will not so simply suppress a smoldering fire of sawdust, as effect of water channeling phenomena in the bulk of the combustible dust, impacting the real wet surface.

As result of all above issues, the fire suppression inside a huge silo of pulverised combustible dust is a very challenging goal, particularly in a short time scale, with the consequence that the emergency emptying attempt could result in a very risky range of accident scenarios, including internal explosion of saw dust mixed with air over the lower explosion limit concentration, internal explosion of carbon monoxide resulting from Boudoir equilibrium as effect of the internal smoldering combustion with insufficient oxygen, collapse of the silo as effect of the explosion, external confined dust explosion, external flash fire of released burning dust etc.

The accident literature on dust silo is however quite rich, but, on the other hand, the awareness of the danger associated with this storage technology, unfortunately, isn't still as diffused between operators and workers as it's between risk analysts and academia.

7.5.2 How it Happened (Incident Dynamics)

The accident involved the miller and storage unit, (Figure 7.5.1) as the sawdust was supplied from wood waste of other companies, but to meet the burner specific it must be finely grind, to pulverised dimension. Also, the sawdust was including, inevitably, a metallic residual (e.g. nails, screws etc.) to be separated before the miller unit, by a screening unit.

Just a malfunction or an error of the screening unit sent some metallic fragments to the miller, with result of sparks and the ignition of sawdust in the in conveyance and silos discharge. Although the spark detection system detected the events, closing the charging system, the feared scenario really happened. A smoldering combustion was detected inside the silo.

At this point, various attempts started, to stop the fire spread on the pulverised saw dust bulk, both venting with the carbon dioxide reserve and, probably, opening the water sprinkler deluge system inside the silo, but without immediate result.

The decision to operate an emergency emptying is still under inqury, at the date of this publication, but we are more interested on the scientific evaluation on the event, so the chain of orders leading to this decision is overall the goal of our analysis in this chapter.

The available literature related to this kind of fire reports of fire events for as long as weeks, with various suppression stategies undertaken, from an oxygen depletion

Figure 7.5.1 Area involved in the accident.

estinguishment, as effect of the combustion to inertization by mean of nitrogen, vented in large quantities, or to water flooding attemps. Each of those solutions is not completely free from risks of escalation, just considering the collapse of the silo as consequence of sudden fractures caused by the material behavior under the nihil ductility temperature after liquid refrigerated nitrogen feed into the bulk, or as conquence of the internal water pressure in a total flooding approach, over the design specific. Also the wet expansion of the contained burning product could be an issue of concern for the shell elements.

As a result, to resolve the "problem", after any other solution worked with the time expectation on this kind of event, the company decided on emptying by the bottom opening, behind the crawl space. This attempt totally undervalued the physical behavior of the pulverised sawdust, in terms of the mechanics of dusts, packing, and kinetic energy of release of packed fractions of the bulk of the silo.

Precisely few hours after the start of discharge operations, with few people involved and without protective clothing nor a water dedusting fire hose placed nearby the area, a massive drop of pulverised sawdust from a highter position in the silo originated an external dust cloud, as the result of the kinetic energy at the discharge hole. (Figure 7.5.2).

The cloud suddenly ignited in a flash fire on the area in front of the silos (Figure 7.5.3), severely burning the two workers, one of whom died few days after in consequence of burns, and igniting fires in one of the two operating machines used to move the sawdust. The footprint of the flash fire was clearly evident in nearby shed.

The evidence of a smoldering combustion inside the silo tank is given from burning areas in the ejected saw dust, see Figure 7.5.4, as well as from the data collected by firefighters, with high carbon monoxide value close to the discharge hole.

The dimension and the impact of the flash fire on the concrete wall of the warehouse was clearly visible.(Figures 7.5.5 to 7.5.6).

Figure 7.5.2 The bottom crawl space, with a discrete part of the sawdust bulk collapsed, generating a dust cloud ignited probably from a pool of burning sawdust inside the silo. The water is spayed by fire service after the flash fire event.

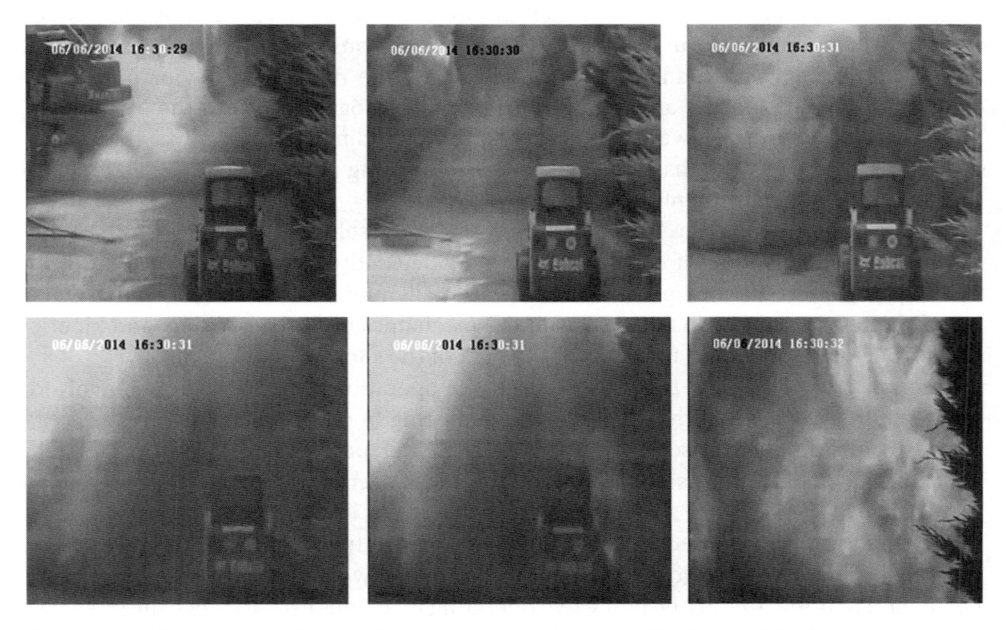

Figure 7.5.3 The sequence of the underestimated and unespected hight speed discharge event, generating the saw-dust cloud, with the flash fire ignited in the last image.

Figure 7.5.4 The smouldering combustion in the saw dust discharged by the silo, in the occurrence of the event.

Figure 7.5.5 Footprint of the flash fire on the front wall of the shed in front of the discharge hole.

7.5.3 Why it Happened

The analysis of the accident is strongly helped by the video sequence from the company video recording system, acquired by the fire service law enforcement office, immediately after the rescue operation. Furthermore, that kind of event isn't unknown, as various previous accidents, some involving firefighters in various countries, are reported in the specialised literature.

The root analysis of the accident should start at the control of the process, starting with the saw-dust screening and after the miller, the transfer system and the storage inside a silo (Figure 7.5.7).

Also, an insight on the origin of the accident chain is clearly related to the safety devices, the carbon dioxide manual system and the water sprinkler inside the silo, and to the awareness of the risk related to saw-dust milling and storage, with regard of the accident scenarios.

The first failure in the events chain leading up to the final flash fire must be clearly indicated in the iron fragment screening system, based on sieve technology. The second

Figure 7.5.6 The development of the flash fire could be deducted by the burned trees. The parked bobcat resulted in being ignited.

Figure 7.5.7 The silo with the baghouse filter at its top. See the vents.

failure in the events chain is the ineffective intervention of the fire shutter, between the miller and the silo feed system. In fact, the sparks were, probably, correctly detected, but the characteristics of an extreme dry and small amount of ignited dust spread smoldering combustion inside the storage.

At this point, anything but an active protection system could be able to stop the fire, obviously with an impact on the plant operations.

The internal sprinkler, probably also for the short time of use, was not able to suppress the internal fire. indeed, challenges of put out a smoldering silo fire by water are well known. An intensive flood may be more effective , compatibly with structural specific of the silo, and in a time of various days, probably various weeks, including a prolonged out of service related to the emptying of wet saw-dust, reduced at this point to slurry.

The characterization of the saw-dust in term of explosion risk must consider the step process suggested by OSHA. Mainly step 3 *"Once a determination of the dust material stability has been established, then the 3rd step should focus on a determination of the particle size and particle size distribution"* and step 5 *"in the risk assessment should be a determination of whether a portion of the dust is smaller than 500 microns. A careful review of the particle size data, especially the particle size distribution for sampled dusts (see step 3), should be made in order to assess the existence of fines and also to see whether bimodal particle size distributions exist which extend into the particle size range that poses a potential hazard."*

An evaluation in term of Minimum Ignition Energy (MIE)and Minimum Explosive Concentration (MEC) could confirm the dry pulverised saw dust as extremely dangerous with regard to the risk of the formation of a cloud and its ignition resulting a premixed flame, both in the form of a flash fire or, with confinement, of an explosion (Figure 7.5.8).

This case study is of some evidence as the dust cloud was not an effect of a primary explosion, but the energy for the cloud generation was provided by the release of an amount packed high enough in the silo to be released with an amount of kinetic energy.

Another important question could be "What could be happen, is it really the worst scenario?"

The answer is that a worst scenario could be possible, as effect of the instauration on a Boudoir Equilibrium ($2CO \leftrightarrow CO_2 + C$) on lack of oxygen concentration for the combustion, and with enough air entrainment, the mixture of carbon monoxide and air could enter in its flammability/explosivity field. The result could be an internal explosion. If its energy had not been dissipated by emergency venting, the collapse of the silo, and the almost instantaneous release of the dust hold up would cause a secondary explosion.

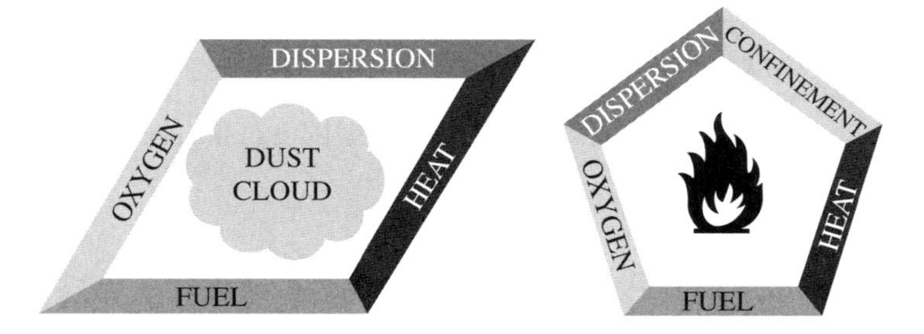

Figure 7.5.8 Elements of a Flash Fire and the Explosion Pentagon.

7.5.4 Findings

The accidents which occurred evidence a scenario of an external flash fire, on an unprotected area with regard to the risks of explosion, as result of a sudden and heavy discharge of pulverised dry combustible dust and the formation of a cloud in consequence only of its kinetic energy at the exit of the silo.

The literature normally refer to this as a secondary effect after a first energy release of a primary dust explosion, but this case study is also of some interest with regard to the design specifics and standards of the active protection system, and with regard to firefighting doctrine for silos with an internal smouldering combustion.

The accident confirms the largely diffused risk underestimation of silo storage of dry pulverised combustible dusts, also with regard to emergency operations and emptying.

7.5.5 Lessons Learned and Recommendations

The main recomandation after this accident is that the the emergency emptying of a huge silo of combustible dust must be the last hope in term of emergency management, and this attempt must be conducted with all available mitigation layers, starting from protective clothings and airbreather with a large use of fire hose, to water dedusting as well as to firefighting and fire radiation mitigation.

Although silo accidents are not a new kind of fire scenario, and the behaviour of the combustible dust cloud is a well known fire and explosion scenario, its underestimation in terms of perceived risk from workers and plant manager is always raising concern between fire service and law enforcement teams.

The research of more effective suppression system inside the confined space of a silo must be conducted together with a better understanding of dust mechanics and dust cloud behaviour.

7.5.6 Forensic Engineering Highlights

The investigation was strongly helped by the evidence of the video record for the reconstruction of the chain of events leading to the tragedy. Without this evidence, the inquiry approach, and the event dynamics would be only partially known, on the basis of the witness.

The footprint of the cloud flashfire on the warehouse wall in front of the silo door is an evidence of its real scale, as well as, obviously, the burnt tress affected by thermal radiation for more than 12 meters from the release point. Albeit the area results were rather confined, any overpressure was however generated.

Last but not least, the event clearly evidenced a lack on the design of active protection system, mainly on the aim of their design with regard to the actual suppression power, in each accident scenario, and, once again, a deep lack in terms of risk analysis in terms of occupational safety and fire safety, with a heavy consequence on human life and health.

7.5.7 Further Readings

Further readings

Amyotte, P., Khan, F., Demichela, M., and Piccinini, N. (2008) (PSAM 9) N. Paper 0114. Ninth International Probabilistic Safety Assessment and Management Conference. Hong Kong, China.

Amyotte, P.R., Eckhof, R.K. (2010) Dust explosion causation,prevention and mitigation: An overview. *Journal of Chemical Health & Safety,* January/February.
Britton, L. (1999) *Avoiding Static Ignition Hazards in Chemical Operations.* Center for Chemical Process Safety of the American Institute of Chemical Engineers.
CCPS (Center for Chemical Process Safety) (2014) *Guidelines for Determining the Probability of Ignition of a Released Flammable Mass.* American Institute of Chemical Engineers.
CCPS (Center for Chemical Process Safety) (2005) *Guidelines for Safe Handling of Powders and Bulk Solids.* New York: American Institute of Chemical Engineers.
Cheremisinoff, N. (2014) *Dust Explosion and Fire Prevention Handbook.* New York, NY: John Wiley & Sons.
Cote, A. (2008) *Fire Protection Handbook.* Quincy, Mass: National Fire Protection Association.
Eckhoff, R. (2003) *Dust Explosions In The Process Industries.* Amsterdam: Gulf Professional Pub.
Gummer, J. and Lunn, G. (2003) Ignitions of Explosive Dust Clouds by Smouldering and Flaming Agglomerates. *Journal of Loss Prevention in the Process Industries,*16:27–32.
Hedlund, F.H. (2018) Carbon dioxide not suitable for extinguishment of smouldering silo fires:Static electricity may cause silo explosion. *Biomass and Bioenergy,* 108:113–119.
Mannan, S. and Lees, F. (2012) *Lee's loss prevention in the process industries.* 4th ed. Boston: Butterworth-Heinemann.
Mannan, S. (2014) *Lees' process safety essentials.* 1e. Kidlington, Oxford, U.K.: Butterworth-Heinemann.
NFPA (National Fire Protection Association). (2007) NFPA 664: Standard for the Pre NFPA 664: Standard for the Prevention ention of Fires and Explosions in Wood Processing and Woodworking Facilities. *NFPA (National Fire Protection Association).*
NFPA (National Fire Protection Association). (2017) NFPA 921: Guide for Fire and Explosion Investigations. *NFPA (National Fire Protection Association).*
OSHA. (2013) *Combustible dust: firefighting precautions at facilities with combustible dust.* Washington: US Occupational Safety and Health Administration.
OSHA. (2014) *Hazard Alert: Combustible Dust Explosions.* Washington: Occupational Safety and Health Administration.
Woodward, J. (1999) *Estimating the Flammable Mass of a Vapor Cloud.* Center for Chemical Process Safety of the American Institute of Chemical Engineers.

7.6 Explosion of a Rotisserie Van Oven Fueled by an LPG System

7.6.1 Introduction

The general information about the case study is shown in Table 7.6.1.

On March 9, 2013 at around 12.15 p.m., during the weekly market in Piazza della Repubblica at Guastalla (RE), a rotisserie van was impacted by a fire and several explosions (Figure 7.6.1). The event involved two other vans, the stalls of a number of traders and several cars parked near the square, causing the death of three persons, relatives of the owner of the van, trapped inside the vehicle, as well as several injuries among the patrons of the market, including the owner himself.

Table 7.6.1 General information about the case study.

Who	Rotisserie van oven power LPG system
What	Explosion (BLEVE [boiling liquid expanding vapour explosion] and fireball)
When	March 9th, 2013
Where	Italy
Consequences	3 victims and 11 injured
Mission statement	Determine the causes of the explosion
Credits	Luca Marmo (Politecnico di Torino)

Figure 7.6.1 The van after the accident.

7.6.2 How it Happened (Incident Dynamics)

The police reported an explosion of the van which occurred at around the time of 12.15 p.m. on March 9, 2013 where three people lost their lives, while the owner of the vehicle was hospitalised for bursts at the Maggiore di Parma hospital. The police reported that during the same episode 10 people suffered burns, all fairly serious, in particular to the face and hands.

The police referred to SIT [Italian body assigned to collect witness summary information] numerous persons able to report on the event as well as acquiring video files from various Internet sources or provided spontaneously by witnesses and relating to the moments following the event.

From statements collected, it appeared that an extensive flash fire developed which mainly affected the van, on the sales counter side, followed by two-three explosions.

The provincial Fire Brigade of Reggio Emilia and the Guastalla police who were immediately on the scene secured the area, ascertained the extent of the damage, identified the various elements of evidence and reconstructed the course of events. The bodies were subjected to medical-legal investigations to ascertain the cause of death. Following the investigation the police reconstructed the facts according to a hypothesis that involved four phases as follows:

1. Phase 1: structural failure of one of the components of the LPG systems fuelling the oven which caused significant leaking of LPG;
2. Phase 2: the LPG was ignited, possibly in contact with hotspot in the oven, causing a flash fire followed by a local fire;
3. Phase 3: the fire engulfed the LPG bottles causing further gas leaks; and
4. Phase 4: the fire caused the explosion of a fourth bottle, which projected fragments up to 17 m away.

The provincial Fire Brigade of Reggio Emilia reported that at approximately 12.15 on 09/03/2013, the local teams intervened on the scene of the accident, to extinguish a fire following a gas leak and subsequent exploding of LPG cylinders inside a rotisserie van. Upon arrival on site, the situation described was one of a large column of thick, black smoke wafting into the sky and the rotisserie van in question completely engulfed in flames with sudden blazes below it. Two vehicles parked on via Pisacane, on the left side of the van (on the side opposite to the carriageway) had burnt parts while on the opposite side of the van, towards the centre of the square, the goods displayed on some of the market stalls were alight. One of two vans parked near these counters had their engine compartment on fire. The intervention involved the rotisserie van, in order to verify whether there were any persons still alive inside, and at the same time to contain the flames, and allowed encircling of the blaze that was affecting the merchandise on the counters and which was proceeding towards the centre of the square.

The event resulted in the total destruction of the rotisserie van, damage from irradiation on one side of the van body and inside the two cars parked on via Pisacane at number 18; irradiation damage to the front of another van located on the side of the sales counter selling household items; at the back of another van (located in front of the previous one and to the rear and to the interior of another van (initially located to the side of the burned van and then moved to safety by the owner). There was damage to the façade and to the fixtures of the first floor of via Pisacane, number 18 and of number 22.

The firemen also collected evidences at the fire scene with the aim to describe the structure of the van facilities, included the LPG plant which immediately resulted fuelled by 4 lpg bottles having the capacity of 15 kg each. The pressure reducer that should equip these plants was not found. The investigation also demonstrated that the van owner used to refuel the 4 LPG bottles at a car gas station (prohibited in Italy) with no control on the amount of LPG refilled. One LPG bottle blew up catastrophically. the shell broke into three main fragments. One landed 17 meters away from the van, the other two (the shell and the bottom) were found close to the van. Two other bottles underwent a partial rupture of the shell. The fourth was find unbroken.

On March 11, 2013 a report was drawn up of the events of March 9, 2013 at the trading area of the Municipality of Guastalla in combination with the Operating Core of the Gustalla police. Accompanying the report were photographic examinations of the floor plan of this area (from Figures 7.6.2 to 7.6.4).

Following the investigation, the Fire Brigade reconstructed the facts as follows.

The morning of the event the LPG fuelled cooking system had been in use for hours. The 4 LPG cylinders thus contained a different fuel level (see Figure 7.6.5) in relation

Figure 7.6.2 Gas cylinders removed as exhibits.

Figure 7.6.3 Valve P.R. TA - W brev. DN 1/4".

to the quantity already supplied to fuel the burner "*therefore two of the three had not yet been involved in the provision and, considering the refill system, contained liquid gas exceeding 80%, probably filling the volume of the ogive*".[1]

The heating device of the LPG cylinders which conveyed warm air from the oven to the bottle closet heat up the LPG with progressive dilatation the liquid that, finding no headspace in the overfilled bottle (for this purpose the reference standard defines the maximum filling ratios, identified as mass of LPG per cylinder capacity unit), to guarantee safe storage at ambient temperature and at higher temperatures, resulted in "*an unbearable mechanical stress*" to the structure of the containers.

1 This condition is known as "*overfilling*" and it is a common cause of LPG bottles catastrophic explosion.

Figure 7.6.4 Copper pipe and fittings found on the ground behind the van.

The overfilled bottle ruptured at the bottom welding, resulting in instant release and evaporation of the liquid LPG with isoenthalpic flash and forming a pool of supercooled gas fraction inside the bottle housing, from which there was further evaporation. The release produced an explosive phenomenon, known as BLEVE, resulting in a shock wave that spread through the passenger compartment of the van *"violently pushing the three women present towards the rear of the driving compartment and the* owner *outwardly in the opposite direction"*. The gas released formed an evaporative cloud that expanded rapidly, spreading through the wide opening of the sales counter, in the direction of the front area of the market square and spreading even into the van, where was ignited by the naked flame present in the oven.

The ignition generated a fire-ball and put the van on fire, involving all the fuel elements therein present. The persistence of high temperatures then triggered *"a domino effect on the other gas receptacles with inevitable tragic repetition of the dynamics"*. There then followed the *"outbreak (or physical explosion)"* of the cylinder, serial number 58137, due to dilation of the liquid phase *"which was also filled beyond the permitted limits"*, with breakage of the same into three large fragments.

The breech, as a reaction, unhinged the metal frame of the underlying compressor; the plating was hurled onto the right side of the square against the flower stall (17 meters from the vehicle), passing through the passenger compartment and breaking the frame metal elements (found approximately 40 meters from the van on the same trajectory); the ogive, as a reaction, shot towards the vehicle roof then fell into the neighbouring area to the rear of the vehicle.

The overpressure wave generated following the explosion destroyed some structural parts of the vehicle, projecting them against the building behind including its roof. It is possible that one of the fragments had violently struck the nearby gas bottle, serial number 17519, causing, at the same time, tearing of the plating with a release of the liquid contained therein, as well as causing its fall to the ground to the rear of the van where the liquid phase of the LPG produced two jet fires, that burned the outside part and the area below the vehicle, turning, as a reaction, the container itself. Only one

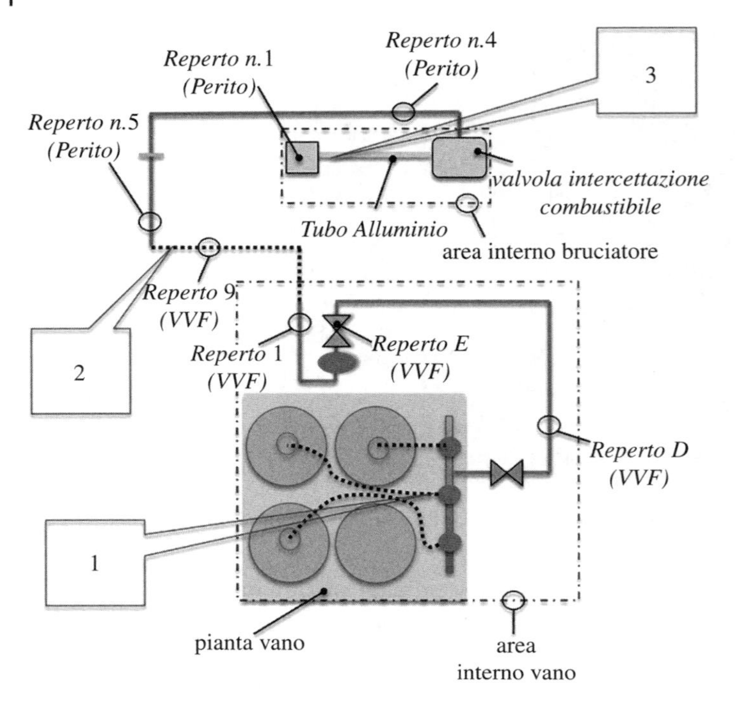

Figure 7.6.5 LPG system diagram indicating the 3 points of possible catastrophic rupture hypothesised during simulations.

bottle, serial number 58137, remained unscathed and was found *"in a horizontal position in the central compartment of the vehicle". This is due to the fact that the same was not connected to the ramp and also because within it the gas was almost finished.*

The reader should keep in mind that the accident dynamic described above (taken from the Fire Brigade report) was not confirmed by the accident investigation team hired by the Judge during the trial. The team came came to different conclusions as described in the following paragraphs.

Figures 7.6.6 to 7.6.9 show the four cylinders taken as exhibits and their labelling.

7.6.3 Why it Happened

The court hired one of the Authors of the book as technical expert to clarify causes and dynamics of the accident. Further investigations were made and more evidences gathered. The evidence which was gathered, the examinations conducted, and the assessments together generate the following conclusions:

It can be stated with reasonable confidence that the starting event was a catastrophic rupture of an LPG supply line to the oven, which caused the release of a substantial quantity of LPG within an extremely short space of time, counted as a few seconds.

This scenario will be defined "scenario a)" (paragraph 7.6.6). It must be considered that at the time of the release the system was likely being operated in such a way as to fuel the LPG oven from a single bottle in liquid form. The bottle in use must therefore have been laid flat or upside down.

Figure 7.6.6 Cylinder A with details of the Fire Brigade labelling, top photo, and of the Expert, photo below.

It cannot be excluded that the initiator event was the mechanical failure of a cylinder resulting from excessive and/or incorrect filling. Two cylinders suffered rupturing of the outer casing (Cylinder A and Cylinder B) but with no BLEVE. The tests conducted on the outer casing of these cylinders, to the extent permissible given the state of preservation, showed evidence of ductile fracture at the bottom of the cylindrical part of the container, essentially indicating that the outer casing ruptured as a result of the combined effect of increased temperature, increased inner pressure and decrease of the mechanical resistance of the material. Bottle A showed signs of violent impact that may have been caused by missiles launched by the explosion of other bottles.

Cylinder D was catastrophically damaged with BLEVE evidences. It can also be inferred that:

1. Cylinder D underwent BLEVE following overheating with nucleation temperature having been reached. This occurred following exposure to fire for several minutes. Moreover, the BLEVE was the last explosion visible in film clips of the event, the most violent of those observed. It happened once the fire had been blazing for some time, to the point that it is accepted that cylinder D was the last to explode; and
2. ruptures of the outer casing of cylinders A and B are typical of a cylinder exposed to radiant heat. This type of rupture takes place where the shell is not wet by the liquid, which keeps the temperature low. Both bottles showed signs of rupturing of the lower part of the cilindrical shell. This is explained, for cylinder B, admitting that it contained a modest quantity of liquid (i.e. a low level that left the fracture zone uncovered) or that it was upside down. Similar considerations can be made for cylinder A. Some CTP have put forward the thesis that cylinder A was on the ground outside of

Figure 7.6.7 Cylinder B with details of the Fire Brigade labelling, top photo, and of the Expert, photo below.

Figure 7.6.8 Cylinder C with details of the Fire Brigade labelling, top photo, and of the Expert, photo below.

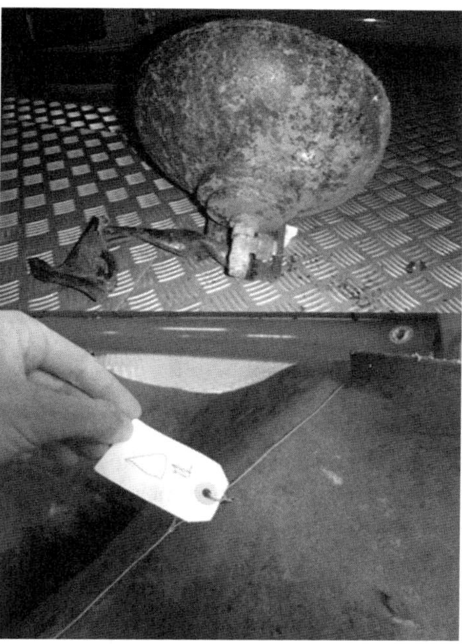

Figure 7.6.9 Cylinder D, in particular the base (in the background cylinder A), the ogive and the coating with labelling of the Expert.

the van before the initial event, but that view clashes with the visible impact marks on the cylinder itself.

The LPG released formed a cloud that mainly extended into the area in front of the van and inevitably ignited in the oven, exploding in the form of a violent flash fire.

The possible breakage points of the van's gas circuit are indicated in Figure 7.6.5. In order of probability with respect to the damage and the dynamics observed these are placed as 3 > 2 > 1.

Both in the presence and in the absence of the pressure reducer, following the catastrophic rupture of an LPG system pipe, a release compatible with the dynamic observed could have been possible but far most likely in the case of supply of LPG at the liquid phase. The catastrophic rupture of the system shall be considered more probable in the absence of pressure reducer due to the greater mechanical and thermal stresses of the system itself (greater pressure and the presence of possible biphasic mixture from incipient evaporation of the LPG with extreme subcooling). With regard to the presence or absence of the pressure reducer, the evidences collected do not allow to formulate a definitive hypothesis.

It is likely that at the time of the events, cylinder A was in use, although the evidence does not explain with certainty how it was connected to the gas system of the van. A brass quick coupler, not foreseen in the original configuration of the system may have been used to this purpose.

One of the party-appointed technical consultants advanced the hypothesis that the initiator event might be found in the spontaneous combustion of the animal fat deriving from cooking of the chickens and which had accumulated in the collection tray and possibly in the area below the oven. The thesis must be taken into consideration, as it is not considered impossible even by important international researchers. The fire hazard of vegetable and animal fats is also known from other publications but there are very few data regarding these types of substances. In general, it can be stated that the flash point is around 265–270 °C and the auto-ignition temperature is between 300 and 450 °C.

In this regard, two considerations may hold: the first is that the temperature reached by the hot fat during several tests runs of the oven is far less than the flash point (approximately 190 °C instead of 265 °C), the second is that the ignition of the fat tank could not have generated the scenario observed, particularly the first violent blaze, without the concomitant release of LPG, as previously stated. These considerations makes the ignition of animal fats far less probable than an LPG leak.

7.6.4 Findings

The analyses carried out on the gas recovered from cylinder C (the only one not broken) reveal that it was a commercial LPG, with no characteristics that would discriminate it between an LPG for automotive use. Cylinder C was probably filled with an A1-type mixture according to Ministerial Decree 15/05/1996 no. 159 (50-50 Propane-Butane).

The same type of LPG can be used for filling containers such as those in use on the vehicle analysed.

It must be emphasised that the illicit practice of unauthorised filling of LPG cylinders at a car gas station that is not adequately equipped for the purpose, even though it is not in a causal connection with the facts of this case, is considered to be a high-risk operation in terms of explosion due to reaching of full hydraulics, with potential consequences equal to if not more than those of the events here discussed.

It can be stated with reasonable confidence that the starting event was a catastrophic rupture of an LPG supply line to the oven, which caused the release of a substantial quantity of LPG within an extremely short space of time, counted as a few seconds.

This scenario will be defined "scenario a)" (paragraph 7.6.6). It must be considered that at the time of the release the system was likely being operated in such a way as to fuel the LPG oven from a single bottle in liquid form. The bottle in use must therefore have been laid flat or upside down.

It cannot be excluded that the initiator event was the mechanical failure of a cylinder resulting from excessive and/or incorrect filling.

Both in the presence and in the absence of the pressure reducer, following the catastrophic rupture of an LPG system pipe, a release compatible with the dynamic observed could have been possible but far most likely in the case of supply of LPG at the liquid phase. The catastrophic rupture of the system shall be considered more probable in the absence of pressure reducer due to the greater mechanical and thermal stresses of the system itself (greater pressure and the presence of possible biphasic mixture from incipient evaporation of the LPG with extreme subcooling). With regard to the presence or absence of the pressure reducer, the evidences collected do not allow to formulate a definitive hypothesis.

The LPG bottles heating system should not be considered as one of the direct or indirect causes of the event, having shown that the likely incidental dynamics did not originate from a rupture of a cylinder due to excessive heating. In the course of the operational tests carried out during the investigation, cylinder temperature conditions were confirmed as being below the limits indicated by the relevant technical standard. It should be noted that the incidental dynamics considered more likely are that indicated as "scenario a" "during the expert evaluation, and that can be summarised as follows:

Scenario a): rupture of an LPG supply system pipe to the oven with fuel release, formation of a flammable cloud, ignition and consequent hot flash followed by general fire. Then the heat stress caused by the fire generated the explosion of three of the four gas cylinders with release of the contents and further propagation of the fire.

One important aspect is the sudden closing of the front door of the van, which in fact trapped the three victims in the work compartment after the rear exit had become impassable due to the fire blast. The causes of this behaviour are not entirely clear. The tests carried out at the manufacturer's premises on one of the vehicles in the workshop, whose door lifting hydraulic circuit was fitted with restraint valves, showed a very short closing time of 25 seconds, a sufficiently long time to evacuate the vehicle in normal conditions. Assuming that it is not known what was the closing time in the case of the vehicle in question, it can be safely stated that the effect of the fire could possibly only have been to increase the speed of closure. This aspect, coupled with the general configuration of the vehicle, with the grill and relative LPG systems located near the exit, without a doubt does not represent a favourable evacuation configuration in case of a serious incident or accident such as the one that occurred at Guastalla.

A different configuration, with the grill located not at the bottom of the vehicle but near the driving compartment, might have allowed the occupants to escape safely.

With reference to the van gas system, the most relevant reference technical standard is UNI CIG 7131 in its various editions of 1972 and 1999. In relation to the ways in which the containers were housed, the vehicle did not fully meet the requirement of non-combustibility of the compartment construction materials. In addition, the design capacity slightly exceeded the standard (75 kg vs. the 70 permitted for multiple container installations).

7.6.5 Lessons Learned and Recommendations

A different configuration, with the grill not at the bottom of the vehicle but near the driving compartment, might have allowed the occupants of the van to escape safely.

7.6.6 Forensic Engineering Highlights

In order to respond fully to the questions that may arise in court, it is necessary to outline all the possible incident scenarios, that is, those sequences of failures, malfunctions and/or human errors that may have generated the observed incidental dynamics. It goes without saying that it is equally important to adequately describe the incidental dynamics associated with each scenario. In doing so, due account will be taken of what the parties have written in their technical consultant reports and what was discussed during the investigation.

From witnesses it is important to acquire information about the status of the location, the visual effects of the incident dynamics, etc.

With reference to the witnesses, it seems clear how the two closest witnesses talked about a white cloud followed almost immediately by a much-extended blaze.

A third witness also described a flare not preceded by explosions.

All these witnesses then refer to explosions following the initial flames.

Other witnesses who were located far away from the van, identify less precisely the start of the event, stating that it coincided with the first of a series of explosions.

In principle the witnesses itself did not allow to distinguish between the two incidental scenarios.

1. Scenario a): rupture of an LPG supply system pipe to the oven with fuel release, formation of a flammable cloud, ignition and consequent hot flash followed by general fire. Then the heat stress caused by the fire generated the explosion of three of the four gas cylinders with release of the contents and further propagation of the fire; and
2. Scenario b): rupture of a cylinder due to mechanical failure, probably caused by unsuitable heating of the compartment and of the cylinder itself, with the concomitance of reaching of the cylinder of overfull hydraulic capacity.

The forensic engineer must aim to distinguish, at least in probabilistic terms, between the two scenarios, based on the available elements.

Going back to the witnesses, but also to the findings about the early stages of the fire, there is evidence of a sudden release of LPG that formed a whitish, suffocating cloud that almost instantly exploded. While there is no direct evidence on the extension of the cloud and the subsequent blaze regarding the rear area of the van, something can be said for the area in front of it. Two witnesses that were in front of the van fled but were enveloped by flames within moments. They managed to escape the wall of flames by running then stop approximately 15 meters away. The third was hit by a blaze while near the van.

Conversely this first flash fire did not appear to have significantly affected the rear of the van (with reference to the sales counter). This can be clearly seen from some footage of the events, in particular from one of the movies collected by the investigators. For clarity of interpretation all the frames of the part of interest of the clip were extracted, that make up Figure 7.6.10. It is clear that in the phase represented by the clip, the cars parked behind the van had not yet been damaged or affected by the flames while they were at a later time.

The elements reported by witnesses who were closer to the van appear to describe a gas release and the almost simultaneous flash fire, causing a phenomenon more akin to a blaze that to an explosion. It should be considered that scenario b), involves an event that is undoubtedly violent and evident as such. The amount of energy developed by the explosion of a pressurised container and by the contextual flash evaporation of a part of the content can be calculated knowing the thermodynamic properties of the content itself.

Data in Table 7.6.2 can be assumed:

In case of a bottle bursting the energy released is 346.55 kJ. Using the known method of equivalent TNT, it is possible to estimate at a scaled distance of 1.99, corresponding to the real distance of two meters, a peak overpressure of 303 kPa.

This value is extremely high and undoubtedly destructive. Such an event would certainly have resulted in extremely significant damage to the back of the van. The destruction of the cylinder compartment can easily be assumed with the projection of objects over a great distance, a scenario similar to the third explosion (that of cylinder D), that occurred with a developed fire, that projected parts of the cylinder to a distance of 17 m.

Figure 7.6.10 Series of frames from "Guastalla tragedia al mercato.avi".

Table 7.6.2 Reference parameters for scenario b).

Cylinder volume	40 l
Cylinder content	20 kg
Propane weight fraction	0.283
Butane weight fraction	0.717
Liquid density	500 kg/m^3
Initial temperature	25 °C
Flash fraction	0.3
Liquid propane specific energy	81 kJ/kg
Vapour propane specific energy	32 kJ/kg
Liquid butane specific energy	40 kJ/kg
Vapour butane specific energy	11 kJ/kg

Such an event would have been perceived by those persons near the van as a violent, if not extremely violent, explosion and would probably have caused serious injuries to anyone nearby, not only in terms of the burns that the flare would have caused but especially as a result of the mechanical effects of the explosion.

It should also be added that the mechanical explosion would certainly and almost immediately have been followed by exploding of the released gas, resulting in considerably large fireballs.

Again, with reference to scenario b), it is appropriate to point out that the examinations carried out on the remains of the cylinders, to the extent possible given the advanced state of corrosion, did not provide any element supporting a brittle failure. On the contrary, all the cylinders clearly showed the results of a ductile rupture, identifiable by the evident local thinning of the thickness of the material. This is certainly evident for cylinders A and B, which exhibited the typical breakage caused by overheating of the flame-affected outer casing. It is less obvious instead for cylinder D which undoubtedly underwent a violent and catastrophic rupture. In any case, some plastic deformation was also observed on the outer casing of cylinder C. In addition, it seems likely that cylinder D was the last to explode, resulting in a BLEVE, thus causing the most violent explosion and projecting the outer casing to a significant distance.

The witnesses and the evidence collected do not allow definition of the size of the LPG cloud formed by the release. However after the ignition no immediate flame propagation to the adjacent vehicles was observed. only later the flames propagated by radiation from the van to the closest vehicles and to the cars parked behind the van. A reasonable order of magnitude of the dimension of the cloud caused by the release at the moment of the ignition may be a hemicylinder of radius of the order of several meters and of height slightly above the average person's height (some of the witnesses stated being enveloped and reported a sense of suffocation, demonstrating that their head was inside the cloud). Under these assumptions, it is possible to estimate the quantity of the LPG released prior to the ignition (Table 7.6.3).

Such a release is fully compatible with scenario a). To identify the location of the leak, a number of C-Phast code simulations were carried out in accordance with certain plausible rupture hypotheses, which should be intended as the sub-scenario of scenario a). These were identified starting from the evidences and the findings of the investigations.

Table 7.6.3 Scenario a), release characteristics.

Cloud range [m]	5
Cloud height [m]	1.8
Butane mole fraction [–]	0,342
Propane mole fraction [–]	0,658
Propane LEL [%vol]	2.2
Butane LEL [%vol]	1.9
Mixture LEL [%vol]	1,993
Mixture stoichiometric concentration [%vol]	0.0444
Mixture stoichiometric concentration [kg/m^3]	0.0966
LPG mass [kg]	1.71
Release time [s]	10
Flow rate [kg/s]	0,171

The starting point was the LPG plant layout included in the technical dossier of the van. However, it should be noted that some evidences were in disagreement with the technical dossier:

1. The presence of four 20 kg cylinders instead of three 25 kg ones;
2. the presence of an element that can be traced back to a coupling between two pipes, damaged by flames;
3. the pressure reducer was not found; and
4. the remains of a brass component, melted on top of on one of the gas bottles, which may have been a quick coupler with a pressure reducer.

The simulations were conducted considering:

1. A catastrophic rupture of the pipe at the points shown in figure 7.6.5;
2. one dispensing cylinder, the other closed by a head valve;
3. gas phase dispensing, or liquid phase dispensing; and
4. presence of a pressure reducer, or absence of pressure reducer.

The chemical-physical conditions of the LPG contained in the bottle in use were derived from the results of the chemical analyses carried out on the gas recovered from bottle C. some experimental runs were carried out to measure the liquid phase temperature in the bottle in use at the oven working conditions (20°C). The presence or absence of the pressure reducer substantially introduces a differential pressure in the circuit which reflects on the release pressure and therefore on the flow rate. The sub-scenarios of scenario a) are resumed in Table 7.6.4, chemical and physical variables, and the results are summarised in Table 7.6.5.

Beyond the evident approximations that arise from the quality of the data available, the following can be observed: in the case of gas spill from the bottle in use (sub-scenarios no. 1, 3, 4, 7, 8), catastrophic break in any of the points 1 to 3 in Figure 7.6.5 results in outflows of the order of 0.003 -0.03 kg/s depending on the rupture location and on the presence of the pressure reducer (in case of rupture at point 1, which is upstream of the reducer, obviously only the case without reducer is considered). These flow rates are to low to be compatiblewith the scenario observed.

Table 7.6.4 Identification of simulations related to scenario a) indicating the breaking point and of the released phase.

simulation ID	Scenariotype	Release phase
1	Breakage at point 1	G
2	Breakage at point 1	L
3	Breakage at point 2, with pressure reducer	G
4	Breakage at point 2, without pressure reducer	G
5	Breakage at point 2, with pressure reducer	L
6	Breakage at point 2, without pressure reducer	L
7	Breakage at point 3, with pressure reducer	G
8	Breakage at point 3, without pressure reducer	G
9	Breakage at point 3, with pressure reducer	L
10	Breakage at point 3, without pressure reducer	L

Table 7.6.5 Results of simulations with C-Phast code.

Simulation	Diameter (mm)	Length (cm)	Actual pressure used to achieve the desired phase	Temperature (°C) and quantity released (kg)	Phase	Flow rate (kg/s)
1	10	150	2.9 barg	20 - 62	G	0.05
2	10	150	16 barg	20 - 62	L	1.05
3	10	485	0.09 barg	20 - 62	G	0,003
4	10	485	2.9 barg	20 - 62	G	0.03
5	10	485	3 barg	20 - 62	L	0.12
6	10	485	16 barg	20 - 62	L	0.64
7	10	660	0.09 barg	20 - 62	G	0,003
8	10	660	2.9 barg	20 - 62	G	0,025
9	10	660	0.09 mbarg	20 - 62	L	0.10
10	10	660	3 barg	20 - 62	L	0.55

As can be seen in Table 7.6.5, all the sub-scenarios that involve liquid spill from the bottle (hence inclined or upturned bottle) produce leak flow rates compatible with the observed scenario (Table 7.6.3).

The most significant flow rates are generated by scenarios 2, 6 and 10, without pressure reducer. From the above considerations, it follows that scenario a) is to be considered much more likely than scenario b), in particular due to breakage at one of the three points indicated in Table 7.6.4, with liquid phase release. Liquid phase release is possible if the cylinder in use at the moment of breakage was turned upside down (i.e. with the valve facing down) or at least horizontally.

Scenario a) is also compatible with the film clips of the events, all of which are related to the already advanced phases of the fire. Of these, in order of time one shows in the early instants a gas bottle on fire at ground level (Figure 7.6.10). Another film clip refers to images in Figure 7.6.11 where the burst of a gas bottle is represented. This film clip

Figure 7.6.11 Still image from "video0054.mp4".

is certainly taken after the one cited above as the van visible in the previous movie has gone. In the film clip, there are clearly two violent explosions, one immediately followed by a second, causing an extended fireball. The following are important elements of this film clip:

1. The first explosion occurs once the fire has started. The van on the right of the one that is burning has already been removed. This film clip is therefore definitely subsequent to the one discussed above; and
2. the two explosions are the result of the explosion of two cylinders. The violent blast at the end of the film clip is probably a consequence of the BLEVE of cylinder D.

A violent explosion is also visible in a third film, whose the still image are represented in Figure 7.6.12. There are no useful elements for temporal placement of this explosion, which is likely the second explosion visible in the previous video.

In summary, the exam of the film clips shows beyond any doubt that at least two explosions (i.e. the explosion of two cylinders, the last of which with violent BLEVE) occurred once the fire had started. There are no elements in the film clips that allow identification of the initial event with certainty.

7.6.7 Further Readings

Further Readings

CCPS (Center for Chemical Process Safety). (1989) *Guidelines for use of vapor cloud dispersion models*. New York: American Institute of Chemical Engineers; 1989.

CCPS (Center for Chemical Process Safety). (1994) *Guidelines for evaluating the Characteristics of Vapor Cloud Explosions, Flash Fires, and BLEVE*. New York: American Institute of Chemical Engineers; 1994.

Figure 7.6.12 Still image from "Untitled.avi".

CCPS (Center for Chemical Process Safety). (1999) *RELEASE: a model with data to predict aerosol rainout in accidental releases*. New York: American Institute of Chemical Engineers; 1999.
CCPS (Center for Chemical Process Safety). (2000) *Guidelines for chemical process quantitative risk analysis*. New York: American Institute of Chemical Engineers; 2000.
Gugan, K. (1979)*Unconfined vapor cloud explosions*. Houston: Gulf Pub. Co., Book Division; 1979.
Hattwig, M. (2004) *Handbook of explosion prevention and protection*. Weinheim: Wiley-VCH; 2004.
Pasquill, F. (1974) *Atmospheric diffusion*. 2nd ed. Chichester: Ellis Horwood; 1974.
Payman, W., and Wheeler, R. (1923) LIV.—The effect of pressure on the limits of inflammability of mixtures of the paraffin hydrocarbons with air. *J. Chem. Soc., Trans.*, 1923;123(0):426–434.

7.7 Fragment Projection Inside a Congested Process Area

7.7.1 Introduction

The general information about the case study are shown in Table 7.7.1.

An incident at a crude oil refinery occurred in which a steel box filled with steam ruptured after a repair activity on a leaking steam line causing the top cover plate to be ejected. Box has been constructed by workers in order to mitigate the leakage, without any specific design. It is unsure if the box has been fixed to the ironworks of the plant ("supported") or not ("unsupported"). So both cases have been investigated. The plate came to rest at a known distance from the origin without any consequences for operators, pipes, vessels, even if the event occurred at a very congested process area. Even in absence of consequences a specific assessment has been carried out to investigate the event and other possible scenarios where the steel plate has different trajectories to determine possible outcomes. Assessment have been conducted to demonstrate the Public Prosecutor's Office that accident scenarios described in the current Safety Case (Safety Report) of the Refinery coming from the requirements of the Seveso EU Directive on major accidents prevention have consequences in terms of hazardous chemicals

Table 7.7.1 General information about the case study.

Who	Congested process area
What	Fragment projection
When	2010
Where	Italy
Consequences	None
Mission statement	Verify the consistency of the safety case describing the industrial risk of the plant with for an eventual additional threat posed by fragments
Credits	Ernst Rottenkolber (Numerics GmbH) Stefan Greulich (Numerics GmbH)

loss of containment with effects (consequences distances) greater than those that could have been raised by the fragment hitting surrounding process lines involving hazardous chemicals. Safety Case consistency had to be demonstrated. Several main topics have been investigated and implemented. First, the conditions inside the steam filled steel box needed to be understood. Finite Element (FE) simulations have been used to determine the conditions of the internal pressure which caused the rupture of the welds, and they have been employed to determine possible trajectories and velocities of the plate. Second, damage to surrounding pipes have been investigated. In the specific framework additional FE simulations have been conducted to estimate damage to various pipes under several impact conditions and impact velocities. Based on the simulations results, damage criteria have been developed for the pipes. Given the damage degree estimated the calculation of the effects of the loss of containment have been assessed and compared to the damage areas reported in the refinery Safety Case to determine the consistency of it regarding fragment impacts accidents that as observed by the Public Prosecutor's Office where not sufficiently developed by the Owner as requested by the Seveso EU Directive. Scope of the numerical assessment conducted is understanding if the incidents described in the refinery Safety Report show effects having the same order of magnitude as those deriving from containment losses associated with impacts of projectiles as the one that caused the accident described in this chapter. In particular results have been compared with the loss of containment events (jet fires and flammable/toxic substances caused by random ruptures of pipes and flanges) described in the Safety Report. Investigation led to a modification of the code FI-BLAST© by Numerics GmbH (DE), a calculation tool specifically design to assess projectiles trajectories and related effects due to the residual energy at impact location.

7.7.2 How it Happened (Incident Dynamics)

A loss of containment has been observed on a steam process line. Maintenance activities have been conducted and repair has been provided with the installation of a box around the leak. The box suddenly ruptured (Figure 7.7.1) and a fragment (a plate) was launched at a very high velocity due to the steam pressure at significant distance with no consequences for operators and items (pipes, vessels, process items or machinery) processing hazardous chemicals.

7.7.3 Why it Happened

Investigation activities has been conducted by mean of specific working packages to understand:

1. The plate launch velocity;
2. development of piping damage criteria based on finite elements simulations;
3. implementation of the damage criteria;
4. estimation of the consequences coming from the damages observed on the process lines; and
5. comparison with the existing scenarios in the current Safety Case (EU Directive) of the process plant where the accident took place.

Figure 7.7.1 Ruptured steel box.

7.7.4 Findings

In first instance, the failure of the steel box is investigated through hydrocode simulations (Figure 7.7.2). The aim of the simulations is to reconstruct the failure process that affected the box in order to assess the launch velocity of the top plate. During the study the influence of steam temperature and corresponding pressure, support conditions and weld strength were examined.

The assessment then investigates the possible damage to five critical nearby pipes assuming the launched plate would impact them. FE simulations are conducted to determine damage under several impact conditions and a range of impact velocities.

Based on a validation study, damage criteria are developed and a damage evaluation methodology is developed.

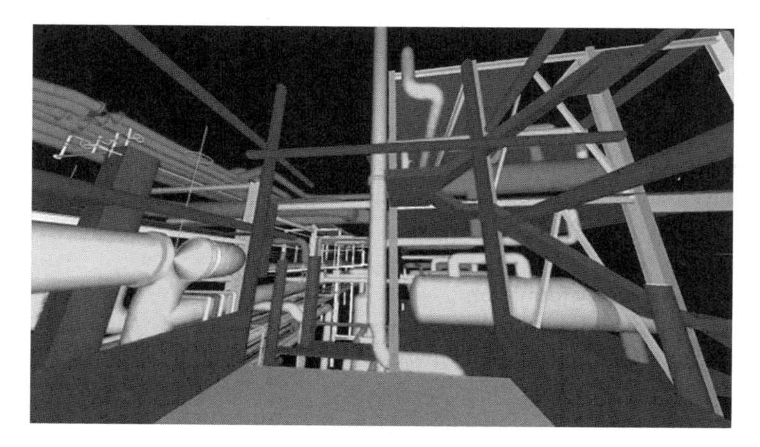

Figure 7.7.2 Process unit tridimensional layout.

Results took into account:

1. Additional features for calculating plate trajectories (including obstacles):
   1. variation of initial velocity;
   2. variation of initial launch angle;
   3. variation of drag coefficient;
   4. user defined impact point of pipes; and
   5. calculation of impact velocity and impact angle;
2. damage functions for estimating pipe damage based on the results of the FE simulations.

The tools employed allowed the assessment of an array of possible scenarios involving the failure of the steel box. In the following paragraphs the main assessment activities have been summarised in the following single assessment packages:

1. Collection of evidences and data;
2. initial plate velocity calculation and box deformation;
3. development of a damage criteria for the impacted pipes;
4. validation;
5. evaluation of damage;
6. results for impacts;
7. FI-BLAST© adaptation to perform a parametric study; and
8. results of the parametric study.

7.7.4.1 Collection of Evidences and Data

Activities started from the collection of the evidences from the real accident. Those evidences were the characteristics of the damaged steel box and the plate that has been launched by the steam pressure inside the box caused by the leakage of the process pipe.

Documentation and data about the plant and process conditions at the time of the event have also been collected. Among those:

1. Process unit plot plan;
2. process unit detailed tridimensional layout; and
3. lines list along with operating/design conditions, piping class, hazardous chemicals.

Tridimensional layout of the process unit have been prepared starting from a 3D laser scanning activity (Figures 7.7.2 and 7.7.3); within the layout the hazardous chemicals (Seveso related) processing lines have been identified on the basis of the lines list document together with the steel box original position on the leaking pipe.

Steel box evidences have been used to build the model to be used in finite elements simulations (Figure 7.7.3).

7.7.4.2 Initial Plate Velocity and Box Deformation

Forensic activities have been started with a specific assessment to estimate the initial plate velocity taking into account the box deformation. The following activities were conducted:

1. Determination of influence of steam temperature and box support conditions ("supported" and "unsupported") on plate velocity;

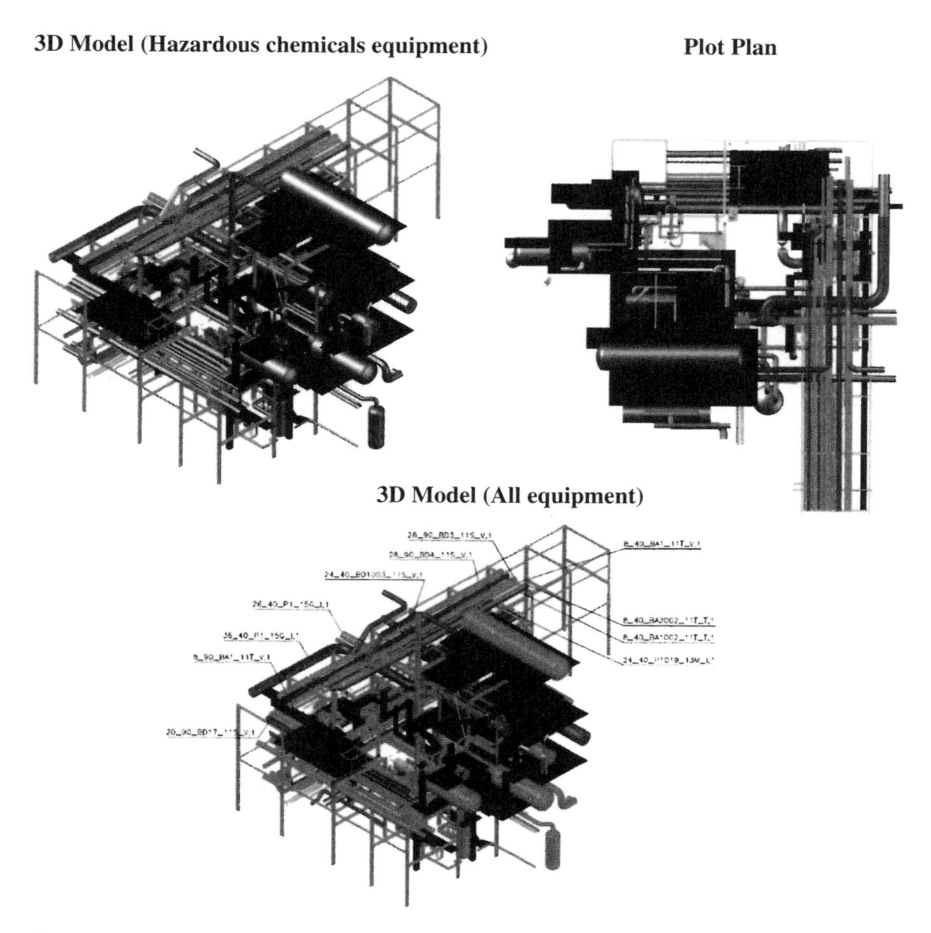

Figure 7.7.3 Process unit involved in the incident tridimensional layout from the 3D laser scanning of the area and the identification of the piping containing hazardous substances.

2. estimation of steam pressure – plate velocity relation for sudden complete failure (considered as "worst case");
3. assessment of maximum stress in the weld with respect to steam pressure and consequent estimation of internal pressure at failure;
4. assessment of box deformation with respect to steam pressure and consequent estimation of internal pressure at failure; and
5. simulations with time delay between initial and final failure of the welds of the box.

Below, some figures about the performed simulations are shown (Figures 7.7.4 to 7.7.6).

The simulated velocities for pressure and temperature variations are in Table 7.7.2.

Simulations helped in identifying two specific issues (Figures 7.7.7 to 7.7.8):

1. Influence of steam temperature can ben assumed to be marginal; and
2. no support ("box unsupported" case) leads to ~15% less plate velocity.

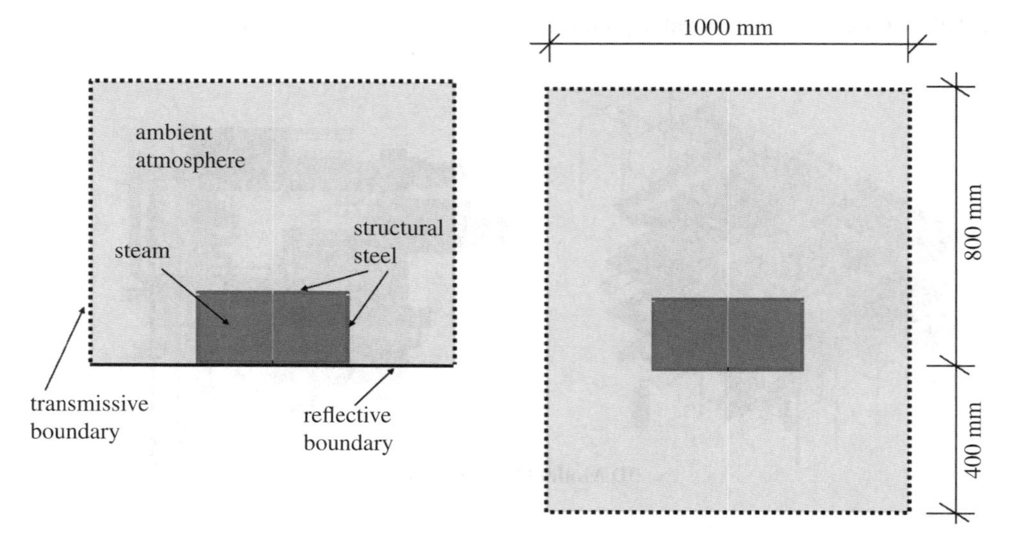

Figure 7.7.4 Model setup of 2D simulations in rotational symmetry: supported box (left) and free box (right).

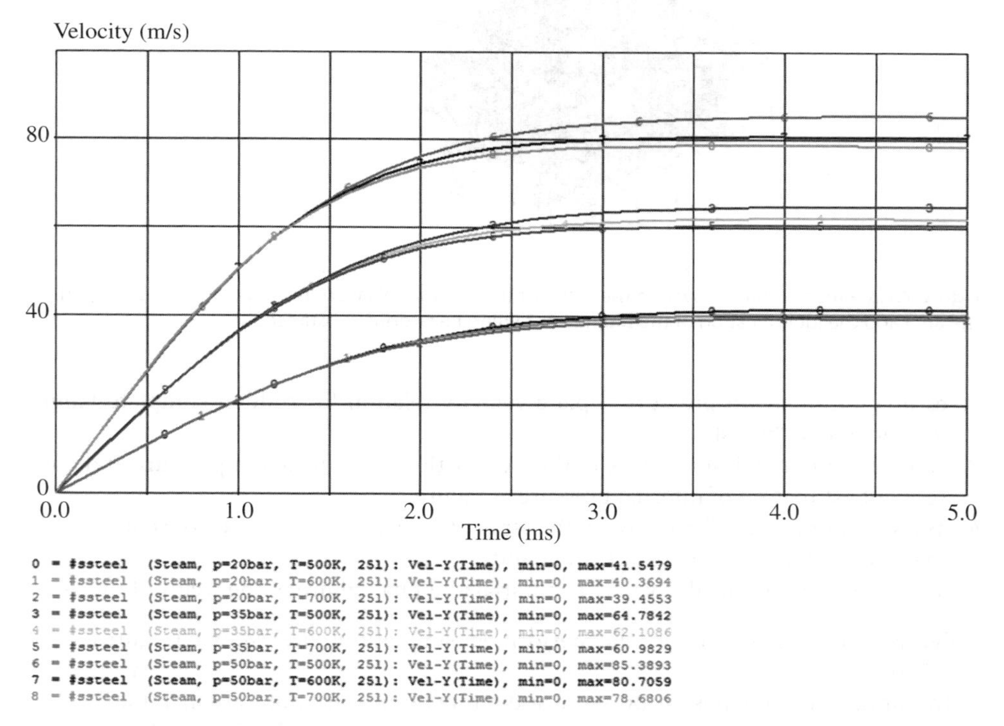

Figure 7.7.5 Velocity profiles of the top plate for different pressure-temperature combinations (case "box supported").

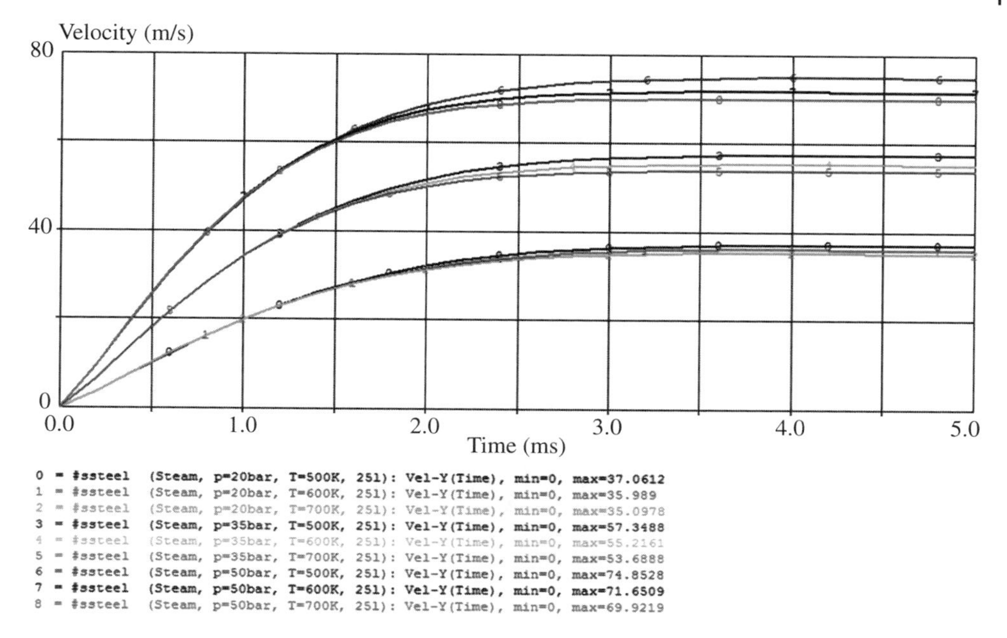

Figure 7.7.6 Velocity profiles of the top plate for different pressure-temperature combinations (case "unsupported box").

Table 7.7.2 Simulation results for steam pressure and temperature variation.

	V_{launch} [m/s] case "box supported"			V_{launch} [m/s] case "box free"		
p [bar]	**500 K**	**600 K**	**700 K**	**500 K**	**600 K**	**700 K**
20	41.5	40.4	39.5	37.1	36.0	35.1
35	64.8	62.1	61.0	57.3	55.2	53.7
50	85.4	80.7	78.7	74.9	71.7	69.9

Hydrocode simulations showed a nearly linear relation for the launch velocity versus the internal pressure of the box in the investigated pressure interval.

Maximum weld stress investigation in relation with steam pressure has been assessed with Autodyn© (Figure 7.7.9). Considering a maximum allowed stress of 470 N/mm^2 it is possible to estimate a failure pressure of nearly 40 bar. On the basis of the total deformation of the box equal to 120 mm the estimated failure pressure is 41,6 bar that determines a launch velocity, for the "worst case" of 62,4 m/s (Figure 7.7.10).

A number of simulations considering delayed failure of the box plate have been conducted in order to verify the observed deformation of the box. Those simulations, shown in Figures 7.7.11 to 7.7.14, allowed some specific considerations about the box failure:

1. Asymmetric deformation can only be due to delayed failure;
2. pressure at initial failure can be determined (~42 bar);

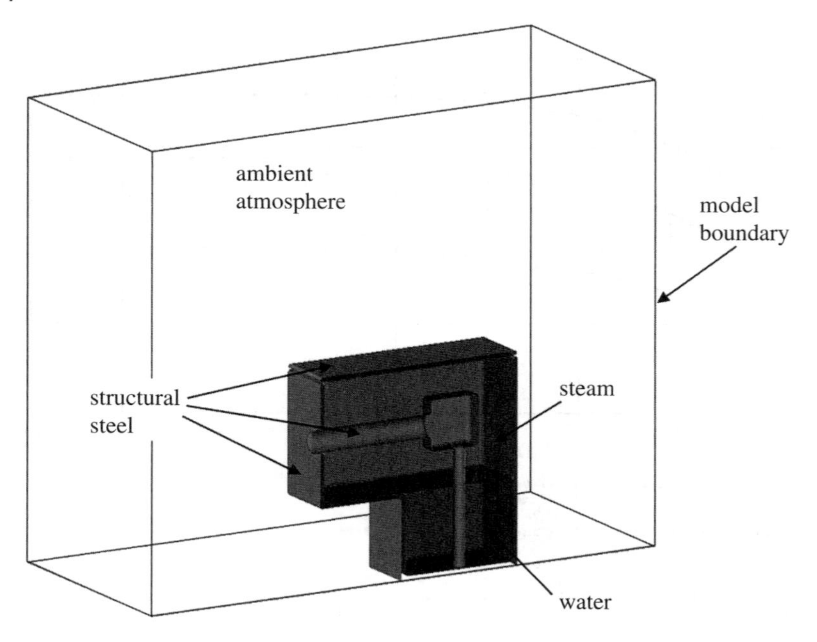

Figure 7.7.7 Numerical model for launch velocity investigation (steam pressure – plate velocity relation has been studied with a specific hydrocode named SPEED© by Numerics GmbH).

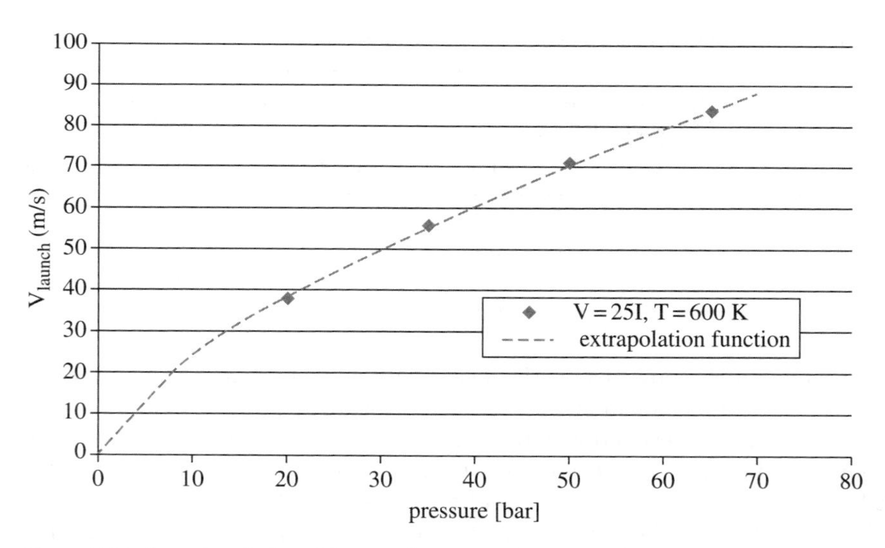

Figure 7.7.8 Launch velocity of the top plate versus box internal pressure.

3. pressure at final failure estimated from box deformation vs. steam pressure results (~20–25 bar); and
4. time between the two pressure levels estimated from intermediate results of steam pressure – plate velocity simulations (~1,5 – 2,0 ms).

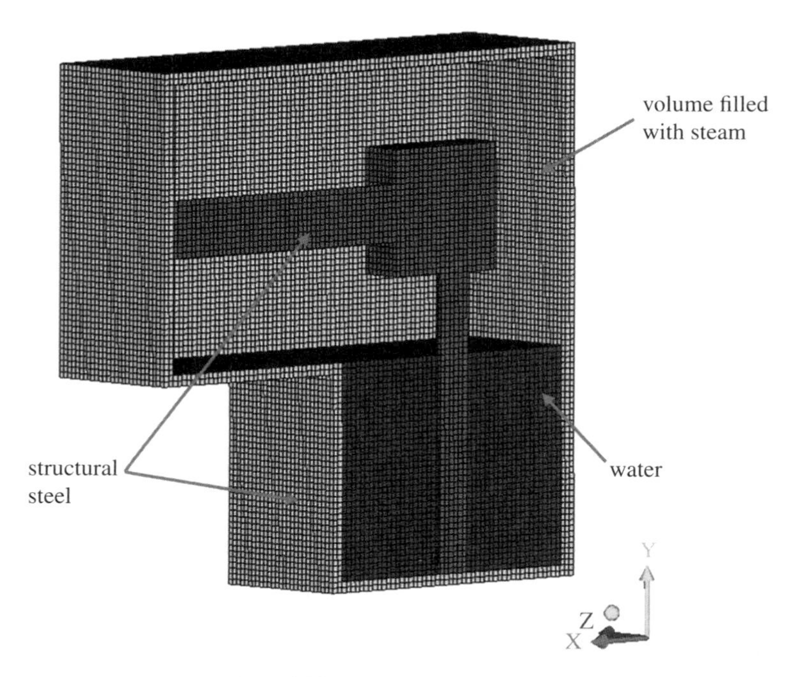

Figure 7.7.9 Numerical model for stress investigation.

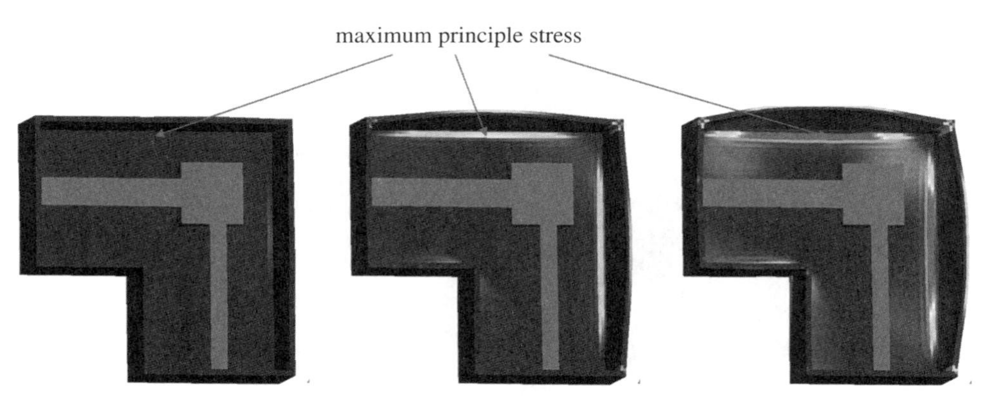

Figure 7.7.10 Plastic deformation of intact box at different internal pressures: 35 bar (left), 50 bar (middle) and 65 bar (right).

Deformations show a good agreement at 2,0 ms failure delay: Considering a plate initial velocity ≈ 42,5 m/s, and an ejection angle a ≈ 23.5° (from vertical) the horizotal and vertical components of the velocity have been identified (Figures 7.7.15 to 7.7.17).

7.7.4.3 Development of a Piping Damage Criteria

Estimation of the pipe damage have been then conducted with FE simulations using the LS-DYNA© tool (specialised in impact analysis). Feasibility has been demonstrated with validation example taken from the literature [1]. See Figure 7.7.18 to 7.7.20.

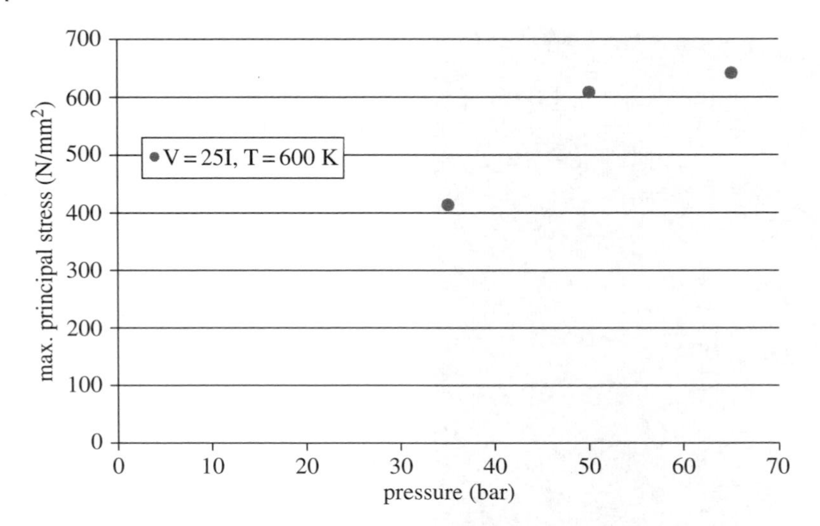

Figure 7.7.11 Main stresses in the weld determined from the numerical simulations.

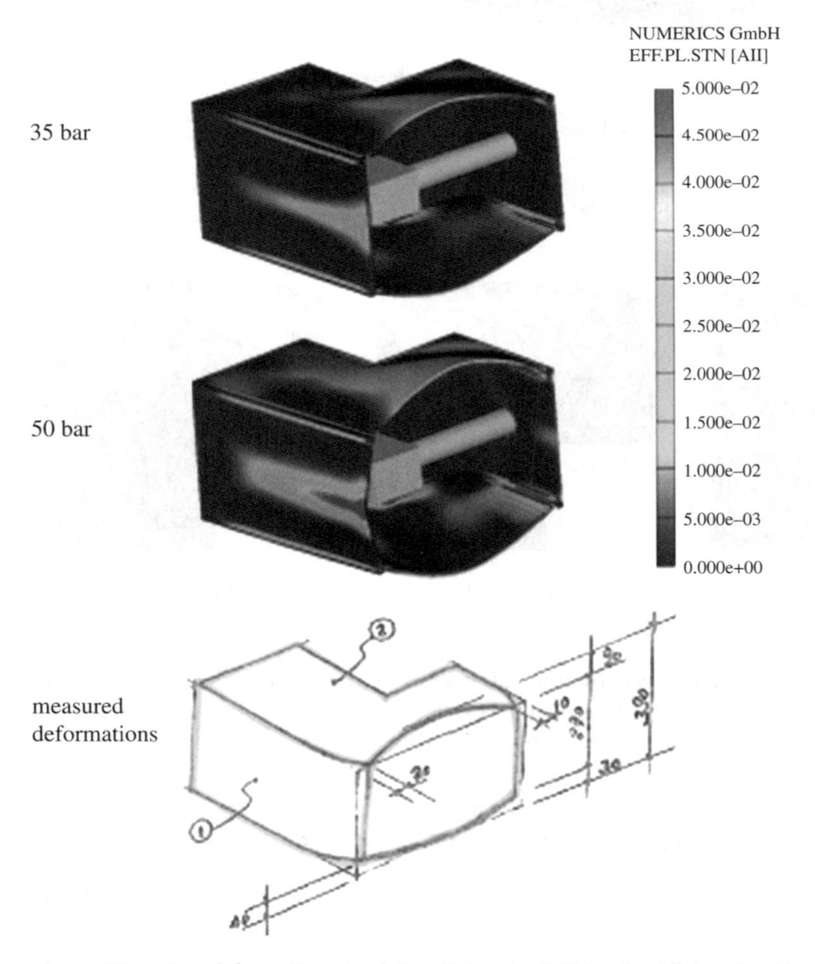

Figure 7.7.12 Box deformation: simulation 35 bar (top), 50 bar (middle) and real box measurements (bottom).

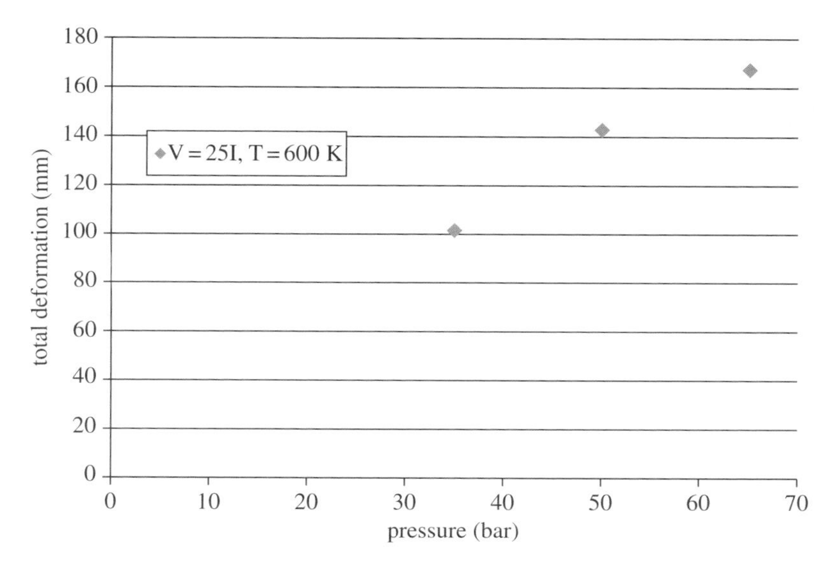

Figure 7.7.13 Total box deformation versus internal pressure.

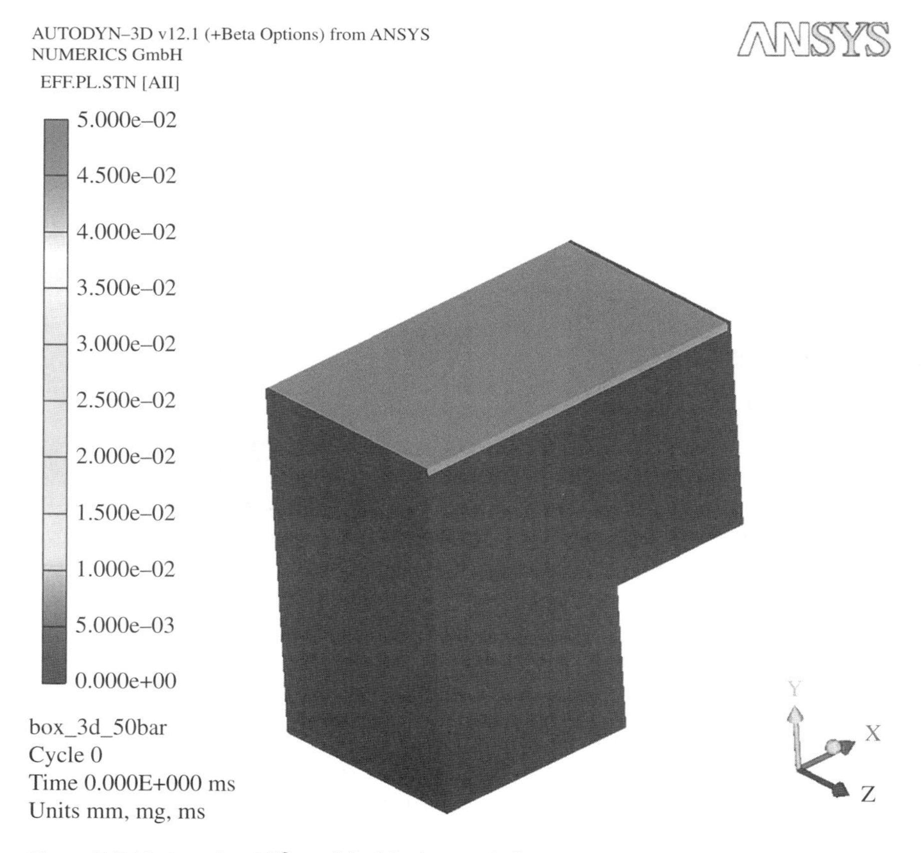

Figure 7.7.14 Autodyn-3D© model of the box and plate.

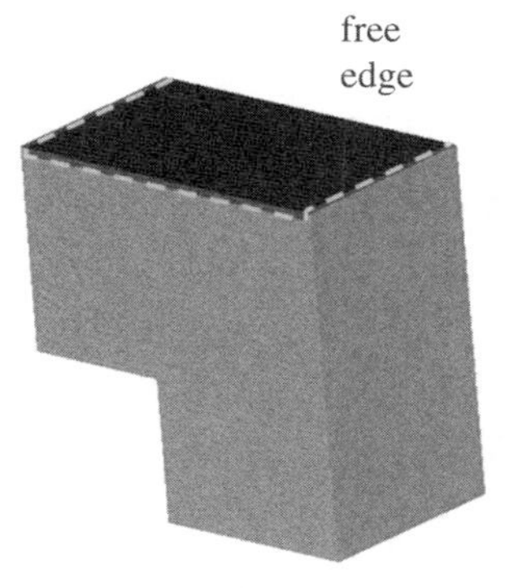

Figure 7.7.15 Numerical model with partly connected top plate, representing the delay condition observed during the box rupture.

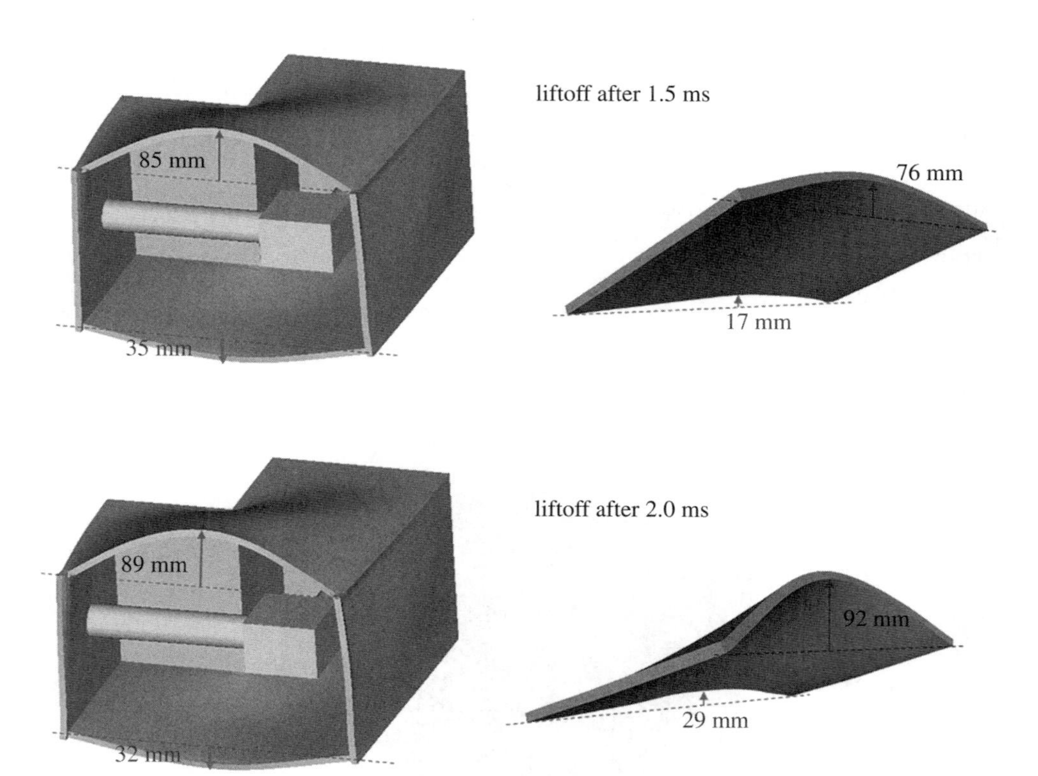

Figure 7.7.16 Results of simulations with delayed failure of the welds (in the pictures 1,5 ms delay and 2,0 ms delay).

FE simulations are compared to experimental results in terms of indentation (pipe deformation) and perforation information (in the form y/n). these simulations demonstrated that the simulated indentation is always higher than real, since the material model does not include the strength increase due to strain rates.

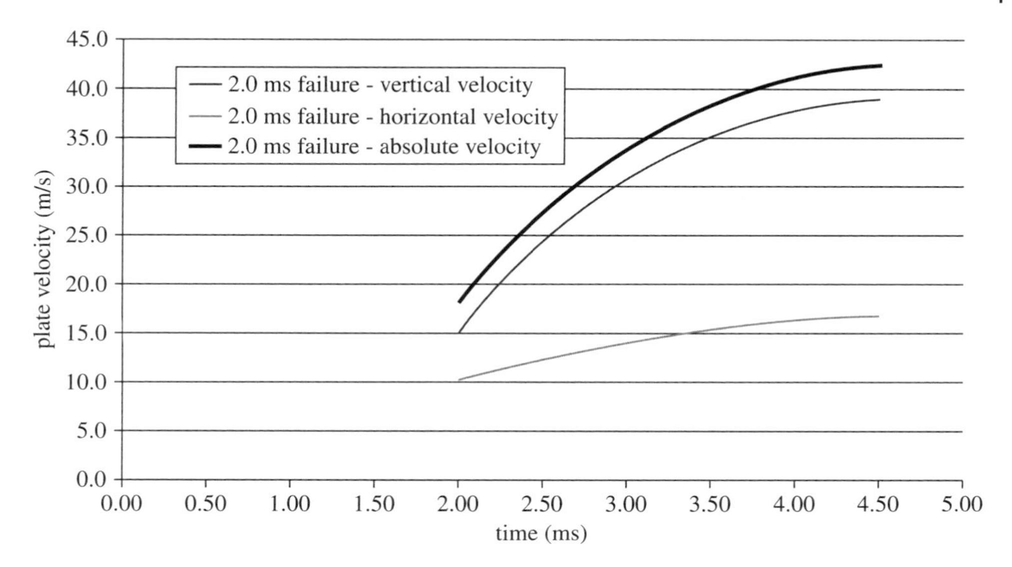

Figure 7.7.17 Evaluation of top plate velocity from simulation with 2.0 ms failure delay.

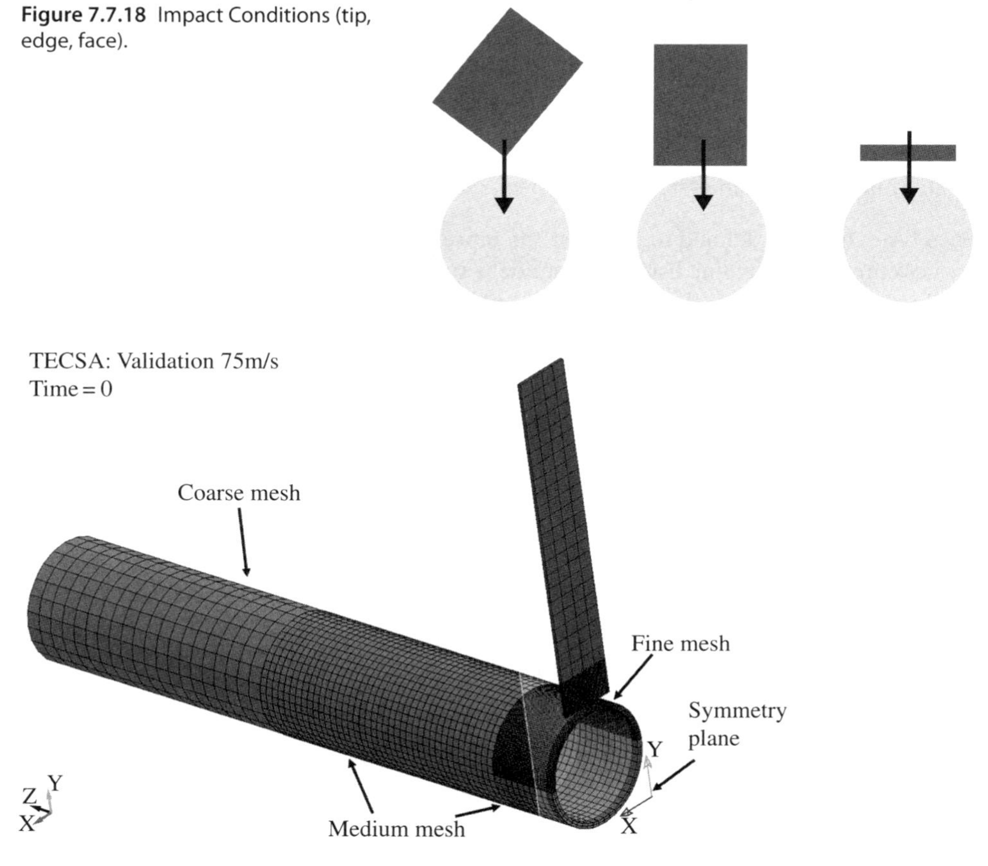

Figure 7.7.18 Impact Conditions (tip, edge, face).

Figure 7.7.19 FE Model showing symmetry along the shotline.

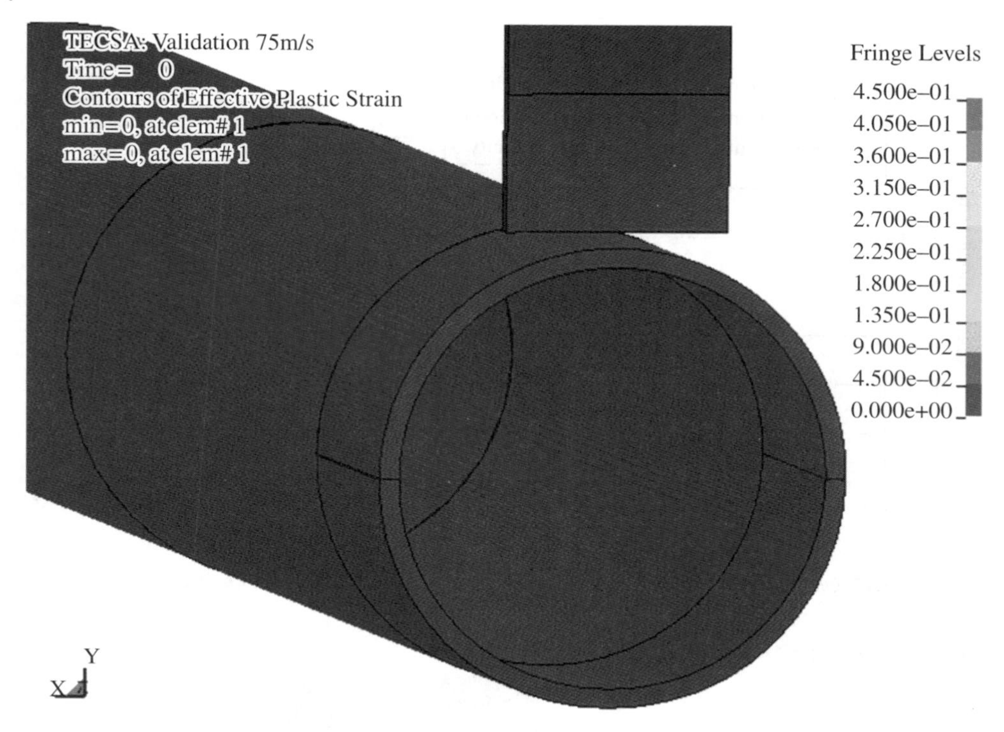

Figure 7.7.20 Validation activity.

On this basis a parametric study has been conducted considering:

1. 5 types of pipes defined to consider the representative main lines of the nearby process unit given the line list document data and the tridimensional model;
2. 3 impact conditions (tip, edge and tip) with various initial velocities to develop the damage criterion; and
3. assumption: empty pipes ("worst case").

A "Dynamic Increase Factor" (DIF) has been introduced to account for increase in material strength at high strain rate. Initial simulations used standard static material properties of API5L Steel and the incorporation of DIF led to calculations characterised by material strength increase with increase strain rate (50%, 75%, 100% strenght increase values have been selected on the basis of data from specialised literature (ASCE: Design of Blast Resistant Builidngs in Petrochemical Facilities, ASCE, Reston VA, 1997, pp 5-10 - 5-20). The resulting simulations are shown in Table 7.7.3.

Considering a 75% strenght increase for the DIF the following pictures show the maximum plastic strains in the pipes at 99 m/s and 143 m/s impact with the fringe level held constant for both simulations. In the simulation with a 99 m/s impact, the max plastic strain does not reach 0,45 and based on the experimental results, this pipe is not damaged. In the simulation with a 143 m/s impact, the maximum plastic strain exceeds 0.45 across more than 50% of the thickness of the pipe (2 elements). According to the experiments, this pipe is perforated. This indicates selecting a plastic strain of 0,45 for

Table 7.7.3 Simulations characterised by a Dynamic Increase Factor.

Strength Increase	Vs [m/s]	Simulation Indentation[mm]	Report Indentation [mm]	Max Plastic Strain in Simulation (Pipe)
0%	75	56	36	0.48
50%	75	42	36	0.39
75%	75	36	36	0.34
100 %	75	31	36	0.30
0%	84	66	46	0.51
50%	84	49	46	0.43
75%	84	42	46	0.38
100 %	84	38	46	0.33
0%	99	80	55	0.56
50%	99	60	55	0.49
75%	99	53	55	0.43
100 %	99	46	55	0.38
0%	143		Perforation	0.68
50%	143		Perforation	0.61
75%	143		Perforation	0.56
100 %	143		Perforation	0.52

material failure is appropriate. This has then been considered the basis of the damage criteria (See Figure 7.7.21).

7.7.4.4 Evaluation of Damages

Simulation results were evaluated to determine the elements at or above the max plastic strain:

1. Damage evaluation is investigated in cut planes made at 1 or 2 mm increments in pipe length direction;
2. decision for a perforation / hole in the FE model → 50% of the thickness have plastic strain at or above 45%;
3. decision for a crack initiation in the FE model → max 25% of the thickness have plastic strain at or above 45%; and
4. the hole length is then the length over the cut planes (Figure 7.7.22).

Damage assessment has been made considering the pipes defined as representative (5 pipes selected from the process unit line list), several impact conditions (3 conditions, see Figure 7.7.18) with various initial velocities from 20 to 80 m/s. Pipes have been assumed empty, with supports every 6 m of length and with an impact point in the middle between two supports (See Figure 7.7.23)

7.7.4.5 Results for Impacts for Some Pipes

Evaluations for pipe having tag number "8_40_B" are shown in Table 7.7.4 as an example of the assessment conducted on pipes selected as representative for the case.

NA means that the pipe is subject to deformation and not crack (Figure 7.7.24).

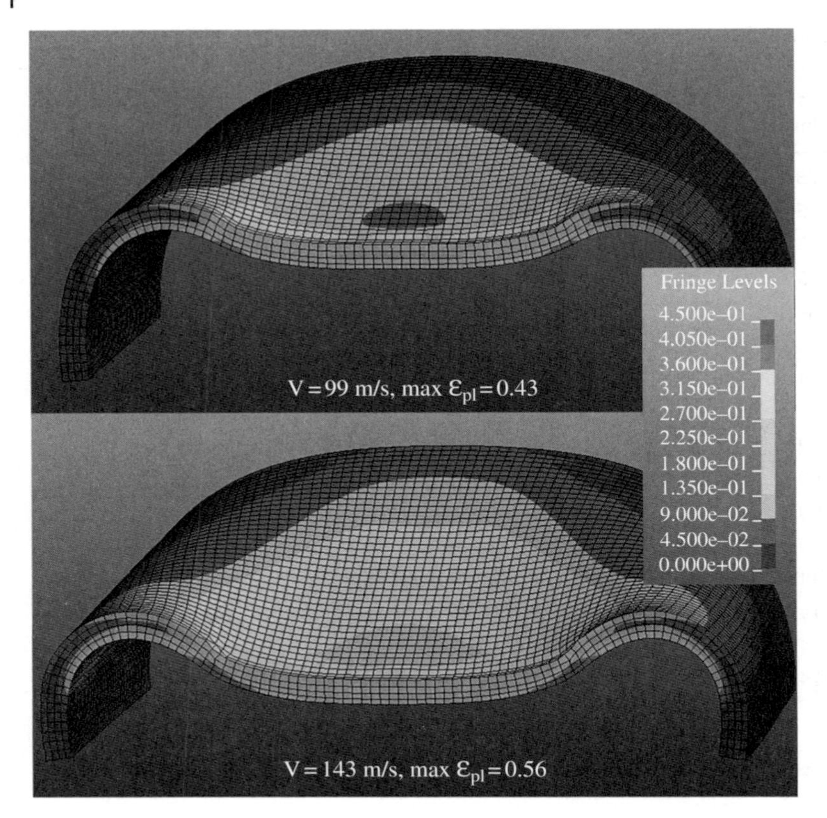

Figure 7.7.21 Maximum plastic strain. Top picture- 99 m/s impact into the pipe. Bottom picture – 143 m/s impact into the pipe. Plastic Strain level held constant in both simulations.

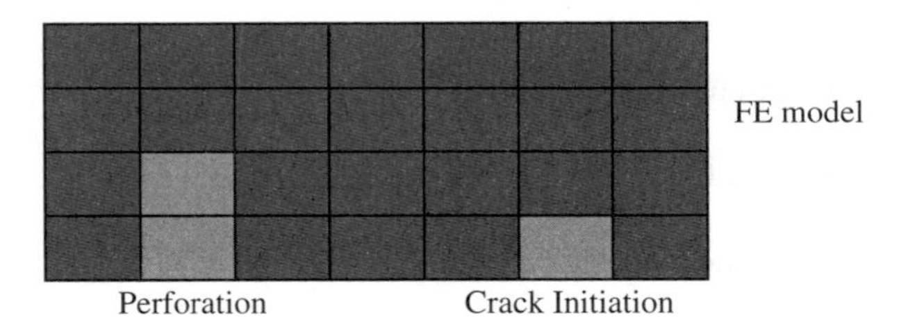

Figure 7.7.22 Crack / perforation criteria in the FE method.

7.7.4.6 FI-BLAST© Adaptation to Perform a Parametric Study

FI-BLAST© simulation code by Numerics GmbH have been modified to include some features able to conduct a parametric assessment all over the involved process unit (Figure 7.7.25). In the new code the following features have been then added:

1. Tools for calculating plate trajectories (including obstacles) in a parametrical study (Figure 7.7.26):
    1. Variation of initial velocity;
    2. Variation of initial angle;

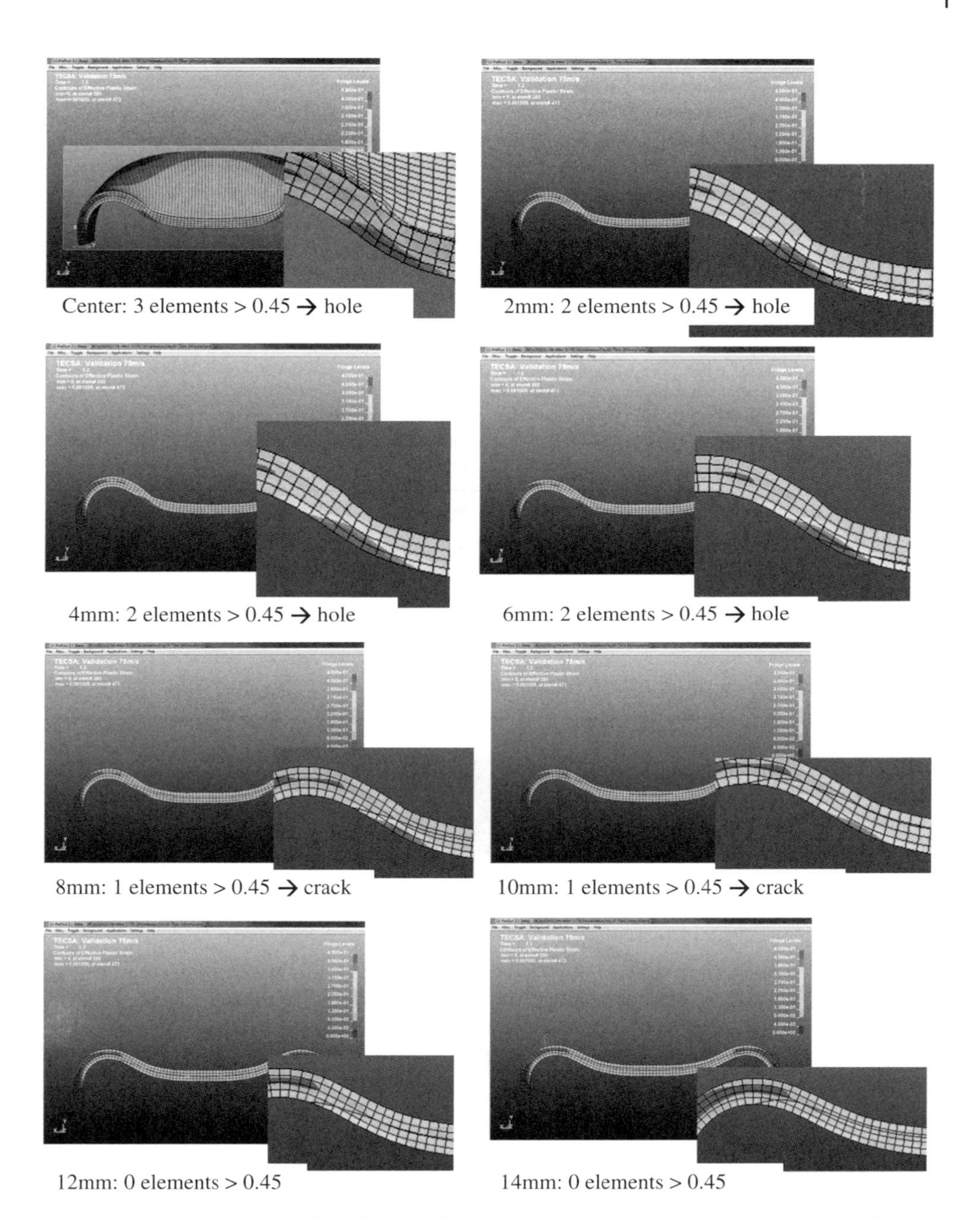

Figure 7.7.23 Damage evaluation using cut planes at 2mm increments: 6 mm long hole as per the damage criteria described in paragraph 7.7.4.4.

Table 7.7.4 Results for impacts.

Pipe	V_S [m/s]	Only when tensile stresses Max Plastic Strain	Damage	Hole Size [mm]	Possible max crack length [mm]
8_40_ba	80	0.72	Hole	61 × 20	-
8_40_ba	70	0.66	Hole	54 × 18	-
8_40_ba	60	0.62	Hole	44 × 14	-
8_40_ba	50	0.56	Hole	32 × 12	-
8_40_ba	40	0.49	Crack initiation	NA	12
8_40_ba	30	0.43	Deformation	NA	-
8_40_ba	20	0.37	Deformation	NA	-

Pipe	V_S [m/s]	Max Indentation [mm]	End/orig	Outer diameter end	Outer diameter orig	Evaluation at [ms]
8_40_ba	80	112.3	0.45	90.8925	203.2	10
8_40_ba	70	96.7	0.52	106.477	203.2	10
8_40_ba	60	81.1	0.60	122.085	203.2	10
8_40_ba	50	66.5	0.67	136.697	203.2	10
8_40_ba	40	48.4	0.76	154.801	203.2	10
8_40_ba	30	33.4	0.84	169.81	203.2	10
8_40_ba	20	19.2	0.91	183.96	203.2	10

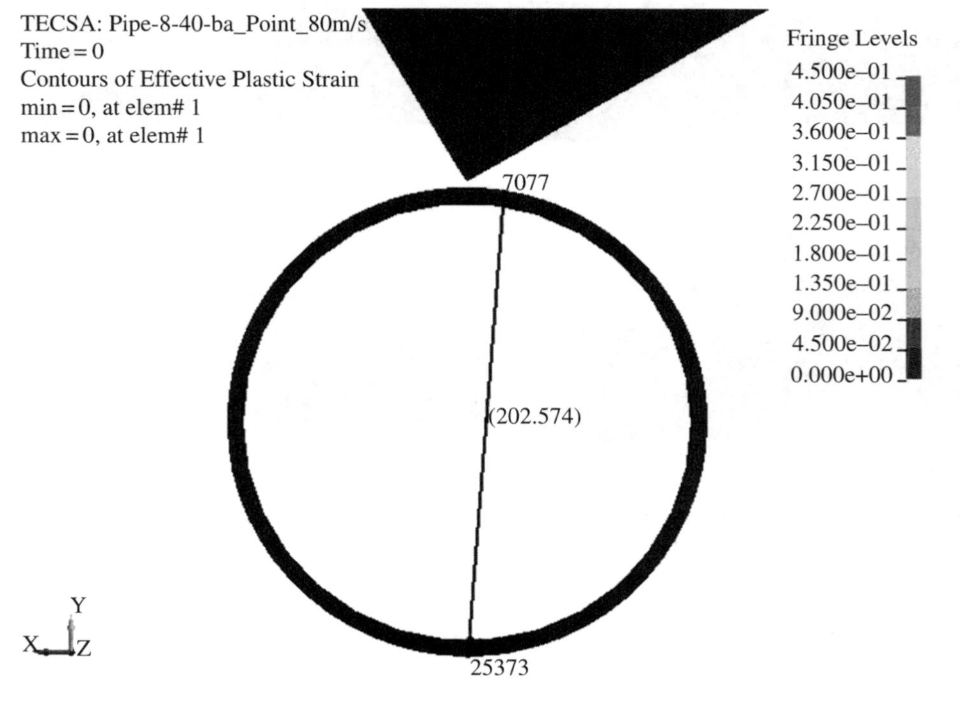

Figure 7.7.24 Plastic strains & deformation: 8_40_BA1002.

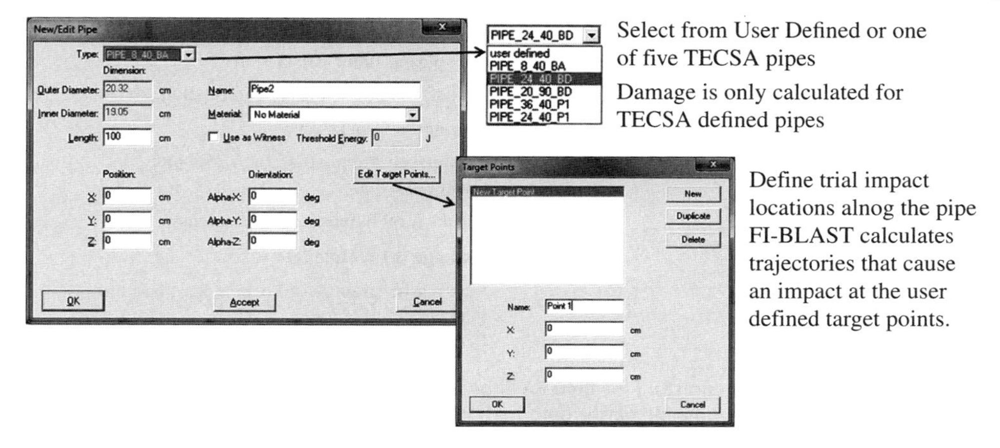

Figure 7.7.25 Modifications of the FI-BLAST© code.

Figure 7.7.26 FI-BLAST© tool: impacting trajectories and pipe damage indicated in grey as shown inside the calculation code to the user.

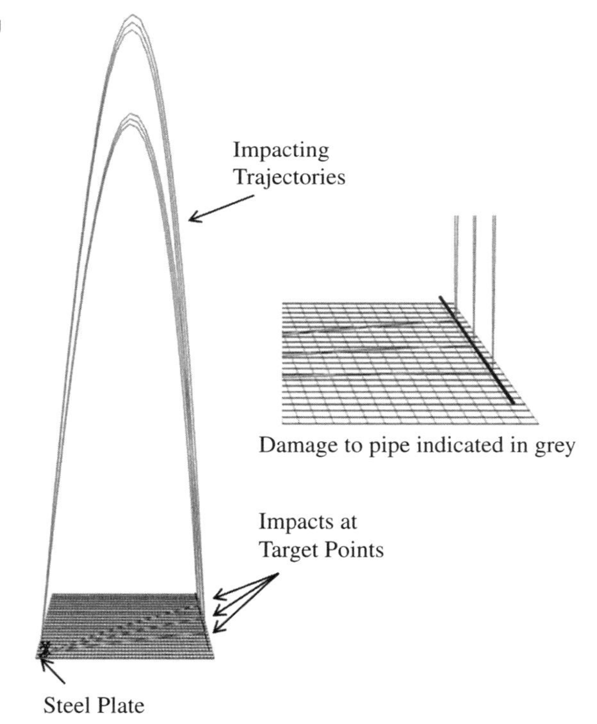

3. Variation of drag coefficient;
4. User defined impact points; and
5. Calculation of impact velocity and impact angle;

Damage results based on the FE simulations (step functions have been implemented between "no damage", "crack" and "hole", linearly interpolated between simulation results) see Figures 7.7.27 and 7.7.28;

2. Pipe objects;

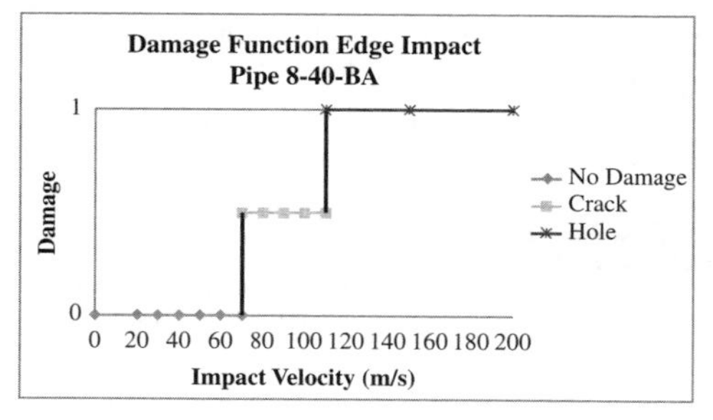

Damage = 0 = No Damage (N)

Max plastic strain < 0.45

No deformation to plastic deformation, no cracks

Damage = Crack (C)

Max plasitc strain ≥ 0.45 for $0.25 \leq$ pipe thickness < 0.5

Cracks form in tensile strain regions

Damage = 1 = Hole (H)

Max plastic strain ≥ 0.45 over at least half the thickness of the pipe
Perforation - Hole

Figure 7.7.27 Damage Function for Pipe 8-40-BA Edge Impact. Damage = 1 indicates a hole in the pipe. Damage = 0 indicates possible plastic deformation but no holes and no cracks. Cracks begin to form, but they do not create a hole.

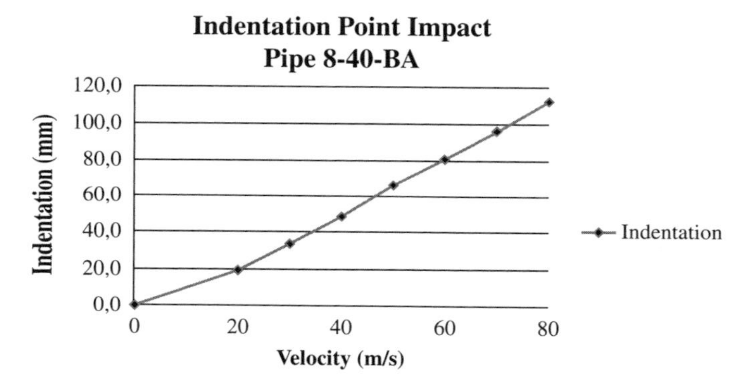

Figure 7.7.28 Indentation function (crack depth due to loss of material from the impact) for Pipe 8-40-BA in the impact location. Black diamonds indicate simulation results. Linear interpolation is used between know points.

3. Graphical display of:
   1. Impacting trajectories
   2. Damage results (worst case)
      1. Red – hole
      2. Yellow – crack
      3. Green – no damage
4. Reporting features in order to show the results in terms of:
   1. impact velocity;
   2. impact angle; and
   3. damage.

7.7.4.7 Results of the Parametric Study

Results of the parametric study on the lines processing hazardous chemicals have been employed to estimate the consequences of an accidental release of toxic due to the failure of the line (Figure 7.7.29). These results have been studied with DNV Phast

Init.Vel. [m/s]	Elev.Angle [deg]	Cw [-]	Imp.Vel. [m/s]	Imp.Angle [deg]	Damage Tip [-]	Damage Edge [-]	Damage Plane [-]
70	3.8	0.68	60.56	77.6	H	N	N
70	84.2	0.68	45.09	89.2	H	N	N
80	3	0.68	69.22	77.6	H	N	N
80	85	0.68	47.51	89.4	H	N	N
70	3.8	0.68	60.7	83.2	H	N	N

User input (Init.Vel., Elev.Angle, Cw) — FI-BLAST Output (Imp.Vel. to Plane)

Note: Damage results do not consider the calculated impact angle. Damage is only reported at the idealized impact conditions.

Tip Impact Hole Area [m²]	Tip Impact Max.Pl.Strain [-]	Tip Impact Max.Indent. [cm]	Edge Impact Hole Area [m²]	Edge Impact Max.Pl.Strain [-]	Edge Impact Max.Indent. [cm]	Plane Impact Hole Area [m²]	Plane Impact Max.Pl.Strain [-]	Plane Impact Max.Indent. [cm]
0.000636	0.62	8.2	0	0.36	5.92	0	0.12	2.42
0.000196	0.53	5.76	0	0.28	4.2	0	0.08	1.5
0.000944	0.66	9.55	0	0.42	6.98	0	0.15	2.94
0.000289	0.54	6.2	0	0.29	4.44	0	0.08	1.64
0.000641	0.62	8.22	0	0.36	5.94	0	0.12	2.43

Figure 7.7.29 Incident Effects Results.

Professional© consequence effects estimation tool generally used in the oil and gas industry for the assessment of the safety report requested by the EU Seveso Directive. It has been demonstrated that scenarios in the current Safety Report of the facility have greater consequences areas in comparison with the effects coming from the accidental release of toxic and or flammable chemicals following a fragment projection (See Figure 7.7.30).

Safety report assessment by the Owner can then be considered sound, consistent, and effective.

7.7.5 Lessons Learned and Recommendations

FI-BLAST© has been improved to assist in the analysis of an incident that occurred at a refinery involving a steel box used to mitigate the steam loss of containment and the reduces pressure in the pipe Two essential adaptations were implemented in the tool to calculate trajectories of the ejected top plate, and to assess the damage to five critical nearby pipes should an impact occur. The adaptations were based on two specific investigations.

In the first study, the failure of the steel box was investigated. The pressure at initial failure and the launch velocity at final liftoff of the top plate of the investigated box were assessed in numerical simulations. In these simulations, it was possible to reconstruct the failure process with rather good accuracy. A relation between internal pressure and launch velocity of the top plate was found. Assuming a sudden complete failure of the weld, a launch velocity of 62,4 m/s was calculated. Assuming a delay of 2 ms, the simulations showed a launch velocity of the top plate of 42,5 m/s at an ejection angle of about 23 degrees.

In the second study, possible nearby pipes damage was assessed. FE simulations are conducted to determine damage under several impact conditions and a range of impact velocities. Based on a validation study, damage criteria are established and a damage evaluation methodology is developed. Simulation results showed the tip impact to be the most critical. Holes resulted in three of the five pipes at velocities as low as 40 m/s. Edge impacts were not as critical. In the velocity range of 20-80 m/s, none of the pipes were perforated. Only the smallest pipe exhibited crack initiation at 80 m/s. Face impacts did not cause damage in the investigated velocity range. Only minor indentations were noticed for the thinner pipes.

With these additional features, FI-BLAST© can be used to analysis an array of possible scenarios involving the failure of the steel box at the refinery.

The parametric study conducted allowed to conclude that if the plate launched by the failure of the steel box would have hit a pipe in the surrounding the eventual loss of containment would have not created consequences in terms of damage ares (for toxic and/or flammable dispersion of an hazardous chemical processed at the facility) greater than those already evaluated by the Owner in its Safety Report (EU Seveso Directive). Safety Report assessment can be considered consistent with the effective industrial risk of the facility.

7.7.6 Forensic Engineering Highlights

The presented case study highlights the benefits deriving from the use of special tools for the analysis of evidence and the definition of hypotheses. In the specific case, in fact, the event did not cause damages to the pipes in the area. However, there was a potential for damage and domino effects. This potential damage could have put into question the safety analysis developed by the plant for its plant in accordance with the requirements of the Seveso Directive which provides for considering both the direct

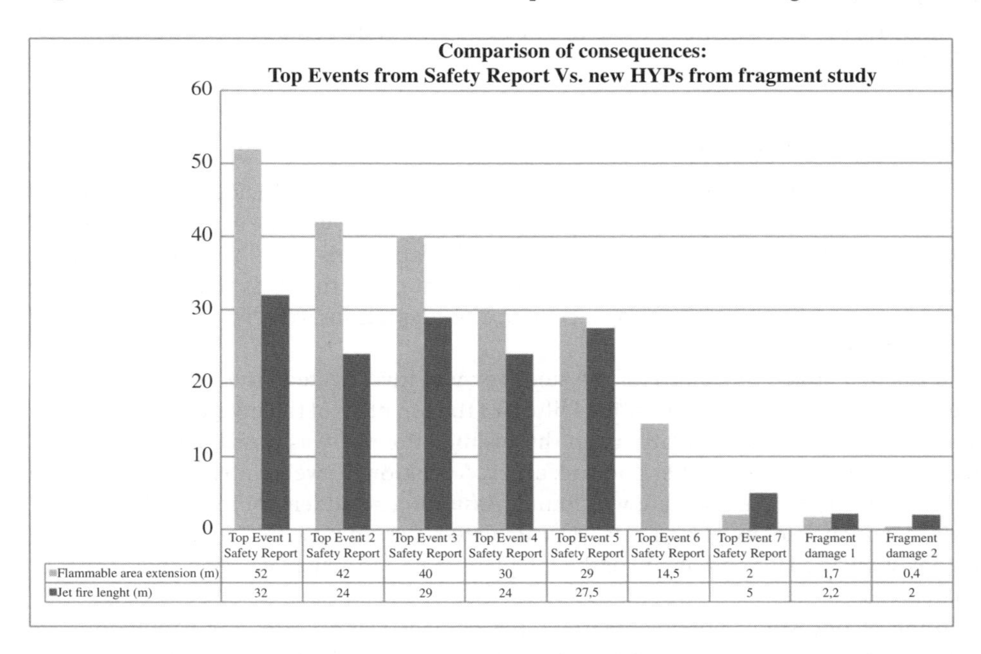

	Top Event 1 Safety Report	Top Event 2 Safety Report	Top Event 3 Safety Report	Top Event 4 Safety Report	Top Event 5 Safety Report	Top Event 6 Safety Report	Top Event 7 Safety Report	Fragment damage 1	Fragment damage 2
Flammable area extension (m)	52	42	40	30	29	14,5	2	1,7	0,4
Jet fire lenght (m)	32	24	29	24	27,5		5	2,2	2

Figure 7.7.30 Comparison of consequences: Top Events from Safety Report Vs. new HYPs from fragment study. Flammable top events comparison.

events and the effects connected to secondary events or dominoes, among which are also included the containment losses associated with the projection of fragments. In order to verify the consistency of the contents of the Safety Report, a detailed analysis of the environment and of the evidences recorded in the incident event was then carried out. On the basis of the estimate of the energy available to the projectile it was therefore assumed a damage to the lines containing dangerous chemicals and the degree of damage to them was studied. The approach, conducted through the use of advanced simulation, has been validated with respect to the available technical literature that recalls real and recognised experiments. In the event that the damage results in a potential loss of containment with consequent incidental scenario (dispersion of toxic or flammable vapors), an estimate was made of the consequent effects in terms of damage distances and the comparison with the damage distances associated with the incidental accident scenarios derived from the current risk analysis (Safety Report) for the refinery.

7.7.7 References and Further Readings

Reference

1 Palmer, A., Neilson, A., Sivadasan, S. (2006) Pipe perforation by medium-velocity impact. *International Journal of Impact Engineering*, 32(7):1145–1157.

Further Readings

ASCE. (1997) *Design of Blast Resistant Building in Petrochemical Facilities*. Reston, Virginia: American Society of Civil Engineers; 1997.

Century Dynamics Inc. (2002) *AUTODYN-Theory Manual*. Century Dynamics Inc; 2002.

NUMERICS GmbH. (2010) "SPEED-Product Description". http://numericsgmbh.de/en/speed.html. NUMERICS GmbH; 2010.

Schmidt, E. (1963) *Thermodynamik*. Springer Publishing; 1963.

Xiaoqing, M., and Stronge, W. (1985) Spherical missile impact and perforation of filled steel tubes. *International Journal of Impact Engineering*,. 1985;3(1):1–16.

7.8 Refinery Process Unit Fire

7.8.1 Introduction

The general information about the case study is shown in Table 7.8.1.

Some of the contents of this example are taken from the MARS report [1]. EU official MARS database on major accidents (Seveso Directive related) can be consulted online for free at https://emars.jrc.ec.europa.eu.

7.8.2 How it Happened (Incident Dynamics)

On 1 December 2005 at 21:40 a fire broke out affecting the heavy fuel purification plant of the refinery.

The company made following statements concerning the accident.

Table 7.8.1 General information about the case study

Who	Crude oil refinery
What	Fire
When	September, 2nd 2005
Where	Province of Genova, Italy
Consequences	No safety consequences. Economic losses = 7 M€
Mission statement	Reconstruct the fire dynamics
Credits	Gianfranco Peiretti (IPLOM S.p.A.)

The shift personnel present testified that a strong hiss was heard lasting a few seconds which was followed with the ignition of the released product and that there had not been any noticeable pressure changes (like pressure waves etc.). The shift personnel present onsite and in particular the shift foreman who was staying on the main access ramp to the atmospheric distillation plants, testified that he saw the jet-fire ignite from the top and propagate down (from approx 18m above ground down to approx 14m in the area comprised between the exchangers E1718 A/B/C and the reactor R1702). Immediately the shift foreman activated the onsite emergency plan and informed the gate guard in order to alert the fire brigade. At 21:40 the fire brigade was alerted. The jet fire affected the quench-line with the 3" hydrogen pipe, which ruptured after 6 min. exposure with the consequent ignition of the hydrogen. The fire took a cylindrical form from the bottom to the top starting at 14 m height affecting above located plant parts, comprising under other the pre-heater of the diathermic oil. Approx 30 minutes after the fire ignited an 8" fuel pipe of the diathermic oil system ruptured and subsequent ignition of the product. The fire was kept under control and evolved without noticeable changes until consumption of the fuel once the pipes were shut off according to the emergency response plan which had been activated.

The fire was extinguished at 1:20 a.m. on September 2005 (3 hours and 40 minutes after the fire initiated) and the state of emergency was called off by the fire brigade at 1:45 a.m.

No damages to persons has been reported consequent to the accident.

According to damage evaluations performed, there has not been any environmental damage, this evaluation has been confirmed also by an environmental indicators assessment performed by ARPA of Genoa.

The accident affected the catalytic hydro-treatment plant named Unit 1700. The Block diagrams relating to this are shown in Figures 7.8.1 and 7.8.2. The main characteristics of the plant are:

- Capacity 1650 t/d light fuel treatment section;
- 145o t/d heavy fuel treatment section Design and construction Tecnimont Construction January - August 1997 Operation start September 1997; and
- Unit 1700 has been designed to improve the characteristics of light and heavy fuel oil produced in the refinery by treating the fuel fractions with high pressure hydrogen.

The technology employed consists essentially in treating the fuel oil with high pressure hydrogen, on a specific catalyst, such to eliminate the sulphur in the fuel oil,

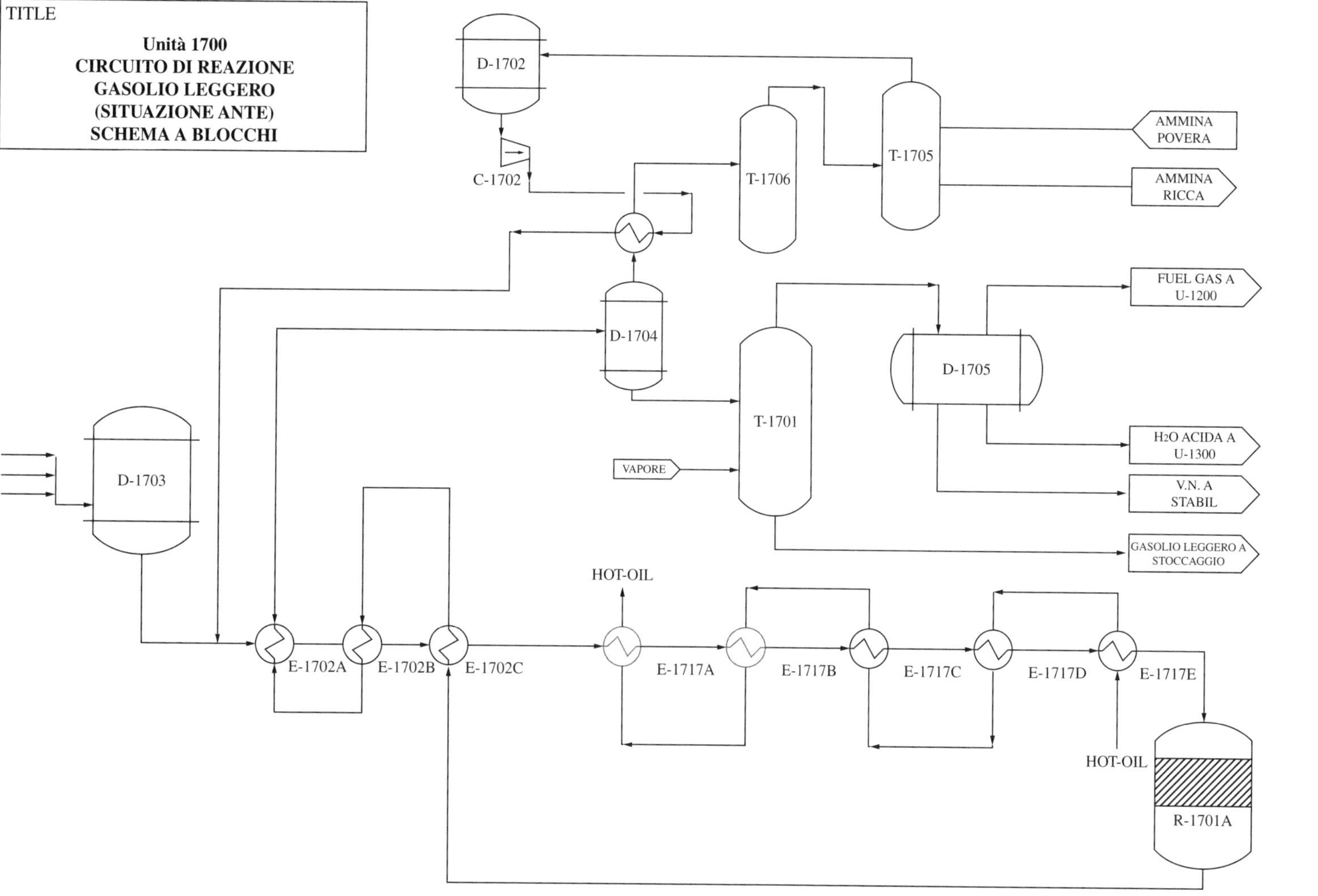

Figure 7.8.1 Block Flow Diagram of the light fuel treatment section, before the incident Source: (Courtesy of IPLOM S.p.A.).

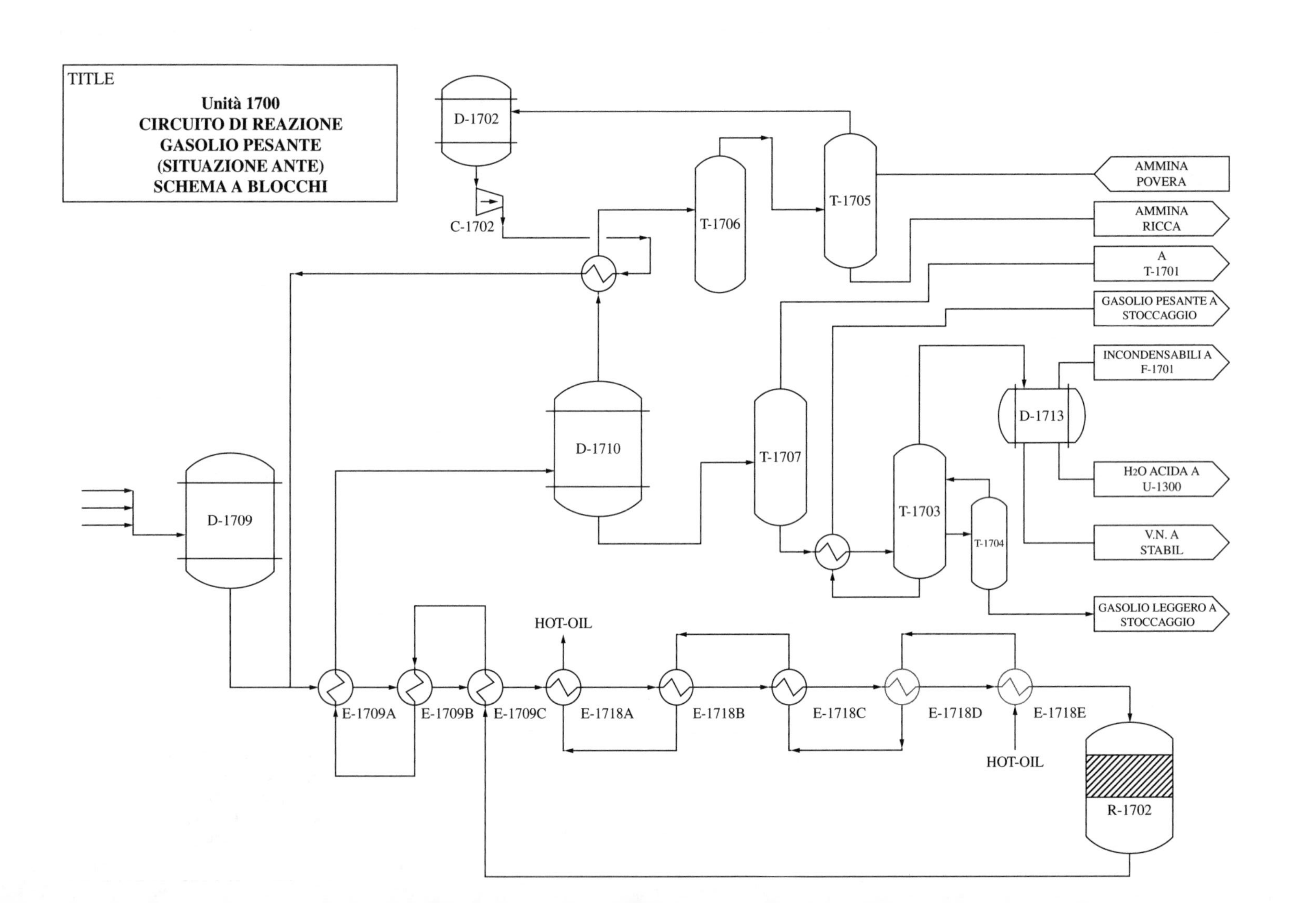
TITLE
Unità 1700
CIRCUITO DI REAZIONE
GASOLIO PESANTE
(SITUAZIONE ANTE)
SCHEMA A BLOCCHI
D-1702
C-1702
T-1706
T-1705
AMMINA POVERA
AMMINA RICCA
A T-1701
GASOLIO PESANTE A STOCCAGGIO
INCONDENSABILI A F-1701
D-1713
H2O ACIDA A U-1300
V.N. A STABIL
GASOLIO LEGGERO A STOCCAGGIO
D-1710
T-1707
T-1703
T-1704
D-1709
HOT-OIL
E-1709A
E-1709B
E-1709C
E-1718A
E-1718B
E-1718C
E-1718D
E-1718E
HOT-OIL
R-1702

produce hydrogen sulphide, hydrogenate the hydrocarbons and improve the other characteristics.

The plant was designed with two heating - reaction - fractionating sections in consideration of the different characteristics of the charges to be treated, respectively one for light fuel oil mixtures (unit 1700) from the Topping and one for the heavy fuel oil mixtures from Vacuum, while foreseeing one single gas purification and compression section for the recycled gas to be reintegrated in the circuit.

The charge, made of a heavy fuel oil mixtures coming from the vacuum distillation plant, is sent to Unit 1700b through 3 pipelines equipped with flow control systems. Once the water contained is eliminated, the charge is pre-heated and then transferred to the reaction section working at 60 bars of pressure.

Before the charge enters the reactor the charge is mixed with the hot recycled hydrogen and then heated up to optimal temperature for the catalytic reaction (on average 360°) through a closed circuit of heating oil. An appropriate catalyst inside the reactor facilitates the desulphuration reactions inside the reactor (hydrogenation of sulphur in H2S).

The reactor effluents , constituted by desulphurised fuel oil and a gas mixture, is cooled down and sent to a liquid-gas separator. The gas constituted principally of hydrogen is washed and purified from the hydrogen sulphide before being recycled, while the desulphurised heavy fuel oil feeds a pre-strippingcolumn (T- 1707), the head flow of this column is then sent to the stripping column of the light fuel-oil section (T- 1701).

The pre-stripped heavy fuel fraction (bottom T-1707) is heated and sent to the main fractionating column (T-1703) in which following products are separated:

- Uncondensable gas;
- virgin nafta; and
- light desulphurised fuel oil heavy desulphurised fuel oil.

The light fuel oil, is sent from the fractionating column to a stripper (T-1704) in which hydrocarbon tails and water are eliminated. The heavy fuel oil is cooled down and transferred to storage.

The area affected by the accident corresponds to the heavy fuel oil purification circuit comprising reactor R1702, the charge /hot oil circuit heat exchange train, the hydrogen injection circuit (quench) and reactor R1702 comprising the control instrumentation.

The incident put at risk about 20 onsite people and 6 emergency personnel (off site). The material losses have been quantified in 5 millions of Euros, while the response, cleanup and restoration costs in 7.6 millions of Euros. The severity of the incident required the interruption of the road and railway crossing nearby. The incident had large media coverage. Figure 7.8.3 shows some photos of the incident.

7.8.3 Why it Happened

Following the results of the investigation performed by the company and the analysis of an amateur video, the company has formulated following assumptions concerning the accident.

The accident, considering the products processed, could have originated by the failure of one of the following plant components:

- Pipes leading to the pressure gauges of reactor R-1702;

Figure 7.8.3 Photos of the incident Source: (Courtesy of IPLOM S.p.A.).

- recycled gas pipe at the bottom of the reactor R-1702 having a quench function;
- diathermic oil pipe (hot oil) entering or exiting the exchanger; and
- flanged joints exchanger E-1718, E-1709 and connection lines.

The company excludes a release from the hot oil circuit as triggering factor of the fire, based on the evidence gathered from the records on the pressure in the circuit which demonstrate the failure 30 min after fire start. Also the video confirms the pipe rupture 30 min after the fire began.

For the same reason a release from the hydrogen pipes it is not considered likely, the records demonstrate that the hydrogen pipe failed 7 min after fire began.

Concerning the flange joints of exchangers E-1718, E-1719 experts requested the dismounting of the exchanger flanged joints, the joint gaskets resulted in not being damaged.

For this reason the company considers the failure of a pipe from the pressure measurement gauges of reactor R-1702 as the most likely accident triggering factor, this assumption is supported by the following facts:

- This part is located in the area corresponding to the epicentre of the fire;
- the area corresponds to the area visually identified by the witnesses;
- the product release (hydrogen an fuel oil) from one of this pipes can cause a 6 m long jet flame as occurred;
- the product supposedly released would have had a high enough temperature and pressure to self-self- ignite or ignite against a hot spot of the plant like the hot oil circuit;
- the damages recorded are caused by overheating (flame exposition) and were not caused by overpressure or explosion.

The pressure measurement records confirm significant pressure changes at the beginning of the event. The company does not have any element allowing to identify the failure cause of that pipe.

Table 7.8.2 lists the events in a tabular timeline.

7.8.4 Findings

The onsite emergency response service of that shift was composed of six persons having the following functions:

- Shift manager in charge of the emergency management;
- Gate guard responsible for the external communication;
- 1st product transfer operator in charge of fire fighting;
- 2nd product transfer operator in charge of fire fighting;
- 3rd distillation plant worker in charge of fire fighting; and
- 5th processing plant worker in charge of fire fighting;

All the team assisted by the other shift personnel forming the operative team, and participated from the beginning of the event in implementing the emergency response.

The operative team of the shift constituted 8 workers with the following functions:

- Shift foreman coordinator responsible for securing the installations;
- Distillation plant Q1 control-room operator responsible for securing the plant from the control-room;
- Processing plants Q2 control-room operator responsible for securing the plant from the control-room;
- 1st distillation worker responsible for securing the installation equipment;
- 3rd services operator responsible for securing the installation equipment;
- Processing plant operator responsible for securing the installation equipment;

Table 7.8.2 Tabular timeline of the main events.

Progression	Time	Event
00.00"	21.40	Event takes place
00.30"	21.40	Shift supervisor declares the emergency
01.00"	21.41	Night porter phones National Fire Brigade and alerts refinery managers
01.00"	21.41	CTI operator activiates foam monitor to protect E1701/E1702/E1709 heat exchangers
01.00"	21.41	4° Field operator activates steam barriers on E1717/E1718 heat exchangers
01.00"	21.41	Shift supervisor activates foam pump to pressurize foam network
01.00"	21.41	5° Field operator activates foam monitor loated near the U1100 process unit in order to prepare a foam bed under the unit
02.00"	21.42	1° operator activates a monitor between conversion and distillation units
02.00"	21.42	1° and 2° operators from storage area activate foam monitors to prevent U1700 gound from hydrocarbons pool formation at the basis of the process unit
02.35"	21.42	From control room panel operator Q2 activates the process shut-down of the main involved process unit (U1700B) together with the trip of C1701/2 compressors. Furthermore he activates the emergency depressurization of the hydro desulphurization process units (HDS). Depressurization is completed in 15'.
03.00"	21.43	I vigili del fuoco di Busalla entrano in raffineria, prendono il comando delle operazioni e iniziano il raffreddamento dell'impianto in ciò coadiucati dalla squadra d'emergenza IPLOM Arrival of the National Fire Brigades. Cooling actions directed to the plant structures with the support of the internal emergency team.
05.00"	21.45	1° storage area field operator activates cooling rings on S48 and S49 storage tanks.
06'.14"	21.46	Probable failure of 3" diameter line with H2 quenching to R1702 reactor.
08'.00"	21.48	2° storage area field operator activates cooling rings on S88, S89, S90 and S33 storage tanks.
10'.00"	21.50	Town Major arrives at the refinery
11'.00"	21.51	Refinery director arrives at the refinery
18'.00"	21.58	Bolzaneto National Fire Brigade team arrives at the refinery
25'.00"	22.05	Genova National Fire Brigade team arrives at the refinery
28'.00"	22.08	Novi Ligure National Fire Brigade team arrives at the refinery
30'.30"	22.10	8" hot oil processing line catastrophic failure. BLEVE from 25 m to 65 m.
33'.00"	22.13	Topping Unit general shut-down
39'.00"	22.19	Vacuum Unit general shut-down

(Continued)

Table 7.8.2 (Continued)

Progression	Time	Event
50' .00"	22.30	Fire becomes limited in extension
1h 50'	23.30	Genova Airport National Fire Brgade team arrives at the refinery with Perlini trucks equipped with very high flow/rate – pressure monirts
3h 35'	01.15	Extremely reduced flames lenght due to nitrogen purge
3h 40'	01.20	Fire completely extinguished
	01.45	Emergency end declaratio
	02.00	Plant purging
	02.00	National Fire Brigade team remains at the site for the night
	02.00	Fax communication to the authorities having jurisdiction to notify the major accidents event according to EU Seveso Directive

- Shift foreman in charge with product transfer responsible for the securing of the storage facilities; and
- 3rd product transfer operator responsible for securing the Boccarda storage.

After approx. 3 minutes from the beginning of the event the fire brigade of Busalla (a team with 6 fire fighters) arrived on-site, subsequently arrived fire brigades from Bolzaneto, Genoa and Novi Ligure with a total of 50 fire fighters.

The off-site emergency plan in force has a temporary value and has a provisory character and was drawn up by the prefecture of Genoa in 1998. The updating activity of the off-site emergency plan has been requested by the prefecture and is still under elaboration. The new offsite emergency plan is under evaluation by the local authorities. The evaluation of the emergency response measures has to be considered preliminary awaiting the results of the technical assessment requested by the competent authorities, the results of the investigation may help to identify the accident causes and indicate organizational measures to improve safety.

7.8.4.1 Examination of the Effects of the Fire

The high pressure of the circuit and the temperature of the leaked product (64 barg and 245 °C) determined initial flame lengths of approx. 6 meters estimated by the witnesses.

Verifications carried out by Tecsa spa using mathematical models confirm the possibility of flame lengths for similar events.

After inspection after the fire, damage to the structures and equipment from 14 meters up to 25 m was detected on an area of 10x10 square meters.

In particular in a smaller area of approx. 5 meters by 5 meters (between 15 and 20 meters) the temperatures reached resulted in the release of the product from the flanged couplings, the collapse of the hot oil and hydrogen pipes, the deformation of the support beams of the E1718 exchangers at an altitude 18 m, the melting of the aluminum sheet and the roasting of the rock wool cushion of the thermal insulation of the equipment.

The main damage area coincides with the point of origin of the fire visually identified immediately by the CTP and other shift staff.

In this area there are the reactor, the E1718 ABCDE exchangers, the hydrogen quench line and the measurement instruments of the Delta P of the R1702 reactor.

The fire extended after approx. 30 minutes to the hot oil circuit, determining the failure of the 8 "pipe at a 25 m altitude for simultaneous yield and internal pressure with the release of the product at a pressure of about 11 barg and at a temperature of about 385 °C (data recorded at DCS) generating a flare ray of about 20 meters with a progress from bottom to top in the direction of the topping column damaging the paint of the insulation and damaging some light points of the 1100 unit.

7.8.4.2 Water and Foam Consumption

During the emergency, approx. 10,000 cubic meters of water with an average flow rate higher than 3,000 cubic meters per hour and 16 cubic meters of foaming liquid.

The reduced consumption of foaming is due to the fact that the fire has developed in altitude without consequences on the ground. It was therefore sufficient to firstly create a foam mat at the base of the plants affected by the fire and keep it constant by the operation of a single foam dispenser. All the other fire-fighting equipment was used to supply only water in order to enhance the cooling of the plants resulting in this pre-eminent action to combat the fire.

After verification of the fire, it was verified that no damage was reported to the structures and equipment below 10 m.

7.8.4.3 Damages

The damages caused by the fire are limited to the unit of purification of the diesel fuel in the deck containing the light and heavy diesel fuel exchangers and in the R1702 reactor deck for an area of about 100 square meters starting from a height of 14 m towards the high.

It was noted in particular:

- Damage to the support structure between 14 and 21 m (Figure 7.8.4);
- damage to the piping between 14 and 25 m;
- damage to electrical and instrumental equipment between 14 and 25 meters;
- effects of overheating and sudden cooling (water jets) to heat exchangers between 14 and 25 meters;
- damage to the insulation of the r1702 reactor;
- loss of the catalyst characteristics of r1702 and r1701a reactors due to prolonged plant shutdown;
- damage to light points and electrical instrumentation cables on adjacent systems; and
- damage to paint and insulation of nearby systems.

7.8.5 Lessons Learned and Recommendations

The operator decided to rebuild the plant with a new executive project, in consideration of the damage caused to the plant, maintaining the same production layout.

The new executive project foresees essentially:

- The complete separation of the light fuel oil section and the heavy fuel section such to avoid for example the possibility of domino effects;
- lowering the maximum height for the installation of exchangers from 25 m to 15 such to facilitate fire extinguishing operations;

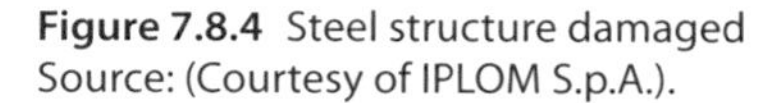

Figure 7.8.4 Steel structure damaged Source: (Courtesy of IPLOM S.p.A.).

- reconstruction of the plant in compliance with the PED directive (CE n° 97/23); and
- rationalising the piping system to minimise adjacencies, relocate valves on the hydrogen quench line in R1072 to maintain the line depressurised, reduce the number of measurement gauges and insertion of valves in a safe area for depressurising the hot oil circuit.

In the following Figures 7.8.5 to 7.8.10, the *ante* and *post* incident situations are shown.

7.8.6 Forensic Engineering Highlights

The investigation required a preliminary onsite inspection, with the main task of evidence collection. Figure 7.8.11 shows some forensic engineering highlights about evidence collection, tagging, and movement. Forensic activities were conducted both in order to discover the causes and the fire dynamics as well as to identify specific areas for improvement. Fire dynamics assessment was supported by both the use of digital data recorded by the distributed control system (DCS) of the refinery (process control system), of amateurs video and of specific simulation carried out with quantitative risk assessment tools. In Figure 7.8.12 some screenshots of the evaluations carried out by TECSA S.r.l. with DNV Phast Professional are shown.

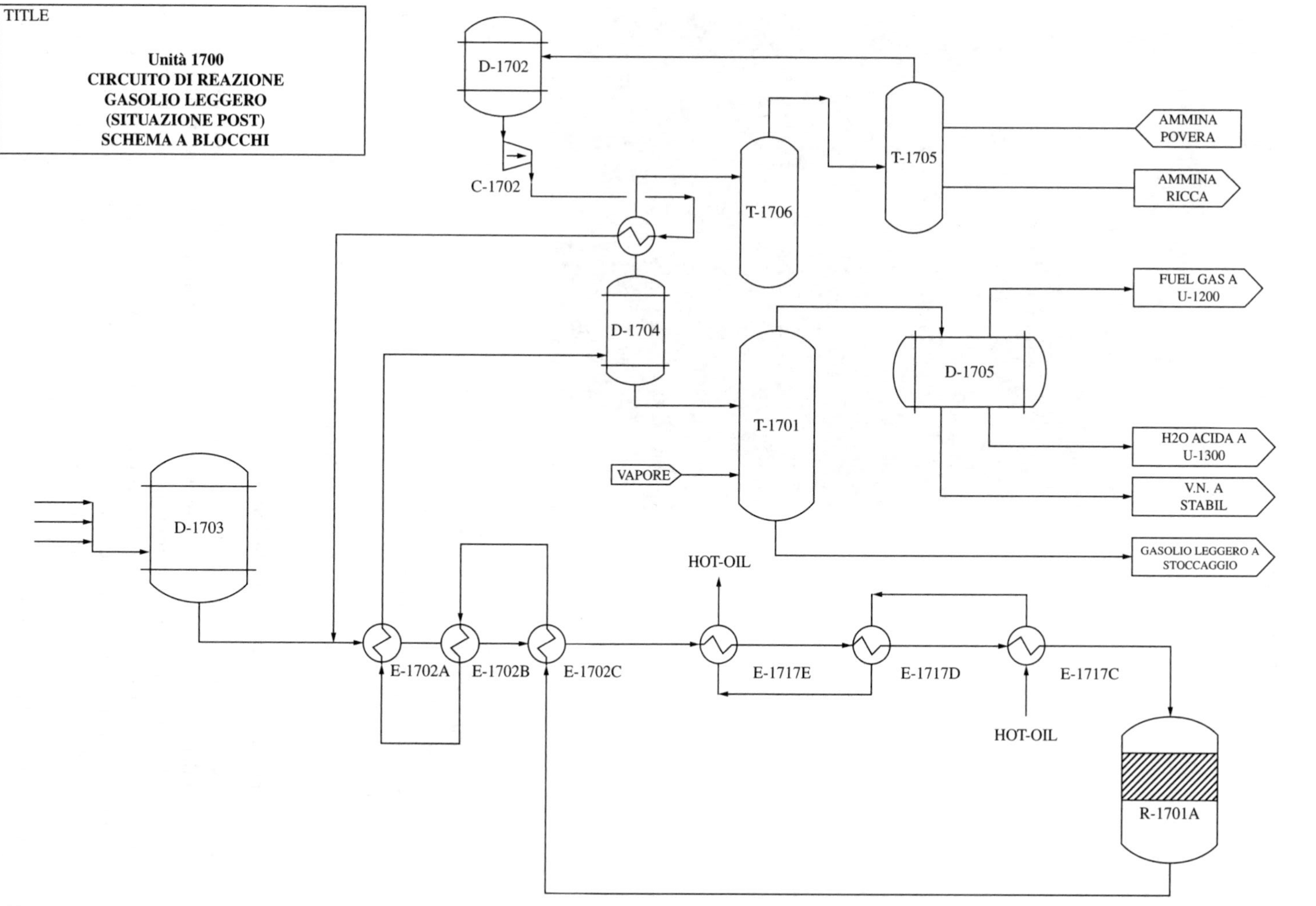

Figure 7.8.5 Block Flow Diagram of the light fuel treatment section, after the incident Source: (Courtesy of IPLOM S.p.A.)

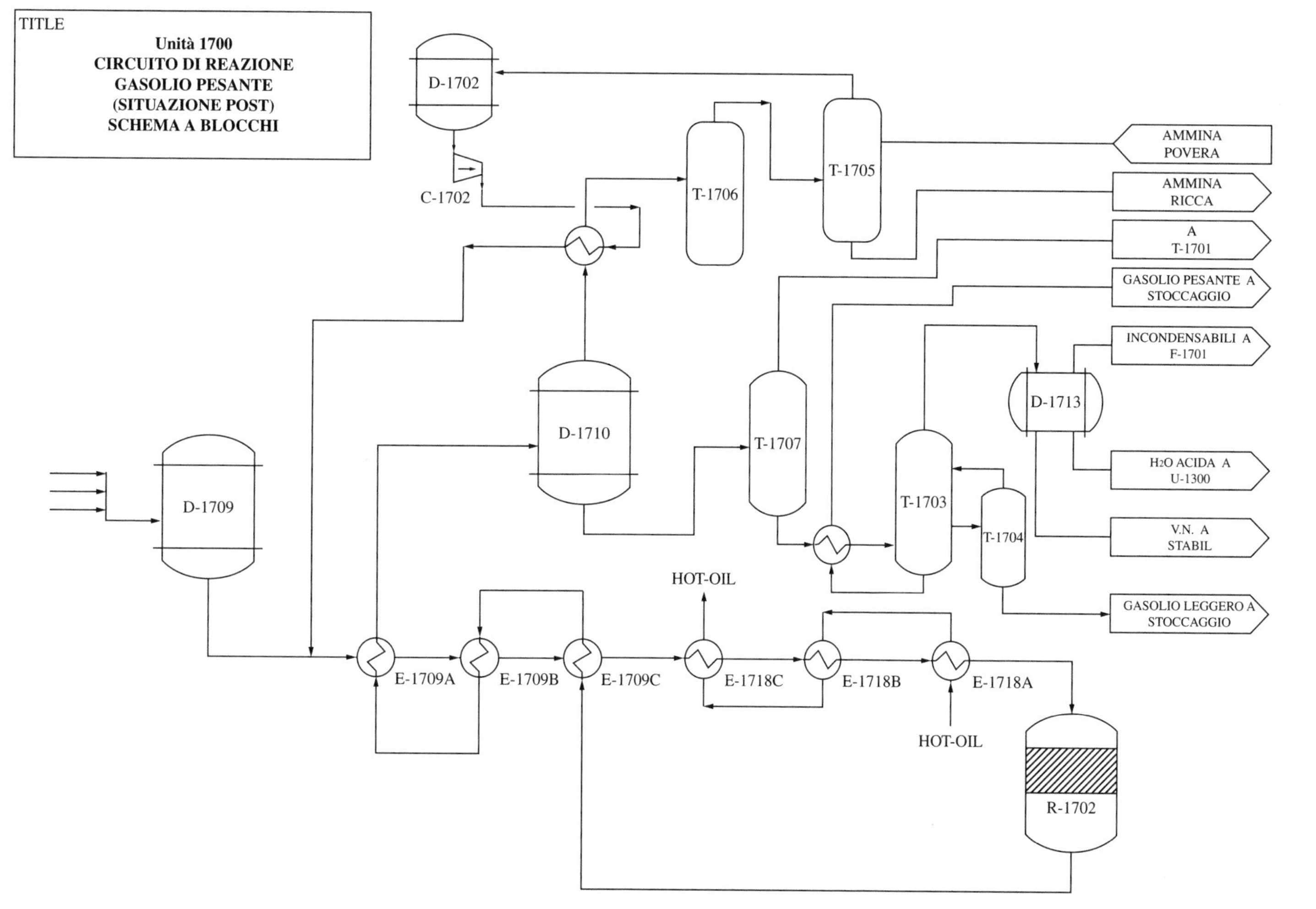

Figure 7.8.6 Block Flow Diagram of the heavy fuel treatment section, after the incident Source: (Courtesy of IPLOM S.p.A.).

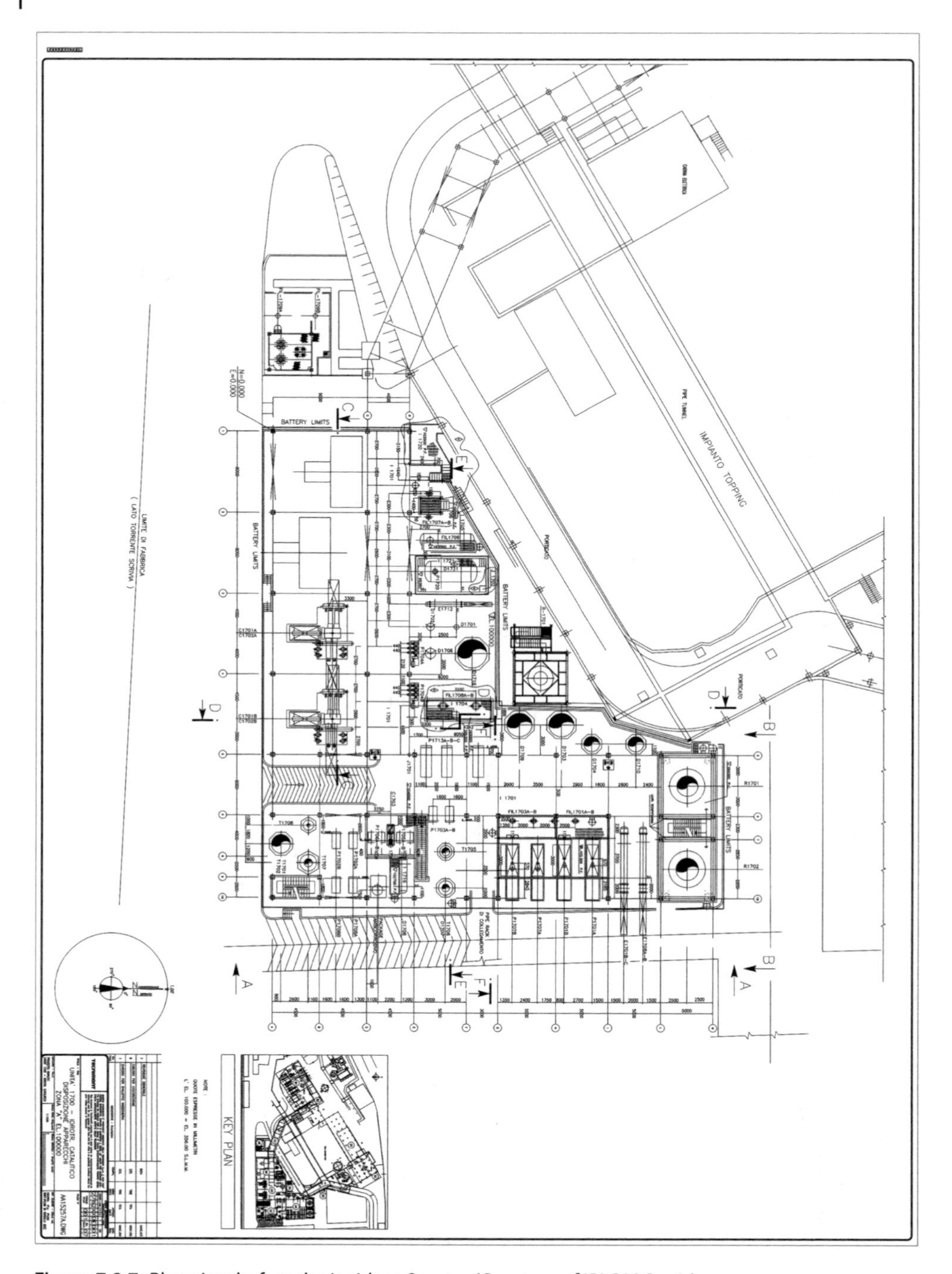

Figure 7.8.7 Plan view before the incident Source: (Courtesy of IPLOM S.p.A.).

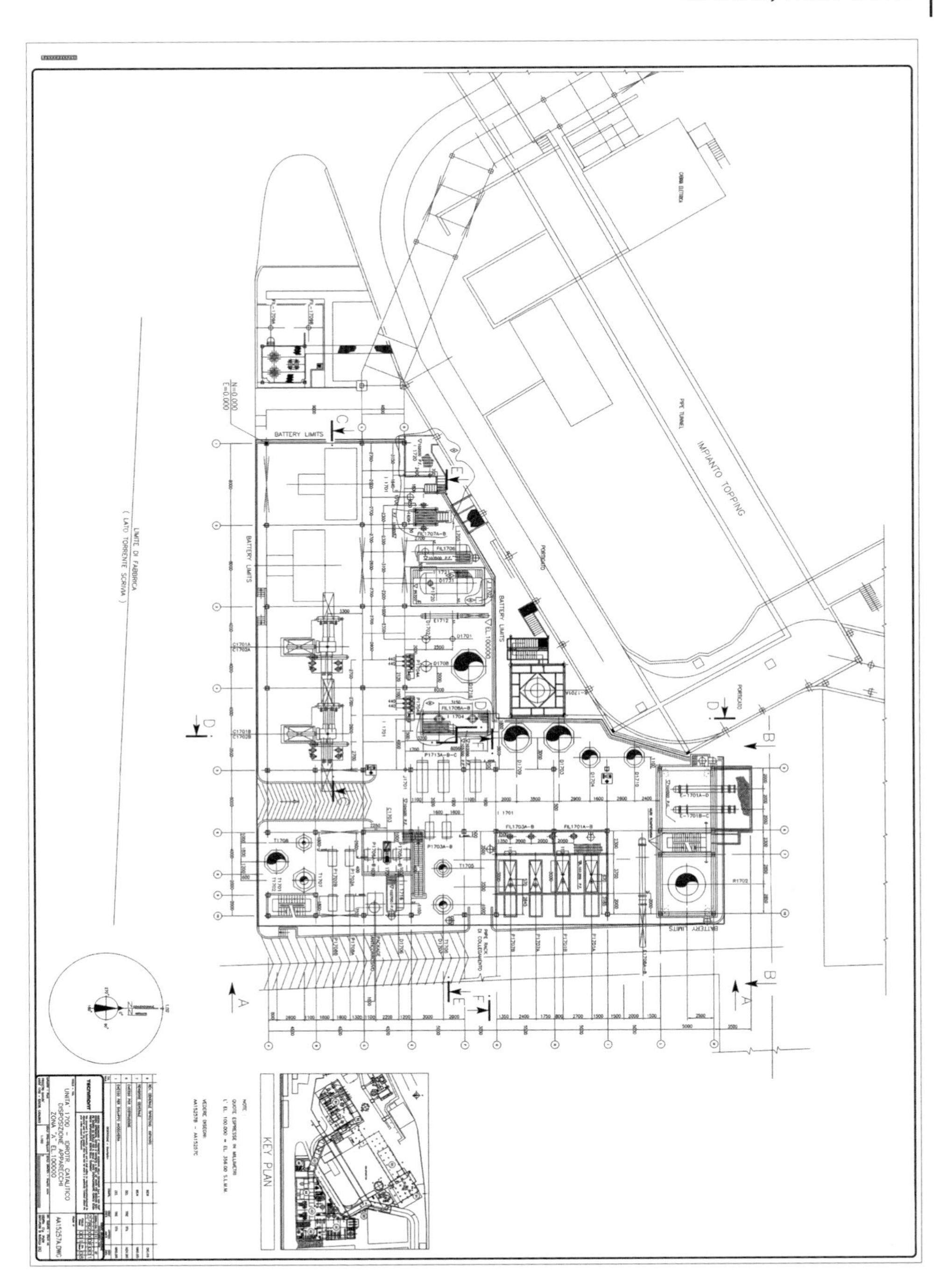

Figure 7.8.8 Plan view after the incident Source: (Courtesy of IPLOM S.p.A.).

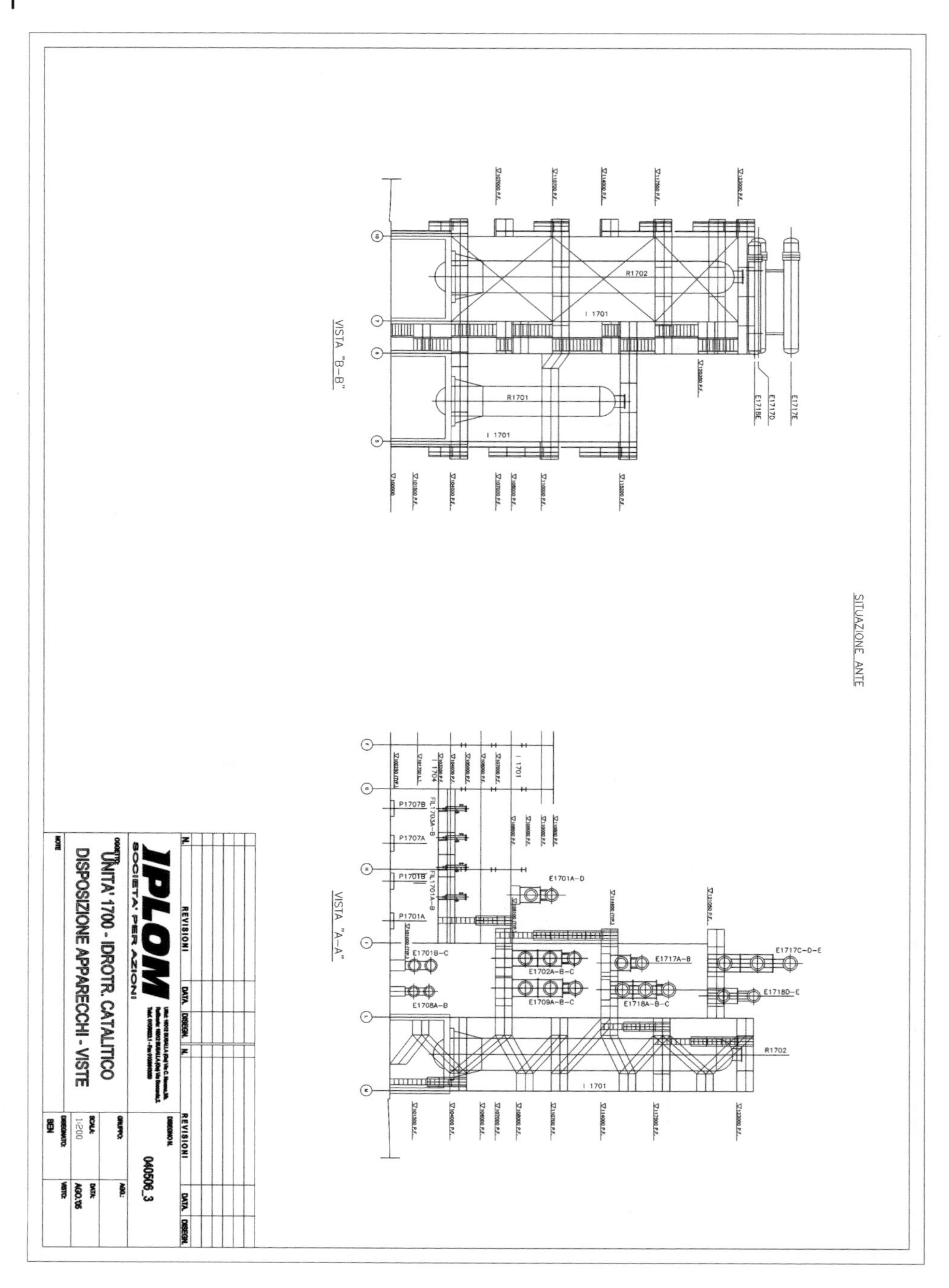

Figure 7.8.9 Unit 1700. Arrangement of equipment before the incident Source: (Courtesy of IPLOM S.p.A.).

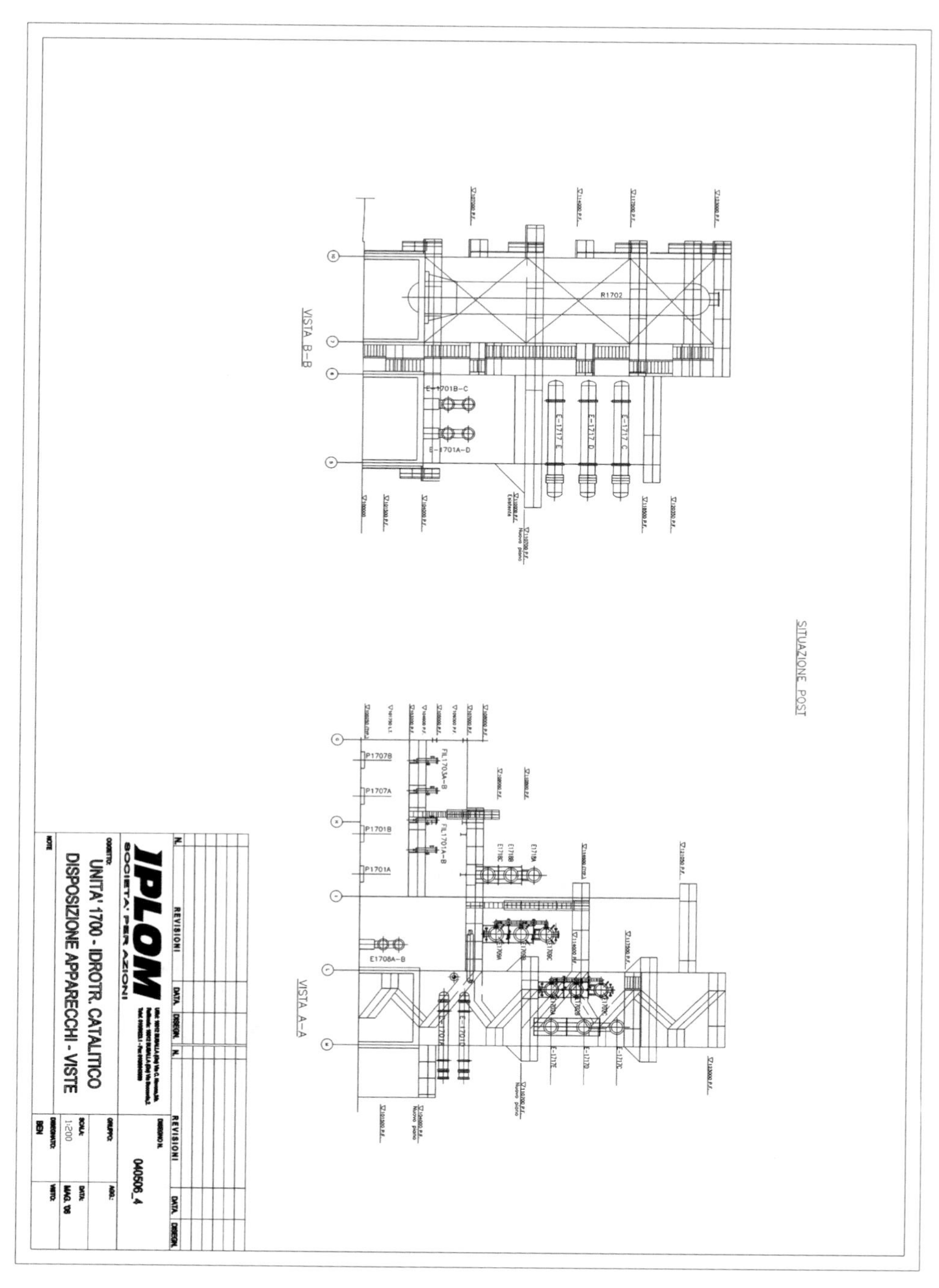

Figure 7.8.10 Unit 1700. Arrangement of equipment after the incident Source: (Courtesy of IPLOM S.p.A.).

Figure 7.8.11 Forensic engineering highlights about evidence collection, tagging, and movement Source: (Courtesy of IPLOM S.p.A.).

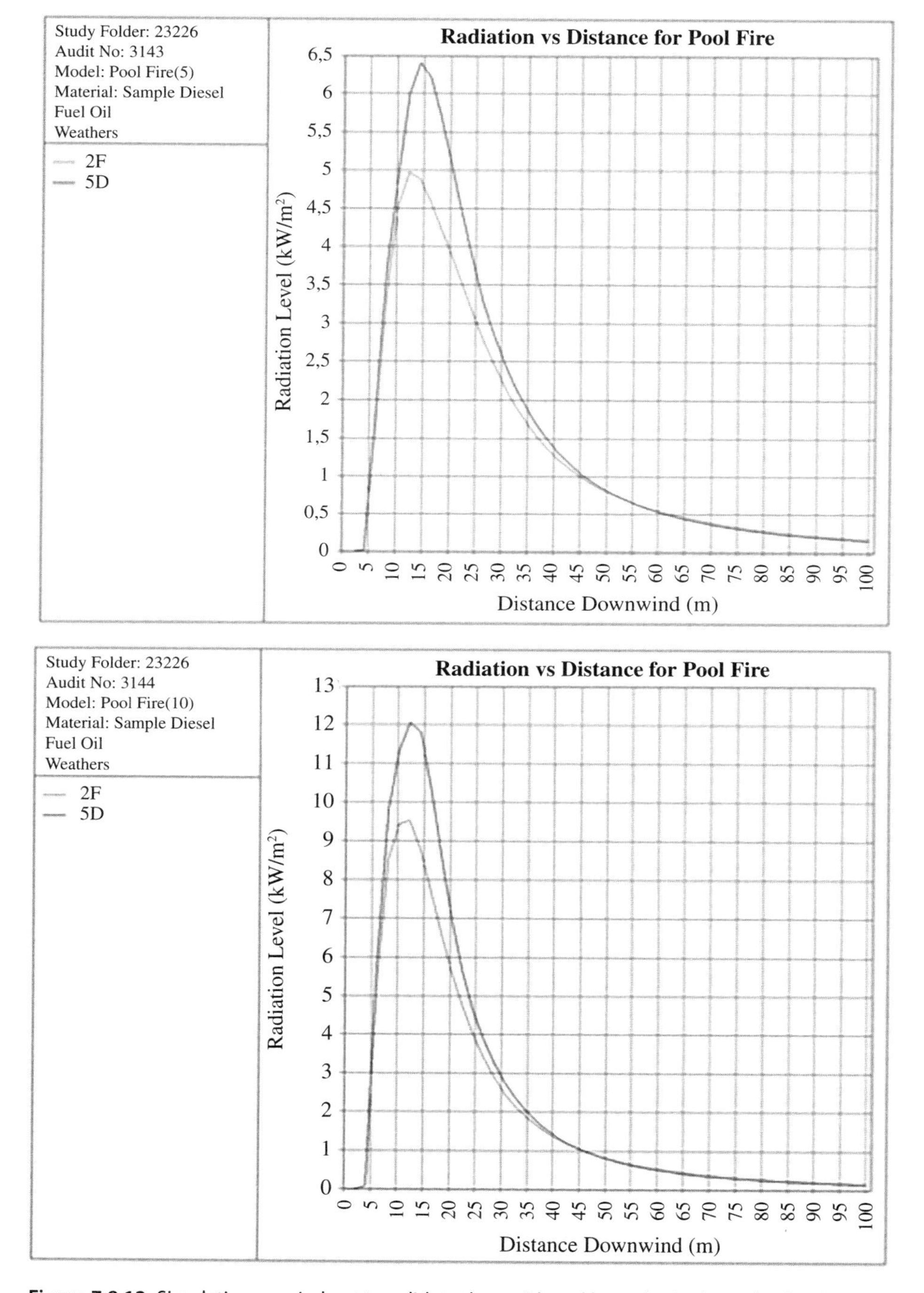

Figure 7.8.12 Simulations carried out to validate the accidental hypothesis about the fire dynamics Radiation at 5 (top) and 10 meters (bottom) by pool fire, in different weather condition (2F and 5D). Source: (Courtesy of TECSA S.r.l.).

Following all the investigations carried out, it was possible not only to identify the specific causes of the incidental event but also to define ways of reconstructing the plant to minimise the potential for damage following a similar event. The observation both of the dynamics of the incident and of the emergency management procedures, with particular reference to the fire-fighting operations carried out by both the internal teams and the external teams, have shown that through a series of specific plant and layout arrangements it is possible to minimise the degree of damage connected with a series of direct and indirect incidental events, which involve both the secondary effects and the domino effects. These layout adjustments, alongside the adoption of active fire protection systems, can improve response to emergencies while minimizing the risk for operators involved in the management of the fire emergency. The study activities of this specific case resulted in the definition of significant improvements aimed at achieving a greater level of inherent safety whose basic criteria can not only obviously be applied in the creation of new process plants but also and perhaps above all to the major revamping activities of existing plants in order to adapt their performance in the field of process safety with respect to foreseeable accident events.

7.8.7 References and Further Readings

Reference

1 EUROPA-eMARS Accident Details - European Commission [Internet]. Minerva.jrc.ec.europa.eu. [cited 15 November 2017]. Available from: https://minerva.jrc.ec.europa.eu/en/emars/accident/view/19158d8a-2bb2-4ea5-dd8a-b0e5a9b7cb0a

Further readings

Ahmad, S., Hashim, H., and Hassim, M. (2016) A graphical method for assessing inherent safety during research and development phase of process design. *Journal of Loss Prevention in the Process Industries,* 2016;42:59–69.

CCPS (Center for Chemical Process Safety). (2009) *Inherently Safer Chemical Processes: A Life Cycle Approach.* 2nd ed. Hoboken: John Wiley & Sons; 2009.

Cozzani, V., Tugnoli, A., and Salzano, E. (2009) The development of an inherent safety approach to the prevention of domino accidents. *Accident Analysis & and Prevention.,* 2009; 41(6):1216–1227.

Kidam, K., Sahak, H., Hassim, M., Shahlan S, Hurme M et al. (2016) Inherently safer design review and their timing during chemical process development and design. *Journal of Loss Prevention in the Process Industries,.* 2016;42:47–58.

Kletz, T., and Amyotte, P. (2009) *Process plants: A Handbook for Inherently Safer Design.* 2nd ed. Boca Raton, FL: CRC Press/Taylor & Francis; 2010.

Mansfield, D., Turney, R., Rogers, R., Verwoerd M, Bots P et al. (1995) How to integrate inherent SHEshe in process development and plant design. ICHEME Symposium Series No. 139.

Rathnayaka, S., Khan, F., and Amyotte, P. (2014) Risk-based process plant design considering inherent safety. *Safety Science,* 2014;70:438–464.

Shariff, A., and Leong, C. (2008) Inherent risk assessment—A new concept to evaluate risk in preliminary design stage. *Process Safety and Environmental Protection,* 2009;87(6):371–376.

Shariff, A., Leong, C., and Zaini, D. (2012) Using process stream index (PSI) to assess inherent safety level during preliminary design stage. *Safety Science*, 2012;50(4): 1098–1103.

Shariff, A., Wahab, N., and Rusli, R. (2016) Assessing the hazards from a BLEVE and minimizing its impacts using the inherent safety concept. *Journal of Loss Prevention in the Process Industries*, 2016;41:303–314.

7.9 Crack in an Oil Pipeline

7.9.1 Introduction

The general information about the case study are shown in Table 7.9.1.

The probability of a spill occurring along a pipeline lies at the core of risk management for pipeline operators. Thus, a look at historical accident trends may provide some insight into this probability [1].

Analyses of data for U.S. petroleum product pipelines operating between 1982 and 1991 indicate that such pipelines of short-to-moderate lengths (for example, 50 miles) are likely to have at least one reportable spill within a 20-year period. Longer lines (as much as 1,000 miles, for example) may suffer a reportable spill within 1 year.

These are major conclusions of analyses by EFA Technologies Inc., Sacramento, of statistics compiled by the U.S. Department of Transportation (DOT) on liquid pipelines operated under the Code of Federal Regulations (CFR) Title 49D, Part 195 Transportation of Hazardous Liquids by Pipeline [2].

In 1992, moreover, the data show that 52.5% of the oil spilled in the U.S. in accidents of more than 10,000 gal each came out of pipelines. Worldwide, pipelines caused 51.2% of the leaks involving this magnitude.

The most important problem in oil pipeline is corrosion: as mentioned in Corrosion Control In Oil And Gas Pipelines March 2010 Vol. 237 No. 3, Gas and Pipeline Journal [3], in the United States, the annual cost associated with corrosion damage of structural components is greater than the combined annual cost of natural disasters, including hurricanes, storms, floods, fires and earthquakes. Similar findings have been made by studies conducted in the United Kingdom, Germany, and Japan.

Typical corrosion mechanisms include uniform corrosion, stress corrosion cracking, and pitting corrosion, as a consequence it is recommended to consider the implications

Table 7.9.1 General information about the case study.

Who	Oil Pipeline
What	Rupture
When	2016
Where	Borzoli, Genoa, Italy
Consequences	Crude Oil Spill
Mission statement	Determine the cause of failure
Credits	Bernardino Chiaia (ARCOS Engineering s.r.l.) Stefania Marello (ARCOS Engineering s.r.l.)

of corrosion damage and failure in the design and, even if corrosion is considered, unanticipated changes in the environment in which the structure operates can result in unexpected corrosion damage. Moreover, the combined effects of corrosion and mechanical damage, and environmentally assisted material damage can result in unexpected failures due to the reduced load carrying capacity of the structure. It follows that an integrated approach based on the use of inspection, monitoring, mitigation, forensic evaluation, and prediction, is fundamental, *a fortiori* in old structures, by considering that this method can provide information about past and present exposure conditions but, in general, they do not directly predict residual life. This goal can be achieved by validated computer models, in which the accuracy is strongly dependent on the quality of the computer model and on associated inputs.

This case study concerns a rupture and spill which occurred in the Oil Pipeline running from Porto Petroli in Multedo to Oil Refinery in Busalla, Genoa (Italy); it was built in the early 1960s, becoming operational in 1963. The total length of the buried oil pipeline is 24.5 km, of which 19.8 km with the outer diameter equals to 16", and the other part has an outer diameter equals to 12". Wall thickness ranges between 7.14 and 12.7 mm, with 7.14 mm being the predominant thickness and MAOP is variable along the pipeline. Steel grade is API 5L X52, characterised by nominal (minimum) tensile yield stress fy = 358 MPa, and nominal tensile rupture stress ft = 455 MPa.

Specifically, the segment of the pipeline affected by the rupture is located between weld labelled as 8420 and weld labelled as 8430 in which wall nominal thickness is 8.74 mm. For our purposes it is important to observe that, at the time of construction, this kind of steel pipe allowed a ± 12.5% thickness tolerance and the pipe was not coated on the inside, whereas the outside was protected by a coating named Protector B.

As a consequence of the accident, the Genoa Public Prosecutor's Office opened a case in order to investigate the causes of rupture considering different aspects, i.e. geology, maintenance, steel, operations.

7.9.2 How it Happened (Accident Dynamics)

On April, 17th 2016, a segment of the Oil Pipeline 16" near Genoa, on the left bank of the Rio Pianego, was affected by a sudden rupture, during the pumping crude oil operation (Figure 7.9.1), from the oil tanker to the refinery in Busalla. Then, after preliminary investigation, oil tanker could leave the harbour and no detailed information about pumping operation has been acquired: it is very important to observe this, because the right explanation of the rupture requires the knowledge of all parameters playing a crucial role, i.e. pressure and back pressure from oil tanker, sequence of the pumping operations, precise nature of fluids pumped and their density. In particular, in this case, the fluids pumped were oil and water (so, fluids with different density, viscosity, weight, etc.), alternate.

Pipeline leaked approximatively 700,000 l of oil, a part of which reached the nearby Rio Pianego, until its confluence with Rio Fegino, reaching Torrente Polcevera and then flowing up to the seafront.

In particular, as mentioned above, the segment of the pipeline affected by the rupture, is located between weld labelled as 8420 and weld labelled as 8430, 6.7 m in length in which wall nominal thickness is 8.74 mm. This segment presents a MAOP equals to 83 MPa.

Figure 7.9.1 Oil Pipeline near Genoa, affected by the rupture. It is evident the crater formed in the soil due to leaked oil pressure.

The figures below (Figure 7.9.2) represent the oil pipeline, after its excavation in order to sampling the segment affected by the rupture. Pipeline has a double line, which runs parallel and the stretch of line here in discussion, is almost vertical. As a consequence of steeper slope, filling geomaterial tends to be unstable. This is one of the most important aspects to consider in order to maintain the structure: over time, filling geomaterials are eroded, flowing downhill towards Rio Pianego, and discovering the pipeline which remains exposed and, as a consequence, more vulnerable.

In order to mitigate this risk, the Owner company realised some geo-engineering works and carried frequent geological surveys to detect problems related to erosion, deformations, etc. which can generate additional stresses in the pipes. In addition to this survey, in the last 25 years, the pipeline was regularly inspected by ILI – In Line Inspection, made by ultrasonic or magnetoscopic scan in order to detect size of and locate metal loss features, also estimating the corrosion rate. In this way, pipeline usage is carried correctly, thanks to a reliability-based pipeline integrity management program to assist engineers in selecting suitable maintenance strategies.

In this period, six inspections were carried out, by using different tools, following the technological development and, as a consequence of the data obtained, maintenance works (segments pipe substitutions) were carried out. It is important to note that this kind of monitoring is not mandatory by law, and these measures have been taken on owner initiative.

In this context, the accident took place with fracture occurring in the lower part of the pipe (Figure 7.9.3), approximatively in the centre of the pipe, in the longitudinal direction, 62 cm long with maximum opening approximately equal to 15 cm. The fluid

Figure 7.9.2 Oil pipeline formed by two pipes with different diameter: 16″ pipeline was affected by the rupture. Images show the pipeline after the excavation to sample the broken segment.

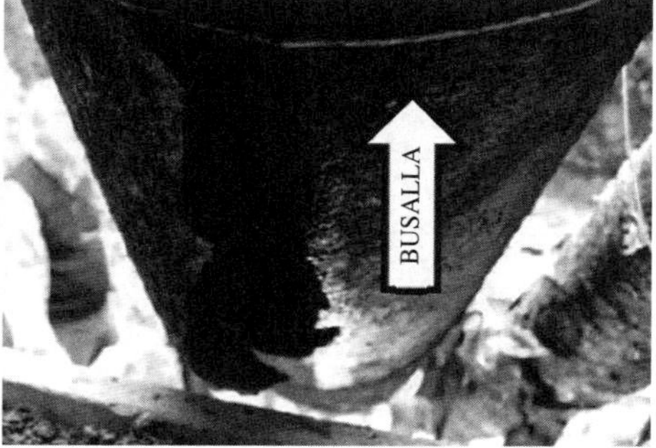

Figure 7.9.3 Detail of the segment affected by fracture and fluid (oil and water, alternate) direction when the accident occurred.

pressure created a sort of crater on the geomaterials surrounding the pipe and fluid flowing downhill was very fast. In the following days an extensive cleansing was activated for environmental purposes.

7.9.3 Why it Happened

Different scenarios were considered to understand the cause of the accident, taking into account different aspects, from a geological-geotechnical, metallurgic, operational point of view, employing various disciplines and frameworks.

First of all, geological aspects were investigated, to evaluate the potential role of a soil movement. Accurate in situ inspections had shown that the crater surrounding the breakpoint (Figure 7.9.1) as described above, was a consequence of the rupture, and not its cause. Although the in situ conditions were complicated, no evidences of significant previous soil erosion or soil slip were detected. To completely understand the geological aspects and to evaluate their role in the accident, rainfall data were also analysed, considering that the main causes of landslides in this region are prolonged and/or severe rain, which provoke a significant decay of the mechanical properties of soils, causing the soil movements. Hence, 75 days before the accident were considered, by using dataset from the nearest meteorological station, located in Bolzaneto: this allowed to observe that in this period no significant rainfall occurred, confirming that rains and geological aspects played no role in this case.

This evidence is further corroborated by the complete integrity of the upstream part of the breaking point in the pipeline, where vegetation and natural engineering work (gabion mats) were undisturbed and still lied vertical.

7.9.4 Experimental Campaign on the Pipeline Segment

A number of specific studies have been conducted to understand the causes of the rupture were conducted on the segment of the pipeline and on the fracture area, from a metallurgical point of view, to verify the exact geometry of the pipe (in particular its thickness, to compare the direct experimental measures with the estimated thickness predicted by ILI) its chemical and mechanical characteristics, the type and mode of fracture. To achieve this, a segment 6.7 m long was sampled (Figure 7.9.2),

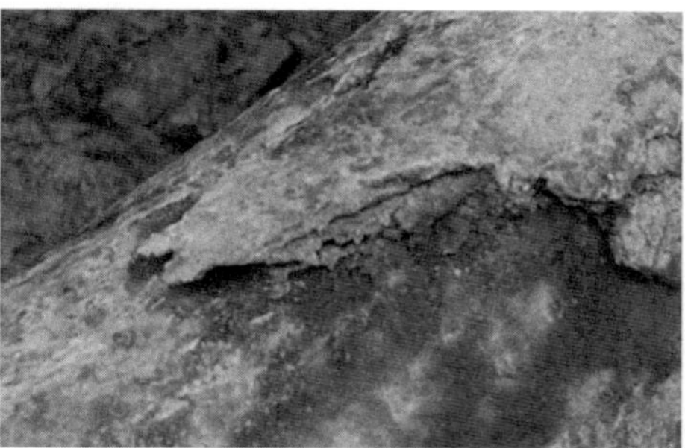

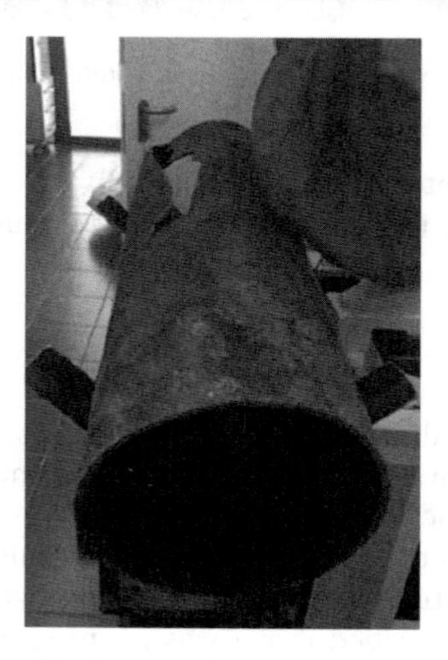

Figure 7.9.4 The segment affected by fracture after sampling and details of external corrosion, related to the age of the pipe.

cleaned and transported to a mechanical University laboratory in Bergamo, for its preparation.

First of all, as shown in Figure 7.9.4, it is possible to put into evidence the corrosion on the outer part of the pipe. Despite of the external protection, due to the age of the pipe which, in this area, is still the original one put in place in the '60s, the corrosion attack seems to be quite generalised.

The following analyses were carried out: chemical analysis, mechanical tests (traction and resilience), metallographic analysis and hardness tests, SEM and EDS analysis. Before the tests, the fracture area was measured with ultrascan and with a mechanical comparator along the crack faces.

Figure 7.9.5 shows the sampling zones for the different mechanical tests and analysis.

In the following, the most important results are detailed, in particular those concerning tensile tests and geometry measurements.

It is important to underline that technical standards do not foresee transversal traction tests for steel pipes, because that would require the straightening of the steel pipe, an operation that causes a disturbance which can affect results. Hence, traction tests were performed on 8 tensile specimens, made in accordance with technical standard

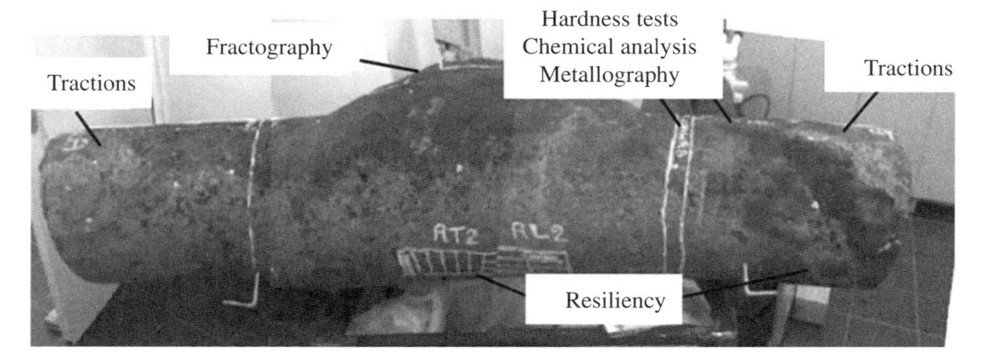

Figure 7.9.5 Pipeline portions destined to mechanical tests and chemical analysis.

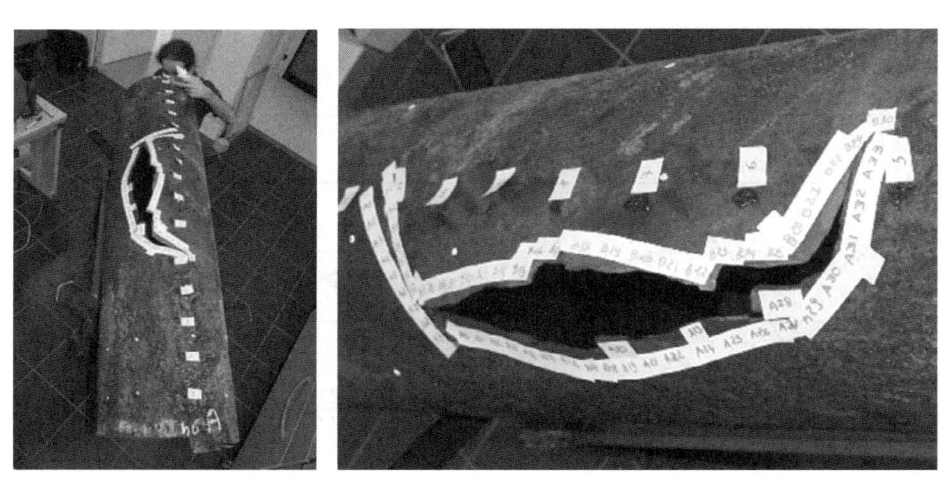

Figure 7.9.6 Pipe segment in which the fracture along the longitudinal line "h 6" and the letter "A" identifying one of the two edges of the pipe (the other one is called "B") are shown. Along the length of the fracture, different positions named from A1 to A33 are marked.

API 5L and with different thicknesses, as a function of the pipe thickness, not being constant along the segment due to the corrosion. Both tensile yield strength (mean value = 410 MPa) and tensile rupture strength (561 MPa) obtained are consistent with nominal characteristics of this type of steel, even if tests shown a certain degree of dispersion about tensile yield strength (nominal tensile yield stress fy = 358 MPa).

Before cutting operations to prepare samples, the pipeline segment was accurately measured, along 4 longitudinal lines, located in the upper part (called "h 12"), in the eastern part (called "h 3"), in the lower part in which is placed the fracture (called "h 6") and in the western part (called "h 9"), as shown in Figure 7.9.6.

Figure 7.9.7 shows thickness measured along 4 lines as described above: it is possible to observe that minimum values concern the lower part of the pipe, while the maximum values are in the upper part, and on the horizontal diameter (h 3 and h 9) thickness is similar to each other. Thickness strongly changes on different lines, from a

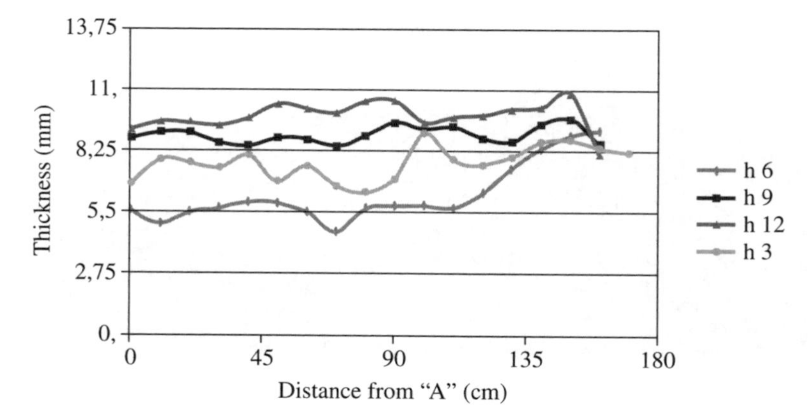

Figure 7.9.7 Thickness measured with ultrascan along four longitudinal lines on the pipe.

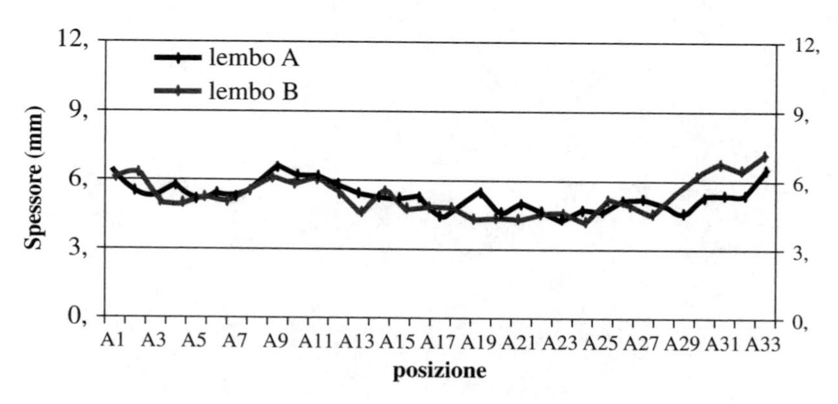

Figure 7.9.8 Crack face thickness measured by ultrascan. Similar data were obtained with a mechanical comparator.

minimum value equal to approximately 5 mm up to a maximum value equal to 11 mm: this variation certainly depends on generalised corrosion but could be also related to the pipe manufacturing process. As mentioned above, in fact, a quite large tolerance in the geometry was allowed at the time of pipe construction. Detailed measures of thickness were carried out along the crack faces: the mean value is approximately equal to 4.5 mm, as shown in Figure 7.9.8, consistent with the values measured along the line "h 6".

The measurements made evidenced the deformation of pipe, which is not only affected by irregular thickness, but also the diameters are not constant. In Figure 7.9.9 the outer diameter measured on the pipe edge is shown in red and the corresponding thickness in white.

In addition to mechanical tests and measurements, micrographic analyses were carried out, in order to obtain information about crack formation. These analysis in particular revealed that the region affected by fracture had suffered a plastic deformation, while there is no evidence of brittle crack propagation. SEM analysis does not evidence any defects or localised anomalies which could originate the crack triggering. Hence,

Figure 7.9.9 Outer diameters (in light grey, in mm) and corresponding thickness (in white, in mm).

the fracture mechanism was reconducted to a plastic one, depending on stresses and strains suffered by pipeline during its life.

7.9.5 Findings

Before the accident, the thickness of the pipeline segment 8420–8430 could be estimated thanks to the ILI inspections. The last inspection carried out in 2012–2013 period showed in this segment, a loss of thickness with a residual thick equal to 44%t, where t is the nominal wall thickness, in this case 8.74 mm. Hence, residual thickness is 4.89 mm, by neglecting manufacturing tolerance in thickness. As a consequence of ILI inspection, a corrosion growth rate was estimated in 0.17 mm/y, hence, at the time of the accident (39 months after the ILI inspections), the actual wall thickness should have been 4.34 mm. This estimated value is comparable with direct measures made in laboratory after the accident.

Based on technical operational instrumentations and pressure monitoring along the pipeline, a pressure equal to 74 barg (7.5 MPa) was active at the moment of the accident. This pressure provokes a stress equal to 344 MPa acting on the pipeline, obtained by applying Mariotte's law:

$$\sigma = D \cdot p/2t \tag{7.9.1}$$

where σ is the stress, D is the pipe diameter, p is the fluid pressure, t is the wall thickness, that in this case taking the following values: D = 397,60 mm, p = 7,5 MPa, t = 4,34 mm. The obtained value of pressure is close to the tensile yield stress fy.

To refine this result, a FEM Model was created and the pipeline segment was modeled by LUSAS software.

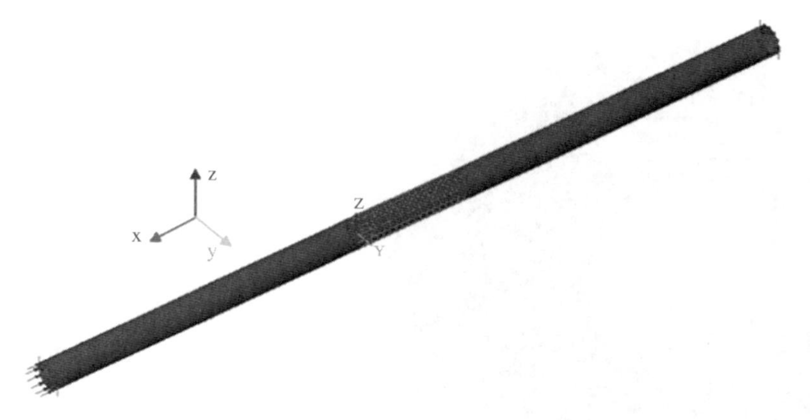

Figure 7.9.10 FEM Model – Global view.

FEM Model description:

- Geometry

The pipe geometry is taken into account according with the geometry acquired by laboratory measurements, by considering the deformation shown in Figure 7.9.9 and the different thickness. The model (Figure 7.9.10) considers a pipeline 11.7 m length, where the central part (1.7 m length) is the segment in which diameter and thickness are variable, while two parts at the edge (each 5 m length with constant thickness) are designed to minimise boundary effects.

- Mesh
  The mesh is formed by FEM elements *Thick Shell*, with longitudinal step equal to 10 cm and radial step equal to 30°.
- Constraints/supports

Both the edges are constrained and translation in the longitudinal direction is not allowed, in accordance with the real pipeline where the length does not permit deformations in this direction. Additional supports make the pipe statically determined in the YZ plane and indetermined in XZ plane. No interaction with the surrounding soil is taken into account.

- Material

The material here considered is steel X52 which nominal Young's Modulus E is equal to 210 GPa while Poisson's ratio ν is equal to 0.3.

- Loads

Only fluid pressure equal to 7.5 MPa acting on pipeline is considered.

- Results analysis

Figure 7.9.11 shows the pipeline deformed mesh, subjected to an internal radial pressure equal to 7.5 MPa. Because the longitudinal elongation was not allowed, pipe homogeneously deforms where the thickness is constant, while in the inner part, where diameter and thickness are variable, the pipe shows buckling effects.

Figure 7.9.11 Deformed Mesh – Global view.

Similarly, the stress state represented in Figure 7.9.12 (Von Mises stresses) is constant to the edges, while in the inner part (the most interesting pipe portion), it shows strong variations accordingly with variations in geometry. By specifically analysing this portion, it is possible to observe that the portions with lower thickness present higher stresses, in particular along the region surrounding the line "h 6", where the crack actually occurred.

The following figures (Figures 7.9.13 and 7.9.14) show respectively principal stresses σ_1 (circumferential), σ_2 (longitudinal), $\sigma 3$ (radial), where the last one, as expected, is negligible compared to the others, hence the pipe shows a membrane behaviour. Principal stresses σ_1 and σ_2 are both tensile stress, whereas circumferential stress is the prevailing stress (330 MPa, value close to that obtained by Mariotte's formula) while longitudinal stress is equal to 85 MPa.

Figure 7.9.12 Von Mises stresses and deformed mesh – Global view.

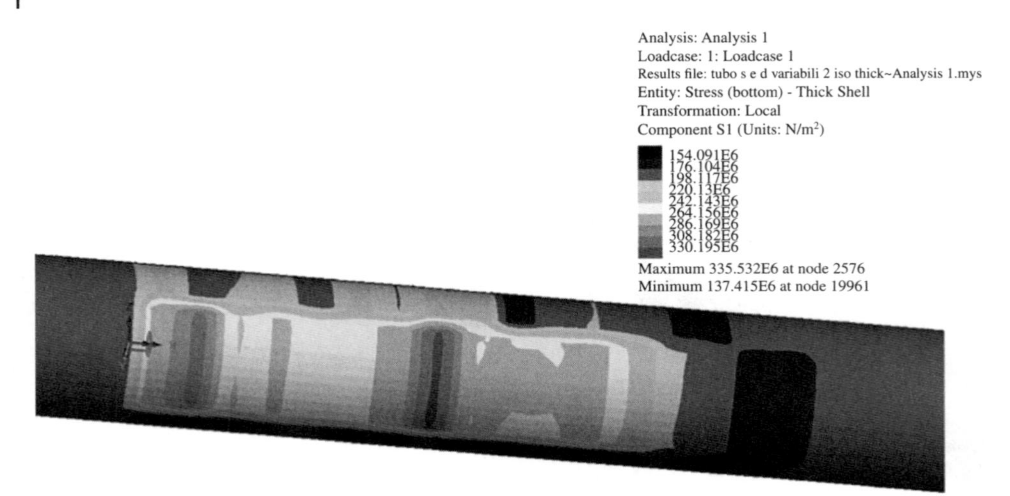

Figure 7.9.13 Principal stress σ1 (circumferential) along generator "h 6".

Tensile longitudinal stress, induced by prevented deformation in this direction, has a positive effect as shown by the Von Mises stresses in Figure 7.9.15, where the maximum stress is equal to 286 MPa.

By comparing the value of Von Mises stress obtained by FEM model with the value of nominal tensile rupture stress ($f_t = 455$ MPa), ore tensile stress obtained by the experimental campaign (560 MPa), it can be concluded that the pipeline, under the estimated pressure acting at the moment of the accident, was subjected to an admissible stress state. The crack triggering hence is due to external causes, not related to the ordinary operational conditions of the pipeline and it is possible that an excess pressure (not recorded by the pipeline monitoring system) occurred, probably caused by the pumps of the oil tanker.

7.9.6 Lessons Learned and Recommendations

First of all, it is important to note that the structure analysed here is aged and was designed with old technical rules. Anyway, the Owner has invested considerable resource to monitoring the pipeline to maintain it. In particular, ILI inspections seem to be the key to manage this kind of structure, where typical industry strategies for reliability-based corrosion management include high-resolution inline inspections (ILI) to measure defects on the pipeline body and estimate failure probabilities based on the very inspection results. Usually, Bayesian probabilistic approaches are the most credible way to calibrate models given observation data and have been commonly employed in energy pipelines' literature over the past decade. The analytical estimation of the high-dimensional integrals involved in the Bayesian updating is not feasible in pipeline problems though and therefore, Markov Chain Monte Carlo (MCMC) sampling techniques are commonly adopted to numerically perform this task. The limitations of these methods include the uncertainty as to whether the final samples have reached the final distribution or not and also a difficulty in ultimately quantifying small probabilities of rare events; for example, rupture due to metal-loss corrosion in the case of energy pipelines.

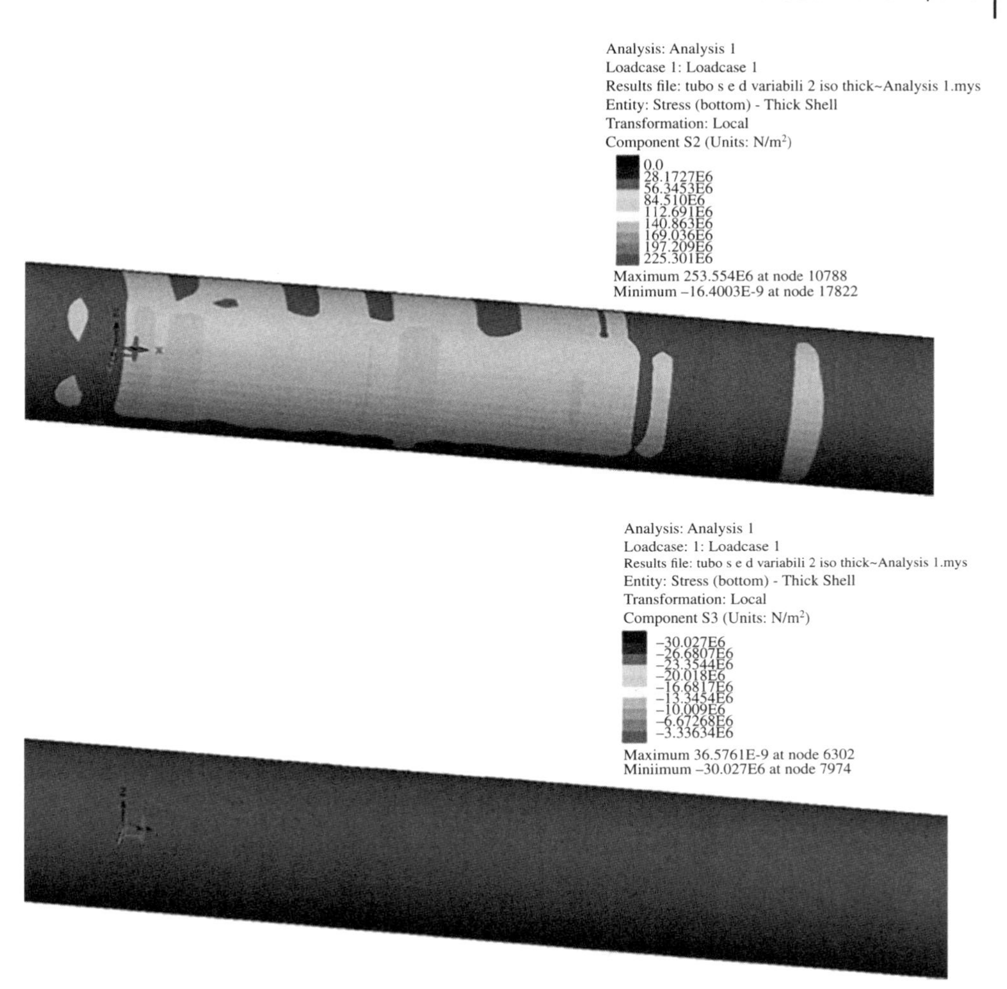

Figure 7.9.14 On the left: Principal stress σ2 (longitudinal) along line "h 6" – On the right: Principal stress σ3 (radial) along line "h 6". It is noted that maximal values are on the edge, at the external supports (so they are fictitious), here not visible.

Recently, an alternative method has been proposed that sets an analogy between Bayesian updating and a reliability problem [4]. This formulation is termed BUS (Bayesian Updating with Structural reliability methods) and enables the use of established structural reliability methods (SRM) to conduct the Bayesian updating. It also facilitates the estimation of small posterior failure probabilities, directly within the same analysis framework [5].

In this case, the Owner should be improving the pipeline management and is evaluating the possibility to apply BUS on the inspection data of this crude oil pipeline, by using last ILI carried out with magnetic flux leakage (MFL) tools in February 2013, and the previous (November 2007 and January 2002) using similar technology's inspection tools. The growth of multiple active metal-loss corrosion defects is modelled through a hierarchical Bayesian framework, whilst the updating is realised based

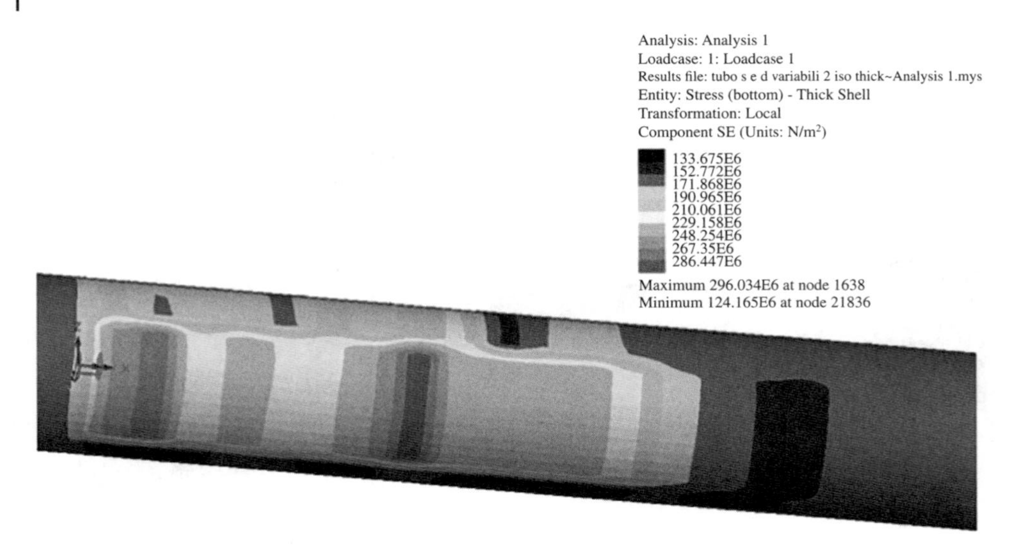

Figure 7.9.15 Von Mises stresses calculated along the longitudinal line "h 6".

on the 2002, 2007 and 2013 ILI data, with the associated measurement errors being comprehensively accounted for. The defects are subjected to internal pressure, which is modelled through a Ferry-Borges stochastic process. The growth of depths of the metal-loss corrosion defects is conducted by means of a homogeneous gamma process. This model also considers the corrosion initiation time. Finally, the BUS-SuS updates the growth model for the defect depth, based on the data collected from the three ILIs. The BUS-SuS also conducts the evaluation of the reliability of the pipeline, in terms of rupture.

The main output of this technique is the corrosion growth rate prediction and the proposed methodology can be incorporated in a reliability-based pipeline integrity management program to assist engineers in selecting suitable maintenance strategies, by elaborating the correct maintenance protocol for the next 5–10 years, where further ILI data can be used to update the maintenance strategies from time to time.

The main challenge of this approach is that the data obtained from two different ILIs with different methodologies normally are not exactly on the same locations. This will cause some difficulties in comparison. Despite this drawback, misalignment of the data position provides more complete information (in terms of more locations) of the whole pipeline over the specified time interval. Interpolation will be used to estimate the thickness reduction if it is a uniform corrosion. More careful interpretation will be carried out if there is a dramatic difference of thickness reduction around the area.

7.9.7 Forensic Engineering Highlights

The key points of the technical investigations carried out for this forensic case are related to the necessity to deploy different and multidisciplinary skills, like geology and geotechnics, material science, metallurgy, structural analysis, fracture mechanics and process engineering.

The data and the information acquired have been managed due to the strong interaction between several experts, also considering the fundamental implications of the in situ investigations. Inverse analyses were carried out and integrated in a multi-criteria framework leading to the final interpretation of the accident.

All this competence therefore contributed to understand the causes of the accident and, especially, to elaborate a new feasible and more robust maintenance protocol, as explained above, to safety manage the pipeline for the future.

7.9.8 References and Further Readings

References

1 Hovey, D.J. and Farmer, E.J. (1993) Pipeline accident, failure probability determined from historical data. *Oil and Gas Journal.*
2 Koch, G.H., Brongers, M.P.H., Thompson, N.G. et al. (2002) *Corrosion costs and preventive strategies in the United States.* McLean, VA (US): Federal Highway Administration.
3 Corrosion Control In Oil And Gas Pipelines [Internet]. *Pipeline and Gas Journal,* [cited 9 December 2017]. Available from: https://pgjonline.com/2010/03/05/corrosion-control-in-oil-and-gas-pipelines/.
4 Straub, D. and Papaioannou, I. (2015) Bayesian Updating with Structural Reliability Methods. *Journal of Engineering Mechanics,* 141(3):04014134.
5 Straub, D., Papaioannou, I., and Betz, W. (2016) Bayesian analysis of rare events. *Journal of Computational Physics,* 314:538–556.

Further Reading

Bigi F. B. (2017) “Le normative fondanti della sicurezza nell’industria di processo, le metodologie di analisi del rischio e la messa in sicurezza di un oleodotto”.

7.10 Storage Building on Fire

7.10.1 Introduction

The general information about the case study are shown in Table 7.10.1.

Table 7.10.1 General information about the case study.

Who	Storage building inside a storage buildings area
What	Fire on roof and on floor under roof
When	April 2011
Where	Nola, Italy
Consequences	Building damages, no fatalities
Mission statement	Determine the cause of the fire, the state of the area involved in the fire, the damages, with a quantification, of structures, plants, equipment, goods, automotive
Credits	Giovanni Manzini

Figure 7.10.1 Photo of the burned roof and the installed PV system.

The fire developed on 20 April 2011 had a large spread on the roof and, in particular, on waterproofing and insulating layers, thin film photovoltaic system installed on the roof and on some portions of the interior of the building (Figure 7.10.1). Later, another fire happened on 26 March 2012 on another building of the same storage center and it involved the roof layers (waterproofing and insulating) and still some portions of the building interiors.

The activities usually carried out in the storage center were in the list of those checked out by the Ministry of Home Affairs because of fire safety concerns, but, at the time of main event (April 2011), without formal authorization of that Ministry to be operating.

The main event (22 April 2011) happened when the work to install the waterproofing and insulating layers and the thin film photovoltaic system was carrying out on the building roof (a temporary consortium of companies was in charge about that).

7.10.2 How it Happened (Accident Dynamics)

On 22 April 2011 (sunny and windy day) 15h00' instant in which, probably, the present staff has noticed the fire, 15h10' alert to company emergency squad, 15h20' alert to fire fighters.

7.10.3 Why it Happened

One of the first conclusions of the investigation was that the fire had started on the building roof and then spread on the roof and later to the building inside (floor below the roof) through the skylights. Another version was that the fire started inside the building (last floor, the one under the roof) and later it was spreading from inside to outside (roof) through the skylights. The compatibility between PV system and roof (according to the Italian regulations) was OK with some "distinguo" about the interaction between skylights and roof (PV system and other layers on the roof).

Fire spread rate (horizontal, on the roof) was about 2 m/min (value calculated/evaluated from real event observation) (See Figure 7.10.2).

After the event some experiments were carried out, together with the collaboration of Fire fighter national corps, and the fire spread rate (horizontal) on a "sandwich" of layers with the same features of the one on the real event roof was about of 5 ÷ 6 cm / min. Obviously the HRR was much less than the real case value, but the difference between the fire spread values (experiments vs. real case) was really big and this feature may have occurred due to a strong heating on the bottom roof covering, especially in its initial steps. Later the great value of HRR could justify the great value of the fire spread rate.

Other possible causes of that rate were analyzed, in particular: the effects of age of XPS (old polystyrene) layer; effects of air in the roof layers (and effects of wind). Because of chemical features, the first one feature was considered unable to improve the rate to that value and the second one even. But a combination of wind (the fire occurred in a windy day) and of roof bottom heating was, on many opinions, able to improve so much the fire spread rate.

In particular, a first analytical modeling based on chemical kinetics and a more "coarse" simulation carried out by the (thermal) network theory, both approaches have led to conclusions according to which that great rate was possible due to a strong heating under the roof.

So, it is really possible that in the initial stages of the fire, that fire rose up to the last floor of the building (the one under the roof) imposing a strong heating on the lower surface of the building roof and later the fire was spread to the roof through the skylights.

7.10.4 Findings

With regard to the fire behaviour of the roof covering materials, from the experimentation carried out after the event it was noted that the PV thin film was able to slow

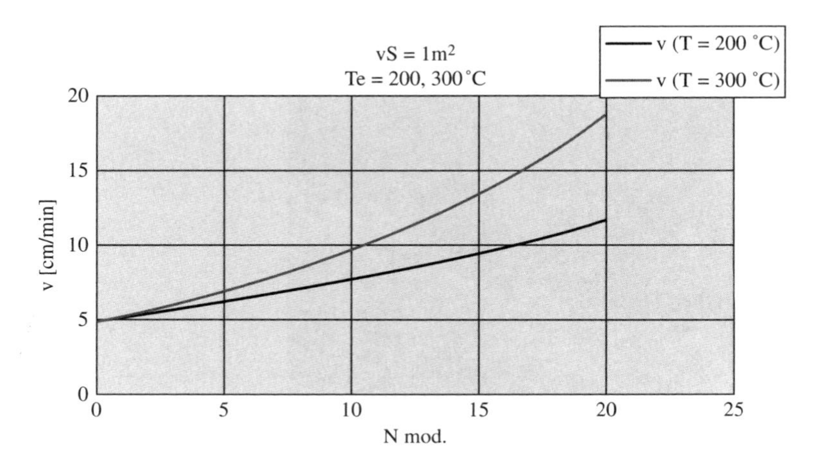

Figure 7.10.2 Curve of the maximum fire spread rate values v on roof surface (surface composed of modules of area equal to 1 m^2 placed continuously one to another one). Cases with bottom surface temperature - Te - equal to 200 °C and 300 °C. The case with more heating (300 °C) is clearly with a bigger rate.

Figure 7.10.3 The PV thin film.

down the fire spread rate with respect to the covering package without such a film as the last/upper layer (Figures 7.10.3 and 7.10.4).

7.10.5 Lessons Learned and Recommendations

From the event under consideration we can also draw some lessons for the future such as:

- It would have been better if the activities had the formal authorization of Ministry of Home Affairs to be operating;

Figure 7.10.4 The burned layers of the roof.

- a more effective fire detection system (e.g. focused on the roof) could have alerted to the fire earlier;
- the skylights are privileged ways to fire spread, so keeping proper distances between them and combustible layers and PV systems and limiting the fire load on the roof would be a good idea;
- in general, the assembly of combustible layers and skylight systems is a factor that can amplify the risk of fire in coverage in a non-negligible way.

7.10.6 Forensic Engineering Highlights

The severity of the incidental event referred to in the present case study was determined by a specific fire behaviour of the materials present in the roof and constituting the substrate of the photovoltaic system, induced in turn by the severe fire that affected the building. Given this relationship, the use of advanced simulation, combined with a series of experimental tests in the field, has quantitatively verified the increase in the speed of propagation of the fire in coverage given the increase in temperature in the underlying compartments due to a fire. This made it possible to exclude the possibility that the ignition could have occurred in coverage by the photovoltaic system itself with the consequent involvement of the entire building: preliminary hypothesis taken into consideration by the technicians in charge of the survey.

7.10.7 Further Readings

Further Readings

Cancelliere, P.G., Manzini, G., and Mazzaro, M. (2017) A review of the photovoltaic module and panel fire tests. IFireSS 2017 – 2nd International Fire Safety Symposium. Naples.

Fiorentini, L., Marmo, L., Danzi, E. et al. (2014) Fire risk analysis of photovoltaic plants. A case study moving from two large fires: from accident investigation and forensic engineering to fire risk assessment for reconstruction and permitting purposes. Warsaw, Poland: 47th ESReDa Seminar on "Fire Risk Analysis".

Fiorentini, L., Marmo, L., Danzi, E., and Puccia, V. (2016) Fire risk assessment of photovoltaic plants. A case study moving from two large fires: From accident investigation and forensic engineering to fire risk assessment for reconstruction and permitting purposes. *Chemical Engineering Transactions*, 48:427–432.

Fiorentini, L., Marmo, L., Danzi, E., and Puccia, V. (2016) *Fire risk assessment of photovoltaic plants. A case study moving from two large fires: from accident investigation and forensic engineering to fire risk assessment for reconstruction and permitting purposes*. Konzerthaus Freiburg (Germany): ICHEME.

Fiorentini, L., Marmo, L., Danzi, E., and Puccia, V. (2015) Fires in Photovoltaic Systems: Lessons Learned from Fire Investigations in Italy. *SFPE Magazine*, 99.

Russo, P., Coccorullo, I., and Russo, G. (2017) Incendio di un capannone durante l'incendio dell'impianto fotovoltaico. IF CRASC 17. Milano: Dario Flaccovio Editore, pp. 179–188.

8

Conclusions and Recommendations

Industrial accidents, especially the most severe ones, are not the result of one single cause; rather different interrelated causal factors are involved. Accident scenarios may be also influenced, directly or indirectly, by all the actors involved in the ordinary work process. The incident investigation reflects the complexity which subsequently arises. The final goal of incident investigations is to identify the temporal sequence of what led to the undesired event, and to find all the causal factors that affected the incident evolution from a latent condition to its actual occurrence. Once the causal factors have been found, corrective actions should be suggested to reduce current risks into an acceptable range, in order to prevent future incidents. This requires the analysis of all the actors involved, from technical systems and front-line operators to procedures and technical standards [1]. A wide set of accident investigation methods is available. Each method has different purposes, with its own pros and cons. Graphical illustrations of the event sequence are highly recommended since they provide, in a single sheet, visual information and give help in collecting the key information and identifying eventual gaps, thus driving the evidence gathering and analysis. When the incident event is really complex, more than one investigation can be set up for each sub-problem areas. In the investigation team, at least one member should have a sound knowledge about the investigation methods and enough experience to choose the ones that best fit the context, providing a proper tool for the challenging objectives.

The diversity of the investigation approaches reflects the different cultural, historical and institutional background in each country, regarding the risk acceptability and its management [2]. Indeed each country has its own history, with experience of severe catastrophes, pollution and major losses; every industry experienced successes and failures and each company has been shaped by them. The subsequent institutional, organisational and cultural processes partially explain the diversity among the perception of incidents and the modalities they are investigated. Nowadays, the harmonisation processes, based on information sharing and integration, tend to converge towards a unitary global approach. Indeed, efforts in improving the accessibility of information and its sharing are recorded globally. From an economic and political standpoint, the birth of permanent investigation boards at the European and International level are another signal of this process of convergence, together with the standardization of safety directives and technical standards. At the industry level, the new approach is reflected in

Principles of Forensic Engineering Applied to Industrial Accidents, First Edition.
Luca Fiorentini and Luca Marmo.

the development of standards and procedures related to accident investigation, and the even larger sharing of the lessons learned from experience. After an accident has occurred, various actors could start multiple investigations, with common purposes but different focus (e.g. organization's management, customers, subcontractors, control authorities, investigation boards, researchers, insurers, justice, and so on), inquiring with different aspects such as safety, engineering, legal, economic, or contractual issues. These different contexts generate the diversity above mentioned. Regardless the possible harmonization on some issues, the diversity is likely to manifest being strongly correlated to the different stakeholders' context and objectives, reflecting on findings and suggested corrective actions. This diversity might erroneously bring the actors in refusing a formal training about the forensic discipline and the investigation methods. The eventuality must be avoided, by teaching them how important is to be trained on the related subjects, especially on the forensic engineering, as the complex discipline presented in this book. The harmonisation process should be encouraged since it will probably enhance the incident investigation process, directly and indirectly favouring the prevention of incidents in the future. The LFE process today remains sometimes ineffective. The asymptotic trend to the improvement of safety and the occurrence of major incidents reveal the failures of the LFE and underline the importance of the organizational factors.

It emerged that an effective prevention lies far from the top event [3]: "Going beyond the widget!" is a key concept and the importance of root causes has been underlined. The control of plant modifications, the necessity to test and inspect protective equipment, the adoption of a user-friendly design are all needs to create a solid ground for a prevention policy. The importance of hazard and operability studies has been stressed out. The outcomes of the analysis usually revealed a need for a better management and critical questions have been posed, trying to go deeper into how human factors bring to an undesired event or what is the involvement of the line, site or corporate level management in the incident. It has been underlined that an investigator looks for causes not for the blame (blaming is fruitless and supports the myth that incidents cannot be foreseen and are unavoidable, which is not true). And once causes are found, the emphasis is moved on to the corrective actions, no more on the causes. The reduction of the risk is the priority of the developed recommendations. Reducing risk means to protect the people [4]. We are not talking only of people who were directly killed as a result of incidents, but also of those who suffered major injuries (and, as a consequence, were absent from work for a certain period, with the additional economic loss for the company). The investigation findings are intimately correlated with the risk assessments and they identify the areas of intervention where the risk assessment need to be reviewed. The interested reader can consult [5] for further reasoning about the lack of business driver to change, the necessity of an organization to find someone to blame, and some critiques about why recommendations are rarely implemented and an organization performs incident investigations only on large incidents.

In conclusion, this book intends to present forensic engineering as a discipline, assigning it specific boundaries and identifying the approach standing at the base: the scientific method. The necessity to look for root causes has been underlined from the very beginning of the book; only by exploring the managerial context, can effective recommendations developed to avoid further incidents. The diversity of industrial accidents has been discussed in Chapter 2, where basic notions on combustions, and the most common incidental scenarios (including the basic concepts on near misses) were given. The process safety has been presented as a fundamental knowledge to possess in order to carry out the analysis through methods discussed in Chapter 5. The importance of accidents, as a source of information to improve the safety levels, was pointed out by presenting different real incidents. The role of the performance indicators revealed as essential to monitor the evolution towards an unsafe conditions that could transform into a real incident. The key concepts of "Uncertainty" and "Risk" have been presented, taking inspiration from the probabilistic approach that is in opposition to the Newtonian standpoint soundly based on the full knowledge, the certainty of the events and the time reversibility. The discussion on investigation and forensic engineering has been enriched with some considerations about legal, ethic, and insurance issues. Some methods for Hazard Identification and Risk Assessment have been presented, including the HAZOP technique, one of the most widely recognised. Chapter 3 ended with the presentation of the most diffused technical standards among the industrial sector, related to the topics faced by the book, with a special mention for NFPA 550, NFPA 921, IEC 61508, and IEC 61511. The forensic engineering workflow was deeply discussed in Chapter 4. It starts with team composition and planning activity. The collection of evidence was stressed, also detailing the sampling process and the types of evidence to be considered to be further recognised and organised. Finally, it is the turn of their analysis and the investigation path has been contextualised within a specific method named "conic spiral". How to report and effectively communicate the outcomes of the investigations is faced in the last Paragraph on Chapter 4. A significant set of investigation methods has been presented in Chapter 5. Starting from the causes and causal mechanism analysis, and the time and events sequence preliminary approach (STEP method), methods have been discussed in depth. A distinction between structured and not structured approach has been carried out. MORT, BSCAT™, Tripod Beta, BFA, RCA, TapRoot®, Apollo RCA™, FTA, ETA, and LOPA have been presented, clarifying their purpose, structure, and context of application, sometimes providing examples. How human factors affect the occurrence of incidents was also among the objective of the discussion. Finally, the derived lessons have been presented in Chapter 6. The attention has been paid on both the pre- and post-accident management and how to develop effective recommendations, also taking into account the communication issues. The necessity of continuous safety (and risk) management and training was claimed and the concept of "safety culture" has been presented as well as the BBS. The necessity to treat near misses, as free source of information to enhance the risk management, was pointed out. Finally, some case studies has been presented, to see how to apply what learned from the reading of the book to real incidents.

References

1 Sklet, S. (2002) *Methods for accident investigation*. 1e. Trondheim: Norwegian University of Science and Technology.
2 Dechy, N., Dien, Y., Funnemark, E. et al. Results and lessons learned from the ESReDA's Accident Investigation Working Group. *Safety Science*, 50(6):1380–1391.
3 Kletz, T. (2001) *Learning from accidents*. 3e. Oxford: Gulf Professional Publishing.

4 Health and Safety Executive. (2004) HSG245: Investigating accidents and incidents: a workbook for employers, unions, safety representatives and safety professionals. 1e. Health and Safety Executive.
5 ABS Consulting (Vanden Heuvel, L., Lorenzo, D., Jackson, L. et al.). *Root cause analysis handbook: a guide to efficient and effective incident investigation*. 3e. Brookfield, Conn.: Rothstein Associates Inc.

9

A Look Into the Future

In the past, the most frequent cause of accidents was lack of knowledge. Today, economic competition and time pressure have a significant impact on the decision-making level, leaving insufficient time to think about what can be wrong and considerations taken are mainly cost-based. It results in wrong decisions, or too late decisions, or no decisions at all. Having this concept in mind and having presented a panoramic view of the forensic engineering discipline, an extra effort is required to promote the Authors' vision about its future. In order to carry out a satisfactory investigation, in the context of an advanced forensic engineering, it is auspicable to have:

- A multidisciplinary team, whose members are experts in the involved disciplines;
- a coordinator who drives the team members to have a unique objective, the same investigative direction, a well-balanced report where every part fits with each other in a congruent final thesis;
- proper investigation methods, to discover the real root causes. It is especially recommended when human factors are supposed to be involved;
- advanced tools (drones, special equipment);
- computer simulations, to evaluate specific aspects (like Finite Element Method, Computational Fluid Dynamics, and so on); and
- a multi-level approach, starting from the general to the particular (like the conic spiral method).

The growing public perception of the risks related to such disasters is one of the driving force towards the challenge of the investigation. Data about incidents, including the immediate and root causes, are collected by several investigation bodies, in order to develop appropriate measures to reduce the risk of similar occurrences. The ESReDA working group collected the opinions of the experts, detecting the challenges during the conduct of investigations [1]. They concern both external and internal conditions, dealing with issues related to methodology, training, competence, scope, independence. The combination of natural disasters (like earthquakes, flooding, wildfires) with a technological incident (e.g. in power plants) is a new type of incident whose interactions does not make its investigation easy, since the resulting major disaster could not be assessed with the conventional methods currently used.

Recently, a clear distinction between the technical investigation and the police investigation has been confirmed. The search for root causes and preventive actions is definitely separated from the blame assignation and the search for accountabilities, responsibilities and culprits, even if sometimes the borderlines can be quite vague. As

Principles of Forensic Engineering Applied to Industrial Accidents, First Edition.
Luca Fiorentini and Luca Marmo.

a consequence, today incident investigation has to deal with several challenges. Mainly they are the lack of integration on different institutional levels and between sectors, the defeats in organisational and methodological level.

The ESReDA members performed a SWOT analysis to discover the major issues to be faced with investigation [2]:

- The broad use of investigation methods has shifted the attention on individuals rather than on systems, on legal aspects instead that on root causes, on the allocation of blame instead that on promoting preventive measures. Therefore, the real objective of an investigation must not be lost;
- the rapid technological and organizational innovation, especially in the ict context, contributed in shifting the failures from hardware components to software. more frequently, a failure occurs at the functional level, not in the mechanical, tangible level. a new approach to model the systems complexity is therefore necessary to perform an effective investigation, reserving a special focus on functional safety and its specialists;
- even if remaining in the same sectors (aviation, rail, maritime, industrial), the complexity of modern society makes the investigation vulnerable. we assist to a fragmentation of competencies, responsibilities and duties among different levels in both public and private authorities. a main challenge is an international harmonization of the investigation approaches within the same sectors, in order to comply with high level technical and quality standards; and
- in a large sociological perspective, governments are retreating from active participation in safety assessment, since this subject is becoming the prerogative of the legislative and punitive power only. however, the continuous involvement in public-private partnerships requires the necessity to cover the role of public safety auditor. the safety investigation agencies may fulfil this role, replacing the national government. but if this role is not recognised, then political sectors and private companies risk being conquered by judicial regulation on behalf of political and administrative bodies.

The details above provides four challenges that may be classified as external challenges. A look into the future shows also four internal challenges, yet unresolved. They are:

- Independence. investigation reports declare to be independent, by composition, proposals, discovered findings and developed recommendations. however, a closer look reveals that some institutional and administrative restrictions affect the investigation independency. examples are: the appointment of national committee members by the political authorities; the legal requirements to access a public committee, which may advantage judges or other professionals; the subordination of some public investigation commissions to a specific ministry. the challenge is to reach a real independency, with organisational, legal, and financial freedom for the investigative body;
- scope. the sectoral approach limits the learning process from experience. some sectors accumulate much more experience than others, and so they have a higher number of developed recommendations to follow in order to prevent the reoccurrence of a similar incident. therefore they have a greater possibility to be inherently safer. a transversal approach is instead recommended, to ensure a useful spread of knowledge and let that also the others' incidents may help the risk reduction in a different context;

- methodology. the survey carried out by the esreda [3, 4] outlined how several organizations did not use a standard method for investigation, and a simple approach is adopted by those that did use a method. moreover, many companies have developed their own methods. the lack of standard methods produce fragmentation of the developed recommendations, which are based on those findings coming from different non-standard methods. this may conflict with the achievement of the professional, public and political consensus that is necessary to implement the lessons learnt during the investigation; and
- training and competence. the rising complexity asks for qualified and expert investigation team members. there is a high demand for qualified academic experts, to cooperate with stakeholders during major investigations. investigators internal to a company may experience a lack of competences because of the low exposure to accidents in their company. the challenge is to develop and implemented shared high-level training courses for investigators.

These are the future challenges asked by the investigation community. As reminded by [4], these major challenges need to be met in order to align with the recent developments demanded for investigation quality and public credibility.

In conclusion, as also reminded in [5], the future investigations must be an effective part of a proactive approach to industrial safety. If we think about the IT solutions to manage incident investigation, today the users are active, while the databases are passive. The Authors believe that the situation will be reversed in the future: users will be passive and databases will be active, automatically and proactively providing useful information for risk assessment studies. Some software houses already provide tools performing this coupling, using the bowtie methodology. The real challenge will the cultural change required to carry out such complex, and complete, approach to industrial safety.

An interesting look into the future is given by the application of virtual reality during the onsite inspection (Figures 9.1 and 9.2). Indeed, if we consider the destructive nature of a fire or an explosion, we understand how the investigation of their causes is

Figure 9.1 Virtual recognition of some signs due to the heat.

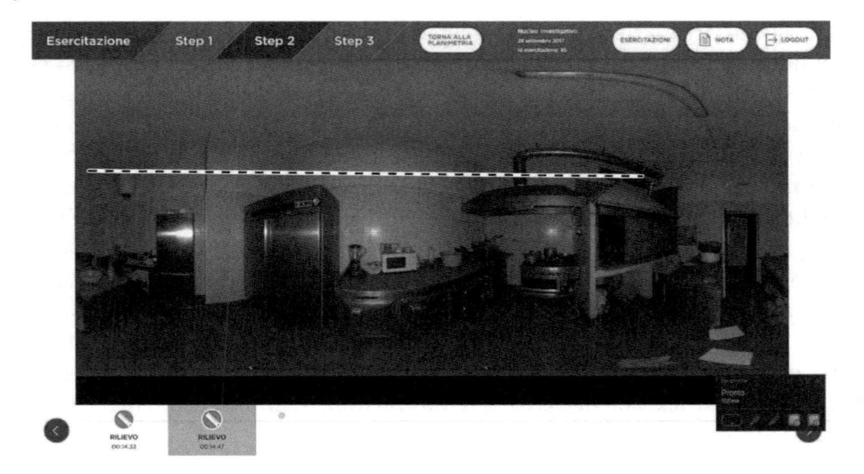

Figure 9.2 Record on the timeline of the performed actions during the geometric survey.

an incredibly complex activity. It often sees investigators working in scenarios characterised by levels of damage to structures and materials that do not allow a reconstruction, if not partial, of the site conditions ex-ante the fire. The higher the ability of the investigators to collect information on the scenario under investigation, interpreting, for example, the signs left by the fire, and the higher the probability that the analysis of the event leads to the identification of the cause, through an objective reconstruction of the facts that have happened. Following this objective, the Investigative Fire Fighting Unit of the Italian National Fire Brigade has decided to train its personnel on the fire and explosion investigation using a high technology system that, through the use of virtual reality, simulates the execution of the inspection activity on real fire scenarios. The application is aimed at the interpretation of the signs of thermal damage left by the fire on materials or structures inside the environment under inspection. The user, concerning the signs of thermal damage observed on the scene, will be able to practice identifying the area of origin of the fire and in understanding the path of propagation of the fire.

References

1 Valvisto, T., Harms-Ringdahl, L., Kirchsteiger, C., and Roed-Larsen, S. (2003) Accident investigation practices - results from a European inquiry. *European Safety Reliability and Data Association (ESReDA).*

2 Roed-Larsen, S. and Stoop, J. (2012) Modern accident investigation – Four major challenges. *Safety Science,* 50(6):1392–1397.

3 ESReDA Working Group on Accident Investigation. (2009) *Guidelines for Safety Investigations of Accidents.* 1e. European Safety and Reliability and Data Association.

4 Dechy, N., Dien, Y., Funnemark, E. et al. (2012) Results and lessons learned from the ESReDA's Accident Investigation Working Group. *Safety Science,* 50(6):1380–1391.

5 Kletz, T. (2001) *Learning from accidents.* 3e. Oxford: Gulf Professional Publishing.

Appendix A

Principles on Probability

A.1 Basic Notions on Probability

Basic notions on probability are defined on the basis of the possible outcomes of a test. All the possible outcomes of a test are denoted by the set $\boldsymbol{\Omega}$. A generic event $\boldsymbol{A}$ is a subset of $\boldsymbol{\Omega}$, therefore the probability that the event $\boldsymbol{A}$ occurs is denoted by $\boldsymbol{P}(\boldsymbol{A})$ and is a positive number between 0 and 1. In particular, the following expression are true:

$$\boldsymbol{P}(\boldsymbol{\Omega}) = 1 \tag{A.1}$$

$$\boldsymbol{P}(\boldsymbol{A} \cup \boldsymbol{B}) = \boldsymbol{P}(\boldsymbol{A}) + \boldsymbol{P}(\boldsymbol{B})\,\textbf{if}\,\boldsymbol{A} \cap \boldsymbol{B} = \emptyset \tag{A.2}$$

If $\boldsymbol{P}(\boldsymbol{A}) = 0$ the event is called impossible.

The addition rule of probability is a theorem claiming that for any two events $\boldsymbol{A}$ and $\boldsymbol{B}$:

$$\boldsymbol{P}(\boldsymbol{A} \cup \boldsymbol{B}) = \boldsymbol{P}(\boldsymbol{A}) + \boldsymbol{P}(\boldsymbol{B}) - \boldsymbol{P}(\boldsymbol{A} \cap \boldsymbol{B}) \tag{A.3}$$

The basic properties of probability can be visually represented using Venn diagrams. The sample space $\boldsymbol{\Omega}$ is represented by a rectangle and each event, i.e. subset of $\boldsymbol{\Omega}$, is often represented by a circle. For example, Figure A.1 shows the addition rule of probability graphically.

The relative probability of an event $\boldsymbol{A}$ respect to an event $\boldsymbol{X}$, that has already taken place, is defined as conditional probability of $\boldsymbol{A}$ knowing $\boldsymbol{X}$. In formula:

$$\boldsymbol{P}(\boldsymbol{A}|\boldsymbol{X}) = \frac{\boldsymbol{P}(\boldsymbol{A} \cap \boldsymbol{X})}{\boldsymbol{P}(\boldsymbol{X})} \tag{A.4}$$

The expression can be easily understood looking at Figure A.2.

From A.4 derives that:

$$\boldsymbol{P}(\boldsymbol{A} \cap \boldsymbol{X}) = \boldsymbol{P}(\boldsymbol{A}|\boldsymbol{X})\cdot\boldsymbol{P}(\boldsymbol{X}) = \boldsymbol{P}(\boldsymbol{X}|\boldsymbol{A})\cdot\boldsymbol{P}(\boldsymbol{A}) \tag{A.5}$$

If two events $\boldsymbol{A}$ and $\boldsymbol{X}$ are independent, i.e. the occurrence of one does not affect the occurrence of the other, then $\boldsymbol{P}(\boldsymbol{A}|\,\boldsymbol{X}) = \boldsymbol{P}(\boldsymbol{A})$. Consequentially, using A.5, we have:

$$\boldsymbol{P}(\boldsymbol{A} \cap \boldsymbol{X}) = \boldsymbol{P}(\boldsymbol{A})\cdot\boldsymbol{P}(\boldsymbol{X}) \tag{A.6}$$

When A.6 is not true, it means that the events $\boldsymbol{A}$ and $\boldsymbol{X}$ are said to be dependent.

When the logic AND-gates have been presented in this book, it has been claimed that the probability associated with the event after an AND-gate was the product of the probabilities of the events feeding the gate. Expression (A.6) explains clearly the

Principles of Forensic Engineering Applied to Industrial Accidents, First Edition.
Luca Fiorentini and Luca Marmo.

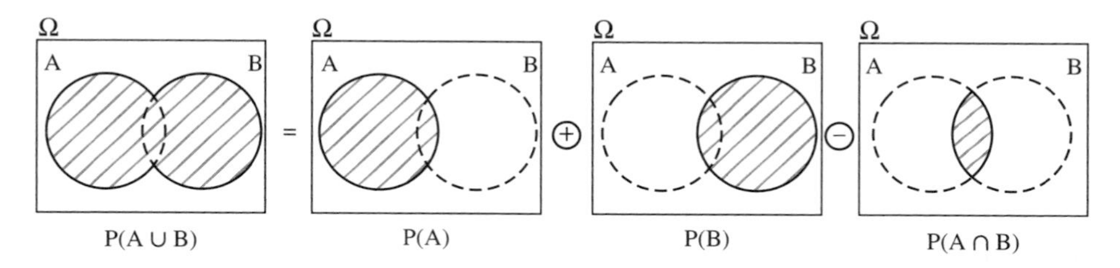

Figure A.1 Visual explanation of the addition rule of probability, through Venn diagrams.

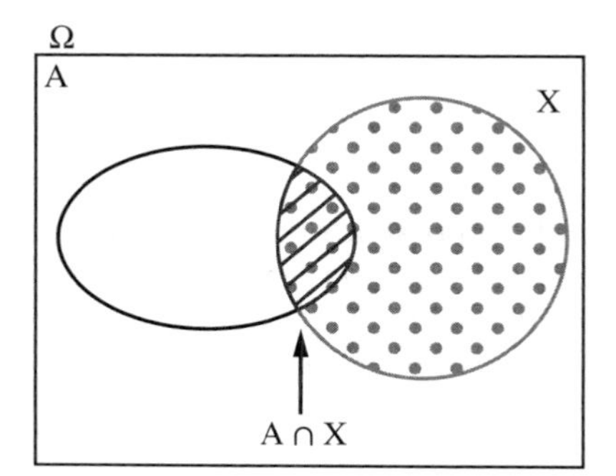

Figure A.2 Visual explanation of the conditional probability, through Venn diagrams.

hypothesis at the base of that assumption. Similarly, when the logic OR-gates have been presented, it was claimed that the probability associated with the event after an OR-gate was the sum of the probabilities of the events feeding the gate. Expression (A.2) shows the mathematic formulation at the base of that assumption. The investigator must be always aware of the hypothesis about events independency at the base of the reasoning and if the hypothesis is no longer true, more complex formulations must be used, like the ones in A.5 and A.3.

Index

Principles of Forensic Engineering Applied to Industrial Accidents, First Edition.
Luca Fiorentini and Luca Marmo.

d

e

g

h

q

r

t

u

v

w